ENCYCLOPEDIA OF MATHEMATICS AND ITS APPLICATIONS

EDITED BY G.-C. ROTA

Volume 57

Skew fields

ENCYCLOPEDIA OF MATHEMATICS AND ITS APPLICATIONS

1. L. A. Santalo *Integral Geometry and Geometric Probability*
2. G. E. Andrews *The Theory of Partitions*
3. R. J. McEliece *The Theory of Information and Coding: A Mathematical Framework for Communication*
4. W. Miller, Jr *Symmetry and Separation of Variables*
5. D. Ruelle *Thermodynamic Formalism: The Mathematical Structures of Classical Equilibrium Statistical Mechanics*
6. H. Minc *Permanents*
7. F. S. Roberts *Measurement Theory with Applications to Decisionmaking, Utility, and the Social Sciences*
8. L. C. Biedenharn and J. D. Louck *Angular Momentum in Quantum Physics: Theory and Applications*
9. L. C. Biedenharn and J. D. Louck *The Racah–Wigner Algebra in Quantum Theory*
10. J. D. Dollard and C. N. Friedman *Product Integration with Applications in Quantum Theory*
11. W. B. Jones and W. J. Thron *Continued Fractions: Analytic Theory and Applications*
12. N. F. G. Martin and J. W. England *Mathematical Theory of Entropy*
13. G. A. Baker, Jr and P. Graves-Morris *Padé Approximants, Part I, Basic Theory*
14. G. A. Baker, Jr and P. Graves-Morris *Padé Approximants, Part II, Extensions and Applications*
15. E. G. Beltrametti and G. Cassinelli *The Logic of Quantum Mechanics*
16. G. D. James and A. Kerber *The Representation Theory of Symmetric Groups*
17. M. Lothaire *Combinatorics on Words*
18. H. O. Fattorini *The Cauchy Problem*
19. G. G. Lorentz, K. Jetter and S. D. Riemenschneider *Birkhoff Interpolation*
20. R. Lidl and H. Niederreiter *Finite Fields*
21. W. T. Tutte *Graph Theory*
22. J. R. Bastida *Field Extensions and Galois Theory*
23. J. R. Cannon *The One-Dimensional Heat Equation*
24. S. Wagon *The Banach–Tarski Paradox*
25. A. Salomaa *Computation and Automata*
26. N. White (ed) *Theory of Matroids*
27. N. H. Bingham, C. M. Goldie and J. L. Teugels *Regular Variations*
28. P. P. Petrushev and V. A. Popov *Rational Approximation of Real Functions*
29. N. White (ed) *Combinatorial Geometries*
30. M. Phost and H. Zassenhaus *Algorithmic Algebraic Number Theory*
31. J. Aczél and J. Dhombres *Functional Equations in Several Variables*
32. M. Kuczma, B. Choczewski and R. Ger *Iterative Functional Equations*
33. R. V. Ambartzumian *Factorization Calculus and Geometric Probability*
34. G. Gripenberg, S.-O. London and O. Staffans *Volterra Integral and Functional Equations*
35. G. Gasper and M. Rahman *Basic Hypergeometric Series*
36. E. Torgersen *Comparison of Statistical Experiments*
37. A. Neumaier *Interval Methods for Systems of Equations*
38. N. Korneichuk *Exact Constants in Approximation Theory*
39. R. A. Brualdi and H. Ryser *Combinatorial Matrix Theory*
40. N. White (ed) *Matroid Applications*
41. S. Sakai *Operator Algebras in Dynamical Systems*
42. W. Hodges *Model Theory*
43. H. Stahl and V. Totik *General Orthogonal Polynomials*
44. R. Schneider *Convex Bodies*
45. G. Da Prato and J. Zabczyk *Stochastic Equations in Infinite Dimensions*
46. A. Björner *et al. Oriented Matroids*
47. G. Edgar and L. Sucheston *Stopping Times and Directed Processes*
48. C. Sims *Computation with Finitely Presented Groups*
49. T. W. Palmer *C*-algebras I*
50. F. Borceux *Handbook of Categorical Algebra 1, Basic Category Theory*
51. F. Borceux *Handbook of Categorical Algebra 2, Categories and Structures*
52. F. Borceux *Handbook of Categorical Algebra 3, Categories of Sheaves*

Skew fields

Theory of general division rings

P. M. COHN, FRS

University College London

CAMBRIDGE UNIVERSITY PRESS
Cambridge, New York, Melbourne, Madrid, Cape Town, Singapore, São Paulo

Cambridge University Press
The Edinburgh Building, Cambridge CB2 8RU, UK

Published in the United States of America by Cambridge University Press, New York

www.cambridge.org
Information on this title: www.cambridge.org/9780521432177

First published 1995
This digitally printed version 2008

A catalogue record for this publication is available from the British Library

ISBN 978-0-521-43217-7 hardback
ISBN 978-0-521-06294-7 paperback

To my grandson
James Abraham Aaronson

CONTENTS

Preface   xi

From the preface to *Skew Field Constructions*   xiii

Note to the reader   xv

Prologue   1

1   **Rings and their fields of fractions**   3
1.1   Fields, skew fields and near fields   3
1.2   The general embedding problem   8
1.3   Ore's method   14
1.4   Necessary conditions for a field of fractions to exist   19
1.5   Stable association and similarity   24
1.6   Free algebras, firs and semifirs   34
1.7   The matrix reduction functor   41
    Notes and comments   45

2   **Skew polynomial rings and power series rings**   **47**
2.1   Skew polynomial rings   47
2.2   The ideal structure of skew polynomial rings   56
2.3   Power series rings   66
2.4   Group rings and the Malcev–Neumann construction   71
2.5   Iterated skew polynomial rings   78
2.6   Fields of fractions for a class of filtered rings   83
    Notes and comments   91

3   **Finite skew field extensions and applications**   **93**
3.1   The degree of a field extension   93
3.2   The Jacobson–Bourbaki correspondence   96
3.3   Galois theory   100
3.4   Equations over skew fields and Wedderburn's theorem   111
3.5   Pseudo-linear extensions   118

3.6	Quadratic extensions	126
3.7	Outer cyclic Galois extensions	133
3.8	Infinite outer Galois extensions	140
3.9	The multiplicative group of a skew field	143
	Notes and comments	150
4	**Localization**	**152**
4.1	The category of epic R-fields and specializations	153
4.2	The matrix representation of fractions	157
4.3	The construction of the localization	162
4.4	Matrix ideals	171
4.5	Universal fields of fractions	177
4.6	Projective rank functions and hereditary rings	184
4.7	Normal forms for matrix blocks over firs	196
	Notes and comments	200
5	**Coproducts of fields**	**202**
5.1	The coproduct construction for groups and rings	203
5.2	Modules over coproducts	210
5.3	Submodules of induced modules over a coproduct	214
5.4	The tensor ring on a bimodule	224
5.5	HNN-extensions of fields	231
5.6	HNN-extensions of rings	239
5.7	Adjoining generators and relations	242
5.8	Derivations	250
5.9	Field extensions with different left and right degrees	259
5.10	Coproducts of quadratic field extensions	268
	Notes and comments	275
6	**General skew fields**	**278**
6.1	Presentations of skew fields	279
6.2	The specialization lemma	283
6.3	Normal forms for matrices over a tensor ring	292
6.4	Free fields	301
6.5	Existentially closed skew fields	308
6.6	The word problem for skew fields	317
6.7	The class of rings embeddable in fields	326
	Notes and comments	329
7	**Rational relations and rational identities**	**331**
7.1	Polynomial identities	331
7.2	Rational identities	335
7.3	Specializations	340
7.4	A particular rational identity for matrices	345
7.5	The rational meet of a family of X-rings	348
7.6	The support relation on generic division algebras	355
7.7	Examples of support relations	362
	Notes and comments	365

8	**Equations and singularities**	**366**
8.1	Algebraically closed skew fields	367
8.2	Left and right eigenvalues of a matrix	374
8.3	Normal forms for a single matrix over a skew field	379
8.4	Central localizations of polynomial rings	386
8.5	The solution of equations over skew fields	395
8.6	Specializations and the rational topology	401
8.7	Examples of singularities	408
8.8	Nullstellensatz and elimination	411
	Notes and comments	418
9	**Valuations and orderings on skew fields**	**420**
9.1	The basic definitions	421
9.2	Abelian and quasi-commutative valuations	427
9.3	Matrix valuations on rings	435
9.4	Subvaluations and matrix subvaluations	442
9.5	Matrix valuations on firs	449
9.6	Ordered rings and fields	457
9.7	Matrix cones and orderings on skew fields	462
	Notes and comments	469
	Standard notations	473
	List of special notations used throughout the text	475
	Bibliography and author index	478
	Subject index	495

When *Skew Field Constructions* appeared in 1977 in the London Mathematical Society Lecture Note Series, it was very much intended as a provisional text, to be replaced by a more definitive version. In the intervening years there have been some new developments, but most of the progress has been made in the simplification of the proofs of the main results. This has made it possible to include complete proofs in the present version, rather than to have to refer to the author's *Free Rings and their Relations*. An attempt has also been made to be more comprehensive, but we are without a doubt only at the beginning of the theory of skew fields, and one would hope that this book will offer help and encouragement to the prospective builders of such a theory. The genesis of the theory was described in the original preface (see the extract following this preface); below we briefly outline the subjects covered in the present book.

The first four chapters are to a large extent independent of each other and can be read in any order, referring back as necessary. Ch. 1 gives the general definitions and treats the Ore case as well as various necessary conditions for the embedding of rings in skew fields. From results in universal algebra it follows that necessary and sufficient conditions for such an embedding take the form of quasi-identities. Later, in Ch.4, we shall find the explicit form of these quasi-identities, and in Ch. 6 we shall see that this set must be infinite. The rest of Ch. 1 gives the definition and basic properties of free algebras and free ideal rings, which play a major role later. It also includes some technical results on the association of matrices and it introduces an important technical tool: the matrix reduction functor.

Ch. 2 studies skew polynomial rings and the fields formed from them, as well as power series rings and generalizations such as the Malcev–Neumann construction, and the author's results on fields of fractions for a class of filtered rings. Ch. 3 is devoted to the Galois theory of skew fields, now almost classical, with applications to (left or) right polynomial equations over skew fields, and special cases of extensions, such as pseudo-linear extensions and cyclic Galois extensions.

Ch. 4 is in many ways the central chapter. The process of forming fields of fractions or more generally epic R-fields for a ring R is described in terms of the *singular kernel*, i.e. the set of matrices that become singular over the field. It is shown how any epic R-field can be constructed from its singular kernel, while the latter has a simple description as prime matrix ideal. This leads to explicit conditions for the existence of a field of fractions. In particular, the rings with a fully inverting homomorphism to a field are characterized as Sylvester domains and it is shown that every semifir has a universal field of fractions. Of the earlier sections only 1.6 is needed here.

Ch. 5 describes the coproduct construction and the results proved here are basic for much that follows. It is also the most technical chapter and the reader may wish to postpone the details of the proofs in 5.1–3 to a second reading, but he should familiarize himself with the results. They are applied in the rest of the chapter to give the HNN-construction for fields and for rings, to study the effect of adjoining generators and relations, particularly matrix relations, and to construct field extensions with different left and right degrees (Artin's problem). Ch. 6 deals with some general questions. There is a study of free fields; here the specialization lemma is an essential tool. Other topics include the word problem and existentially closed fields.

Ch. 7 on rational identities is mainly devoted to Bergman's theory of specializations between rational meets of X-fields; it is independent of most of the rest and can be read at any stage.

In Ch. 8 the rather fragmentary state of knowledge of singularities (which in the general theory take the place of equations in the commutative theory) is surveyed, with an account of the problems to be overcome to launch a form of non-commutative algebraic geometry. Ch. 9 deals with valuations and orderings on skew fields from the point of view of the general construction of Ch. 4 and it shows for example how to construct valuations and orderings on the free field.

The exercises are intended for practice but serve also to present additional developments in brief form, as well as some open problems. Some historical background is given in the Notes and comments.

The theory of division algebras (finite-dimensional over a field) is very much further advanced than the general theory of skew fields, and a comprehensive account including a full treatment of division algebras would have thrown the whole out of balance and resulted in a very bulky tome. For this reason that topic has largely been left aside; this was all the more reasonable as the subject matter is much more accessible, and no doubt will be even more so with the forthcoming publication of the treatise by Jacobson and Saltman.

Nearly all the material in this volume has been presented to the Ring Theory Study Group at University College London and I am grateful to the members of this group for their patience and help. I would like to thank Mark L. Roberts for his comments on early chapters and George M. Bergman for his criticism of *Skew Field Constructions*, which has proved most useful. My thanks also go to the staff of the Cambridge University Press for their help in transforming the manuscript into a book, with a particular word of thanks to their copy editor Mr Peter Jackson, who corrected not merely grammatical but also mathematical slips. As always, I shall be glad to receive any constructive criticism from readers, best of all, news of progress on the many open problems.

London, February 1995   P. M. Cohn

From the preface to *Skew Field Constructions*

The history of skew fields begins with quaternions, whose discovery (in 1843) W. R. Hamilton regarded as the climax of his far from ordinary career. But for a coherent theory one has to wait for the development of linear associative algebras; in fact it was not until the 1930's that a really comprehensive treatment of skew fields (by Hasse, Brauer, E. Noether and Albert) appeared. It is an essential limitation of this theory that only skew fields finite-dimensional over their centres are considered.

Although general skew fields have made an occasional appearance in the literature, especially in connexion with the foundations of geometry, very little of their properties was known until recently, and even particular examples were not easy to come by. The first well known case is the field of skew power series used by Hilbert in 1899 to illustrate the fact that a non-archimedean ordered field need not be commutative. There are isolated papers in the 1930's, 1940's and 1960's (Moufang, Malcev, B. H. Neumann, Amitsur and the author) showing that the free algebra can be embedded in a skew field, but the development of the subject is hampered by the fact that one has no operation that can be performed on skew fields (over a given ground field) and again produces a skew field containing the original ones. In the commutative case one has the tensor product, which leads to a ring from which fields can then be obtained as homomorphic images. The corresponding object in the general case is the free product and in the late 50's the author tried to prove that this could be embedded in a skew field. This led to the development of firs (= free ideal rings); it could be shown (1963) that any free product of skew fields is a fir, but it was not until 1971 that the original aim was achieved, by proving that every fir is embeddable in a skew field, and in fact has a universal field of fractions. Combining these results, one finds that any free product or 'coproduct' of fields has a universal field of fractions, or a *field coproduct*, as we shall call it. It is this result which forms the starting point for these lectures.

As the name indicates, these really are lecture notes, though not for a single set of lectures. For this reason they may lack the polish of a book, but it is hoped that they have not entirely lost the directness of a lecture. The material comes from courses I have given in Manchester and London; some parts follow rather closely lectures given at Tulane University (1971), the University of Alberta (1972), Carleton University (1973), Tübingen (1974), Mons (1974), Haifa Technion (1975), Utrecht (1975) and Ghent (1976). It is a pleasure to acknowledge the hospitality of these institutions, and the stimulating effect of such critical audiences.

NOTE TO THE READER

The reader of this book is expected to have a fair background in algebra, particularly ring theory and commutative field theory. Any standard results needed are usually quoted from the author's *Algebra* (referred to as A.1, 2, 3, see the Bibliography).

All theorems, propositions, lemmas and corollaries are numbered consecutively in a single series in each section; thus Th. 2.2 is followed by Prop. 2.3 in Section 1.2, and outside Ch. 1 they are referred to as Th. 1.2.2, Prop. 1.2.3 respectively. Occasional results needed but not proved are usually given letters, e.g. Th. 6.A. The end (or absence) of a proof is indicated by ■. Most sections have exercises; open-ended (or open) problems are marked °. Unexplained notations can be found in the list of standard notations on p. 473.

References to the bibliography are by the author's name and the last two digits of the year of publication if after 1900, e.g. Ore [31], Cramer [1750], with primes to distinguish publications by the same author in the same year.

PROLOGUE

> O glücklich, wer noch hoffen kann,
> Aus diesem Meer des Irrtums aufzutauchen!
> Was man nicht weiss, das eben brauchte man,
> Und was man weiss, kann man nicht brauchen.
>
> Goethe, *Faust I*

One of the principal aims of this book is to describe some methods of constructing skew fields. The case most studied so far is that of skew fields finite-dimensional over their centres. But a finite-dimensional k-algebra, where k is a commutative field, is a field whenever it has no zero-divisors. On the one hand this enormously simplifies their study, while on the other hand it puts many constructions out of bounds (because they produce infinite-dimensional algebras). The study of fields that are not necessarily finite-dimensional over their centres is still in its early stages, and the methods needed here are not very closely related to those used on finite-dimensional algebras – the relation between these subjects is rather like the relation between finite and infinite groups.

There are some ways of obtaining a field directly, for example Schur's lemma tells us that the endomorphism ring of a simple module is a field, and the coordinatization theorem shows that when we coordinatize a Desarguesian plane, the coordinates lie in a field. But these methods are not very explicit, and we shall have no more to say about them. For us the usual way to construct a field is to take a suitable ring and embed it in a field. What is to be understood by 'suitable' will transpire later.

There are five methods of interest to us; they are

(1) Ore's method (Ch. 1),
(2) The method of power series (Ch. 2),
(3) Inverse limits of Ore domains (Ch. 2),

1

(4) A general criterion (Ch. 4),

(5) An application of the specialization lemma (Ch. 6).

As a test ring we shall use the free algebra on a set X over a commutative field k, written $k\langle X \rangle$. All five methods can be used on $k\langle X \rangle$, and each has its pros and cons. (1) is particularly simple, but not in any way canonical, (2) and (3) provide a convenient normal form, while (4) gives, at least in principle, a complete survey over all possible embeddings, indeed over all homomorphisms of our ring into fields. Finally (5) applies only to free algebras, where it gives an easy existence proof of the universal field of fractions.

The main applications are to the construction of the field coproduct, which shows that the class of skew fields possesses the amalgamation property and allows a form of HNN-construction. The consequences are described here, but it is clear that the existing range of constructions is still rather limited, mainly because a good specialization theory is still lacking (see 8.8 below). One would hope that the present work will offer encouragement to others working towards that goal.

1

Rings and their fields of fractions

Fields, especially skew fields, are generally constructed as the field of fractions of some ring, but of course not every ring has a field of fractions and for a given ring it may be quite difficult to decide if a field of fractions exists. While a full discussion of this question is left to Ch. 4, for the moment we shall bring some general observations on the kind of conditions to expect (mainly quasi-identities) in 1.2 and give some necessary conditions relating to the rank of free modules in 1.4, as well as some sufficient conditions. On the one hand there is the Ore condition in 1.3, generalizing the commutative case; on the other hand and perhaps less familiar, we have the trivializability of relations, leading to semifirs in 1.6, which include free algebras and coproducts of fields, as we shall see in Ch. 5. Some general relations between matrices over rings, and the applications to the factorization of elements over principal ideal domains (needed later) are described in 1.5.

Although readers will have met fields before, a formal definition is given in 1.1 and is contrasted there with the definition of near fields, which however will not occupy us further. The final section 1.7 deals with the matrix functor and its left adjoint, the matrix reduction functor, which will be of use later in constructing counter-examples.

1.1 Fields, skew fields and near fields

By a *field* we understand a set K with two binary operations, *addition*, denoted by a plus sign: $+$, and *multiplication*, denoted by a cross, $\times$, a dot, $\cdot$, or simply by juxtaposition, with two distinguished elements, zero: 0 and one: 1, such that

  (i) K is a group under addition, with 0 as neutral element,

3

(ii) $1 \neq 0$ and $K^{\times} = K \backslash \{0\}$ is a group under multiplication, with 1 as neutral element,
(iii) the two operations are related by the *distributive laws*:

$$x(y + z) = xy + xz, (x + y)z = xz + yz \text{ for all } x, y, z \in K.$$

The groups K in (i) and $K^{\times}$ in (ii) are the *additive group* and the *multiplicative group* of K respectively.

Our first observation is that the additive group is always abelian. For, using first the left and then the right distributive law, we have

$$(x + 1)(y + 1) = (x + 1)y + (x + 1) \cdot 1 = xy + y + x + 1,$$

while an expansion on the other side gives

$$(x + 1)(y + 1) = x(y + 1) + 1 \cdot (y + 1) = xy + x + y + 1.$$

Equating the results and cancelling xy on the left and 1 on the right, we find that $y + x = x + y$, as claimed.

If the multiplicative group of K is abelian, K is a *commutative field*; when commutativity is not assumed, K is called a *skew field* or also a *division ring*. Since skew fields form the topic of this book, we shall use the term 'field' to mean 'not necessarily commutative field' and only occasionally add 'skew', when emphasis is needed.

Let K be a field. Any *subfield* of K (i.e. a subset of K admitting all the operations of K) contains 1 and hence the subfield generated by 1. This least subfield, often denoted by Π, is called the *prime subfield* of K. It is either the rational field $\mathbf{Q}$ or $\mathbf{Z}/p$, the integers mod p, for some prime p. Accordingly K is said to have *characteristic* 0 or p; this characteristic is also written char K.

Given a field K and a subset X of K, the *centralizer* of X in K is defined as the set

$$\mathscr{C}_K(X) = \{y \in K | xy = yx \text{ for all } x \in X\}.$$

This set is easily seen to be a subfield of K. In the special case $X = K$ we obtain the *centre* C of K:

$$C = \{y \in K | xy = yx \text{ for all } x \in K\}.$$

Clearly the centre is a commutative subfield containing the prime subfield Π.

Just as rings arise naturally as the endomorphism sets of abelian groups, or more generally, of modules, i.e. groups with operators, so fields arise as endomorphism sets of simple modules. Their importance stems from the fact that linear algebra, first developed over the real numbers, can be carried out over any field. This applies even to skew fields, as long as we do not try to form determinants. In fact there is a form of determinant over skew fields, the Dieudonné determinant, but this will play only a limited role here. The main difference is that whereas the structure of commutative fields is fairly well known since the fundamental paper of Steinitz [10], information on skew fields is much more fragmentary. The theory is best developed for fields finite-dimensional over their centres (division algebras), but we shall mainly be concerned with fields infinite-dimensional over their centres, where a full classification is not to be expected.

It is a natural question to ask what can be said about endomorphism sets of non-abelian groups. Let G be a group written multiplicatively and consider the set $\mathcal{M}(G)$ of all mappings preserving 1 of G into itself. On $\mathcal{M}(G)$ we have two operations, the multiplication arising by composition of mappings and addition arising from the group operation:

$$x(\alpha\beta) = (x\alpha)\beta, \, x(\alpha + \beta) = x\alpha \cdot x\beta \text{ for all } x \in G, \, \alpha, \beta \in \mathcal{M}(G).$$

It follows that the left distributive law holds,

$$\alpha(\beta + \gamma) = \alpha\beta + \alpha\gamma, \quad \alpha, \beta, \gamma \in \mathcal{M}(G),$$

but the right distributive law fails to hold in general. In fact we have

$$(\alpha + \beta)\gamma = \alpha\gamma + \beta\gamma$$

for all $\alpha, \beta \in \mathcal{M}(G)$ only when γ is an endomorphism of G. However, if we restrict ourselves to endomorphisms we no longer have an addition, because the sum of two endomorphisms need not be an endomorphism. Thus $\mathcal{M}(G)$ fails to be a ring only in that it lacks the right distributive law (except for the vestige $0\alpha = 0$) and the commutativity of addition. It forms an example of a *near ring*; a subring whose non-zero elements all have inverses is a *near field*. Near fields have been used in the study of permutation groups, in geometry, as the rings coordinatizing certain translation planes and in the classification of finite subgroups of skew fields (see Amitsur [55] and for the results, 3.9 below), but they will not occupy us further in this volume. For a detailed account of near fields see Wähling [87].

In any field K the addition can be expressed in terms of the multiplication xy and the operation $x + 1$. For we clearly have

$$x + y = \begin{cases} (xy^{-1} + 1)y & \text{if } y \neq 0, \\ x & \text{if } y = 0. \end{cases}$$

This observation leads to a definition of fields which emphasizes the multiplicative structure. Let G be any group, written multiplicatively; by the *group with* 0 on G we understand the set $G_0 = G \cup \{0\}$ with multiplication xy as in G for x, $y \neq 0$, while $x0 = 0x = 0$ for all $x \in G_0$.

LEMMA 1.1.1. *Let G be a group and G_0 the group with 0 on G. Suppose that $\sigma\colon G_0 \to G_0$ is a map such that $e\sigma = 0$ for some $e \in G$ and further,*
(i) *$0\sigma = 1$, where 1 is the neutral element of G,*
(ii) *$(y^{-1}xy)\sigma = y^{-1} \cdot x\sigma \cdot y$ for all x, $y \in G$,*
(iii) *$[(xy^{-1})\sigma \cdot y]\sigma = ([x\sigma \cdot y^{-1}]\sigma)y$ for all $x \in G_0$, $y \in G$.*
Then G_0 is a field with respect to its multiplication and the addition

$$x + y = \begin{cases} (xy^{-1})\sigma \cdot y & \text{if } y \neq 0, \\ x & \text{if } y = 0. \end{cases} \tag{1}$$

Proof. By (1), $x + 0 = x$, $0 + x = (0x^{-1})\sigma \cdot x = 1 \cdot x = x$ for $x \neq 0$. Now with the help of (1), (iii) may be written as

$$(x + y)\sigma = x\sigma + y.$$

Further, (1) shows that $x\sigma = x + 1$, hence

$$(x + y) + 1 = (x + 1) + y. \tag{2}$$

Now the definition (1) shows that for $yz \neq 0$,

$$xz + yz = [(xz(yz)^{-1})\sigma]yz = (xy^{-1} \cdot \sigma)yz = (x + y)z,$$

hence

$$(x + y)z = xz + yz. \tag{3}$$

This has been shown to hold for y, $z \neq 0$. If $z = 0$, both sides reduce to 0, while for $y = 0$, both become xz, so (3) holds identically in G_0.

Next we have to prove

$$z(x + y) = zx + zy. \tag{4}$$

If one of x, y, z is 0, this is clear; otherwise we have by (ii),

$$zx + zy = (zx \cdot y^{-1}z^{-1})\sigma \cdot zy = z(xy^{-1}\sigma)z^{-1} \cdot zy$$
$$= z(xy^{-1}\sigma)y$$
$$= z(x + y)$$

and (4) follows.

Next we have, by (2),

$$(xz^{-1} + yz^{-1}) + 1 = (xz^{-1} + 1) + yz^{-1};$$

multiplying on the right by z and using (3) twice, we find

$$(x + y) + z = (x + z) + y, \tag{5}$$

at least when $z \neq 0$, but for $z = 0$ it holds trivially. Taking $x = 0$, we find that $y + z = z + y$, hence addition is commutative and so (5) can be rewritten to give the associative law:

$$(x + y) + z = x + (y + z).$$

Finally, for $x \neq 0$, we have

$$ex + x = (ex \cdot x^{-1})\sigma \cdot x = 0 \cdot x = 0.$$

Thus x has the additive inverse ex, and this is true even when $x = 0$ and $e0 = 0$. This shows G_0 to be a group under addition, with neutral element 0. In particular, e is the additive inverse of 1 and writing -1 for e, we obtain the usual notation for a field. ∎

Exercises

1. Show that every near field with fewer than nine elements is a field (for a near field on nine elements, see Ex. 4).

2. Show that if in Lemma 1.1, (ii) is omitted, we obtain a near field.

3. Let K be any field with a subgroup P of index 2 in $K^\times$ and with an automorphism of order 2, $x \mapsto x'$, mapping P into itself. Define a new multiplication on K by the rule

$$x \circ y = \begin{cases} xy & \text{if } x \in P, \\ xy' & \text{if } x \notin P. \end{cases}$$

Verify that K with this multiplication is a near field which is not a field.

4. (Dickson [05]) Apply Ex. 3 to construct a near field on any field of p^2 elements, where p is an odd prime.

5. (Ferrero [68]) Let Γ be an additive group with a group G acting on it by fix-point-free automorphisms (i.e. $\alpha g = \alpha$ for $\alpha \in \Gamma$, $g \in G$ implies $\alpha = 0$ or $g = 1$). Let Δ_i ($i \in I$) be a family of orbits $\neq \{0\}$ in Γ, with representatives δ_i and on Γ define a multiplication by putting $\alpha \circ \beta = \beta g_\alpha$ if $\alpha \in \Delta_i$ and g_α is the unique element of G satisfying $\delta_i g_\alpha = \alpha$; otherwise, i.e. if $\alpha \notin \Delta_i$ for all i, put $\alpha \circ \beta = 0$. Verify that except for lacking a one, Γ is a near ring. (This shows that every group Γ is the additive group of some near ring, possibly lacking a one.)

6. Show that in any ordered field (see 9.6) the set of all non-negative elements, with the operation $x\sigma = x + 1$, satisfies the conditions (i)–(iii) of Lemma 1.1 (this shows that the condition $e\sigma = 0$ cannot be omitted).

7. Show that any element of a ring having both a left inverse and a right inverse has a unique two-sided inverse.

8. (Kohn and Newman [71]) Show that in any field K of characteristic $\neq 2$ the following identity holds:

$$[(x + y - 2)^{-1} - (x + y + 2)^{-1}]^{-1} - [(x - y - 2)^{-1} - (x - y + 2)^{-1}]^{-1} =$$

$$\tfrac{1}{2}(xy + yx).$$

Why cannot xy be expressed in this way unless K is commutative?

9. In any ring show that if $1 - xy$ is a unit, then so is $1 - yx$. (Hint. Use elementary transformations to transform $\mathrm{diag}\,(1, 1 - xy)$ to $\mathrm{diag}\,(1 - yx, 1)$.)

10. Show that the centralizer of any subset of a field is a subfield.

1.2 The general embedding problem

A basic difference between groups and rings on the one hand and fields on the other is that the former, but not the latter, form a *variety*, i.e. a class defined by identical relations (see A.3, 1.3). In particular, a group may be described by generators and defining relations and any set of generators and relations yields a group; similarly for rings, whereas a given set of (ring) generators and defining relations cannot always be realized in a field. The usual method of obtaining a field, especially a skew field, is as field of fractions of a ring. This makes it important to study methods of embedding rings in fields. In this section we shall make some general observations on the embedding problem, and we begin by introducing some terminology.

Let R be a ring; by a *field of fractions* of R we understand a field K together with an embedding $R \to K$ such that K is the field generated by the image of R. Our task then is to find when a ring has a field of fractions. For commutative rings the answer is easy (and well known). It falls into three parts:

(i) Existence. A field of fractions exists for a ring R if and only if R is an *integral domain*, i.e. the set $R^{\times} = R\backslash\{0\}$ is non-empty and closed under multiplication.

(ii) Uniqueness. When a field of fractions exists, it is unique up to a unique isomorphism, thus given two fields of fractions of R, $\lambda_i: R \to K_i$ $(i = 1, 2)$, there exists a unique isomorphism $\varphi: K_1 \to K_2$ such that $\lambda_1 \varphi = \lambda_2$.

(iii) Normal form. Each element of the field of fractions can be written in the form a/b, where $a, b \in R$, $b \neq 0$, and $a/b = a'/b'$ if and only if $ab' = ba'$.

Of course this is not really a 'normal form'; only in certain cases such as **Z** or $k[x]$ is there a canonical representative for each fraction (see also Ex. 1).

Let us now pass to the non-commutative case. The absence of zero-divisors is still necessary for a field of fractions to exist, but not sufficient. The first counter-example was found by Malcev [37], who writes down a semigroup whose semigroup ring over **Z** is an integral domain but cannot be embedded in a field (see Ex. 3 below). Malcev expressed his example as a cancellation semigroup not embeddable in a group, and it prompted him to ask for a ring R whose set $R^\times$ of non-zero elements can be embedded in a group, but which cannot itself be embedded in a field. This question was answered affirmatively nearly 30 years later, in 1966, and will be dealt with in 5.7 below.

After giving his example, Malcev went on in a remarkable pair of papers (Malcev [39]) to provide a set of necessary and sufficient conditions for a semigroup to be embeddable in a group. This is an infinite set of conditions, and Malcev showed that no finite subset could be sufficient. The first two conditions express cancellability:

$$xy = xz \Rightarrow y = z, \quad yx = zx \Rightarrow y = z; \qquad (1)$$

next came the condition (using $\wedge$ to mean 'and'):

$$ax = by \wedge cx = dy \wedge au = bv \Rightarrow cu = dv. \qquad (2)$$

The other conditions were similar, but more complicated (Malcev [39], or UA, VII. 3), and they were all of the form

$$A_1 \wedge A_2 \wedge \ldots \wedge A_n \Rightarrow B, \qquad (3)$$

where $A_1, \ldots, A_n, B$ are certain equations, with the universal quantifier $\forall$ for all the variables prefixed. Such a condition (3) is called a *quasi-identity* or a *universal Horn sentence*; when the As are missing, we have an *identity*.

As a matter of fact it follows from general principles of universal algebra that the class of semigroups embeddable in groups is a *quasi-variety*, i.e. definable by quasi-identities. For it can be shown to be a *universal class* (definable by sentences with universal quantifiers over all variables, i.e. universal sentences), and one has the following theorem (see e.g. UA, VI. 4):

A class of algebras is a quasi-variety if and only if it is universal and admits direct products, or equivalently, if and only if it admits direct products and subalgebras.

We remark that such a class always contains the one-element subalgebra, as the product of the empty family. With the help of this result it is not hard to check that the class of semigroups embeddable in groups is a quasi-variety. At the same time we see that integral domains do not form a quasi-variety, since they do not admit direct products, and neither do rings embeddable in fields. Nevertheless they come very close to being a quasi-variety. To be precise, if $\mathcal{D}$ denotes the class of integral domains, $\mathcal{F}$ the class of fields and $s\mathcal{F}$ the class of subrings of fields, then there is a quasi-variety $\mathcal{Q}$ such that

$$s\mathcal{F} = \mathcal{D} \cap \mathcal{Q}. \tag{4}$$

To find $\mathcal{Q}$, we recall some definitions. A ring R is called *regular* (in the sense of von Neumann) if for any $a \in R$ there exists $x \in R$ such that $axa = a$. If for each $a \in R$ there exists $x \in R$ such that $a^2x = a$, R is said to be *strongly regular*. Despite its appearance, the condition of strong regularity is left–right symmetric, as the next lemma shows. We recall that a ring is said to be *reduced*, if it contains no nilpotent elements $\neq 0$, i.e. $x^2 = 0$ implies $x = 0$.

LEMMA 1.2.1. *A ring is strongly regular if and only if it is regular and reduced.*

Proof. Assume that R is strongly regular. If $a^2 = 0$, take x to satisfy $a^2x = a$; then $0 = a^2x = a$, so R is reduced. Moreover, for any $a \in R$ and for $x \in R$ such that $a^2x = a$, we have

$$(axa - a)^2 = axa^2xa - axa^2 - a^2xa + a^2$$
$$= axa^2 - axa^2 - a^2 + a^2 = 0,$$

hence $axa - a = 0$ and this shows R to be regular.

Conversely, if R is regular and reduced, let $a \in R$ and take $x \in R$ such that $axa = a$. Then

$$(a^2x - a)^2 = a^2xa^2x - a^2xa - a^3x + a^2$$
$$= a^3x - a^2 - a^3x + a^2 = 0,$$

hence $a^2x - a = 0$ and so R is strongly regular. ∎

Now any regular ring R is *semiprimitive*, i.e. its Jacobson radical $\mathfrak{J}$ is zero. For if $a \in \mathfrak{J}$ and $axa = a$, then $a(xa - 1) = 0$ and $xa - 1$ is a unit, by the definition of $\mathfrak{J}$, so $a = 0$. It follows that R is a subdirect product of primitive rings, which as homomorphic images of R are again regular (see A.3, Th. 10.4.1, p. 405). Now any primitive ring is clearly prime (A.3,

Prop. 10.6.1, p. 413), and there is little more that can be said about a regular prime ring. But a prime ring which is strongly regular is a field. To see this we first show that R has no idempotents apart from 0 and 1. For if e is an idempotent and $x \in R$, then $(ex(1 - e))^2 = 0$; since R is reduced, we have $ex(1 - e) = 0$ for all $x \in R$, and by primeness, either $e = 0$ or $e = 1$. Now given $a, x \in R$ satisfying $axa = a$, it follows that ax is idempotent, hence $ax = 0$ or 1, and $ax = 0$ only if $a = axa = 0$. Hence R is a field and we have proved

THEOREM 1.2.2. *A strongly regular primitive ring is a field. Any strongly regular ring is a subdirect product of fields.* ∎

It is clear that the class of rings embeddable in strongly regular rings is a quasi-variety. We shall show that this can be taken to be the class $\mathfrak{2}$, but for the proof we shall need a result which is best proved by using ultraproducts. We briefly recall the background, referring to A.3, 1.6 for details.

Let I be a non-empty set. A *filter* on I is a set $\mathscr{F}$ of subsets of I which is non-empty, admits intersections and contains with any subset X of I all subsets $\supseteq X$. Since $\mathscr{F}$ is non-empty, it follows that $I \in \mathscr{F}$ and we shall exclude the family of all subsets by requiring that $\varnothing \notin \mathscr{F}$. If the set $\mathscr{P}(I)$ of all subsets of I is regarded as a Boolean algebra, a filter on I may be regarded as the dual of a proper ideal in $\mathscr{P}(I)$. As for ideals one can show that any filter is contained in a maximal one; such a maximal filter is called an *ultrafilter* and may also be characterized by the fact that for any subset X of I, either X or its complement (but not both) belongs to the ultrafilter. For example, given $c \in I$, the set of all subsets containing c forms an ultrafilter, the *principal filter* generated by c. When I is finite, every ultrafilter has this form, as is easily seen, but on infinite sets there are non-principal ultrafilters, as we shall see in a moment.

Let $\{X_\lambda\}$ be a family of subsets of I with the following 'finite intersection property':

(FIP) *Any finite subfamily of the X_λ has a non-empty intersection.*

Then the set $\mathscr{F}$ of all subsets containing some finite intersection $\bigcap X_\lambda$ is a filter, as is easily checked; it is the filter *generated* by the family $\{X_\lambda\}$. To give an example, let I be an infinite set; then the *cofinite* sets, i.e. the subsets with a finite complement, have the finite intersection property and so generate a filter $\mathscr{F}_0$. Take an ultrafilter containing $\mathscr{F}_0$; this is an example of a non-principal ultrafilter on I.

Filters are used in the construction of certain homomorphic images of direct products. Let R_i $(i \in I)$ be a family of rings and let $P = \prod R_i$ be their direct product, with the natural projections $\pi_i : P \to R_i$. For any filter

$\mathcal{F}$ on the index set I we define the *reduced product* $P/\mathcal{F}$ as a homomorphic image of P by the rule:

> *For any* $x \in P$, $x = 0$ *if and only if* $\{i \in I \,|\, x\pi_i = 0\} \in \mathcal{F}$.

If the subsets of I belonging to $\mathcal{F}$ are called *$\mathcal{F}$-large*, then $P/\mathcal{F}$ may be described as the homomorphic image of P by the ideal of all elements vanishing on an $\mathcal{F}$-large index set. When $\mathcal{F}$ is an ultrafilter, the reduced product $P/\mathcal{F}$ is called an *ultraproduct*. It is clear that the reduced product is again a ring, though the reduced product of fields need not be a field. But an ultraproduct of fields is again a field; more generally, any elementary sentence of logic applying to rings holds in an ultraproduct precisely if it holds in all the factors of an $\mathcal{F}$-large set (see A.3, Th. 1.6.4, p. 30).

The next proposition (whose statement does not mention ultra-products) shows how these ideas may be applied:

PROPOSITION 1.2.3. *Let R be a ring and $f_\lambda : R \to K_\lambda$ ($\lambda \in \Lambda$) a family of homomorphisms from R to fields, such that for any finite subset X of $R^\times$, some f_λ maps no element of X to zero. Then R is embeddable in a field.*

Proof. For each $x \in R^\times$ define $\Lambda(x) = \{\lambda \in \Lambda \,|\, x f_\lambda \neq 0\}$; by hypothesis,

$$\Lambda(x_1) \cap \ldots \cap \Lambda(x_n) \neq \varnothing,$$

for any finite family $x_1, \ldots, x_n$ of non-zero elements of R. Hence the sets $\Lambda(x)$ generate a filter, and we can find an ultrafilter $\mathcal{F}$ containing all the $\Lambda(x)$. We can combine the f_λ to a homomorphism $f : R \to P = \prod K_\lambda$; let $g : P \to P/\mathcal{F}$ be the natural homomorphism to the reduced product. As an ultraproduct of fields $P/\mathcal{F}$ is a field and for each $x \in R^\times$, $x f_\lambda \neq 0$ on an $\mathcal{F}$-large set, namely $\Lambda(x)$. Hence $xfg \neq 0$ and so fg is the desired embedding. ∎

Let us return to the equation (4). In order to identify $\mathfrak{Q}$ let us define $\mathfrak{R}$ as the class of strongly regular rings. Clearly $\mathfrak{R}$ admits direct products, for if R_λ is a family of strongly regular rings and $P = \prod R_\lambda$ is their direct product, take any $a = (a_\lambda) \in P$. Then there exists $x_\lambda \in R_\lambda$ such that $a_\lambda^2 x_\lambda = a_\lambda$; hence on putting $x = (x_\lambda)$, we have $a^2 x = a$, showing that P is again strongly regular. Thus $\mathfrak{R}$ admits direct products; in the same way we see that the class $s\mathfrak{R}$ of subrings of strongly regular rings admits direct products; moreover it admits subrings and so it is a quasi-variety. We assert that $\mathfrak{Q}$ may be taken to be $s\mathfrak{R}$, thus we have

$$s\mathcal{F} = \mathfrak{Q} \cap s\mathfrak{R}. \qquad (5)$$

For clearly any ring embeddable in a field is an integral domain and is embeddable in a strongly regular ring. Conversely, if R is an integral domain and embeddable in a strongly regular ring, then by Th. 2.2, R is embeddable in a direct product of fields; hence for each $x \in R^\times$ we can find a homomorphism to a field which is non-zero on x. It follows that each finite family $\{x_1, \ldots, x_n\}$ can be mapped to non-zero elements, since we can apply the argument to the product $x_1 \ldots x_n$. So the condition of Prop. 2.3 is satisfied and R is embeddable in a field. Thus we have proved (5); the result may be stated as

THEOREM 1.2.4. *A subring of a strongly regular ring is embeddable in a field if and only if it is an integral domain.* ∎

As a consequence we have

COROLLARY 1.2.5. *A subring of a direct product of fields is embeddable in a field if and only if it is an integral domain.* ∎

These results show that the class of integral domains embeddable in fields can be defined by quasi-identities. Later, in 6.7, we shall find an explicit set of such quasi-identities; this set is infinite and we shall see that it cannot be replaced by a finite set of quasi-identities, or indeed by any finite set of elementary sentences.

Exercises

1. Let R be a commutative Bezout domain, i.e. an integral domain in which every finitely generated ideal is principal. Show that every element of its field of fractions can be written in the form a/b, where a, b are coprime, and a, b are unique up to multiplication by a unit.

2. Let $R = k[x, y, z, t; xz = yt]$. Show that there is no way of writing the fraction x/y so that it is defined and either finite or ∞ under every homomorphism of R to a field.

3. Verify that (2) holds in every subsemigroup of a group, but not in every cancellation semigroup. (Hint. Try writing the conditions in terms of 2×2 matrices; cf. also 1.4.)

4. Show that every strongly regular ring is semiprime, i.e. $xRx = 0$ implies $x = 0$.

5. Let R be a ring with a unary operation $x \mapsto x'$ such that $x^2 x' = x$, $x'^2 x = x'$. Show that x' is uniquely determined by x.

6. Let R be a non-trivial ring such that for all a, $c \neq 0$ in R, either $xa = c$ or $ax = c$ has a solution in R. Show that R is a field.

7. Show that every identity can be expressed as a quasi-identity.

1.3 Ore's method

We shall now look at the embedding problem in more detail and in particular treat an important special case, first described by Ore [31]. But we begin with some general remarks on the problem of constructing inverses in rings.

Let R be a ring and S a subset of R. A homomorphism $f: R \to R'$ to another ring R' is called *S-inverting* if for each $s \in S$, sf is an invertible element of R', i.e. an element with a two-sided inverse; such an invertible element is also called a *unit*. The following result, although trivial to prove, is useful in considering S-inverting maps.

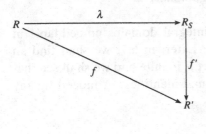

PROPOSITION 1.3.1. *Given a ring R and a subset S of R, there exists a ring R_S with an S-inverting homomorphism $\lambda: R \to R_S$ which is universal S-inverting, in the sense that for each S-inverting homomorphism $f: R \to R'$ there is a unique homomorphism $f': R_S \to R'$ such that $f = \lambda f'$.*

As with all universal constructions, the universal property determines R_S up to isomorphism (see A.2, 1.3).

Proof. To construct R_S we take a presentation of R and for each $s \in S$ adjoin an element s' with defining relations $ss' = s's = 1$. The map λ is defined by assigning to each element in R the corresponding element in the given presentation. Then $s\lambda$ is invertible, the inverse being s'. Thus we have a ring R_S and an S-inverting homomorphism $\lambda: R \to R_S$, and it remains to verify the universal property of λ. Given any S-inverting homomorphism $f: R \to R'$, we define $f': R_S \to R'$ by mapping $a\lambda$ (for $a \in R$) to af and s' to $(sf)^{-1}$, which exists in R', by hypothesis. Any relation in R_S must be a consequence of the relations in R and the relations expressing that s' is the inverse of $s\lambda$. All these relations still hold in R', so f' is well-defined and it is clearly a homomorphism. It is unique because its values on $R\lambda$ are prescribed, as well as on $(S\lambda)^{-1}$, by the uniqueness of inverses. ∎

The ring R_S constructed here is called the *universal S-inverting ring* or the *localization at S* of R. We have in fact a functor from pairs (R, S) consisting of a ring R and a subset S of R, with morphisms $f: (R, S) \to (R', S')$ which are homomorphisms from R to R' mapping S into S', to the category Rg of rings and homomorphisms.

All this is easily checked, but it provides no information about the structure of R_S. In particular we shall be interested in a normal form for the elements of R_S and an indication of the size of the kernel of λ, which could be the whole of R (e.g. if $0 \in S$). This general question will be treated in Ch. 4; for the moment we shall make some simplifying assumptions, which will allow us to obtain a complete answer in an important special case.

Let us look at the commutative case first. To get a convenient expression for the elements of R_S we shall take S to be *multiplicative*, i.e. $1 \in S$ and if $a, b \in S$, then $ab \in S$. In this case every element of R_S may be written as a fraction a/s, where $a \in R$, $s \in S$, and $a/s = a'/s'$ if and only if $as't = a'st$ for some $t \in S$. This is not exactly what one understands by a normal form, but it is sufficiently explicit to allow us to determine the kernel of λ, viz.

$$\ker \lambda = \{a \in R | at = 0 \text{ for some } t \in S\}. \tag{1}$$

Ore's idea consists in asking under what circumstances the elements of R_S have this form, when commutativity is not assumed. We must be able to express $s^{-1}a$, where $a \in R$, $s \in S$, as $a_1 s_1^{-1}$ ($a_1 \in R$, $s_1 \in S$), and on multiplying up, we find $as_1 = sa_1$. More precisely, we have $as_1 1^{-1} = sa_1 1^{-1}$, whence $as_1 t = sa_1 t$ for some $t \in S$. This is the well known *Ore condition* and it leads to the following result:

THEOREM 1.3.2. *Let R be a ring and S a subset, such that*
(D.1) *S is multiplicative,*
(D.2) *For any $a \in R$, $s \in S$, $sR \cap aS \neq \varnothing$,*
(D.3) *For any $a \in R$, $s \in S$, $sa = 0$ implies $at = 0$ for some $t \in S$.*
Then the universal S-inverting ring R_S may be constructed as follows. On $R \times S$ define the relation

$$(a, s) \sim (a', s') \text{ whenever } au = a'u', \ su = s'u' \in S \text{ for some } u, u' \in R. \tag{2}$$

This is an equivalence on $R \times S$ and the quotient set $R \times S/\sim$ is R_S. In particular, the elements of R_S may be written as fractions $a/s = as^{-1}$ and $\ker \lambda$ is given by (1). ∎

The full proof is a lengthy but straightforward verification, which may be left to the reader (see A.3, Th. 9.1.3, p. 350). It can be simplified a little by observing that the assertion may be treated as a result on monoids; once the 'universal S-inverting monoid' R_S has been constructed, by the method of this theorem, it is easy to extend the ring structure of R to R_S. A multiplicative subset S of R satisfying (D.2) is called a *right Ore set*; if $sa = 0$ or $as = 0$ for $a \in R$, $s \in S$ implies $a = 0$, S is said to be *regular*. More generally, if S satisfies (D.3), it is said to be right *reversible* and if (D.1–3) hold, S is sometimes called a *right denominator set*. Of course a corresponding construction can be carried out on the left, leading to left fractions $s^{-1}a$, when the left Ore condition holds.

When S is contained in the centre of R, in particular, when R is commutative, (D.2–3) are automatic and can be omitted. If R is an integral domain, (D.3) can be omitted and if moreover $S = R^\times$ then (D.2) reads $aR \cap bR \neq 0$ for $a, b \neq 0$. In that case $R_{R^\times}$ is a field, for when $a, b \in R^\times$, then ab^{-1} has the inverse ba^{-1}. Thus we have

COROLLARY 1.3.3. *Let R be an integral domain such that*

$$aR \cap bR \neq 0 \quad for \ a, b \in R^\times. \tag{3}$$

Then the localization of R at $R^\times$ is a field K and the natural homomorphism $\lambda: R \to K$ is an embedding.

Only the last part remains to be proved, and this is clear because ab^{-1} for $a, b \neq 0$ has the inverse ba^{-1} and $\ker \lambda = 0$, by (1) for an integral domain. ∎

An integral domain R satisfying condition (3) is called a *right Ore domain*, so the corollary tells us that every right Ore domain has a field of fractions.

It is important to observe that the localization at a right Ore set is essentially unique. Let us first note that the construction is functorial. Thus, given a map between pairs $f: (R, S) \to (R', S')$, i.e. a homomorphism f from R to R' such that $Sf \subseteq S'$, we have the diagram shown, and by universality there is a unique homomorphism $f_1: R_S \to R'_{S'}$ such that the resulting square commutes. In other words, λ is a natural transformation. It follows in particular that if f is an isomorphism, then so is f_1.

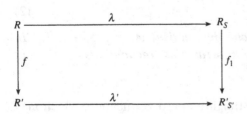

So far R, R' have been quite general; suppose now that R is a right Ore domain and $R' = D$ is any field of fractions of R, thus we have an embedding $f: R \to D$. Putting $S = R^{\times}$, we have a homomorphism $f_1: R_S \to D$, which is an embedding, because R_S is a field. The image of R_S under f_1 is a field containing R and hence equal to D, because D was a field of fractions. Thus f_1 is an isomorphism, and we have proved

PROPOSITION 1.3.4. *The field of fractions of a right Ore domain is unique up to isomorphism.* ∎

The result is of particular interest because it ceases to hold for general rings. We shall soon (in 2.1) meet rings which have several non-isomorphic fields of fractions.

If in the above diagram we take $R' = R$ to be a right Ore domain and $S = S' = R^{\times}$, then f is an endomorphism of R and the condition $Sf \subseteq S$ expresses the fact that f is injective. This sufficient condition for extendability is clearly also necessary. Moreover, f_1 is uniquely determined by f, because we have

$$(ab^{-1})f_1 \cdot bf = af.$$

This provides a criterion for the extendability of endomorphisms:

COROLLARY 1.3.5. *Let R be a right Ore domain with field of fractions K. Then an endomorphism of R can be extended to an endomorphism of K if and only if it is injective; its extension is then unique.* ∎

This result also fails to extend to general rings, as we shall see in 4.5.

We conclude this section with a sufficient condition for a ring to be an Ore domain which is often useful.

PROPOSITION 1.3.6. *Let R be an integral domain; then either R is a right Ore domain or it contains a right ideal which is free of infinite rank as R-module. In particular, every right Noetherian domain is right Ore.*

Proof. Suppose that R is an integral domain which is not right Ore. Then there exist $a, b \in R^{\times}$ such that $aR \cap bR = 0$; we claim that the elements $b, ab, a^2b, \ldots$ are right linearly independent over R, so the right ideal generated by them is free, of infinite rank.

If this were not so, then there would be a relation $\sum a^i b c_i = 0$, where the c_i are not all 0. Let c_r be the first non-zero coefficient; then we can cancel a^r and obtain the relation

$$bc_r + abc_{r+1} + \ldots + a^{n-r}bc_n = 0.$$

Hence

$$a(bc_{r+1} + \ldots + a^{n-r-1}bc_n) = -bc_r \neq 0,$$

and this contradicts the assumption on a and b. Now the last sentence follows because a free right ideal of infinite rank clearly cannot be finitely generated. ■

Since a principal right ideal domain is right Noetherian, we have

COROLLARY 1.3.7. *Every principal right ideal domain is a right Ore domain and hence has a field of fractions.* ■

Exercises

1. Give a proof of Th. 3.2; in particular, verify the expression (1) for the kernel of λ.

2. Let $R = \varinjlim R_\lambda$ be the direct limit of a directed system of Ore domains. Show that R is an Ore domain.

3. Show that the isomorphism in Prop. 3.4 is unique.

4. Let M be a monoid with a submonoid S, satisfying the following conditions: (i) for all $m \in M$, $s \in S$, $mS \cap sM \neq \varnothing$, (ii) if $sm = sm'$ for $m, m' \in S$ and $s \in S$, then there exists $t \in S$ such that $mt = m't$. Show that the universal S-inverting monoid M_S of M (defined as for rings) can be constructed as a set of fractions mt^{-1} such that $m_1 t_1^{-1} = m_2 t_2^{-1}$ if and only if $m_1 t = m_2 t'$, $t_1 t = t_2 t' \in S$ for some $t, t' \in M$. Find the conditions for the natural homomorphism from M to M_S to be an embedding.

5. Let R be a right Ore domain with field of fractions K and let $f: R \to D$ be a homomorphism to a field D such that D as a field is generated by im f (thus D is an epic R-field in the sense of 4.2). Show that K contains a subring L with a single maximal ideal $\mathfrak{m}$ such that $L/\mathfrak{m} \cong D$.

6. Let R be a commutative ring, A an $n \times n$ matrix over R and $S = R\langle x_{ij}; AX = XA = I\rangle$, where $X = (x_{ij})$, thus S is the universal A-inverting ring. Show that S is commutative. (Hint. Verify first that x_{ij} centralizes R.)

7. Show that for a principal ideal domain R and a right Ore set S in R the localization R_S is again principal.

1.4 Necessary conditions for a field of fractions to exist

The most obvious condition for the embeddability of a ring in a field is the absence of zero-divisors (where $\vee$ stands for 'or'):

$$xy = 0 \Rightarrow x = 0 \vee y = 0, \quad 1 \neq 0. \tag{1}$$

As is well known, in the commutative case it is also sufficient, but for general rings, (1) does not really go to the heart of the matter, for here we need conditions involving addition as well as multiplication. In this section we shall discuss a number of such conditions and their interrelation; some of them will play a role in what follows. In any case, all are easily stated and of some independent interest.

Let R be any ring and F a free left R-module on a basis B. If B is infinite, then any two bases of F have the same cardinal (see A.2, Prop. 4.4.4, p. 142), but for finite B this need not be the case. We shall say that F *has unique rank* if any two bases of F have the same number of elements, and the number of these elements will be called the *rank* of F, written $\mathrm{rk}(F)$. Over a field, every module is free, of unique rank, as is well known; even for subrings of fields the rank shows good behaviour that we shall now consider in more detail.

DEFINITION 1. A ring R is said to have *invariant basis number* (IBN) if every free R-module has unique rank.

By the earlier remark this condition is automatic for free modules that are not finitely generated, and the duality $^nR = \mathrm{Hom}_R(R^n, R)$ shows that it is enough to assume unique rank for finitely generated free *left* R-modules. Most rings arising naturally have IBN; in particular, any field or more generally, any ring with a homomorphism to a field has IBN. Examples lacking IBN will occur later when we come to construct integral domains not embeddable in fields.

Next we have a strengthening of IBN:

DEFINITION 2. A ring R is said to have *unbounded generating number* (UGN) if for every $n = 1, 2, \ldots$ there is a finitely generated left R-module which cannot be generated by fewer than n elements.

It is equivalent to require that for all n, R^n cannot be generated by fewer than n elements. For if this holds, R satisfies the condition; conversely, when it is false, suppose that R^m can be generated by $m - 1$ elements. Since every m-generator left R-module is a homomorphic image of R^m, it follows that any module on m generators can be generated by $m - 1$ elements, and by induction it follows that every finitely generated module can be generated by $m - 1$ elements, so UGN fails to hold.

If IBN fails for R, say $R^m \cong R^n$ for $m > n$, then R^m has a basis of $n < m$ elements, so UGN fails to hold; thus UGN implies IBN.

Next we come to an even stronger condition.

DEFINITION 3. A ring R is said to be *weakly n-finite* if any generating set of n elements of R^n is free. If R is weakly n-finite for all n, it is called *weakly finite*.

Weakly 1-finite rings are sometimes called 'von Neumann finite', 'inverse symmetric' or 'directly finite'.

Let R be a non-trivial ring. If R^n can be generated by fewer than n elements, we can find n-element generating sets for R^n that are not free, so any weakly finite (non-trivial) ring has UGN. Thus for any non-trivial ring we have the implications

$$\text{WF} \Rightarrow \text{UGN} \Rightarrow \text{IBN}, \tag{2}$$

where WF stands for 'weakly finite'. The trivial ring 0 is weakly finite but does not have UGN or IBN. Moreover, neither of the implications in (2) can be reversed (see Cohn [66]).

To restate the conditions, let us suppose that R^n has a generating set of m elements. Then we have an exact sequence

$$0 \to K \to R^m \to R^n \to 0,$$

which splits, because R^n is free. So we obtain the relation

$$R^m \cong R^n \oplus K. \tag{3}$$

This isomorphism allows us to restate the three conditions in the following form:

(IBN) *For all m, n, $R^m \cong R^n$ implies $m = n$,*

(UGN) *For all m, n, $R^m \cong R^n \oplus K$ implies $m \geqslant n$,*

(WF) *For all n, $R^n \cong R^n \oplus K$ implies $K = 0$.*

By taking bases for our modules, we obtain a restatement in terms of matrices:

(IBN) *For $A \in {}^nR^m$, $B \in {}^mR^n$, if $AB = I_n$, $BA = I_m$, then $m = n$,*

(UGN) *For $A \in {}^nR^m$, $B \in {}^mR^n$, if $AB = I_n$, then $m \geqslant n$,*

(WF) *For A, $B \in R_n$, if $AB = I$, then $BA = I$.*

These conditions again make it clear that (2) holds for any non-trivial ring. Further, it can be shown that a ring R has UGN if and only if some non-zero homomorphic image of R is weakly finite (see FR, Prop. 0.2.2, p. 8 or Th. 4.6.8 below). Any field (even skew) has all three properties; more generally all three hold for any subring of a field, for any Artinian or Noetherian ring (A.2, 4.4, p. 144) and for any commutative ring (always excluding the trivial ring). In fact, a ring lacking any of these properties is usually regarded as pathological.

Let R be a non-trivial ring without IBN. Then for some $m \neq n$,

$$R^m \cong R^n. \tag{4}$$

We take $(r, r + d)$ to be the first such pair of distinct integers (m, n) in the lexicographic ordering. Then it is not hard to verify that (4) holds precisely when $m = n$ or $m, n \geq r$ and $d | m - n$. In this case R is said to have *free module type* $(r, r + d)$. For example, if V is an infinite-dimensional vector space over a field k, then $V \cong V \oplus V$, and it follows easily that $\mathrm{End}_k (V)$ is a ring of free module type $(1, 2)$.

Next take a non-trivial ring R without UGN. Then for some n, R^n can be generated by $n - 1$ elements. If the least such n is m, then every finitely generated R-module can be generated by $m - 1$ elements. Let us call such a ring non-UGN of type m. From the matrix form of the conditions we see that there is a universal non-UGN ring of type m, namely the k-algebra generated by the entries of an $m \times (m - 1)$ matrix A and an $(m - 1) \times m$ matrix B subject to the defining relations (in matrix form) $AB = I$. We shall denote it by U_m. Similarly we can form a universal non-IBN ring of free module type $(r, r + d)$ by taking the entries of an $r \times (r + d)$ matrix A and an $(r + d) \times r$ matrix B with defining relations (in matrix form) $AB = I$, $BA = I$. This ring will be denoted by $V_{r,r+d}$. Every non-UGN ring is a homomorphic image of an appropriate universal non-UGN ring, and similarly for non-IBN rings.

We can also form a non-WF ring by taking the k-algebra W_n on the entries of two $n \times n$ matrices A, B with defining relations $AB = I$, but now it is no longer the case that all homomorphic images are non-WF: we obtain a weakly finite ring by imposing the relations $BA = I$.

Let us record the effect of homomorphisms on these conditions.

PROPOSITION 1.4.1. *Let $R \to \bar{R}$ be a homomorphism between rings.* (i) *If R is non-IBN of free module type $(r, r + d)$, then $\bar{R}$ is non-IBN of free module type $(r', r' + d')$, where $r' \leq r$, $d' | d$.* (ii) *If R is non-UGN of type m, then $\bar{R}$ is non-UGN of type m', where $m' \leq m$.*

Proof. (i) By hypothesis, $R^r \cong R^{r+d}$; on tensoring with $\bar{R}$ we obtain $\bar{R}^r \cong \bar{R}^{r+d}$, from which the result follows. Similarly, (ii) follows on tensoring the isomorphism $R^{m-1} \cong R^m \oplus K$. ∎

It follows that the conditions IBN and UGN are reflected by isomorphisms:

COROLLARY 1.4.2. *Given a homomorphism* $R \to \bar{R}$, *if* $\bar{R}$ *has IBN or UGN, then so does* R. ∎

There is another way to express UGN, which requires a definition that is much used later. Let A be a square matrix, say $n \times n$, over a ring R, and consider the different ways of writing A as a product:

$$A = PQ, \text{ where } P \text{ is } n \times r \text{ and } Q \text{ is } r \times n, \tag{5}$$

for varying r. If in every representation (5) of A we have $r \geq n$, A is said to be *full*. The matrix formulation of UGN above shows that it holds if and only if every unit matrix is full. More generally we have the following criterion for UGN:

PROPOSITION 1.4.3. *A ring R has unbounded generating number if and only if every invertible matrix is full. In particular, over a (non-trivial) weakly finite ring every invertible matrix is full.*

Proof. If every invertible matrix is full, then I_n is full for all n and so UGN holds. Conversely, suppose that A is invertible, but not full, say (5) holds with $r < n$. Then $I = P \cdot QA^{-1}$, so I is not full and hence UGN fails to hold for R. The last part follows because every non-trivial weakly finite ring has UGN. ∎

We now turn to a condition of a different kind. We recall that for any square matrix A over a field K (even skew) the following four conditions are equivalent:

(S.1) *A has no left inverse,*
(S.1°) *A has no right inverse,*
(S.2) *A is a left zero-divisor,*
(S.2°) *A is a right zero-divisor.*

A square matrix over a field with these properties is called *singular*, any other square matrix is called *non-singular*; the latter holds (by (S.2, 2°)) precisely when the matrix is regular, or equivalently, when A is full, as is easily verified.

When A is singular, there is a non-zero column u such that $Au = 0$. If U is an invertible matrix with u as its first column, then we have

$$U^{-1}AU = \begin{pmatrix} 0 & c \\ 0 & A_1 \end{pmatrix}, \tag{6}$$

for some row c and $(n-1) \times (n-1)$ matrix A_1. In particular, if A is nilpotent, say $A^r = 0$, then A is singular and the matrix A_1 in (6) is again nilpotent. An induction shows that when A is nilpotent, then there is an invertible matrix T such that $T^{-1}AT$ is upper triangular, with zeros on the main diagonal. It follows that an $n \times n$ nilpotent matrix A over a field satisfies $A^n = 0$, so we have proved

PROPOSITION 1.4.4. *An $n \times n$ nilpotent matrix A over a field satisfies $A^n = 0$.* ∎

Clearly this condition must also hold for any matrix over a subring of a field. The condition of Prop. 4.4 is known as *Klein's nilpotence condition*, since Klein [69] showed, by an ingenious use of Malcev's conditions:

If R is an integral domain such that every nilpotent $n \times n$ matrix C over R satisfies $C^n = 0$, then $R^{\times}$ is embeddable in a group.

Klein showed also that his condition is not necessary for $R^{\times}$ to be embeddable in a group. As we have seen, it is necessary for embeddability in a field, but not sufficient, as counter-examples of Bergman [74'] show (see 5.7, Ex. 4). Klein shows further that his condition implies weak finiteness:

PROPOSITION 1.4.5. *Any ring satisfying Klein's nilpotence condition is weakly finite.*

Proof. Let $A, B \in R_n$ satisfy $AB = I$. Then $A^r B^r = I$ for all $r \geqslant 1$, by an easy induction. Moreover, $A(I - B^r A^r) = (I - B^{r-1}A^{r-1})A$, hence the matrix $C = A(I - B^r A^r)$ satisfies $C^i = (I - B^{r-i}A^{r-i})A^i$; in particular, $C^r = 0$. Choosing $r = n + 1$, $i = n$, we find that

$$I - BA = (I - BA)A^n B^n = C^n B^n = 0,$$

because $C^n = 0$, by Klein's condition. Hence $BA = I$, as we had to show. ∎

The converse does not hold, as an example by Klein [70'] shows.

Exercises

1. Show that a matrix (not necessarily square) over a ring, which has a right inverse, is left regular; verify that the converse holds over any Artinian ring. Deduce that over an Artinian ring the four conditions (S.1–2°) are equivalent for any square matrix.

2. Verify that a ring satisfying Klein's nilpotence condition is reduced, and deduce that a prime ring satisfying Klein's nilpotence condition is an integral domain.

3. Show that in a weakly finite ring, if A is full, then so is $A \oplus I$.

4. Let R be a weakly finite ring and A, B square matrices over R. If $\begin{pmatrix} A & C \\ 0 & B \end{pmatrix}$ for some block C is invertible, show that A and B are both invertible. Show that weak finiteness is necessary as well as sufficient.

5. Show, by taking a suitable homomorphic image, that the universal non-WF ring satisfies UGN.

6. (Jacobson [50]) Let R be a ring with elements a, b such that $ab = 1 \neq ba$. Show that the elements $e_{ij} = b^{i-1}(1 - ba)a^{j-1}$ form an infinite set of matrix units and the e_{ii} form an infinite family of pairwise orthogonal idempotents.

7. (Kirezci [82]). Let $V_{n,m}$ be the universal non-IBN k-algebra of free module type (n, m). Show that $V_{1,m}$ for $m > 1$ is simple.

8. (After Kirezci [82]) If the defining matrices for $V_{n,m}$ are A, B and those for $V_{n',m'}$ are A', B', show by mapping A' to $A(A \oplus I)$ and B' to $(B \oplus I)B$ that there is a homomorphism from $V_{n,n+rk}$ to $V_{n,k}$. Deduce that for all $m > n \geqslant 1$ there is a homomorphism $V_{n,m} \to V_{n,n+1}$.

9. Show that if R has free module type $(r, r + d)$ and $n \geqslant r$, then $\mathfrak{M}_n(R)$ has free module type $(1, 1 + d/(d, n))$.

1.5 Stable association and similarity

This somewhat technical section provides some necessary background for the factorization of elements and matrices. It is convenient to have these results in one place, but the reader may well decide to skip this section and refer back to it later when necessary.

Let R be any ring. Two matrices A, A' over R are said to be *associated* if there exist invertible matrices P, Q over R such that

$$A' = PAQ.$$

If R is a ring with IBN, then every invertible matrix is square, so if A is $m \times n$, the same then holds for A'.

Sometimes a weaker condition is needed. We shall say that A and A' are *stably associated* if $A \oplus I$ is associated to $A' \oplus I$ for some unit matrices (not necessarily of the same size). Thus if A is $m \times n$, then A' will be $(m + r) \times (n + r)$ for some $r \in \mathbf{Z}$. If we define the *index* of an $m \times n$ matrix as $n - m$, we see that over a ring with IBN, stably associated matrices have the same index.

Any finitely presented left R-module M is given by a resolution

$$R^m \xrightarrow{\alpha} R^n \rightarrow M \rightarrow 0,$$

where the map α is given by an $m \times n$ matrix A. It is clear that stably associated matrices give rise to isomorphic modules. Moreover, if A is left regular, so that α is injective, then any left regular matrix A' presenting an isomorphic module must be stably associated to A. This result will not be needed here, so the proof is omitted (see FR, 0.6).

There is a related notion which is sometimes needed. Two matrices A, B with the same number of rows are called *right comaximal* if the matrix $(A \; B)$ has a right inverse. Similarly two matrices A', B' with the same number of columns are *left comaximal* if $(A' \; B')^T$ has a left inverse. An equation

$$AB' = BA' \tag{1}$$

is called *comaximal* if A, B are right comaximal and A', B' are left comaximal. The relation between these concepts is given by

THEOREM 1.5.1. *Let R be a ring and $A \in {}^rR^m$, $A' \in {}^sR^n$. Then the following conditions are equivalent:*
(a) *A, A' satisfy a comaximal relation (1),*
(b) *there is an $(r + n) \times (s + m)$ matrix $\begin{pmatrix} A & B \\ * & * \end{pmatrix}$ with a right inverse of the form $\begin{pmatrix} * & -B' \\ * & A' \end{pmatrix}$.*
In particular, (a) and (b) hold whenever
(c) *A and A' are stably associated,*
and in a weakly finite ring (a)–(c) are equivalent for two matrices A, A' of the same index.

Proof. If (a) holds, say A, A' satisfy (1), then by comaximality, there exist matrices C, D, C', D' such that

$$AD' - BC' = I, \qquad DA' - CB' = I, \tag{2}$$

and on writing

$$P = \begin{pmatrix} A & B \\ C & D \end{pmatrix}, \qquad Q = \begin{pmatrix} D' & -B' \\ -C' & A' \end{pmatrix}, \tag{3}$$

we have $PQ = \begin{pmatrix} I & 0 \\ S & I \end{pmatrix}$ for some S, hence P has the right inverse

$$\begin{pmatrix} D' & -B' \\ -C' & A' \end{pmatrix} \begin{pmatrix} I & 0 \\ -S & I \end{pmatrix} = \begin{pmatrix} * & -B' \\ * & A' \end{pmatrix},$$

and (b) follows. Conversely, if Q in (3) is a right inverse of P, then (1) and (2) hold, and this shows (1) to be a comaximal relation. Thus (a) and (b) are equivalent.

If (c) holds, say we have

$$\begin{pmatrix} A & 0 \\ 0 & I \end{pmatrix}\begin{pmatrix} P_1 & P_2 \\ P_3 & P_4 \end{pmatrix} = \begin{pmatrix} Q_1 & Q_2 \\ Q_3 & Q_4 \end{pmatrix}\begin{pmatrix} I & 0 \\ 0 & A' \end{pmatrix}, \tag{4}$$

then we have the relation

$$AP_2 = Q_2 A'. \tag{5}$$

Now

$$(A \ Q_2)\begin{pmatrix} P_1 & 0 \\ 0 & I \end{pmatrix} = (AP_1 \ Q_2) = (Q_1 \ Q_2),$$

by (4), and $(Q_1 \ Q_2)$ has a right inverse, therefore A and Q_2 are right comaximal. By symmetry (5) shows that P_2 and A' are left comaximal; thus (c) $\Rightarrow$ (a).

Now if R is weakly finite and A, A' have the same index, then $m - r = n - s$, hence $r + n = s + m$ and the matrices in (b) are square; by weak finiteness, the right inverse is a two-sided inverse and we have the equation

$$\begin{pmatrix} I & 0 \\ -C' & I \end{pmatrix}\begin{pmatrix} I & 0 \\ 0 & A' \end{pmatrix}\begin{pmatrix} A & B \\ C & D \end{pmatrix} = \begin{pmatrix} I & B \\ 0 & I \end{pmatrix}\begin{pmatrix} A & 0 \\ 0 & I \end{pmatrix},$$

which shows A and A' to be stably associated. ∎

We now turn to factorizations of elements and begin by introducing a notion weaker than comaximality. The definition could be framed for matrices, but only the case of elements will be needed.

A relation $ab' = ba'$ is called *coprime* if a, b are *left coprime*, i.e. they have no non-unit common left factor, and a', b' are *right coprime*. Clearly any two right comaximal elements are left coprime; in a principal ideal domain the converse holds, for here $aR + bR = dR$ and if a, b are left coprime, then d is a unit and so a, b are then right comaximal.

Let R be an integral domain and $a, a' \in R^\times$. Consider a module homomorphism

$$f: R/Ra \to R/Ra'. \tag{6}$$

If $1 \mapsto b'$, then $x \mapsto xb'$ and $a \mapsto 0$, hence $ab' \in Ra'$, say

$$ab' = ba'. \tag{7}$$

Thus a homomorphism (6) is given by a relation (7), and conversely, every relation (7) leads to a homomorphism (6). Here b' is determined by f up to an element of Ra'; thus if $b_1' = b' + za'$, then by (7), $ab_1' = a(b' + za') = ba' + aza' = (b + az)a'$. So b is determined by f up to an element of aR and f defines a unique R-homomorphism

$$f^*: R/a'R \to R/aR, \quad x \mapsto bx. \tag{8}$$

Let us call a module of the form R/Ra or R/aR, where $a \neq 0$, *strictly cyclic*. What the above argument shows is that the categories of left and right strictly cyclic modules over an integral domain are dual to each other; this is also called the *factorial duality*.

We note that the map (6) is injective if $xb' \in Ra'$ implies $x \in Ra$, i.e. if (7) is a least common left multiple (LCLM) of a' and b'; when this is so, (7) is left coprime. For the surjectivity of (6) we require $c \in R$ such that $c \mapsto 1$, i.e. $cb' - 1 \in Ra'$, so for some $d \in R$ we have

$$da' + cb' = 1.$$

This expresses the fact that (7) is left comaximal, hence right coprime. In particular, over a PID we have an isomorphism (6) if and only if there is a comaximal relation (7). With Th. 5.1 this yields conditions for an isomorphism of strictly cyclic modules over a PID:

PROPOSITION 1.5.2. *Let R be a principal ideal domain. Then for any $a, a' \in R^\times$ the following conditions are equivalent:*
  (a) *there is a coprime relation*

$$ab' = ba'; \tag{7}$$

  (b) *there is a comaximal relation (7);*
  (c) $R/Ra \cong R/Ra'$;
  (c°) $R/aR \cong R/a'R$. ∎

Two elements a, a' satisfying condition (b) are said to be *similar*. Thus in a PID the isomorphism between strictly cyclic modules is described by similarity. It is clear from Th. 5.1 that in a weakly finite ring two elements are similar if and only if they are stably associated, while in the commutative case similar elements are just associated.

An element of a ring will be called an *atom*, or *irreducible*, if it is a non-unit which cannot be written as a product of two non-units. Let R be a PID and $a \in R^\times$; any factorization $a = p_1 \ldots p_r$ into non-unit factors corresponds to a chain of right ideals

$$R \supset p_1 R \supset p_1 p_2 R \supset \ldots \supset p_1 \ldots p_r R = aR, \tag{9}$$

where the inclusions are proper, because the p_i are non-units. If all the p_i

are atoms, then (9) is a composition series, and by applying the Jordan–
Hölder theorem (A.1, 9.3) to two composition series, $a = p_1 \ldots p_r$
$= p_1' \ldots p_s'$, we find that $r = s$ and for some permutation $i \mapsto i'$ of
$\{1, \ldots, r\}$, p_i is similar to $p_{i'}'$. This is expressed by saying that R is a
unique factorization domain (UFD); in the commutative case it reduces
to the usual meaning of this term, because then similar elements are
associated. The main point of distinction for general UFDs is that
corresponding factors are merely similar and not necessarily associated,
and the factors cannot always be rearranged at will. The composition
length of (9) is also called the *length* of a.

Let us return to a general integral domain R and specialize (6) by
taking $a' = a$. We see that the endomorphisms of R/Ra are given by
elements b' such that

$$ab' = ba,$$

for some $b \in R$. Let us define the *idealizer* of the left ideal Ra as the
largest subring of R containing Ra as an ideal:

$$\mathcal{I}(Ra) = \{b' \in R | ab' \in Ra\}.$$

Clearly this is a subring and from what has been said, there is a surjective
homomorphism from $\mathcal{I}(Ra)$ to $\mathrm{End}_R(R/Ra)$ obtained by mapping
$b' \in \mathcal{I}(Ra)$ to the endomorphism $x \mapsto xb'$. The kernel is easily seen to be
Ra, hence we have

$$\mathrm{End}_R(R/Ra) \cong \mathcal{I}(Ra)/Ra.$$

The quotient $\mathcal{I}(Ra)/Ra$ will be denoted by $\mathcal{E}(Ra)$ and called the
eigenring of Ra.

An element c of a ring R is said to be *right invariant* if c is right regular
and $Rc \subseteq cR$; this means that for each $x \in R$ there exists $x' \in R$ such that
$xc = cx'$, and here x' is uniquely determined by x, because c is right
regular. *Left invariant* elements are defined similarly and a left and right
invariant element is said to be *invariant*. Thus an invariant element in a
ring R is a regular element c such that $Rc = cR$.

An element a of R is said to be *bounded* if it is a left factor of an
invariant element: $c = ab$, where c is invariant. It is then also a right
factor, for $cb = b'c = b'ab$, hence $c = b'a$; this shows the notion defined
here to be symmetric. Any invariant element with a as factor is called a
bound for a; in general there may be no least bound, but in a PID we can
form the largest ideal contained in Ra. It will have the form Ra^*, where
a^* is invariant with a as factor, and clearly a^* is a factor of any bound of
a. Thus a^* is the *least bound* of a, unique up to unit factors. This least
bound a^* can also be characterized as the generator of the annihilating

ideal of R/Ra; it follows that any element similar to a is again bounded, with the same bounds. An element with no bounded factors other than units is said to be *totally unbounded*.

For bounded elements in a PID we have the following commutation formula. Two elements a, b are said to be *totally coprime*, if no non-unit factor of a is similar to a factor of b.

LEMMA 1.5.3. *In a principal ideal domain R, if a is bounded and totally coprime to u, then there is a comaximal relation*

$$au = u_1 a_1, \tag{10}$$

and hence

$$R/Rau \cong R/Ra \oplus R/Ru. \tag{11}$$

Proof. If we can find a relation (10), where a_1 is totally coprime to u and u_1 totally coprime to a, then it must be coprime and hence comaximal, so it only remains to find the relation.

Let $a^* = a'a$ be the least bound of a, and $a'au = a^*u = u_0 a^*$. If the LCRM of a' and u_0 is $a'u_1 = u_0 b$, then for some a_1, $a^* = ba_1$ and $au = u_1 a_1$. Here u_1 is similar to a right factor of u_0 and so is totally coprime to a, while a_1 is a factor of a^* and so totally coprime to u. Hence this is the required relation; now (11) follows by comaximality. ∎

An invariant element c in a ring is called an *I-atom* if it is a non-unit and the only factors of c which are again invariant are units or associates of c. One would expect the least bound of a bounded atom to be an I-atom; this is in fact the case, as we shall now show.

THEOREM 1.5.4. *Let R be a principal ideal domain.*

(i) *Every non-zero ideal of R has the form Rc = cR, where c is an invariant element, and R/Rc is a simple ring if and only if c is an I-atom;*

(ii) *every I-atom is a product of similar bounded atoms;*

(iii) *if p is a bounded atom, then its least bound p^* is an I-atom whose atomic factors are precisely all the atoms similar to p; moreover, the eigenring K of Rp is a field and we have*

$$R/Rp^* \cong \mathfrak{M}_n(K), \text{ where n is the length of } p^*. \tag{12}$$

Proof. (i) Let $\mathfrak{a}$ be a non-zero ideal of R. By definition, $\mathfrak{a} = Ra = a'R$, hence $\mathfrak{a} = Rc$ with an invariant generator c. Now $R/\mathfrak{a}$ is simple if and only if $\mathfrak{a}$ is maximal, but any ideal containing Rc has the form Rb, where b is a factor of c and is again invariant. So Rc is maximal precisely when c has no non-trivial invariant factors, i.e. it is an I-atom.

(ii) Let p^* be an I-atom and take any atom p dividing p^*; clearly p^* is the least bound of p. Since p is bounded, any similar atom p' is again bounded. If the least bound of p' is c, then $c(R/Rp) = 0$, hence $Rc \subseteq Rp$ and so $Rc \subseteq Rp^*$. By symmetry, $Rp^* \subseteq Rc$, so c is associated to p^*. Thus any atom similar to p is a factor of p^*, and it follows that

$$Rp^* \subseteq \bigcap \{Rp' \,|\, p' \text{ similar to } p\}. \tag{13}$$

If we can show that the right-hand side is a two-sided ideal, then equality must hold, since p^* is an I-atom. Take a in the right-hand side of (13); given $b \in R$ and p' similar to p, either $b \in Rp'$, and then $ab \in Rp'$, or $Rb + Rp' = R$, and so

$$Rb + Rp' = Rc,$$

where $c = p_1 b = b_1 p'$, and this is a coprime relation. It follows that p_1 is similar to p and so $a \in Rp_1$, and $ab \in Rp_1 b = Rb_1 p' \subseteq Rp'$. This shows the right-hand side of (13) to be an ideal, and so equality holds in (13). It follows that R/Rp^* is a sum of terms R/Rp, hence there is a chain from Rp^* to R whose quotients are all isomorphic to R/Rp, and so p^* is a product of elements similar to p.

(iii) Let p be a bounded atom; its least bound p^* is an I-atom, for if $p^* = cd$, where c, d are invariant non-units, then p is a factor of c or of d, contradicting the fact that p^* was the least bound of p. By (ii) p^* is a product of atoms similar to p and the atoms dividing p^* are precisely the atoms similar to p. The eigenring K of R/Rp, as endomorphism ring of the simple module R/Rp is a field, by Schur's lemma and R/Rp^* is a sum of copies of R/Rp, hence a direct sum of finitely many copies. If p^* has length n, then

$$R/Rp^* \cong (R/Rp)^n.$$

Taking endomorphism rings and bearing in mind that the endomorphism ring of any ring A as left A-module is A, we have the desired conclusion (12). ∎

Over a PID R every cyclic module is either free of rank 1 or of the form R/Ra for some $a \in R^\times$. By a *torsion module* we understand a module M in which every element is annihilated by a non-zero element of R. If no element of M other than zero is annihilated by a non-zero element of R, then M is said to be *torsion-free*. As for abelian groups one easily verifies that any module M (over a PID) has a uniquely determined maximal torsion submodule tM and that M/tM is torsion-free. When M

is finitely generated, then so is M/tM, and being torsion-free, it is free, so that we can write $M = tM \oplus F$, where F is a free submodule, unique up to isomorphism. The following structure theorem can be stated for finitely generated modules, but we limit ourselves to cyclic modules, as that is all we shall need:

THEOREM 1.5.5. *Let R be a principal ideal domain. Any cyclic left R-module is either free of rank 1 or of the form*

$$R/Ra \cong R/Rq_1 \oplus \ldots \oplus R/Rq_k \oplus R/Ru, \qquad (14)$$

where each q_i is a product of bounded similar atoms, while atoms in different q's are dissimilar, and u is totally unbounded. Moreover, u and the q_i are uniquely determined up to order and similarity.

Proof. It is clear that over a PID any submodule of a cyclic module is again cyclic; thus a submodule of R/Ra has the form Rb/Ra, where $a = cb$, and $Rb/Rcb \cong R/Rc$. Consider the bounded atoms that are factors of a; suppose that $p_1, \ldots, p_k$ are pairwise dissimilar bounded atoms such that each bounded atomic factor of a is similar to exactly one of the p_i. If p_i^* denotes the bound of p_i, then the submodule of R/Ra annihilated by p_i^* is again cyclic, of the form R/Rq_i, where q_i is a product of atoms similar to p_i. Clearly the sum $\sum R/Rq_i$ is direct and is equal to the submodule of R/Ra annihilated by $p_1^* \ldots p_k^*$. By Lemma 5.3 we obtain a decomposition (14), where R/Ru has no non-zero submodule with non-zero annihilator. It follows that u is totally unbounded, for if $u = u'pu''$ with a bounded atom p, where (by induction) u' is totally unbounded, then by Lemma 5.3, there is a comaximal relation $u'p = p_0u_0$, hence $u = p_0u_0u''$ and R/Ru has the submodule $Ru_0u''/Rp_0u_0u'' \cong R/Rp_0$, which is annihilated by p_0^*. This contradiction shows u to be totally unbounded, and it establishes the decomposition (14). Here R/Rq_i is unique as the submodule annihilated by p_i^* while R/Ru is unique up to isomorphism, hence q_i and u are unique up to similarity. ∎

We shall also need a result on direct decompositions. A non-zero element a of a PID R is said to be *indecomposable* if R/Ra is indecomposable as left R-module; by the factorial duality this notion is actually left–right symmetric. Now an indecomposable bounded element may be described as follows:

PROPOSITION 1.5.6. *Let R be a principal ideal domain. Then a bounded indecomposable element is a product of similar atoms and has a* *bound of the form p^n, where p is an I-atom. Two bounded indecomposable elements have the same bound if and only if they are similar.*

Proof. Let q be bounded indecomposable and consider the direct decomposition (14) for R/Rq. The term R/Ru is absent because q is bounded and since q is indecomposable there is only one term. Thus q is a product of similar atoms, say $q = p_1 \ldots p_r$; each p_i has the same bound p, an I-atom, hence q has the bound p^r and it follows that the least bound q^* of q also has this form. Here Rq^* is determined by the similarity class of q as the annihilator of R/Rq, while R/Rq is determined by Rq^* as an indecomposable part of R/Rq^*. ∎

If R is an Ore domain with centre Z, then the field of fractions of R has a centre containing Z, but not necessarily generated by it. By how much it fails is described in the next result:

PROPOSITION 1.5.7. *Let R be a right Ore domain and K its field of fractions, and let C be the centre of K.*

(i) *Then any element of C has the form ab^{-1}, where*

$$axb = bxa \quad \text{for all } x \in R. \tag{15}$$

Conversely, if $a, b \in R^\times$ and (15) holds, then $ab^{-1} \in C$.

(ii) *If R is a principal right ideal domain, then any $u \in C$ can be written as $u = ab^{-1}$, where a and b are left and right coprime, and for any representation of u as a quotient of coprime elements, there is an injective endomorphism α of R such that*

$$xa = ax^\alpha, xb = bx^\alpha \quad \text{for all } x \in R. \tag{16}$$

(iii) *If R is a principal ideal domain, then α in (ii) is an automorphism of R.*

Proof. (i) Suppose that $a, b \in R^\times$ satisfy (15). Then $ab = ba$ and on dividing (15) by b on the left and right, we obtain

$$b^{-1}ax = xab^{-1} \quad \text{for all } x \in R, \tag{17}$$

hence $b^{-1}a = ab^{-1} \in C$. Conversely, if $ab^{-1} \in C$, then $a = ab^{-1} \cdot b = bab^{-1}$, hence $ab = ba$, so (17) holds and on multiplying up we find (15).

(ii) Now assume that R is a PRID and take $u \in C$. Write $u = ab^{-1}$, $a = da_1$, $b = db_1$, where d is a HCLF of a, b. Then $ab^{-1} = b^{-1}a = b_1^{-1}a_1$ and $a_1 = b_1 \cdot b_1^{-1}a_1 = b_1^{-1}a_1 \cdot b_1$, hence $b_1a_1 = a_1b_1$. Thus we may assume in (i) that a, b are left coprime. We may also take them to be right coprime, for if $a = a_1d$, $b = b_1d$, then $ab^{-1} = a_1b_1^{-1}$ and now the same argument applies. We claim that

$$aR \cap bR = abR.$$

For we know that the LCRM of a and b has the form $ab' = ba'$. We also have $ab = ba$, hence $a = a'c$, $b = b'c$ and since a, b are right coprime, c must be a unit, so $ab = ba$ is the LCRM. Now $bxa = axb$ is a right multiple of $ab = ba$, hence for some $f \in R$, $bxa = baf$, $axb = abf$. Clearly f depends on x and is uniquely determined by it. Denoting it by x^α, we have

$$xa = ax^\alpha, \ xb = bx^\alpha.$$

Each of these equations determines x^α uniquely and it is easily checked that the map $x \mapsto x^\alpha$ is an injective endomorphism of R. If R is left as well as right principal, we have, by symmetry an injective endomorphism β of R such that $ax = x^\beta a$, $bx = x^\beta b$. Hence $xa = ax^\alpha = x^{\alpha\beta}a$, so $\alpha\beta = 1$ and similarly, $\beta\alpha = 1$, hence α is an automorphism. ∎

Exercises

1. Show that any two associated elements in a ring are similar, and that the converse also holds if one of the elements is invariant.

2. Use Th. 5.1 to show that if $A \in {}^r R^m$ and $A' \in {}^s R^n$ are stably associated over a weakly finite ring R, then $A \oplus I_n$ is associated to $I_m \oplus A'$.

3. Let R be a PRID and I the set of all its right invariant elements. Show that I is a regular right Ore set in R and that the localization R_I is a simple PID. If R is a (left and right) PID, verify that R_I is a field if and only if R is not primitive.

4. An integral domain R is said to be *rigid*, if for any $c \in R^\times$, $c = ab' = ba'$ implies $aR \subseteq bR$ or $bR \subseteq aR$. Show that any Noetherian rigid domain is a valuation ring (see 9.1). Show that in any atomic rigid domain every element not zero or a unit can be written as a product of atoms: $c = p_1 \ldots p_r$ where the factor p_i is unique up to associates.

5. Show that for $a \neq 0$ in a PID R, R/Ra and R/aR have the same annihilator, viz. $Ra^* = a^*R$, where a^* is the least bound of a.

6. Prove the extension of Th. 5.5 to finitely generated modules.

7. A ring will be called *fully right inversive* if every left regular matrix (not necessarily square) has a right inverse. Show that a ring R is fully right inversive if and only if the only finitely presented bound right R-module is 0. (A module M is

bound if $\text{Hom}(M, R) = 0$. It can be shown that every ring R can be embedded in a fully right inversive ring in which all left regular matrices over R remain left regular (and so acquire right inverses), see Cohn [93]).

1.6 Free algebras, firs and semifirs

In this section we shall meet conditions that are sufficient for embeddability in a field, although their sufficiency will only be proved in Ch. 4. They are not necessary, but they hold in a wide class of cases, particularly in free algebras and in the coproduct of fields, as we shall see in Ch. 5.

Let R be any ring. A relation of r terms

$$a \cdot b = a_1 b_1 + \ldots + a_r b_r = 0 \tag{1}$$

is said to be *trivial* if for each $i = 1, \ldots, r$ either $a_i = 0$ or $b_i = 0$. In the zero ring every relation is trivial, but excluding this case, there are always non-trivial relations. Nevertheless there are rings in which all such relations can be trivialized. If there is an invertible $r \times r$ matrix P over R such that $aP^{-1} \cdot Pb = 0$ is a trivial relation, then (1) is said to be *trivialized* by P or also *trivializable*. More generally, a matrix product

$$AB = 0, \tag{2}$$

where A is $m \times r$ and B is $r \times n$ is *trivializable* if there is an invertible $r \times r$ matrix P which *trivializes* (2), i.e. such that for each $i = 1, \ldots, r$ either the ith column of AP^{-1} or the ith row of PB is zero.

For any non-negative integer n we define an *n-fir* as a non-zero ring in which every relation (1) of at most n terms is trivializable. If R is an n-fir for all n, it is called a *semifir*. Let us denote the class of all n-firs by $\mathcal{F}_n$ and the class of all semifirs by $\mathcal{F}$; then it is clear that

$$\mathcal{F}_0 \supseteq \mathcal{F}_1 \supseteq \mathcal{F}_2 \supseteq \ldots, \bigcap \mathcal{F}_n = \mathcal{F}. \tag{3}$$

It is easily checked that $\mathcal{F}_0$ is the class of all non-zero rings, $\mathcal{F}_1$ is the class of all integral domains and $\mathcal{F}_2$ the class of all integral domains in which any two elements with a non-zero common left multiple generate a principal left ideal. In particular, a commutative domain is a 2-fir precisely if every 2-generator ideal is principal; by induction we see that every finitely generated ideal is principal, thus a commutative 2-fir is nothing other than a Bezout domain. As the next theorem will show, every commutative 2-fir is in fact a semifir, thus the chain (3) collapses to two terms. By contrast, in the general case all terms of the chain (3) are distinct (see 5.7).

We now give an alternative description of n-firs and semifirs. First we remark that if we have a relation (1) in an n-fir, where $r \leqslant n$, then we can transform (1) to a relation $a' \cdot b' = 0$, where $a' = (a_1' \quad \ldots$

a'_k 0   ...   0), $b' = (0$ ...   0 b'_{k+1} ... $b'_r)^T$ and $a'_1, \ldots, a'_k$ are right linearly independent. This follows easily by an induction on $r - k$. The transformation from $a \cdot b$ to $a' \cdot b' = aP^{-1} \cdot Pb$ is often called an *internal modification*.

THEOREM 1.6.1. *Let R be a non-zero ring. Then for every integer $n \geq 0$, the following conditions are equivalent:*
  (a) *R is an n-fir,*
  (b) *any left ideal of R generated by $r \leq n$ left linearly dependent elements has a family of fewer than r generators,*
  (c) *any submodule on at most n generators of a free left R-module is again free, of unique rank,*
  (a^o)–(c^o) the left–right duals of (a)–(c).
Further, R is a semifir if and only if every finitely generated left (or equivalently, right) ideal is free, of unique rank.

Proof. (a) $\Rightarrow$ (b). Suppose that $\mathfrak{a}$ is a left ideal generated by $u_1, \ldots, u_r$ $(r \leq n)$ which are left linearly dependent, say $a \cdot u = 0$, where $u = (u_1 \quad \ldots \quad u_r)^T$, $a = (a_1, \ldots, a_r) \neq 0$. Then by (a) there is a matrix P trivializing this relation, say $u' = Pu$, $a' = aP^{-1}$, $a' \cdot u' = 0$. Further, $\mathfrak{a}$ is generated by the components of u'; since $a \neq 0$, we have $a' \neq 0$, so by triviality some component of u' is 0 and hence $\mathfrak{a}$ is generated by fewer than r elements. Thus (b) holds, and by taking a generating set of $\mathfrak{a}$ of least cardinal we see that $\mathfrak{a}$ is free as left R-module.

(b) $\Rightarrow$ (c). Let F be a free left R-module and G a submodule on a generating set $u_1, \ldots, u_r$, where $r \leq n$. If G is not free on $u_1, \ldots, u_r$, suppose that $\sum a_i u_i = 0$, where the a_i are not all zero. By transforming to a'_i, u'_i we may suppose that $a'_1, \ldots, a'_k$ are right linearly independent and $a'_{k+1} = \ldots = a'_r = 0$, where $k \geq 1$. By projecting from F to R we see that all the coordinates of $u'_1, \ldots, u'_k$ must vanish, hence $u'_1 = \ldots = u'_k = 0$ and so G has a generating set of fewer than r elements. Taking a generating set of least cardinal, we see that G is free. If G has a basis of $r \leq n$ elements and another basis, of $s < r$ elements, then $R^r \cong R^s$, so G has a surjective endomorphism with kernel isomorphic to R^{r-s}. Hence a generating set of s elements maps to a set of s non-free generators under this epimorphism, which as before yields a generating set of fewer than s elements. This is a contradiction and it shows that the rank of G is unique.

(c) $\Rightarrow$ (a). Given a relation (1), the vector $b = (b_1 \quad \ldots \quad b_r)^T$ defines a linear map $f: R^r \to R$ by right multiplication and we obtain an exact sequence

$$0 \to \ker f \to R^r \to \operatorname{im} f \to 0. \tag{4}$$

As r-generator submodule of R, im f is free, so this sequence splits and we can change the basis in R^r to one adapted to the decomposition $R^r \cong \ker f \oplus \operatorname{im} f$, where $\dim \operatorname{im} f = t$ say. If this change of basis is described by $P \in \mathbf{GL}_r(R)$, and $a' = aP^{-1}$, $b' = Pb$, then since $a' \cdot b' = 0$, the components of a' lie in $\ker f$. Thus the components of a' after the first t are zero, while the first t components of b' are 0, so (1) has been trivialized. Now (a°)–(c°) follow by the symmetry of (a).

If R is a semifir, (a) holds for all n, hence by (c), any finitely generated left ideal is free of unique rank. Conversely, when this holds, and R contains free left ideals of arbitrary finite rank, then (c) holds and hence (a), for all n. When this is not the case, R is a left Ore domain, by Prop. 3.6, and hence it has IBN; thus (c) holds and hence (a). ∎

Suppose that over an n-fir R we have a matrix relation $AB = 0$, where A is $m \times r$ and B is $r \times s$. Then right multiplication by B defines a linear map $f: R^r \to R^s$, leading to an exact sequence (4). Here im f is an r-generator submodule of R^s, so if $r \leqslant n$, im f is free and as in the proof of the theorem we can find $P \in \mathbf{GL}_r(R)$ to trivialize the relation $AB = 0$. Thus we have

COROLLARY 1.6.2. *Over an n-fir any matrix relation $AB = 0$, where A has at most n columns, can be trivialized by an invertible matrix.* ∎

This result can still be generalized. Suppose that AB merely has a block of zeros, say

$$AB = \begin{pmatrix} C' & 0 \\ C & C'' \end{pmatrix} \begin{matrix} r' \\ r''. \end{matrix} \qquad (5)$$
$$\begin{matrix} s' & s'' \end{matrix}$$

If A' denotes the block consisting of the first r' rows of A and B'' is the block consisting of the last s'' columns of B, then $A'B'' = 0$, hence this product can be trivialized and we obtain

LEMMA 1.6.3 (Partition lemma). *Let R be an n-fir and let $A \in {}^rR^n$, $B \in {}^nR^s$ be such that the product AB has an $r' \times s''$ block of zeros as in (5), where r', r'', s', s'' indicate the numbers of rows and columns respectively. Then there exists $P \in \mathbf{GL}_n(R)$ and a decomposition $n = n' + n''$ such that AP^{-1} has an $r' \times n''$ block of zeros and PB has an $n' \times s''$ block of zeros.* ∎

A (non-zero) ring in which every right ideal is free, of unique rank is called a *right fir*. *Left firs* are defined similarly and a left and right fir is called a *free ideal ring*, or *fir* for short. In the commutative case firs reduce to PIDs, more generally we have

PROPOSITION 1.6.4. *A ring is a principal right ideal domain if and only if it is a right fir and satisfies the right Ore condition.*

Proof. In a right fir R every right ideal is free and if R is also right Ore, then any two elements of R are right linearly dependent, hence no right ideal needs more than one generator and so R is right principal. Conversely, a PRID is right Noetherian and hence right Ore, thus all right ideals are free and we have a right fir. ∎

As an example of a fir we have the free k-algebra (for any commutative field k) on a set X. It is written $k\langle X \rangle$ and consists of all k-linear combinations of words on the set X. If $X = \{x\}$, this algebra is just the polynomial ring $k[x]$, but as soon as X has more than one element, the algebra is non-commutative. That $k\langle X \rangle$ is a fir will be shown in 5.4; for the moment we note that a commutative fir is just a PID (by Prop. 1.6.4), and $k[x]$ is a PID, as is well known and easily proved with the help of the Euclidean algorithm. In fact a generalization of the latter, the weak algorithm, can be used to show that $k\langle X \rangle$ is a fir, but we shall use a different method of proof in Ch. 5.

The homomorphic image of a semifir need not be a semifir; this is clear since the free algebra is a semifir, while most of its homomorphic images are not. However, a retract of a semifir is again a semifir. We recall that a ring S is called a *retract* of a ring R if there is a homomorphism $f: R \to S$ with a left inverse (composing from left to right). The left inverse must then be injective and if we identify S with its image in R, we can describe a retract of R as a subring S of R with a homomorphism from R to S whose restriction to S is the identity map.

PROPOSITION 1.6.5. *Any retract of an n-fir is an n-fir; likewise for semifirs.*

Proof. Let $f: R \to S$ be a retract, where R is an n-fir, and consider an r-term relation in S, where $r \leqslant n$:

$$a \cdot b = \sum a_i b_i = 0. \tag{1}$$

Over R we can trivialize it: $a' \cdot b' = 0$ holds trivially, where $a' = aP^{-1}$, $b' = Pb$ for an invertible matrix P over R. Applying f to P we obtain a matrix over S which trivializes (1) in S. This proves the result for n-firs;

the conclusion for semifirs follows because a semifir is just an n-fir for all n. ∎

Often a generalization of free algebras is needed. Let K be any field and E a subfield of the centre of K. Then the *tensor K-ring* on a set X over E, $K_E\langle X \rangle$, is defined as the K-ring generated by X over K with the defining relations

$$\alpha x = x\alpha \quad \text{for all } x \in X, \alpha \in E.$$

As we shall see in 5.4, this ring is also a fir. In fact, E could be taken to be any subfield of K, not necessarily central, but we have imposed this condition to ensure that substitution can be carried out: any map $X \to K$ extends to a homomorphism $K_E\langle X \rangle \to K$.

Tensor rings can be defined more generally for any bimodules and this construction will also be needed later. Let K and E be as before and let M be a K-bimodule centralizing E, i.e. $\alpha m = m\alpha$ for any $m \in M$, $\alpha \in E$. Define $M^n = M \otimes \ldots \otimes M$ (n factors), where the tensor product is taken over K. Then the direct sum

$$K \oplus M^1 \oplus M^2 \oplus \ldots \tag{6}$$

becomes a ring by using the natural homomorphism $M^r \otimes M^s \to M^{r+s}$ together with linearity to define the multiplication. This ring is often written $K_E\langle M \rangle$, where the context usually indicates whether it is taken over a set or a bimodule. The tensor ring on a set X may in fact be regarded as a special case, by taking M to be the direct sum of copies of $K° \otimes_E K$ indexed by X.

The tensor K-ring $K_E\langle M \rangle$ on a bimodule M can be shown to be a fir by using the weak algorithm (see FR, 2.6), but here we shall merely prove (in 5.4) that it is a semifir, which is enough for our purposes. In particular this will apply to the free K-ring $K_E\langle X \rangle$ on a set X.

Besides the free algebra and the free K-ring we shall also need to consider free power series rings. Let X and $E \subseteq K$ be as before; then we can consider the set $K_E\langle\langle X \rangle\rangle$ of formal series in X as a K-ring in a natural fashion. Formally this ring may be defined as the completion of the free K-ring $K_E\langle X \rangle$ in the topology defined by the powers of the ideal generated by X. This power series ring is again a semifir (see 5.4), but unlike the free K-ring, it is not a fir when $X \neq \varnothing$ (see FR, 3.4).

When X is infinite, it is sometimes convenient to assign degrees to the members of X. The case just described is that where each $x \in X$ has degree 1, but if we assign degrees in such a way that for any integer N only finitely many members of X are of degree less than N, then the resulting ring will include terms such as $\sum_{x \in X} x$. Thus $K_E\langle\langle X \rangle\rangle$ will depend on the degree function that is used for X. In the uses we make of power series, X will be finite and then $K_E \langle\langle X \rangle\rangle$ is independent of the degree function used, so this problem does not arise.

At first sight free algebras seem far removed from the kind of rings that are usually considered, but in fact they can be found in any non-Ore domain.

PROPOSITION 1.6.6. *Let R be an integral domain with centre F. Then either R is left and right Ore or it contains a free algebra of countable rank.*

Proof. Suppose that R is not left Ore and take $x, y \in R^\times$ such that $Rx \cap Ry = 0$, i.e. x and y are left linearly independent over R. We claim that the F-algebra generated by x and y is free. For if not, then there is a polynomial in x and y which vanishes; we choose such a polynomial f of least degree and write it as

$$f = \alpha + f_1 x + f_2 y,$$

where $\alpha \in F$ and f_1, f_2 have lower degree than f. If $\alpha = 0$, then $f_1 x + f_2 y = 0$, which contradicts the hypothesis, so $\alpha \neq 0$ and f_1, f_2 cannot both vanish, say $f_1 \neq 0$. Hence

$$0 = yf = \alpha y + yf_1 x + yf_2 y$$

$$= yf_1 x + (\alpha + yf_2)y,$$

which is again a contradiction. This shows the F-algebra on x and y to be free. Now the subalgebra generated by $z_n = xy^n$ $(n = 0, 1, 2, \ldots)$ is easily seen to be free on the z_n and so it satisfies all the conditions. ∎

To obtain a homological description of firs, we recall that a ring is said to be *left hereditary* if every left ideal is projective, *left semihereditary* if every finitely generated left ideal is projective. It is well known that for a left (semi)hereditary ring, every (finitely generated) submodule of a projective left R-module is again projective (A.3, Th. 3.4.4, p. 97), so a left hereditary ring is just a ring of left global dimension at most 1. Similar definitions apply on the right and give distinct classes of rings (see Ex. 3).

The homological dimension of a module measures how far the module is from being projective. In this sense one can also ask how far a projective module is from being free. We shall not introduce a measure here, but define a ring R to be *projective-free* if every finitely generated projective left R-module is free, of unique rank. The duality $P \mapsto \mathrm{Hom}_R(P, R)$ shows this notion to be left–right symmetric. For example, any local ring is projective-free (FR, Cor. 0.5.5, p. 22). It is easily verified that a ring is a left (semi)fir if and only if it is projective-free and left (semi)hereditary.

In Ch. 4 we shall show that every semifir can be embedded in a field. For the moment we shall merely note that semifirs satisfy Klein's nilpotence condition.

PROPOSITION 1.6.7. *Let R be an n-fir and C a nilpotent $n \times n$ matrix over R. Then there is a conjugate $P^{-1}CP$ which is upper triangular, with zeros on the main diagonal; in particular, $C^n = 0$.*

Proof. Since C is nilpotent, we have $C^{m+1} = 0$ for some $m \geqslant 0$. Choose the least such m and write $B = C^m$; then $B \neq 0$ and $CB = 0$, hence by Cor. 6.2, there is an invertible $n \times n$ matrix U such that CU has its first column equal to 0, and the same holds for $U^{-1}CU$. Omitting the first row and column from $U^{-1}CU$, we obtain an $(n - 1) \times (n - 1)$ matrix which is again nilpotent; by induction it is conjugate to an upper triangular matrix with zeros on the main diagonal, hence the same holds for C. ∎

With the help of Klein's theorem (Klein [69]) this shows that for every semifir R, the monoid $R^\times$ can be embedded in a group. In fact this already holds for 2-firs, by the Gerasimov–Malcolmson localization theorem (FR, Th. 7.11.22, p. 484).

Exercises

1. Give a direct proof that a semifir is weakly finite.

2. Let A be a commutative integral domain. Show that the polynomial ring $A[x]$ is a fir if and only if A is a field.

3. Show that $\mathbf{Q}$ as $\mathbf{Z}$-module is not projective. Deduce that the triangular matrix ring $\begin{pmatrix} \mathbf{Z} & \mathbf{Q} \\ 0 & \mathbf{Q} \end{pmatrix}$ is right but not left hereditary.

4. Prove that a ring is a left (semi)fir if and only if it is left (semi)hereditary and projective-free.

5. Show that a retract of a PID is again a PID.

6°. Is every retract of a fir again a fir?

7. Let R be any ring and $\mathfrak{a}$ a right ideal which is free as right R-module and satisfies $\mathfrak{a}^2 = \mathfrak{a}$. Show that either $\mathfrak{a} = R$ or $\mathfrak{a} = 0$.

8. (G. M. Bergman) Show that if an inverse of xy is adjoined to $k\langle x, y\rangle$, the resulting ring has an idempotent $\neq 0, 1$.

9. Show that a ring R is a semifir and a local ring if and only if $R \neq 0$ and for any $a_1, \ldots, a_n \in R$ satisfying a relation $\sum a_i b_i = 0$, where the b_i are not all zero, there is such a relation with one of the b_i a unit (Cohn [92']).

10. Show that in a semifir R every finitely generated left or right ideal $\mathfrak{a}$ satisfies $\mathfrak{a} \otimes_R \mathfrak{a} \cong \mathfrak{a}^2$.

11. Let R be a semifir and $c \in R$. Show that any right ideal $\mathfrak{a}$ in R satisfies $\mathrm{rk}(\mathfrak{a} \cap cR) \leqslant \mathrm{rk}\mathfrak{a}$, with equality if and only if $\mathfrak{a} + cR$ is principal.

1.7 The matrix reduction functor

We shall now take a closer look at Rg, the category of rings and homomorphisms, and some related categories. Here the unit element 1 is understood to be a constant operator. The category Rg has as its final object 0, the zero ring consisting of 0 alone, for every ring has a unique homomorphism to 0. There is also an initial object, the ring of integers $\mathbf{Z}$, for there is always a unique homomorphism $\mathbf{Z} \to R$, because $\mathbf{Z}$ is generated by the constant operator 1, whose image in R is determined. This amounts to treating rings as $\mathbf{Z}$-algebras.

More generally, if A is any commutative ring, by an A-*algebra* we understand a ring R which is an A-module such that

$$\alpha(xy) = \alpha x \cdot y = x \cdot \alpha y \quad \text{for all } x, y \in R, \alpha \in A. \tag{1}$$

It is easily verified that an A-algebra is simply a ring R with a homomorphism from A to the centre of R, viz. $\alpha \mapsto \alpha \cdot 1$. The expression '$A$-algebra' will be taken to imply that A is a commutative ring.

Often a more general notion is needed. Let A be any ring, not necessarily commutative. By an A-*ring* we understand a ring R which is an A-bimodule such that

$$\alpha(xy) = (\alpha x)y, (x\alpha)y = x(\alpha y), x(y\alpha) = (xy)\alpha,$$

$$\text{for all } x, y \in R, \alpha \in A. \tag{2}$$

Again it is easy to see that an A-ring is nothing other than a ring R with a homomorphism $A \to R$, viz. $\alpha \mapsto \alpha \cdot 1$. If this homomorphism is injective, the A-ring is called *faithful*. We shall write Rg_A for the category of A-rings and homomorphisms, where the homomorphisms are understood to be compatible with A; this will be so if they are ring-homomorphisms and A-bimodule homomorphisms. Clearly Rg_A is a category with initial object A; it is essentially the comma category (Rg, A) (see 5.1 below; A.3, p. 69).

In the category Rg_A we can, for any pair of objects R, S, form the pushout:

$$
\begin{array}{ccc}
A & \longrightarrow & R \\
\downarrow & & \downarrow \\
S & \longrightarrow & P
\end{array}
\qquad (3)
$$

Here P may be obtained by taking a presentation of R as A-ring: $R = A\langle X; R\rangle$, where X is a generating set with defining relations R, and $S = A\langle Y; S\rangle$, and take the joint presentation (assuming the sets X, Y to be disjoint):

$$
P = R \underset{A}{*} S = A\langle X \cup Y; R, S\rangle. \qquad (4)
$$

This is usually called the *coproduct* of R and S *over* A (or sometimes the *free product* of R and S, *amalgamating* A). Its existence is clear by abstract nonsense, but in practice we shall often want to have further information on the structure of the coproduct, e.g. whether the coproduct (4) is *faithful*, i.e. the natural homomorphisms from R, S to P in (3) are injective. These questions will form the subject of Ch. 5.

Another category of interest to us is the category Rg_n of $n \times n$ matrix rings. Its objects are rings with n^2 constant operators e_{ij} (in addition to 1), traditionally called the *matrix units*, although they are not units except in the trivial case $n = 1$. They satisfy the conditions familiar from matrix theory:

$$
e_{ij}e_{kl} = \delta_{jk}e_{il}, \quad \sum e_{ii} = 1, \qquad (5)
$$

where $\delta_{jk} = 1$ if $j = k$ and 0 otherwise (Kronecker delta). The morphisms in Rg_n are ring homomorphisms preserving the matrix units. For any $n \geq 1$ this category is equivalent to the category of rings. To prove this fact we introduce the *matrix functor* $\mathfrak{M}_n$, a covariant functor from Rg to Rg_n. With any ring R it associates the $n \times n$ matrix ring $\mathfrak{M}_n(R) = R_n$ whose elements are $n \times n$ matrices over R, with the matrix units e_{ij} being given by the matrix with (i, j)-entry 1 and all other entries 0. Clearly any ring homomorphism $f: R \to S$ uniquely determines a homomorphism from $\mathfrak{M}_n(R)$ to $\mathfrak{M}_n(S)$ in Rg_n, by applying f to the separate matrix entries. Conversely, any homomorphism $\mathfrak{M}_n(R) \to \mathfrak{M}_n(S)$ arises in this way from a homomorphism $R \to S$, because R can be characterized within $\mathfrak{M}_n(R)$ as the centralizer of all the e_{ij} and it is mapped to the centralizer of the e_{ij} within $\mathfrak{M}_n(S)$, which is S. Thus the functor $\mathfrak{M}_n$ is faithful and full; moreover, every object T in Rg_n is of the form $\mathfrak{M}_n(C)$, where C is the centralizer of the e_{ij} in T. This shows $\mathfrak{M}_n$ to be a category equivalence (see A.2, Prop. 1.3.1, p. 19). Thus we have proved

THEOREM 1.7.1. *The matrix functor* $\mathfrak{M}_n$ *establishes an equivalence between* Rg *and* Rg_n, *for any* $n \geq 1$. ∎

The inverse of the matrix functor is a functor from Rg_n to Rg which associates to any $n \times n$ matrix ring with a specified set of matrix units e_{ij} the centralizer of the e_{ij}. There is another functor from Rg_n to Rg, namely the *forgetful functor*, which forgets the e_{ij}. It is important to distinguish these two functors; we shall not use a special notation for the inverse of $\mathfrak{M}_n$, but denote the forgetful functor applied to a matrix ring $\mathfrak{M}_n(R)$ by $\overline{\mathfrak{M}_n}(R)$.

We now come to another functor, perhaps less familiar than the others. By the *n-matrix reduction functor* $\mathfrak{W}_n$ we shall understand the left adjoint of the $n \times n$ matrix functor. Thus we have

$$\mathrm{Rg}(R, \overline{\mathfrak{M}_n}(S)) \cong \mathrm{Rg}(\mathfrak{W}_n(R), S). \qquad (6)$$

Our first task is to show that such a functor exists. Taking R to be the initial object $\mathbf{Z}$ in (6), we see that $\mathfrak{W}_n(\mathbf{Z})$ is again an initial object in Rg, hence by uniqueness,

$$\mathfrak{W}_n(\mathbf{Z}) \cong \mathbf{Z}.$$

Next, by the equivalence of Rg and Rg_n we obtain from (6),

$$\mathrm{Rg}(R, \overline{\mathfrak{M}_n}(S)) \cong \mathrm{Rg}_n(\mathfrak{M}_n(\mathfrak{W}_n(R)), \mathfrak{M}_n(S)). \qquad (7)$$

Let us put $\mathfrak{F}_n(R) = \mathfrak{M}_n(\mathfrak{W}_n(R))$; (7) shows $\mathfrak{F}_n(R)$ to have the following universal property: There is a homo-

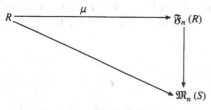

morphism $\mu: R \to \mathfrak{F}_n(R)$ such that every homomorphism from R to a matrix ring $\overline{\mathfrak{M}_n}(S)$ can be factored uniquely by a homomorphism $\mathfrak{F}_n(R) \to \mathfrak{M}_n(S)$. Here μ is obtained by choosing $\mathfrak{M}_n(S) = \mathfrak{F}_n(R)$ in (7) and taking the map on the left corresponding to the identity map on the right.

This description makes it clear how to construct $\mathfrak{F}_n(R)$: we take the ring R and adjoin n^2 elements e_{ij} satisfying the equations (5). If R is an A-algebra, we shall also want the relations

$$\alpha e_{ij} = e_{ij}\alpha \quad \text{for all } \alpha \in A \text{ and } i, j = 1, \ldots, n,$$

to ensure that $\mathfrak{F}_n(R)$ is again an A-algebra. When $R = A$, this construction just gives $\mathfrak{M}_n(A)$, while in general we have a coproduct over A:

$$\mathfrak{F}_n(R) = R \underset{A}{*} \mathfrak{M}_n(A).$$

This is an $n \times n$ matrix ring; the underlying ring, viz. the centralizer of the e_{ij}, is $\mathfrak{W}_n(R)$, or more explicitly, $\mathfrak{W}_n(R; A)$. This mode of formation leads to the following rule-of-thumb for forming $\mathfrak{W}_n(R)$:

Take the elements of R and interpret them as $n \times n$ matrices, with the elements of A as scalars.

More generally, given an A-algebra R with a presentation $R = A\langle X; \Phi \rangle$, we can form another A-algebra by interpreting each element x of X as an $m_x \times n_x$ matrix, where $n_x = m_y$ if a product xy occurs in Φ, $m_x = m_y$ if a sum of terms begins with x and y respectively and $n_x = n_y$ if a sum of terms ending in x, y respectively occurs in Φ.

We give some examples to illustrate these rules, where k is a commutative field:

1. Let $R = k[x]$, the polynomial ring in x. Then $\mathfrak{W}_n(R)$ is the free k-algebra on n^2 generators x_{ij}.

2. $R = k[t, t^{-1}]$, the ring of Laurent polynomials (see 2.1). Here $\mathfrak{W}_n(R)$ is the k-algebra on $2n^2$ generators x_{ij}, y_{ij} $(i, j = 1, \ldots, n)$ with defining relations in matrix form, writing $X = (x_{ij})$, $Y = (y_{ij})$, $XY = YX = I$. More generally, we can interpret t as $m \times (m + d)$ matrix and t^{-1} as $(m + d) \times m$ matrix, and so obtain the universal non-IBN ring of free module type $(m, m + d)$.

3. $R = k[x; x^r = 0]$. Here $\mathfrak{W}_n(R)$ is the k-algebra with n^2 generators x_{ij} and defining relations in matrix form, writing $X = (x_{ij})$, $X^r = 0$.

Later, in 5.7 we shall see that for $n > 1$, $\mathfrak{W}_n(R)$ is an integral domain whenever $R \neq 0$. Applied to the example 3, with $r > n > 1$, this provides an example of an integral domain not embeddable in a field, by Prop. 4.4.

Exercises

1. Show that for a non-commutative ring A, the category of A-algebras (as defined by (1)) is equivalent to the category of $\bar{A}$-algebras, where $\bar{A} = A/\mathfrak{c}$, $\mathfrak{c}$ being the commutator ideal of A (thus $\bar{A}$ is the largest commutative homomorphic image of A).

2. An A-ring R is said to be *augmented* if there is an A-ring homomorphism $R \to A$. Show that an augmented A-ring R can be expressed as $R = A \oplus \mathfrak{a}$, where $\mathfrak{a}$ is the kernel of the above homomorphism $R \to A$ ($\mathfrak{a}$ is called the *augmentation ideal*). Show that the coproduct of augmented A-rings is faithful. If $A = k[x]$, give an example of two A-rings whose coproduct is not faithful.

3. What is the appropriate n-matrix reduction functor for the category of A-rings, where A is a non-commutative ring?

4. Show that the correspondence $R \mapsto e_{11} R e_{11}$ is a functor from Rg_n to Rg which is inverse to $\mathfrak{M}_n$.

5. Show that the matrix functor $\mathfrak{M}_n$ does not have a right adjoint.

6. For any A-rings B, C, show that $B \underset{A}{*} \mathfrak{M}_n(C) \cong \mathfrak{M}_n(R)$, where $R = \mathfrak{W}_n(B \underset{A}{*} C;$ $C)$.

Notes and comments

The first non-commutative field was the ring of quaternions, constructed by Sir William Hamilton in 1843, with the aim of representing vectors in space (in the way vectors in the plane could be represented by complex numbers). Hamilton and his followers built up an elaborate geometrical calculus on this foundation, which eventually led to vector analysis, but which is not relevant to our theme. Skew fields finite-dimensional over their centres (division algebras) were studied intensively in the 1920s and 1930s and there has been a revival of interest since 1970. For an up-to-date account see Jacobson and Saltman [a].

The first skew field infinite-dimensional over its centre was constructed in 1903 by Hilbert in his study of the foundations of geometry (Hilbert [03]). This was the field of skew Laurent series $F((x; \alpha))$, where F is the rational function field $\mathbf{R}(t)$ and α is the automorphism $f(t) \mapsto f(2t)$ of F (see 2.3), to illustrate the fact that an ordered field need not be commutative, when the ordering is non-Archimedean. In principle this example already occurs in the first edition, Hilbert [1899], but there (in a rare mistake) he takes the automorphism α to be $f(t) \mapsto f(-t)$; this is not order-preserving and so the ordering cannot be extended (in fact the resulting field is 4-dimensional over its centre). For a brief history see Cohn [92].

Near fields were introduced by Dickson [05] in studying the axiomatics of field theory; a thorough treatment can be found in Wähling [87]. The definition of fields in terms of their multiplicative group given in Lemma 1.1 is taken from Cohn [61'], with some simplifications suggested by N. G. Greenwood. Similar sets of axioms were used by Dicker [68] and Leissner [71]. A remarkable set of axioms for fields, dispensing with the distributive laws, was found by Pickert [59], who proves that a set K with two binary operations $x + y$, xy is a field provided that (i) K is an abelian group under the operation $x + y$ with neutral element 0, (ii)

$K\backslash\{0\}$ is a group under the operation xy, with neutral element 1, (iii) $K\backslash\{1\}$ is a group under the operation $x \circ y = x + y - xy$, with neutral 0 (where $-$ denotes the inverse of $+$, as usual).

Quasi-identities have been studied in universal algebra, see UA, Ch. VI, where they occur as particular Horn sentences, and Malcev [73]; see also Cohn [71']. Ultraproducts were introduced by Łos [55] and today constitute a basic tool in model theory, see Chang and Keisler [73], UA, V. 5.

O. Ore [31] studied the construction that bears his name and proved Cor. 3.3. In Ore [33] he applied these results to skew polynomial rings, in particular the ring of differential polynomials $k(t)[x; 1, \mathrm{d}/\mathrm{d}t]$ (see 2.1). The concept of invariant basis number was first defined and studied by Leavitt [57]; see also Everett [42]. A comparative study of IBN, UGN and weak finiteness was made by Cohn [66]. That Noetherian domains are Ore was first proved by Goldie [58].

The notion of similarity goes back to Fitting [36], who examines this concept for matrices over any ring. In the form given here it is taken from Cohn [63], see also FR, 0.6 and 3.2. Invariant and bounded elements were studied in the context of pseudo-linear transformations by Jacobson [37], see also the account in Ch. 3 of Jacobson [43]. Our treatment is based on that of Ch. 3 of FR, where Prop. 5.2 and Th. 5.4 are established for semifirs.

Firs and semifirs (then called 'local firs') were introduced in Cohn [64]. Here we have included only a few basic results needed later; for a detailed account see FR Ch. 1, 4 and 5.

The matrix reduction functor arose in conversation between G. M. Bergman and the author in 1969 and is described in Cohn [69]; see also Bergman [74'], Cohn [79] and FR, 2.11. It was also defined in Procesi [73], Ch. IV, but used only in commutative rings, where it leads essentially to the generic matrix ring.

2

Skew polynomial rings and power series rings

The true analogue of a polynomial ring in the non-commutative case is the free algebra, or more generally, the tensor ring. But there is a half-way house, rather like the Ore domain (of which it is an instance), namely the skew polynomial ring, which was introduced by Ore [33]. We examine its elementary properties in 2.1 and make a further study of its ideal theory in 2.2. By forming a completion we obtain in 2.3 from a skew polynomial ring a skew power series ring, whose properties in some senses are simpler than for polynomials. This leads in 2.4 to the Malcev–Neumann construction, a far-reaching generalization allowing the group algebra of any totally ordered group to be embedded in a field. With its help we can iterate the skew polynomial ring construction in 2.5 to form the rings first studied by Jategaonkar, which provide a rich source of counter-examples.

The final section 2.6 applies Ore's method to filtered rings whose associated graded ring is an Ore domain. It is shown that the filtered ring can be embedded in a field, constructed as an inverse limit, and this construction is then used to embed the universal associative envelope of any Lie algebra in a field.

2.1 Skew polynomial rings

Commutative field extensions are usually constructed as residue-class rings of polynomial rings over a field. For skew fields the polynomial construction is not the most general means of forming an extension, but it is an important special case because it allows Ore's construction from 1.3 to be used; in fact Ore [33] was one of the first to consider skew polynomial rings formally.

Let K be a field with a central subfield C. Then the K-ring generated by an indeterminate x over C is the tensor K-ring $K_C\langle x\rangle$ already encountered in 1.6. Its elements do not look like polynomials at all,

$$a + \sum b_i x c_i + \sum d_j x e_j x f_j + \ldots, \quad a, b_i, c_i, d_j, e_j, f_j \in K,$$

and they are no easier to handle than expressions in several variables. We shall make the simplifying assumption that all coefficients can be written on the right, but we shall allow the coefficient ring to be arbitrary, to begin with.

Thus let A be any ring, t a new symbol to represent the indeterminate and consider a ring R whose elements can all be uniquely expressed in the form

$$f = a_0 + ta_1 + \ldots + t^n a_n, \quad \text{where } a_i \in A. \tag{1}$$

We shall call such an expression f a *polynomial* in t and define its *degree*, $\deg f$, as n if $a_n \neq 0$; the zero polynomial 0 is said to have degree $-\infty$. If in (1), $a_n = 1$, f is said to be *monic*. The additive group of R is just the direct sum of countably many copies of A, by the uniqueness of (1). To multiply two elements, say $f = \sum t^i a_i$ given by (1) and $g = \sum t^j b_j$ we have, by distributivity, $fg = \sum t^i (a_i t^j) b_j$ and so it will only be necessary to prescribe $a_i t^j$. We shall also assume that

$$\deg fg \leqslant \deg f + \deg g. \tag{2}$$

Then in particular, at for any $a \in A$ has degree at most 1, so

$$at = ta^\alpha + a^\delta, \tag{3}$$

where $a \mapsto a^\alpha$, $a \mapsto a^\delta$ are mappings of A into itself. This is already enough to fix the multiplication in R, for now we can work out at^r by induction on r:

$$at^r = (ta^\alpha + a^\delta)t^{r-1} = [t^2 a^{\alpha^2} + t(a^{\alpha\delta} + a^{\delta\alpha}) + a^{\delta^2}]t^{r-2} = \ldots.$$

We derive some consequences from (3):

$$(a + b)t = t(a + b)^\alpha + (a + b)^\delta, \quad at + bt = ta^\alpha + a^\delta + tb^\alpha + b^\delta,$$

hence

$$(a + b)^\alpha = a^\alpha + b^\alpha, \quad (a + b)^\delta = a^\delta + b^\delta, \tag{4}$$

and

$$(ab)t = t(ab)^\alpha + (ab)^\delta, \quad a(bt) = a(tb^\alpha + b^\delta) = ta^\alpha b^\alpha + a^\delta b^\alpha + ab^\delta,$$

so

$$(ab)^\alpha = a^\alpha b^\alpha, \quad (ab)^\delta = a^\delta b^\alpha + ab^\delta. \tag{5}$$

Further, $1t = t1 = t$, therefore

$$1^\alpha = 1, \quad 1^\delta = 0. \tag{6}$$

From (4)–(6) we see that α is an endomorphism of A and δ is an α-derivation of A, i.e. a mapping such that

$$(a + b)^\delta = a^\delta + b^\delta, \quad (ab)^\delta = a^\delta b^\alpha + ab^\delta \quad \text{for all } a, b \in A. \tag{7}$$

We note that (7) entails $1^\delta = 0$, by putting $a = b = 1$ in the second equation (7). Conversely, if A is any ring with an endomorphism α and an α-derivation δ, then the set of all expressions (1) can be made into a ring by defining addition componentwise and multiplication by the commutation rule (3). The verifications are straightforward, and they show that in the resulting ring the form (1) for the elements is unique. This proves part (i) of

PROPOSITION 2.1.1. *Let A be a ring, α an endomorphism and δ an α-derivation of A and put $R = A\langle t; at = ta^\alpha + a^\delta, a \in A \rangle$. Then*
  (i) *each $f \in R$ can be uniquely written as $\sum t^i a_i$ $(a_i \in A)$,*
 (ii) *R is an integral domain, provided that A is an integral domain and α is injective,*
(iii) *if A is a field, then R is a principal right ideal domain, which is also a principal left ideal domain whenever α is an automorphism.*

To prove (ii), we note that when A is an integral domain and α is injective, then we have equality in (2), and hence R is then an integral domain. Next, let A be a field and let $\mathfrak{a}$ be a right ideal in R. For $\mathfrak{a} = 0$ there is nothing to prove; otherwise let $p = a_0 + ta_1 + \ldots + t^n$ be the monic polynomial of least non-negative degree in $\mathfrak{a}$. Given $f \in \mathfrak{a}$, we can by the usual division algorithm write $f = pq + r$, where $q, r \in R$ and $\deg r < \deg p$. But then $r = f - pq \in \mathfrak{a}$, and if $r \neq 0$, we obtain a monic polynomial of lower degree than p in $\mathfrak{a}$, which contradicts the minimality of p, so $r = 0$ and $f \in pR$. Hence $\mathfrak{a} = pR$ and R has been shown to be right principal. If α is an automorphism, with inverse β, then, writing $a^\alpha = b$, we have $a = b^\beta$ and we can re-write (3) in the form $b^\beta t = tb + b^{\beta\delta}$ or

$$tb = b^\beta t - b^{\beta\delta},$$

which shows by symmetry that R is left principal. ∎

The ring R formed here is called the *skew polynomial ring* in t over A associated with α, δ and is denoted by $A[t; \alpha, \delta]$. When $\delta = 0$, we also write $A[t; \alpha]$ instead of $A[t; \alpha, 0]$. If moreover $\alpha = 1$, we obtain the polynomial ring in a central indeterminate t over A, written $A[t]$. For a field K the principal right ideal domain $K[t; \alpha, \delta]$ has a field of fractions, by Cor. 1.3.7, which will be denoted by $K(t; \alpha, \delta)$.

In matrix notation the commutation rule (3) may be written

$$a(1 \quad t) = (1 \quad t)\begin{pmatrix} a & a^\delta \\ 0 & a^\alpha \end{pmatrix},$$

and the conditions (4)–(6) may be summed up by saying that the mapping of A into the matrix ring A_2 defined by

$$a \mapsto \begin{pmatrix} a & a^\delta \\ 0 & a^\alpha \end{pmatrix} \tag{8}$$

is a ring homomorphism. More precisely, it is a homomorphism into the ring $\mathbf{T}_2(A)$ of upper 2×2 triangular matrices over A, which in a suggestive notation may also be written $\begin{pmatrix} A & A \\ 0 & A \end{pmatrix}$. More generally, any homomorphism from A to $\mathbf{T}_2(A)$ has the form

$$a \mapsto \begin{pmatrix} a^\alpha & a^\delta \\ 0 & a^\beta \end{pmatrix},$$

and it is easily verified that α, β are endomorphisms of A, while δ satisfies

$$(a + b)^\delta = a^\delta + b^\delta, \quad (ab)^\delta = a^\alpha b^\delta + a^\delta b^\beta. \tag{9}$$

Such a mapping is called an (α, β)-derivation of A; as before, a $(1, \beta)$-derivation is also called a β-derivation. As an example of an (α, β)-derivation we have the mapping

$$a \mapsto ca^\beta - a^\alpha c, \quad \text{where } c \in A. \tag{10}$$

This is called the *inner* (α, β)-derivation induced by c; any derivation not of this form is said to be *outer*.

Let A be a right Ore domain with field of fractions K. In Cor. 1.3.5 we saw that any injective endomorphism of A extends to K; in fact this also applies to derivations. Thus let α be an injective endomorphism of A; its unique extension to K is again written as α. Any α-derivation δ defines a homomorphism (8) from A to $\mathbf{T}_2(A)$ which by functoriality extends to a homomorphism from K to $\mathbf{T}_2(K)$, say

$$u \mapsto \begin{pmatrix} u & u' \\ 0 & u^\alpha \end{pmatrix}.$$

Clearly the map $u \mapsto u'$ is an α-derivation on K extending δ and we shall write u^δ instead of u' for $u \in K$. Thus we have shown that the derivation associated with an injective endomorphism extends to K:

PROPOSITION 2.1.2. *Let A be a right Ore domain with field of fractions K, and let α be an injective endomorphism of A and δ an α-derivation. Then α extends to a unique endomorphism of K and δ extends to a unique α-derivation of K.* ∎

In a skew polynomial ring $R = A[t; \alpha, \delta]$ over any ring A we can change the variable by writing

$$t' = ta + b,$$

where $a, b \in A$ and a is a unit in A. Clearly this leaves the ring R unchanged, but it will in general change α and δ and by a judicious choice of variable it may be possible to reduce α to 1 or δ to 0. Suppose that α is an inner automorphism:

$$c^\alpha = ucu^{-1} \text{ for some } u \in \mathrm{U}(A).$$

Writing $t' = tu$, we find that $ct' = ctu = (tc^\alpha + c^\delta)u = tuc + c^\delta u$, hence

$$ct' = t'c + c^\delta u \text{ for all } c \in A;$$

thus α has been reduced to 1 by a change of variable.

Secondly, assume that δ is an inner α-derivation:

$$c^\delta = cd - dc^\alpha \text{ for some } d \in A.$$

On writing $t' = t - d$, we have $ct' = c(t - d) = tc^\alpha + cd - dc^\alpha - cd = t'c^\alpha$, so

$$ct' = t'c^\alpha.$$

Thus if either α or δ is inner, it can be reduced to 1 or 0 respectively by a change of variable. In fact, when neither α nor δ is inner, R is quite restricted, as the next result shows:

THEOREM 2.1.3. *Let K be a skew field with centre C and let $R = K[t; \alpha, \delta]$ be a skew polynomial ring with endomorphism α and α-derivation δ. Then*
  (i) *if α does not leave C fixed, then δ is inner and by a suitable choice of variable may be taken to be zero,*
 (ii) *if α leaves C fixed but δ does not map it to 0, then α is inner and by a suitable choice of variable may be taken to be 1,*
(iii) *if α leaves C fixed and δ maps it to zero, then C is contained in the centre of R.*

Proof. Assume that α does not leave C fixed, say $\gamma^\alpha \neq \gamma$ for some $\gamma \in C$. Then on writing $t' = \gamma t - t\gamma$ we have, for any $c \in K$,

$$ct' = c(\gamma t - t\gamma) = \gamma tc^\alpha + \gamma c^\delta - tc^\alpha\gamma - c^\delta\gamma = t'c^\alpha,$$

hence we have $R = K[t'; \alpha]$, so δ has been reduced to zero. Next assume that α leaves C fixed but there exists $\gamma \in C$ such that $\gamma^\delta \neq 0$. Then $\gamma^\alpha = \gamma$ and for any $c \in K$ we have $c\gamma = \gamma c$, hence $c\gamma^\delta + c^\delta\gamma = \gamma c^\delta + \gamma^\delta c^\alpha$, i.e. $c\gamma^\delta = \gamma^\delta c^\alpha$ and writing $t' = t(\gamma^\delta)^{-1}$, we have

$$ct' = tc^\alpha(\gamma^\delta)^{-1} + c^\delta(\gamma^\delta)^{-1} = t(\gamma^\delta)^{-1}c + c^\delta(\gamma^\delta)^{-1} = t'c + c^{\delta'},$$

where δ' is a 1-derivation, i.e. an ordinary derivation of K. This shows α to be inner. In the remaining case α reduces to 1 and δ reduces to 0 on C, hence t centralizes C as well as K and so C is contained in the centre of R. ∎

In the special case where $[K:C]$ is finite, all linear endomorphisms and all linear derivations are inner, by the Skolem–Noether theorem:

PROPOSITION 2.1.4. *Let K be a field finite-dimensional over its centre C. Then any C-linear endomorphism of K is an inner automorphism and any C-linear derivation is inner.*

Proof. Any endomorphism α of K must be injective, since the kernel is a proper ideal; hence it is an automorphism by finite dimensionality. Now α is inner, by the Skolem–Noether theorem (A.3, 7.1 or Cor. 3.3.6 below).

To prove the result for derivations we shall need the Skolem–Noether theorem for simple rings: any two homomorphisms from a finite-dimensional simple algebra into a central simple finite-dimensional algebra A have images that are conjugate in A (see A.3, Th. 7.1.6, p. 262). We could take $\alpha = 1$, by what has been shown, but this is not necessary. We have the following two homomorphisms from K to $T_2(K)$:

$$a \mapsto \begin{pmatrix} a & 0 \\ 0 & a \end{pmatrix}, \quad a \mapsto \begin{pmatrix} a & a^\delta \\ 0 & a^\alpha \end{pmatrix}.$$

By Skolem–Noether the images are conjugate, thus there is a non-singular matrix $\begin{pmatrix} p & q \\ r & s \end{pmatrix}$ over K such that

$$\begin{pmatrix} a & 0 \\ 0 & a \end{pmatrix}\begin{pmatrix} p & q \\ r & s \end{pmatrix} = \begin{pmatrix} p & q \\ r & s \end{pmatrix}\begin{pmatrix} a & a^\delta \\ 0 & a^\alpha \end{pmatrix}.$$

Hence

$$ap = pa, \quad aq = pa^\delta + qa^\alpha,$$
$$ar = ra, \quad as = ra^\delta + sa^\alpha.$$

It follows that $p, r \in C$; they cannot both vanish, say $p \neq 0$; then on dividing by p we may assume that $p = 1$ and so $aq = a^\delta + qa^\alpha$, which shows δ to be inner. ∎

For any skew field K we obtain a rational function field $K(t)$ by forming the polynomial ring $K[t]$ in a central indeterminate and taking its field of fractions. Its centre is described by the next result:

PROPOSITION 2.1.5. *Let K be a field with centre C. Then the function field $K(t)$ formed with a central indeterminate t has the centre $C(t)$.*

Proof. Every element of $K(t)$ has the form $\varphi = fg^{-1}$, where $f, g \in K[t]$. We shall use induction on $d(\varphi) = \deg f + \deg g$ to prove that if φ is in the centre of $K(t)$, then $\varphi \in C(t)$, the converse being evident. For $d(\varphi) = 0$ we have an element of K and the result holds by definition. If $d(\varphi) > 0$, we may assume that $\deg f \geq \deg g$, replacing φ by φ^{-1} if necessary. By the division algorithm, $f = qg + r$, where $\deg r < \deg g$, with uniquely determined $q, r \in K[t]$. Let us write $u^c = c^{-1}uc$ for $u \in K(t)$, $c \in K^\times$; then

$$\varphi = fg^{-1} = q + rg^{-1}, \quad \varphi^c = q^c + r^c(g^c)^{-1}.$$

Since φ is in the centre of $K(t)$, we have $\varphi^c = \varphi$, i.e.

$$q - q^c = r^c(g^c)^{-1} - rg^{-1}. \tag{11}$$

Now $v(\varphi) = \deg g - \deg f$ is a valuation on $K(t)$ and the left-hand side of (11) has a value ≤ 0, unless $q = q^c$, while the right-hand side has a strictly positive value, which gives a contradiction. Without using valuations we can say: put $t = s^{-1}$; then $K(t) = K(s)$ and for $s = 0$ the right- but not the left-hand side of (11) vanishes. This shows that both sides must vanish, $q^c = q$ and rg^{-1} is in the centre, but $d(rg^{-1}) < d(fg^{-1})$, so the result follows by induction. ∎

Later, in 2.2, we shall meet another proof of this fact.

Let us examine more carefully the conditions under which a skew polynomial ring over a field is left principal.

PROPOSITION 2.1.6. *Let K be a field with an endomorphism α and an*

α-derivation δ, and let $R = K[t; \alpha, \delta]$. Then the following conditions are equivalent:

(a) α is an automorphism,
(b) R is left principal,
(c) R is left Ore.

Proof. (a) $\Rightarrow$ (b) by Prop. 1.1 and (b) $\Rightarrow$ (c) is clear. To show that (c) $\Rightarrow$ (a), assume (c); it will be enough to show that α is surjective. Let $c \in K$; by hypothesis there exist $f, g \in R$ such that

$$ft = gtc \neq 0.$$

A comparison of degrees shows that $\deg f = \deg g = n$, say. Let $f = t^n a + \ldots$, $g = t^n b + \ldots$, then by comparing highest terms we find $a^\alpha = b^\alpha c$ and so $c = (b^{-1}a)^\alpha$. This shows α to be surjective and hence an automorphism. ∎

This result together with Prop. 1.6.6 leads to an embedding of a free algebra in a field, as was observed by Jategaonkar [69'] and independently, Koshevoi [70]: Let K be a field with a non-surjective endomorphism α, e.g. the rational function field $k(t)$ with endomorphism α: $f(t) \mapsto f(t^2)$, and form the skew polynomial ring $R = K[x; \alpha]$. This is a principal right ideal domain, by Prop. 1.1, and so has a field of fractions D, say. By Prop. 1.6 it is not left Ore, so it contains a free algebra of infinite rank. This then provides an embedding of the free algebra in a field.

In spite (or perhaps because) of its simplicity this construction is of limited use, because not every automorphism of the free algebra can be extended to an automorphism of the field of fractions, constructed here. One application of this construction due to J. L. Fisher [71] is to show that the free algebra has many different fields of fractions, in contrast to the situation in the commutative case, where the field of fractions is unique up to isomorphism, when it exists (Prop. 1.3.4). Let $A = k[t]$ be the polynomial ring in a central indeterminate t over a commutative field k; for $n = 2, 3, \ldots$ it has the endomorphism α_n: $f(t) \mapsto f(t^n)$, which is clearly not surjective because t does not lie in the image (for $n > 1$). Now form $R = A[x; \alpha_n]$ and consider the subring S of R generated by x and $y = xt$ over k. By Prop. 1.6 it follows that $Sx \cap Sy = 0$, hence the subalgebra on x and y is free, and for different n we get distinct, i.e. non-isomorphic, embeddings, because

$$x^{-1}yx = tx = xt^n = x(x^{-1}y)^n = (yx^{-1})^n x,$$

hence

$$x^{-1}y = (yx^{-1})^n.$$

The skew polynomial ring admits the following natural generalization. Let A be any ring and M a monoid. Then the *monoid ring* on M over A consists of all formal sums $\sum sa_s$, where $s \in M$, $a_s \in A$ and almost all the a_s vanish, with addition and multiplication:

$$\sum sa_s + \sum sb_s = \sum s(a_s + b_s), \tag{12}$$

$$\sum sa_s \sum tb_t = \sum sta_s b_t. \tag{13}$$

When A is commutative and M is a group, this is just the well known group algebra on M over A. The homomorphism $\sum sa_s \mapsto \sum a_s$ is called the *augmentation mapping*.

The construction can still be generalized in two ways. Firstly, if we are given an action of M on A by endomorphisms, i.e. a homomorphism f of M into the monoid of endomorphisms of A, indicated by $f_s \colon a \mapsto a^s$ ($a \in A$, $s \in M$), this allows us to define a skew monoid ring, in which (13) is replaced by the formula

$$\sum sa_s \sum tb_t = \sum sta_s^t b_t,$$

which arises by using the commutation rule

$$as = sa^s.$$

Secondly, the multiplication of elements of M in the monoid ring may be modified by a factor set:

$$s \cdot t = stm_{s,t},$$

where $\{m_{s,t}\}$ satisfies the usual factor set conditions. We shall not enter into further details at this stage, since only special cases will be needed where the necessary formalism can be derived *ad hoc*. There is just one case of more frequent occurrence, namely the case of an infinite cyclic group. The group ring of the free group on a single generator t over A is called the ring of *Laurent polynomials*, written $A[t, t^{-1}]$. Its elements are finite sums $\sum t^i a_i$ ($a_i \in A$), where the exponent i may also be negative. This ring can also be described as the localization of the polynomial ring $A[t]$ at t; similarly we obtain a ring of skew Laurent polynomials by localizing the skew polynomial ring $A[t; \alpha, \delta]$ at t.

Exercises

1. Let K be a field with centre C. Show that for any $f \in K[t]$ the highest common left factor of all the afa^{-1} ($a \in K^{\times}$) is either a constant or a polynomial with coefficients in C.

2. Let k be a commutative field and let α be the endomorphism $f(t) \mapsto f(t^2)$ of the rational function field $E = k(t)$. Verify that $D = E(t; \alpha)$ is a field with centre k.

3. Let k be a commutative field with an endomorphism $\alpha \neq 1$. Show that every α-derivation is inner. (Hint. Verify that for all $a, b \in k$ we have $a^\delta(b^\alpha - b) = b^\delta(a^\alpha - a)$; note that this is a special case of Prop. 1.4.)

4. Let K be a field finite-dimensional over its centre and let α be an endomorphism of K. Show that any α-derivation of K is inner unless α itself is an inner automorphism.

5. Let $q = p^r$, where p is a prime number and $r \geq 1$, write $\mathbf{F}_q$ for the field of q elements and denote by T the endomorphism $f \mapsto f^p$ of $\mathbf{F}_q[x]$. Verify that each polynomial $\sum T^i c_i$ defines an endomorphism of $\mathbf{F}_q[x]$, where c_i acts by right multiplication, and that $aT = Ta^p$. Deduce that the endomorphisms form a skew polynomial ring $\mathbf{F}_q[T; \alpha]$, where $\alpha: a \mapsto a^p$ (O. Ore [33']).

6. Let D be a field finite-dimensional over its centre C. Given $f \in D[x]$, show that there exists $g \in D[x]^\times$ such that $fg = gf \in C[x]$.

7. The real quaternions $\mathbf{H}$ may be defined as an $\mathbf{R}$-algebra with basis $1, i, j, k$ and multiplication $i^2 = j^2 = -1$, $ij = -ji = k$. Writing α for complex conjugation in $\mathbf{C}$, show that there is a homomorphism $\mathbf{C}[t; \alpha] \mapsto \mathbf{H}$ defined by $t \mapsto j$ and deduce that $\mathbf{H} \cong \mathbf{C}[t; \alpha]/(t^2 + 1)$.

8. An *involution* of a field K is defined as an antiautomorphism $x \mapsto x^*$ whose square is the identity. In a field with involution * an element u is *unitary* if $u^*u = 1$. Show that for $u \neq 1$, $u^*u = 1$ holds if and only if $v + v^* = 1$, where $v = (1 - u)^{-1}$.

9. An involution * is *unitary-trivial* if $u^*u = 1 \Rightarrow u = 1$. Show that a field with a unitary-trivial involution, not the identity, is of characteristic 2 and any element commuting with its image is fixed under *.

10. Let K be a field with an automorphism α and an involution *. Show that for any $c \in K$ there exists an involution (necessarily unique) of $L = K(t; \alpha)$ extending * and mapping x to xc if and only if $(*\alpha)^2 = I(c)$, the conjugation by c, and $c^{*\alpha} = c^{-1}$. Further, if * is unitary-trivial on K, then so is its extension to L.

     By iterating this process construct a field of infinite degree over its centre with a unitary-trivial involution not the identity (see Cohn [79']).

2.2 The ideal structure of skew polynomial rings

Many division algebras arise as residue-class rings of skew polynomial rings over skew fields, which as we saw in 2.1, are (left or right) principal ideal domains; this makes it of interest to determine the elements generating two-sided ideals. We recall from 1.5 that an *invariant* element in a ring R is a regular element c such that $Rc = cR$. Clearly every unit and every regular element in the centre is invariant. Our first concern is to note that every ideal of an integral domain which is principal as left and as right ideal has an invariant generator.

PROPOSITION 2.2.1. *Let R be an integral domain and $\mathfrak{a}$ a non-zero ideal which is principal as left and as right ideal. Then $\mathfrak{a}$ has an invariant generator, i.e. there exists $c \in R^{\times}$ such that*

$$\mathfrak{a} = cR = Rc. \tag{1}$$

Proof. By hypothesis $\mathfrak{a} = cR = Rc'$, hence $c = uc'$, $c' = cv$ and so $c = uc' = ucv$. Now $uc \in cR$, say $uc = cw$, hence $cwv = ucv = c$, which shows that $wv = 1$. Thus v is a unit, by symmetry so is u and $cR = Rc$. ∎

Next we observe that in a polynomial ring over a field every ideal has a polynomial with central coefficients as generator. Since every non-zero polynomial is associated to a monic polynomial, it is enough to prove the result for the latter.

PROPOSITION 2.2.2. *Let K be any field with centre C and let $K[t]$ be the polynomial ring over K in a central indeterminate t. Then a monic polynomial in $K[t]$ is left or right invariant if and only if all its coefficients lie in C.*

Proof. Clearly any polynomial with coefficients in C is (left and right) invariant. Conversely, assume that $f = t^n + t^{n-1}a_1 + \ldots + a_n$ is right invariant in $K[t]$. Then for any $c \in K$ we have $cf = fc'$ for some c', which must lie in K, by comparing degrees, thus

$$t^n c + t^{n-1} ca_1 + \ldots + ca_n = t^n c' + t^{n-1} a_1 c' + \ldots + a_n c'.$$

Equating coefficients we find that $c' = c$, $ca_i = a_i c'$, hence $ca_i = a_i c$ for $i = 1, \ldots, n$, so $a_i \in C$, as claimed. ∎

Let K be a skew field, α an endomorphism and δ an α-derivation of K. As we saw in 2.1, the skew polynomial ring $R = K[t; \alpha, \delta]$ is then a PRID, so every non-zero ideal of R has the form fR, where $f \neq 0$ and $Rf \subseteq fR$, so f is right invariant. This makes it of interest to find a criterion for right invariance. Let us call f *right K-invariant* if $f \neq 0$ and $Kf \subseteq fK$, thus f is right invariant if and only if f is right K-invariant and $tf \in fR$; explicit criteria for right invariance are derived in Prop. 2.3 and

Th. 2.5 below. We remark that in considering the right ideal fR we may assume f to be monic in t, by dividing by the leading coefficient; then f has the form

$$f = t^n + t^{n-1}a_1 + \ldots + a_n, \quad \text{where } a_i \in K. \tag{2}$$

However, we need not restrict ourselves to fields; in what follows, K may be any integral domain and α an injective endomorphism. The conditions for $f \in R = K[t; \alpha, \delta]$ to be right invariant are

$$cf = fc' \quad \text{for all } c \in K,$$

$$tf = f(tb + a),$$

where c', b, $a \in K$, by a comparison of degrees. Taking f as in (2), we find

$$cf = ct^n + \ldots = t^n c^{\alpha^n} + \ldots,$$

$$fc' = t^n c' + \ldots,$$

where the dots indicate lower terms in t; hence $c' = c^{\alpha^n}$. Similarly we have

$$tf = t^{n+1} + t^n a_1 + \ldots,$$

$$f(tb + a) = t^{n+1}b + t^n a + t^{n-1}a_1 tb + \ldots,$$

$$= t^{n+1}b + t^n(a + a_1^\alpha b) + \ldots;$$

for right invariance we must have $b = 1$, $a + a_1^\alpha = a_1$ and so we obtain the following description of right invariant elements:

PROPOSITION 2.2.3. *Let K be an integral domain with an injective endomorphism α and an α-derivation δ and put $R = K[t; \alpha, \delta]$. Then a monic polynomial f in R of degree n is right K-invariant if and only if*

$$cf = fc^{\alpha^n} \text{ for all } c \in K; \tag{3}$$

f is right invariant if and only if (3) holds, as well as

$$tf = f(t + a), \text{ where } a = a_1 - a_1^\alpha, \tag{4}$$

a_1 being the coefficient of t^{n-1} in f. ∎

If α is an automorphism of K with inverse β, then (3), (4) can be written as $fc = c^{\beta^n}f$, $ft = (t - a^{\beta^n})f$, hence f is then also left invariant, and so we have

COROLLARY 2.2.4. *If in Prop. 2.3, α is an automorphism of K, then any monic right (K-) invariant element of R is (K-) invariant.* ∎

For any monic polynomial f in $K[t; \alpha, \delta]$, given by (2), let us define its *divergence* as

$$\Delta(f) = tf - f(t + a), \quad \text{where } a = a_1 - a_1^\alpha. \tag{5}$$

We note that $\Delta(f)$ has a lower degree than f. For on writing $\deg f = n$ and ignoring terms of degree less than n, we have

$$\Delta(f) = t^{n+1} + t^n a_1 - t^n(t + a) - t^{n-1}a_1(t + a) + \ldots$$
$$= t^n(a_1 - a - a_1^\alpha) + \ldots,$$

and the coefficient of t^n vanishes by the definition of a.

This leads to the following criterion for invariance:

THEOREM 2.2.5. *Let K be a field with endomorphism α and α-derivation δ and put $R = K[t; \alpha, \delta]$. If f is a monic right K-invariant polynomial of degree n, then its divergence $\Delta(f)$ is again right K-invariant of degree less than n and $\delta\alpha^n - \alpha^n\delta$ is an inner (α^n, α^{n+1})-derivation. Morever, f is right invariant if and only if $\Delta(f) = 0$.*

Further, if $u = \sum_0^m t^{m-i}a_i$ $(a_0 = 1)$ is the monic right K-invariant polynomial of least positive degree m, then

$$\Delta(u) = -a_m^\delta - a_m a, \quad \text{where } a = a_1 - a_1^\alpha, \tag{6}$$

and u is right invariant, unless α^{m+1} is an inner automorphism of K.

The element u will be called the *minimal right K-invariant element* of R.

Proof. The right invariance of f is characterized by the vanishing of $\Delta(f)$, by Prop. 2.3. To verify the right K-invariance of $\Delta(f)$, we have, for any $c \in K$, on writing $\lambda = \alpha^n$,

$$c\Delta(f) = ctf - cf(t + a)$$
$$= (tc^\alpha + c^\delta)f - fc^\lambda(t + a)$$
$$= tfc^{\alpha\lambda} + fc^{\delta\lambda} - ftc^{\lambda\alpha} - fc^{\lambda\delta} - fc^\lambda a$$
$$= [tf - f(t + a)]c^{\alpha\lambda} + f[ac^{\alpha\lambda} + c^{\delta\lambda} - c^{\lambda\delta} - c^\lambda a].$$

Hence we find

$$c\Delta(f) - \Delta(f)c^{\alpha\lambda} = f[ac^{\alpha\lambda} + c^{\delta\lambda} - c^{\lambda\delta} - c^\lambda a]. \tag{7}$$

The left-hand side is of degree less than n, while the right-hand side has the factor f of degree n, so both sides must vanish and we obtain

$$c\Delta(f) = \Delta(f)c^{\alpha^{n+1}}, \tag{8}$$

$$c^{\delta\lambda} - c^{\lambda\delta} = c^{\lambda}a - ac^{\lambda\alpha}. \tag{9}$$

Now (8) shows $\Delta(f)$ to be right K-invariant, and (9) shows $\delta\lambda - \lambda\delta$ to be an inner $(\alpha^{n}, \alpha^{n+1})$-derivation.

Suppose now that $u = \sum_{0}^{m} t^{m-i}a_i$ is the monic right K-invariant polynomial of least positive degree; then since $\Delta(u)$ is right K-invariant of lower degree, it must be of degree zero. Now

$$\Delta(u) = \sum t^{i+1}a_{m-i} - \sum t^{i}a_{m-i}t - \sum t^{i}a_{m-i}a,$$

and the constant term is easily seen to be $-a_m^{\delta} - a_m a$. Hence (6) follows, and (8), applied to u, shows that either $\Delta(u) = 0$ and u is right invariant, or $\Delta(u) \neq 0$ and α^{m+1} is an inner automorphism. ■

From this result it is easy to obtain a criterion for the simplicity of skew polynomial rings, using the following observation on the occurrence of invariant elements:

LEMMA 2.2.6. *Let K be a field with an endomorphism α and an α-derivation δ and put $R = K[t; \alpha, \delta]$. If R contains a right K-invariant element of positive degree, then it contains a right invariant element of positive degree.*

Proof. We observe that the divergence satisfies the rule:

$$\Delta(fg) = \Delta(f)g + f\Delta(g), \tag{10}$$

for any monic polynomials f, g such that g is right K-invariant. Of course Δ is not a derivation, since it is not K-linear. To prove (10), let us write $f = t^n + t^{n-1}a_1 + \ldots$, $g = t^m + t^{m-1}b_1 + \ldots$, $a = a_1 - a_1^{\alpha}$, $b = b_1 - b_1^{\alpha}$, and $\mu = \alpha^m$. Then $ag = ga^{\mu}$, hence

$$\Delta(f)g + f\Delta(g) = [tf - f(t + a)]g + f[tg - g(t + b)]$$

$$= tfg - f(t + a)g + ftg - fg(t + b)$$

$$= tfg - fg(t + b + a^{\mu}).$$

Since $fg = t^{m+n} + t^{m+n-1}(b_1 + a_1^{\mu}) + \ldots$, this is just $\Delta(fg)$, and (10) follows.

Let u be the minimal right K-invariant element, of degree m, say. As we saw in Th. 2.5, $\Delta(u) \in K$, hence for any $n \geqslant 0$ we have by a repeated application of (10),

$$\Delta(u^{n+1}) = \sum u^{n-i}\Delta(u)u^i = u^n(\sum_0^n \Delta(u)^{\alpha^{im}}).$$

It follows that $tu^{n+1} \in u^n R$ for all $n \geqslant 1$, hence $t^s u^n \in u^{n-s} R$, and this lies in uR provided that $s < n$. Now any polynomial f can be written as

$$f = uq + r, \quad \text{where } \deg r < m.$$

Therefore $fu^m = uqu^m + ru^m \in uR$, and it follows that

$$Ru^m \subseteq uR.$$

Thus uR contains a two-sided ideal $Ru^m R \neq 0$, which has a right invariant element of positive degree as generator. ∎

We now have the following simplicity criterion:

THEOREM 2.2.7. *Let K be a field with endomorphism α and an α-derivation δ and let $R = K[t; \alpha, \delta]$. If the (α^n, α^{n+1})-derivation $\delta\alpha^n - \alpha^n\delta$ is outer for all $n = 1, 2, \ldots$, then R is a simple ring.*

Proof. If R is not simple, it has a non-unit right invariant element f. If $\deg f = n$, then by Th. 2.5, $\delta\alpha^n - \alpha^n\delta$ is inner. Moreover in the contrary case, R does not even have non-constant right K-invariant elements, by Lemma 2.6. ∎

In the special case when $\delta = 0$ we can get a more explicit result. An automorphism α of a field K is said to have *inner order* r if α^r is the least positive power which is inner, of the form $I(e): a \mapsto eae^{-1}$. If α^r is outer for all $r > 0$ or α is not an automorphism, α is said to have *infinite inner order*.

PROPOSITION 2.2.8. *Let K be a field with an endomorphism α and consider the skew polynomial ring $R = K[t; \alpha]$. (i) If α has infinite inner order, then every right K-invariant element is right invariant and the right invariant elements are all of the form $t^r c$ ($c \in K^\times$). (ii) If α is an automorphism of inner order r, say $\alpha^r = I(e)$, then $u = t^r e$ centralizes K and the invariant elements are of the form $t^m g$, where g is a polynomial in u with coefficients in the centre of K.*

Proof. Let $f = t^n + t^{n-1}a_1 + \ldots + a_n$ be right invariant. Then $tf = f(t + a)$, where $a = a_1 - a_1^\alpha$; comparing lowest terms, we find that $a = 0$, so $tf = ft$ and on writing this out and equating coefficients, we find

$$a_i^\alpha = a_i, \quad i = 1, \ldots, n. \tag{11}$$

Next we have $cf = fc^{\alpha^n}$ for any $c \in K$, hence

$$c^{\alpha^{n-i}} a_i = a_i c^{\alpha^n}.$$

Now α^i is injective; cancelling α^{n-i}, we find that

$$ca_i = a_i c^{\alpha^i}, \quad i = 1, \ldots, n. \tag{12}$$

If α has infinite inner order, it follows that $a_i = 0$ and f reduces to t^n, so in this case the only right invariant elements are $t^n c$.

If α^r is an inner automorphism, say $\alpha^r = I(e)$, let us write $u = t^r e$. Then for $c \in K$,

$$cu = ct^r e = t^r c^{\alpha^r} e = t^r ece^{-1} \cdot e = uc,$$

so u centralizes K. Now any right invariant element is invariant, and if f given as before is invariant, then (12) above shows that $a_i = 0$ unless r divides i; hence we have $f = t^m g$, where g is a polynomial in u. Moreover, g is invariant, so $cg = gc^\mu$ for some automorphism μ. Comparing highest powers of u we see that $\mu = 1$, so $cg = gc$ and if $g = \sum u^{s-i} b_i$, then

$$0 = cg - gc = \sum u^{s-i}(cb_i - b_i c).$$

Hence b_i lies in the centre C of K, thus f equals a power of t times an element of $C[u]$, as claimed. ∎

Next we allow a derivation but restrict α to be an automorphism. An element of the skew polynomial ring over K will be called *K-central* if it centralizes K.

THEOREM 2.2.9. *Let K be a field with an automorphism α and an α-derivation δ and put $R = K[t; \alpha, \delta]$. If R is not simple, then there are non-unit K-invariant elements; let u be a monic one, of least positive degree m, say. Either (i) α has infinite inner order, in which case every K-invariant element has the form $u^r c$, $r \in \mathbf{N}$, $c \in K^\times$, or (ii) α has finite inner order, so an element of the form $u^s c$ is K-central. If $v = u^d e$ is a K-central element of least positive degree, then every K-invariant element has the form $f = u^r f_1 c$, where f_1 is a polynomial in v with coefficients in the centre of K and c induces the automorphism α^{n-rm}, $n = \deg f$, $m = \deg u$.*

Proof. We have seen that R is not simple precisely when there are non-unit K-invariant elements (Lemma 2.6), thus let u be the minimal (monic) K-invariant element, of degree m, say. Any K-invariant element f, taken monic for convenience, may be written in the form

$$f = uq + r, \quad \text{where } q, r \in R, \deg r < m.$$

By Prop. 2.3, $cu = uc^\mu$, where $\mu = \alpha^m$; if f has degree n and we put $v = \alpha^n$, then $fc^v = cf = cuq + cr = uc^\mu q + cr$, hence

$$u(qc^v - c^\mu q) = cr - rc^v. \tag{13}$$

On the left there is a factor of degree m, while the right-hand side has degree less than m; hence both sides vanish and so

$$cr = rc^v.$$

Thus r is K-invariant, of degree less than m, hence $r \in K$. Moreover, by (13) and the fact that α is an automorphism, we have

$$cq = qc^{\alpha^{n-m}}.$$

Applying the same argument to q, we find, by induction on the degree of f,

$$f = u^s + u^{s-1}b_1 + \ldots + b_s, \quad \text{where } b_i \in K, \tag{14}$$

hence $n = sm$. The highest coefficient in (14) is 1, because both u and f are monic. Now for any $c \in K$ we have

$$fc^v = u^s c^v + u^{s-1}b_1 c^v + \ldots + b_s c^v,$$

$$cf = u^s c^{\mu^s} + u^{s-1}c^{\mu^{s-1}}b_1 + \ldots + cb_s.$$

Equating powers of u, we see again that $v = \mu^s$, i.e. $n = sm$ and

$$cb_i = b_i c^{\mu^i}, \quad i = 1, \ldots, s. \tag{15}$$

Suppose first that α has infinite inner order. Then $b_i = 0$ and by (14), every K-invariant polynomial is of the form $u^r c$, where $c \in K$. The alternative is that a power of α is inner, say α^z is the least such power. Then $\mu^i = \alpha^{im}$ is inner if and only if $z | im$, so by (15) $b_i = 0$ unless $z | im$. Hence either $f = u^s$ or f is equal to u^r times a polynomial in u^d, where $d = z/(z, m)$. By hypothesis,

$$cu^d = u^d c^{\alpha^{dm}} = u^d ece^{-1} \text{ for some } e \in K^\times \text{ and all } c \in K.$$

Hence $v = u^d e$ is K-central and $u^{-r}f$ can be written as a polynomial in v, say $f = u^r(\sum v^j c_j)$. Now $g = u^{-r}f$ is K-invariant of degree $n - rm$, so if $\lambda = \alpha^{n-rm}$, then $ag = ga^\lambda$ and

$$0 = ag - ga^\lambda = \sum v^j(ac_j - c_j a^\lambda).$$

Hence $ac_j = c_j a^\lambda$ and it follows that all the cs can be obtained from a single one by multiplying by an element of the centre of K. Thus $f = u^r f_1 c$, where f_1 centralizes K and c induces the automorphism α^{n-mr}, which is what we had to show. ∎

We shall apply this result to determine the centre of the corresponding field of fractions.

THEOREM 2.2.10. *Let K be a skew field with centre C. Given an automorphism α and an α-derivation δ of K, consider $U = K(t; \alpha, \delta)$, the field of fractions of the skew polynomial ring $R = K[t; \alpha, \delta]$.*

(i) *If R is simple or α has infinite inner order, then the centre of U is the set*

$$C_0 = \{a \in C \,|\, a^\alpha = a, \, a^\delta = 0\}. \tag{16}$$

(ii) *If R is not simple and α has finite inner order, so that there exist non-unit invariant elements in R, let v be the monic invariant element of least positive degree inducing an inner automorphism $I(c): a \mapsto cac^{-1}$; then the centre of U is $C_0(vc)$, the field of rational functions in vc over C_0.*

Proof. By Prop. 1.1, R is a PID and it is clear that in any case the centre of U contains C_0. By Prop. 1.5.7, every element of the centre of U, $\mathscr{C}(U)$, has the form $fg^{-1} = g^{-1}f$, where for some automorphism λ of R,

$$pf = fp^\lambda, \quad pg = gp^\lambda \text{ for all } p \in R. \tag{17}$$

If R is simple, then there are no non-unit invariant elements in R, so in that case $\mathscr{C}(U) \subseteq K$. An element a of K centralizes K precisely when $a \in C$, and if $at = ta$, then $ta = ta^\alpha + a^\delta$, hence $a^\alpha = a$, $a^\delta = 0$, so $\mathscr{C}(U) = C_0$.

We may now take R to be not simple and hence to possess non-unit invariant elements. Assume that α has infinite inner order and let u be the minimal K-invariant element, inducing the automorphism β, say, of K. If $fg^{-1} \in \mathscr{C}(U)$, we can write $f = u^r c^{-1}$, $g = u^s d^{-1}$, where $c, d \in K$. For any $a \in K$ we have $af = au^r c^{-1} = u^r a^{\beta^r} c^{-1} = fca^{\beta^r} c^{-1}$, so f induces the automorphism $\beta^r I(c)$, and similarly g induces $\beta^s I(d)$. By (17) we have

$$\lambda = \beta^r I(c) = \beta^s I(d). \tag{18}$$

Now $\beta = \alpha^m$, where $m = \deg u$. Since α has infinite inner order, then so does β, and by (18), $r = s$ and $fg^{-1} = g^{-1}f = dc^{-1} \in K$. Since this element must centralize K, we find as before that it lies in C_0.

There remains the case when α has finite inner order. By hypothesis there are non-unit K-invariant and hence non-unit invariant elements. Let z be the minimal invariant element, say $pz = zp^\mu$ for $p \in R$. By (17) f is invariant and we have

$$f = zq + r, \quad \text{where } \deg r < \deg z.$$

For any $a \in K$, we have $fa^\lambda = af = azq + ar = za^\mu q + ar$, hence

$$z(qa^\lambda - a^\mu q) = ar - ra^\lambda.$$

Here the right-hand side has lower degree than the factor z on the left, so both sides must vanish. By the minimality of z we have $r \in K$ and q is again invariant, so f is a polynomial in z over K, by induction on $\deg f$. Thus we have $f = \sum z^i a_i$, $g = \sum z^i b_i$, where $a_i, b_i \in K$. By (17) we have

$$0 = pf - fp^\lambda = \sum z^i(p^{\mu^i} a_i - a_i p^\lambda),$$

hence all the coefficients vanish and so $\mu^i \lambda^{-1}$ is an inner automorphism whenever $a_i \neq 0$ and similarly for g. Now we may assume f and g to be without a common factor, hence a_0, b_0 are not both 0, say $a_0 \neq 0$ and dividing f, g by a common factor, we may assume that $a_0 = 1$. Since $pa_0 = p = a_0 p^\lambda$, it follows that $\lambda = 1$ and the powers of z occurring in f or g induce inner automorphisms. Let $v = z^r$ be the least such power, inducing $I(c)$ say; then vc is central and f, g can be expressed as polynomials in vc with coefficients in C_0. ■

In the special case $\alpha = 1$, $\delta = 0$ we again reach the conclusion of Prop. 1.5.

Exercises

1. Let K be a field with endomorphisms α, β. Show that the (α, β)-derivations on K form an additive group, and for any (α, β)-derivation δ and any endomorphisms λ, μ, $\lambda\delta\mu$ is a $(\lambda\alpha\mu, \lambda\beta\mu)$-derivation.

2. Let E be a finite extension of the Galois field $\mathbf{F}_p$ with automorphism $\alpha: a \mapsto a^p$ and put $R = E[t; \alpha]$. Find an E-invariant element of positive degree which is not invariant.

3. Let K be a field with an endomorphism α and an α-derivation δ such that $\alpha\delta = \delta\alpha$. Show that α extends to $U = K(t; \alpha, \delta)$ and that δ is induced by an inner α-derivation of U.

4. Let K be a field with a non-surjective endomorphism α. Show that in $K(t; \alpha)$ the union $\bigcup t^{-n} K t^n$ has an automorphism extending α and is the least subfield containing K with this property.

5. Show that in a skew polynomial ring $K[t; \alpha, \delta]$ over a field K, a polynomial f (not necessarily monic) is invariant if and only if f is K-invariant and $tf = f(tb + c)$. Determine b and c in terms of the coefficients of f.

6. Let E be a commutative field with an automorphism α of order n and fixed

field F. Show that for any irreducible polynomial f over F, $f(t^n)$ is an invariant element of $R = E[t; \alpha]$ and the residue-class ring is simple, with one exception. What is its centre? When is it a field? What is the exception?

7. Let E be a field with centre k. Given a polynomial f over k show that over E, f splits into a number of factors all of the same degree. Deduce that an irreducible polynomial of prime degree over k either stays irreducible over E or splits into linear factors. (Hint. Use Th. 1.5.4.)

8. Let $R = K[t; \alpha, \delta]$ be a skew polynomial ring over a field K with endomorphism α and α-derivation δ such that $\alpha\delta = \delta\alpha$. Given f as in (2), verify (by comparing coefficients of t^{n-1} in (3)) that

$$a_1 c^{\lambda\alpha} - c^\lambda a_1 = nc^{\lambda\delta}, \quad \text{where } \lambda = \alpha^{n-1}.$$

Deduce that for char $K = 0$, δ is inner on im λ.

9. (after J.-P. Van Deuren) Show that the field of fractions of the complex-skew polynomial ring $C[t; {}^-]$ contains a central element $x = t^2$ such that for any odd m, $x^m - 1$ is a square. Hence obtain a subfield of given genus g. (Observe that the function field of the curve $y^2 = x^{2g-1} - 1$ is of genus g, see Cohn [91], 4.6.)

10. Let k be a commutative field containing a primitive nth root of 1, ζ say. Define an automorphism σ on $F = k(y)$ over k by the rule $\sigma: y \mapsto \zeta y$ and form the skew function field $F(x; \sigma)$. Find its centre and show that its dimension over the centre is n^2.

2.3 Power series rings

In the commutative case the familiar power series ring $K[[t]]$ may be regarded as the completion of the polynomial ring $K[t]$ with respect to the 't-adic topology', i.e. the topology obtained by taking the powers of the ideal generated by t as a neighbourhood base at zero. No problems arise in extending this concept to the ring $K[t; \alpha]$ for any endomorphism α of K. In this way we obtain the ring $K[[t; \alpha]]$ of all skew power series. Suppose now that α is an automorphism and consider the ring $K((t; \alpha))$ of *skew Laurent series*; they are series of the form $\sum_{-r}^{\infty} t^i a_i$ with componentwise addition and multiplication by the commutation rule:

$$at^n = t^n a^{\alpha^n}. \tag{1}$$

Since n may now be negative, we have had to restrict α to be an automorphism. We remark that $K((t; \alpha))$ is again a field, for any non-zero series can be written $t^{-r}c(1 - \sum_1^{\infty} t^i a_i)$ and this has the inverse $[\sum_n(\sum_i t^i a_i)^n]c^{-1}t^r$. This field of skew Laurent series can be formed even when α is not an automorphism, say by localizing at the powers of t, but now its elements can no longer all be written in the form $\sum t^i a_i$.

Suppose now that we have a field K with an endomorphism α and an α-derivation δ. Here we face a difficulty; the above topology on $K[t; \alpha]$ may be described in terms of the order-function:

$$o(f) = r \text{ if } f = t^r a_r + t^{r+1} a_{r+1} + \ldots + t^n a_n \quad (a_r \neq 0).$$

It turns out that when $\delta \neq 0$, the multiplication is not continuous in the t-adic topology, as the formula

$$a \cdot t = t a^\alpha + a^\delta \tag{2}$$

shows, and any attempt to construct the completion directly will fail. A way out of this difficulty is to introduce $z = t^{-1}$ and rewrite (2) as a commutation formula for z. We then get

$$za = a^\alpha z + z a^\delta z. \tag{3}$$

Owing to the inversion we now have to shift coefficients to the left, and as (3) shows, we cannot usually do this completely in the polynomial ring, but we can do it to any desired degree of accuracy by applying (3) repeatedly:

$$za = a^\alpha z + a^{\delta\alpha} z^2 + z a^{\delta^2} z^2 = \ldots$$
$$= a^\alpha z + a^{\delta\alpha} z^2 + a^{\delta^2\alpha} z^3 + \ldots + a^{\delta^{n-1}\alpha} z^n + z a^{\delta^n} z^n. \tag{4}$$

If δ is locally nilpotent, i.e. each element of K is annihilated by some power of δ, then (4) can be used as a commutation formula in the skew polynomial ring. But in any case, in the power series ring we can pass to the limit and obtain the formula

$$za = a^\alpha z + a^{\delta\alpha} z^2 + a^{\delta^2\alpha} z^3 + \ldots. \tag{5}$$

The ring obtained in this way is clearly an integral domain, and the set consisting of all powers of z is a left Ore set, by (5), so we can form the ring of fractions, which is in effect the ring of skew Laurent series in z. We shall not use a special notation for this ring but remark that it is actually a field:

THEOREM 2.3.1. *Let K be a skew field with an endomorphism α and an α-derivation δ, and consider the following ring:*

$$R = K\langle z; za = a^\alpha z + z a^\delta z \text{ for all } a \in K \rangle. \tag{6}$$

This ring has a completion $\hat{R}$, consisting of all power series $\sum a_i z^i$ and $Z = \{1, z, z^2, \ldots\}$ is a left Ore set whose inversion yields a field, consisting of all power series of the form

$$\sum_{i=0}^{\infty} z^{-r} a_i z^i. \tag{7}$$

When α is an automorphism, the series can be written $\sum b_i z^i$ or also $\sum z^i c_i$.

Proof. Consider the completion $\hat{R}$ consisting of all series $\sum_0^{\infty} a_i z^i$. It is clear that these series form a ring with the commutation rule (5). The order function satisfies the relation $o(fg) = o(f) + o(g)$, because α is injective, and this shows $\hat{R}$ to be an integral domain. Moreover, (5) shows that for any $a \in K$ there exists $f \in \hat{R}$ such that $za = fz$, thus z is left invariant, hence $zg = g^{\lambda} z$ for any $g \in \hat{R}$ and so $z^r g = g^{\lambda^r} z^r$. This shows Z to be a left Ore set, hence the elements $z^{-r} f$ ($r \geqslant 0$, $f \in \hat{R}$) form again a ring. To show that this ring is in fact a field, we first note that every element of order 0 in $\hat{R}$ is a unit. For if $f = \sum a_i z^i$, where $a_0 \neq 0$, we have

$$a_0^{-1} f = 1 - \sum b_i z^i, \quad \text{where } b_i = -a_0^{-1} a_i \ (i > 0).$$

Hence

$$f^{-1} = (1 - \sum b_i z^i)^{-1} a_0^{-1} = \sum_0^{\infty} (\sum b_i z^i)^n a_0^{-1}.$$

By expanding and rearranging this series we obtain an expression of the form $\sum c_i z^i$; this follows because for any m, the terms from $(\sum b_i z^i)^n$ with $n > m$ do not contribute to the coefficient of z^m. This shows f to be a unit.

Now any non-zero power series of the form (7) can be written as $z^{-r} g z^s$, where $r, s \geqslant 0$ and g is of order 0 and hence a unit. It has the inverse $z^{-s} g^{-1} z^r$, and this shows the localization to be a field.

When α is an automorphism, we can clearly pull the coefficients through to one side or the other and so obtain the given form for the series. ∎

For power series the centre is not hard to determine, at least when α has infinite inner order. If K is a field with centre C, and an endomorphism α and α-derivation δ are given, then the (α, δ)-*reduced centre* of K is defined as the subfield

$$C_0 = \{a \in C; a^{\alpha} = a, a^{\delta} = 0\}. \tag{8}$$

PROPOSITION 2.3.2. *Let K be a skew field with an automorphism α of infinite inner order and an α-derivation δ. Let $\hat{R}$ be the completion of the ring (6) and U its field of fractions. Then the centre of U is the (α, δ)-reduced centre C_0 of K.*

Proof. Clearly $C_0 \subseteq \mathscr{C}(U)$; to prove equality, we take any $f \in U$ and write it as a Laurent series: $f = \sum a_i z^i$. If $cf = fc$, then

$$\sum (ca_i - a_i c^{\alpha^i}) z^i = 0,$$

hence $ca_i = a_i c^{\alpha^i}$ and so $a_i = 0$ unless $i = 0$, and $a_0 \in C$. Further, a_0 centralizes z precisely when $az^{-1} = z^{-1}a = z^{-1}a^\alpha + a^\delta$, i.e. $a^\alpha = a$, $a^\delta = 0$, so a_0 lies in C_0, which is therefore the centre. ∎

From this result we obtain another proof of part of Th. 2.10(i) by applying the following criterion, well known in the case of complex power series:

PROPOSITION 2.3.3 (Rationality criterion). *Let K be a field with an automorphism α. Then a formal series $\sum t^i a_i$ in $K((t; \alpha))$ is a rational function of t if and only if there exist integers r, n_0 and elements $c_1, \ldots, c_r \in K$ such that*

$$a_n = a_{n-1}^\alpha c_1 + a_{n-2}^{\alpha^2} c_2 + \ldots + a_{n-r}^{\alpha^r} c_r \quad \text{for all } n > n_0. \tag{9}$$

For this is just the condition that $(\sum t^i a_i)(1 - \sum_1^r t^j c_j)$ should be a polynomial, except for a factor t^{-k}. ∎

We conclude this section with two constructions. The first is a result of Köthe [31], allowing us to construct outer automorphisms of skew fields:

PROPOSITION 2.3.4. *Let K be a commutative field with an automorphism α and put $E = k((t; \alpha))$. Given any automorphism β of k such that $\alpha\beta = \beta\alpha$, extend β to E by the rule $t^\beta = t$. Then β is an inner automorphism of E if and only if $\beta = \alpha^r$ for some $r \in \mathbf{Z}$.*

Proof. If β is inner on E, then there exists $a \in E^\times$ such that

$$u^\beta = a^{-1}ua \text{ for all } u \in E,$$

hence $ua = au^\beta$. Let $a = \sum t^i a_i$ and first take $u = t$. Then $t^\beta = t$, so

$$\sum t^{i+1} a_i = \sum t^i a_i t = \sum t^{i+1} a_i^\alpha,$$

hence $a_i^\alpha = a_i$. Next take $u = b \in k^\times$; then $b(\sum t^i a_i) = \sum t^i a_i b^\beta$ by hypothesis and so $\sum t^i (b^{\alpha^i} - b^\beta) a_i = 0$, hence $b^\beta = b^{\alpha^i}$ whenever $a_i \neq 0$. This holds for some i, so $\beta = \alpha^r$ for some $r \in \mathbf{Z}$. Conversely, if $\beta = \alpha^r$, then β is inner, induced by t^r. ∎

For example, if $k = F(s)$, where F is any field of characteristic 0, and

$\alpha: s \mapsto s + 1$, then $\beta: s \mapsto s + 1/2$ is an outer automorphism of $E = k((t; \alpha))$.

Secondly we shall sometimes need a field with a prescribed field as centre. This is accomplished by the next result:

PROPOSITION 2.3.5. *Let k be a commutative field. Then there exists a field D whose centre is k and which is infinite-dimensional over k.*

Proof. Take the rational function field $k(t)$ and adjoin roots of the equations $x^{2^n} = t$ for $n = 1, 2, \ldots$. The resulting field may be denoted by $E = k(t, t^{1/2}, t^{1/4}, \ldots)$ and may also be obtained as the field of fractions of the group algebra over k of the additive group of all dyadic fractions $m/2^n$. On E we have the automorphism $\alpha: f(t) \mapsto f(t^2)$; we claim that the fixed field of α is precisely k. Any element f of E may be written as a rational function of t^{2^r} for some $r \in \mathbf{Z}$. If $f \notin k$, then the possible values of r are bounded above. Let us choose r as large as possible, so that f is not a rational function of $t^{2^{r+1}}$. Since f^α is such a function, it follows that $f^\alpha \neq f$, and this shows the fixed field to be k.

We now form the field of skew Laurent series

$$D = E((x; \alpha)),$$

and claim that D has centre k. Any element of D has the form $f = \sum x^i a_i$, where $a_i \in E$. If f centralizes D, then $\sum x^{i+1} a_i = xf = fx = \sum x^{i+1} a_i^\alpha$, hence $a_i^\alpha = a_i$ and so $a_i \in k$, by the first part. Further, $\sum x^i a_i t = ft = tf = \sum t x^i a_i = \sum x^i t^{2^i} a_i$, hence $a_i(t - t^{2^i}) = 0$ and so $a_i = 0$ unless $i = 0$. Therefore $f = a_0 \in k$ as claimed. Clearly D is infinite-dimensional over k, since e.g. $1, t, t^2, \ldots$ are linearly independent. ∎

We remark that the field D used here can be formed more simply as follows. Let $F = k(t)$ be the rational function field with endomorphism $\alpha: f(t) \mapsto f(t^2)$. The skew polynomial ring $F[x; \alpha]$ has a field of fractions $F(x; \alpha)$ which is clearly a subfield of the field D of skew Laurent series constructed in the proof of Prop. 3.5, and that proof shows that $F(x; \alpha)$ has all the required properties.

Exercises

1. Find an automorphism α of $K = \mathbf{R}(t)$ and an automorphism β of $K(x; \alpha)$, both of infinite inner order, such that $\alpha\beta = \beta\alpha$ and α, β are not commensurable, i.e. there is no relation $\alpha^r = \beta^s$, $r, s \in \mathbf{N}$.

2. Let K be a skew field with automorphisms α, β such that $\alpha\beta = \beta\alpha$. Put

$D = K((x; \alpha))$ and extend β to D by $x^\beta = x$. Show that β is inner on D if and only if $\beta = \alpha^r I(c)$ for some $r \in \mathbf{Z}$, $c \in K^\times$.

3. Let D be the field generated over a commutative field k of characteristic 0 by x, y with the relation $xy - yx = 1$. Show that the centre of D is k.

4. Let K be a field with a surjective derivation δ and consider the field of Laurent series in x with commutation formula $cx = xc + xc^\delta x$. Verify that $\sum x^{i-1} c_i - \sum x^i c_i x^{-1} = \sum x^i c_i^\delta$ and deduce that the field of skew Laurent series has a surjective inner derivation.

5. (Lazerson [61]) Let k be a commutative field of finite characteristic p and adjoin commuting indeterminates to form $K = k(x_1, x_2, \ldots)$. Verify that K has a derivation δ such that $x_i^\delta = x_{i-1}$ ($i > 1$), $x_1^\delta = 1$. On $L = K(t; 1, \delta)$ show that for $q = p^n$, δ^q is the inner derivation induced by t^q and for any $a \in L$ there exists $q = p^n$ such that $[ax_q, t^q] = a$; thus δ is surjective on L.

6. By forming ultraproducts of the fields in Ex. 5 for different p obtain a field of characteristic 0 with a surjective inner derivation.

7. Let K be a field with an automorphism α and consider the skew power series field $L = K((t; \alpha))$. Show that no element of $L \backslash K$ can be right algebraic over K, in the sense that all its powers are right linearly independent over K (see 3.4). (Hint. If $u = \sum t^i c_i \in L$ is right algebraic over K but $u \neq 0$, show that the order must be zero and then repeat the argument with $u - c_0$.)

2.4 Group rings and the Malcev–Neumann construction

Let M be a monoid and consider the monoid ring KM over a field K (possibly skew). As defined in 2.1, this is the vector space over K with M as basis, made into a ring by means of the multiplication in M. We ask: for which groups G is KG embeddable in a field? Clearly it is necessary for KG to be an integral domain, and for this to hold, G must be torsion-free. For if $u \in G$ is of finite order n, then

$$(1 - u)(1 + u + u^2 + \ldots + u^{n-1}) = 0. \tag{1}$$

When G is abelian, this condition is also sufficient:

THEOREM 2.4.1. *Let G be an abelian group and k any commutative field. Then the group algebra kG is embeddable in a field if and only if G is torsion-free.*

Proof. The necessity follows by (1), so assume that G is torsion-free. Writing G additively for the moment, we can regard it as a $\mathbf{Z}$-module and because G is torsion-free, we can embed it in a $\mathbf{Q}$-module, i.e. a vector space over $\mathbf{Q}$. By taking an ordered basis of this space and using a lexicographic ordering of the coefficients, we obtain a total ordering of G, which makes it into an ordered group. Going back to multiplicative notation, we thus have a total ordering on G such that $s \leqslant s'$, $t \leqslant t'$ imply $st \leqslant s't'$. Now it is easy to see that the group algebra of an ordered group is an integral domain. Let

$$a = a_1 s_1 + \ldots + a_m s_m, \text{ where } a_i \in k^\times, s_1 < s_2 < \ldots < s_m,$$

$$b = b_1 t_1 + \ldots + b_n t_n, \text{ where } b_j \in k^\times, t_1 < t_2 < \ldots < t_n.$$

Hence $ab = a_1 b_1 s_1 t_1 + \ldots$, where the dots represent terms $> s_1 t_1$, and this shows that $ab \neq 0$. Thus kG is a commutative integral domain and so it can be embedded in a field. ∎

For non-abelian groups it is still not known whether kG can have zero-divisors when G is torsion-free, though this has been established in many special cases. When G is totally ordered, kG is an integral domain, as the proof of Th. 4.1 shows, but in that case kG can actually be embedded in a field. This will be proved in Th. 4.5, but some preparation is necessary.

An ordered set is said to be *well-ordered* if every non-empty subset has a least element. This is the familiar definition for totally ordered sets; if a partially ordered set is well-ordered, it must be totally ordered, as we see by applying the definition to pairs of elements. For this reason the definition in the general case has to be modified; it will be convenient to have several equivalent forms of the definition. By an *antichain* we understand a set of pairwise incomparable elements.

LEMMA 2.4.2. *Let S be a partially ordered set. Then the following three conditions are equivalent:*
  (a) *every infinite sequence contains an infinite ascending sequence – given (a_i) in S, there exists a sequence (n') of integers such that $m' < n' \Rightarrow a_{m'} \leqslant a_{n'}$;*
  (b) *every non-ascending sequence (a_i), $a_i \nleqslant a_j$ for $i < j$, is finite;*
  (c) *every strictly descending sequence $a_1 > a_2 > \ldots$ is finite and every antichain is finite.*

Proof. (a) $\Rightarrow$ (b) is clear, and likewise (b) $\Rightarrow$ (c), since both strictly descending chains and antichains are non-ascending. It remains to show that (c) $\Rightarrow$ (a). Assume that (c) holds and let (a_i) be an infinite sequence in S. By (c) this sequence contains minimal elements and these elements form an antichain and so are finite in number; hence we can choose one of them, say b_1 such that

$$b_1 \leq a_n \text{ for infinitely many } n. \tag{2}$$

Omitting all terms a_n not satisfying (2), we obtain an infinite sequence (b_j) say, such that $b_1 \leq b_n$ for all n. Repeating the argument, we obtain by induction on n an infinite ascending sub-sequence of (a_i), so (a) is satisfied. ∎

A partially ordered set satisfying the conditions (a)–(c) of this lemma is said to be *partly well-ordered* (PWO); clearly for totally ordered sets this agrees with the definition of a well-ordered set.

It is clear from the definition that any subset of a PWO set is again PWO, likewise for the image of a PWO set under an order-preserving map. Further, the union of two PWO sets is PWO, and for any two PWO sets S, T their Cartesian product $S \times T$, ordered by the rule

$$(s, t) \leq (s', t') \text{ if and only if } s \leq s' \text{ and } t \leq t',$$

is again PWO. For any infinite sequence in the product contains an infinite sub-sequence in which the first components are in ascending order, and this contains an infinite sub-sequence in which the second components are in ascending order, so that (a) is satisfied. We shall also need conditions for a monoid with a PWO generating set to be PWO. By a *divisibility ordering* on a monoid M we understand a partial ordering '$\leq$' on M such that

(O.1) $s \leq s', t \leq t' \Rightarrow st \leq s't'$ *for all* $s, t, s', t' \in M$,

(O.2) $1 \leq s$ *for all* $s \in M$.

When (O.1–2) hold, then M is *conical*, i.e. $st = 1$ implies $s = t = 1$. For if $st = 1$, then $1 \leq t$, hence $s \leq st = 1$, so $s = 1$, and $t = st = 1$. We shall show now that a monoid with a divisibility ordering is PWO, provided it has a PWO generating set.

LEMMA 2.4.3. *Let M be a monoid with a divisibility ordering. If M is generated by a partly well-ordered set X, then M itself is partly well-ordered.*

Proof. Let X^* be the free monoid on X, with the partial ordering induced by that of X; this means that $x_1 \ldots x_m \leq y_1 \ldots y_n$ holds precisely when integers $1', 2', \ldots, m'$ exist such that $1 \leq 1' < 2' < \ldots < m' \leq n$ and $x_i \leq y_{i'}$. The natural homomorphism from X^* to M is order-preserving, so it will be enough to show that X^* is PWO. If this were not so, we would have a non-ascending sequence (a_n) in X^*, thus

$$a_i \nleq a_j \text{ for } i < j.$$

Choose a non-ascending sequence (a_n) with a_1 of minimal length. Among all non-ascending sequences beginning with a_1 choose one with a_2 of minimal length, and so on: $a_1, a_2, \ldots$. For each i, either $a_i \in X$ or

$$a_i = x_i b_i, \quad x_i \in X, \quad b_i \in X^* \backslash \{1\}. \tag{3}$$

There can only be finitely many a_i in X, because this is PWO, so the a_i in (3) still form an infinite non-ascending sequence. We claim that the set $B = \{b_i\}$ of b_i occurring in (3) is PWO. Otherwise we could find an infinite non-ascending sub-sequence (u_i), where $u_i = b_{i'}$. Let i_0 be least among $1', 2', \ldots$; by omitting finitely many bs we may assume that $i_0 = 1'$. Now consider the sequence

$$a_1, a_2, \ldots, a_{1'-1}, b_{1'}, b_{2'}, \ldots. \tag{4}$$

If $a_i \leqslant b_{j'}$ for some $i < 1'$ and some j, then $a_i < a_{j'}$, which contradicts the fact that (a_i) is non-ascending. This and the fact that $(b_{i'})$ is non-ascending shows that (4) is non-ascending. But this contradicts the choice of $a_{1'}$; so $B = \{b_i\}$ is PWO. Now it follows from (3) that (a_i) as product of two PWO sets is PWO. This shows that X^* and with it M is PWO, as claimed. ∎

Let M be a monoid and k a commutative field, as before. We can form the function space k^M consisting of all functions from M to k. For any $a = (a_s) \in k^M$ we define its *support* as

$$\mathcal{D}(a) = \{s \in M \,|\, a_s \neq 0\}.$$

The functions of finite support form a subspace $k(M)$ of the k-space k^M; on $k(M)$ we can define a multiplication by the rule

$$\text{If } a = (a_s), b = (b_t), \text{ then } ab = c, \text{ where } c = (c_u), c_u = \sum_{st=u} a_s b_t. \tag{5}$$

Since a, b have finite support, the sum on the right of (5) is finite, and it is easily verified that $k(M)$ is isomorphic to the monoid ring kM, and so may be identified with it.

We can think of the elements of k^M as formal series $\sum s a_s$, but there is usually no multiplication, because the number of solutions (s, t) of $st = u$ for a given u may be infinite. Suppose now that M has a divisibility ordering. Then we can consider the subset $k((M))$ of k^M consisting of all functions with partly well-ordered support. It turns out that this is in fact a ring:

THEOREM 2.4.4. *Let M be a monoid with a divisibility ordering, k a commutative field and $k((M))$ the set of all series with partly well-ordered support. Then $k((M))$ is a k-algebra with the monoid algebra kM as subalgebra. The units in this algebra are the series whose support includes 1.*

Proof. Let $a = \sum sa_s$, $b = \sum tb_t$ be elements of $k((M))$. By definition their supports $\mathcal{D}(a)$, $\mathcal{D}(b)$ are PWO; their sum $a + b$ has as support a subset of $\mathcal{D}(a) \cup \mathcal{D}(b)$ and so it is again PWO. Taking next the product ab, consider for a given $u \in M$, the set of all pairs $(s, t) \in \mathcal{D}(a) \times \mathcal{D}(b)$ such that

$$st = u. \tag{6}$$

If this set is infinite, then since $\mathcal{D}(a)$ is PWO, there is an infinite sub-sequence (s_i, t_i) such that $s_1 < s_2 < \ldots$; hence $t_1 > t_2 > \ldots$, and so $\mathcal{D}(b)$ contains an infinite non-ascending sub-sequence, which is a contradiction. Thus (6) has only finitely many solutions and so the product ab is defined; moreover $\mathcal{D}(ab)$ as image of $\mathcal{D}(a) \times \mathcal{D}(b)$ under the map $(a, b) \mapsto ab$, is PWO, therefore ab lies in $k((M))$. The associative and distributive laws are easily checked, hence $k((M))$ is a k-algebra, whose subalgebra consisting of all elements of finite support is the monoid algebra.

If $a = \sum sa_s$ has an inverse, then for some $s \in \mathcal{D}(a)$ there exists t such that $st = 1$, and so $s = 1$. Conversely, let a be a series whose support contains 1; on dividing by its coefficient, we can take a in the form $a = 1 - b$, where $b = \sum sb_s$ $(s > 1)$. We claim that $1 + b + b^2 + \ldots$ lies in $k((M))$; once that is established, it is easily verified that $\sum b^n$ is the inverse of a. Now the monoid generated by $\mathcal{D}(b)$ with the divisibility ordering is PWO by Lemma 4.3, hence $\bigcup \mathcal{D}(b^n)$ is PWO and no element of M can belong to infinitely many of the $\mathcal{D}(b^n)$, because the solutions of $s_1 \ldots s_n = u$, for a fixed n, form an anti-chain in a PWO set. Thus $\sum b^n$ is well-defined with PWO support, and it is the inverse of a. ∎

Let G be an *ordered group*, i.e. a group with a total ordering satisfying (O.1). With the help of Th. 4.4 we can show that the group algebra kG can be embedded in a field.

THEOREM 2.4.5. *Let G be an ordered group and k a commutative field. Then the set $k((G))$ of series over k with well-ordered support in G is a field.*

We remark that the proof goes through even if the field k is skew.

Proof. The subset M of G defined by

$$M = \{u \in G | u \geqslant 1\}$$

is a monoid with a divisibility ordering, almost by definition; hence $R = k((M))$ is a k-algebra by Th. 4.4. It is an integral domain, for we clearly have $1 \neq 0$ and if for $f \in R^\times$ we define its order $o(f)$ as the least element in the support of f, then $o(fg) = o(f)o(g)$ and it follows that the product of non-zero elements is non-zero. We claim that M is a (left and right) Ore set in R. Let $u \in M$, $f \in R$; every element p in $\mathscr{D}(f)$ satisfies $p \geqslant 1$, hence $u^{-1}pu \geqslant u^{-1}u = 1$, so $f_1 = u^{-1}fu \in R$ and

$$uf_1 = fu.$$

This shows M to be a right Ore set; by symmetry it is also left Ore. If L is the localization of R at M, then it is clear that $L \subseteq k((M))$, and here we have equality, for if $f \in K((G))$, $f \neq 0$, say $o(f) = v$, then $v^{-1}f \in R$ and $o(v^{-1}f) = 1$, hence $v^{-1}f$ is a unit in R and so $f = v \cdot v^{-1}f$ is a unit in L. This shows that L is a field and coincides with $k((G))$. ∎

The construction of the field of power series in this theorem is called the *Malcev–Neumann construction*. It applies in particular to free groups, since the latter can be ordered. To verify this fact, let F be the free group on a set X and define the *lower central series* of F recursively as

$$\gamma_1(F) = F, \quad \gamma_{r+1}(F) = (F, \gamma_r(F)),$$

where for subgroups G, H, (G, H) is the subgroup generated by all commutators $(g, h) = g^{-1}h^{-1}gh$, for $g \in G$, $h \in H$. It can be shown that $\gamma_r(F)/\gamma_{r+1}(F)$ is abelian torsion-free, so we can totally order each γ_r/γ_{r+1}. Moreover, $\bigcap \gamma_r = 1$, so for any $a \neq 1$ there is a unique r such that $a \in \gamma_r(F)\backslash\gamma_{r+1}(F)$ (see M. Hall [59], p. 166f. and A.3, 4.6). Now write $a > 1$ if the residue class $a\gamma_{r+1}(F)$ is > 1, and $a < 1$ otherwise.

Another more direct way of ordering F (due to G. M. Bergman [90]) is to take any set X' in bijection with X and define a map from F to the power series ring $\mathbf{R}\langle\langle X'\rangle\rangle$ by the rule:

$$x \mapsto 1 - x', x^{-1} \mapsto \sum_0^\infty (x')^n, \text{ if } x \text{ corresponds to } x'.$$

This provides an embedding; now $\mathbf{R}\langle\langle X'\rangle\rangle$ can be totally ordered by taking any ordering on X', extending it to the lexicographic ordering on the free monoid on X' and ordering $\mathbf{R}\langle\langle X'\rangle\rangle$ by the sign of its lowest term.

Since the free group F on any set X can be ordered, we see by Th. 4.5 that the power series ring $k((F))$ is a field. It contains the group ring kF and we thus obtain

COROLLARY 2.4.6. *The group algebra of any free group can be embedded in a field.* ∎

This provides another embedding of the free algebra $k\langle X\rangle$, since the latter is a subalgebra of kF, where F is the free group on X.

Exercises

1. Let G be any group and K any field. Show that the group ring KG has UGN.

2. If G is a torsion-free abelian group and K a skew field, show that the group ring KG is embeddable in a field, without using Th. 4.5. (Hint. Write G as union of free abelian groups and use Ex. 1.3.2)

3. Prove Th. 4.4 for the monoid ring KM over a skew field K.

4. Let G be an ordered group, acting on a field K by automorphisms, $a \mapsto a^g$ ($g \in G$). Show that the formal series in G over K with skew multiplication: $ag = ga^g$ again form a field.

5. (R. Moufang [37]) Let G be the free metabelian group on a, b and write $u = (a, b) = a^{-1}b^{-1}ab$. Verify that every element of the group algebra kG can be written as a finite sum $\sum a^r b^s u^\varphi \lambda_{rs}(\varphi)$, where r, s are integers, $\lambda_{rs}(\varphi) \in k$ and φ is an element of the free abelian group on α, β, with the commutation rules $u^\varphi a = au^{\varphi\alpha}$, $u^\varphi b = bu^{\varphi\beta}$. Verify that this expression is unique, if each φ is a polynomial in α, β. Show further that the formal series $\sum a^r(b^s f_{rs})$, where the f_{rs} are rational functions in the $u^{\alpha^i \beta^j}$ form a field containing kG.

6. Show that in Ex. 5 the subalgebra generated by a, b is free and hence deduce an embedding of the free algebra of rank 2 in a field. Verify that the least field so obtained is not isomorphic to the field of fractions constructed in Cor. 4.6.

7. Let K be a field and G an ordered group, and let A be the skew group ring of G over K with basis u_α ($\alpha \in G$) and multiplication $u_\alpha u_\beta = u_{\alpha\beta} m_{\alpha,\beta}$, where the $m_{\alpha,\beta} \in K^\times$ satisfy the factor set conditions. Verify that the formal series in G over K with well-ordered support and with the above multiplication rule again form a field, which contains A as a subring.

8. Let G be an ordered group and kG the group algebra over a field k. Show that every unit in kG is *trivial*, i.e. of the form αu, $\alpha \in k^\times$, $u \in G$.

9. Show that a field K can be embedded in a field L such that all automorphisms of K are induced by inner automorphisms of L. (For a generalization see 5.5.)

10°. Find conditions for the Laurent polynomial ring $R[t, t^{-1}]$ over a semifir R to be a semifir. Find extensions to the skew case. Is the resulting ring a fir when R is a fir? (Hint. See the proof of the inertia lemma 6.2.1.)

2.5 Iterated skew polynomial rings

We have seen that a polynomial ring over a field is a principal ideal domain, and it is clear that this condition is necessary, i.e. if a polynomial ring is principal, the coefficient ring must be a field. This is true even for skew polynomial rings relative to an automorphism (see Th. 5.1 below), but for a general endomorphism it need not hold. The precise conditions were determined by Jategaonkar [69]:

THEOREM 2.5.1. *Let A be a ring with an endomorphism α and put $R = A[t; \alpha]$. Then R is a principal right ideal domain if and only if A is a principal right ideal domain and α maps $A^\times$ into $U(A)$, the group of units of A. In particular, if α is an automorphism, then R is right principal if and only if A is a field.*

Proof. If R is right principal, then so is A because it is a retract of R (obtained by putting $t = 0$). Further, for any $a \in A^\times$ we have $aR + tR = cR$, where c is the highest common left factor of a and t. It follows that c has degree 0, as factor of a, so $t = cf$, where f has degree 1, say $f = td + e$. Now $t = ctd + ce = tc^\alpha d + ce$, which shows that $ce = 0$, $c^\alpha d = 1$, so c^α is a unit (A is an integral domain, so every element with a right inverse is a unit). Now we have $au + tv = c$ and putting $t = 0$, we see that a is associated to c, hence a^α is associated to c^α, a unit, so a^α is a unit, as claimed.

Conversely, if the given conditions hold, R is clearly an integral domain. Let $\mathfrak{a}$ be a right ideal in R; when $\mathfrak{a} = 0$, there is nothing to prove; otherwise let n be the least degree of polynomials occurring in $\mathfrak{a}$. The leading coefficients of polynomials of degree n in $\mathfrak{a}$ form with 0 a right ideal in A, which is principal, generated by a say. If $f = t^n a + \ldots \in \mathfrak{a}$, then $a^\alpha \in U(A)$ and hence $ft = t^{n+1} a^\alpha + \ldots$ has a unit as highest coefficient. It follows that $\mathfrak{a}$ contains a monic polynomial of degree $n + 1$ and so also of all higher degrees. Now it is clear that $\mathfrak{a} = fR$, hence R is a principal right ideal domain. The last sentence follows because for an automorphism α, the condition only holds when $A^\times = U(A)$, i.e. A is a field. ∎

This result shows that under favourable circumstances one may iterate the polynomial ring construction and still get a PRID, and it suggests the following definition.

By a *J-skew polynomial ring* we understand a skew polynomial ring $A[t; \alpha]$ such that α is injective and satisfies *Jategaonkar's condition*:

(J.0) $A^\alpha \subseteq U(A) \cup \{0\}$.

For example, this condition holds whenever A is a field; what is of interest is that there are other cases. It is easily verified that a J-skew polynomial ring over A is an integral domain if A is. We shall be interested in direct limits of iterated J-skew polynomial rings, which may be defined as follows. Let τ be an ordinal number. A ring R is called a *J-ring of type* τ, if R has a chain of subrings R_λ ($\lambda < \tau$), such that

(J.1) $R_0 = U(R) \cup \{0\}$ *(hence R_0 is a field)*,

(J.2) $R_{\lambda+1}$ *is a J-skew polynomial ring over R_λ, for all $\lambda < \tau$*,

(J.3) $R_\lambda = \bigcup_{\mu < \lambda} R_\mu$ *for any limit ordinal $\lambda \leqslant \tau$*,

(J.4) $R_\tau = R$.

Explicitly we have $R_{\lambda+1} = R_\lambda[t_\lambda; \alpha_\lambda]$ and it follows from the definition that each element c of R can be uniquely written as

$$c = \sum t_{\lambda_1} \ldots t_{\lambda_r} c_{\lambda_1 \ldots \lambda_r} \quad (c_{\lambda_1 \ldots \lambda_r} \in R_0, \lambda_1 \geqslant \ldots \geqslant \lambda_r). \tag{1}$$

It is easily verified that $U(R_\lambda) = U(R_0)$ for all λ, hence we have

COROLLARY 2.5.2. *Any J-ring (of any type τ) is a principal right ideal domain.* ∎

It turns out that J-rings can be characterized as integral domains with a Euclidean algorithm (usually transfinite) and unique remainder (see Lenstra [74]; FR, 8.8). We shall not carry out this verification but we shall show how to construct J-rings of prescribed type. Skew polynomial rings over fields are just the J-rings of type 1; J-rings of type 2 can be obtained by an *ad hoc* construction (Cohn [67], see also Ex. 3), but beyond this the general case is no harder than the finite case. Moreover, one cannot use induction directly, since the coefficient ring depends essentially on the order type. Jategaonkar uses an ingenious argument involving ordinals; below is a direct proof based on the Malcev–Neumann construction (Th. 4.4).

We observe that to achieve the form (1) we need a commutation rule of the form

$$t_\mu t_\lambda = t_\lambda u_{\mu\lambda} \quad (\mu < \lambda),$$

where $u_{\mu\lambda}$ has to be a unit in R_0. More generally, this must be true for products of ts, which we may as well take in normal form, as in (1). Thus we shall need a formula of the form

$$t_\mu t_{\lambda_1} \ldots t_{\lambda_r} = t_{\lambda_1} \ldots t_{\lambda_r} u_{\mu\lambda_1\ldots\lambda_r} \quad (\lambda_1 \geq \ldots \geq \lambda_r, \lambda_1 > \mu, r \geq 1). \quad (2)$$

It turns out that this is enough to give the required construction. Thus let $T = \{t_\lambda\}$ $(\lambda < \tau)$ be a family of indeterminates, denote by F_T the free group on T and put $E = k((F_T))$. Let K be the subfield of E generated by the elements

$$u_{\mu\lambda_1\ldots\lambda_r} = (t_{\lambda_1} \ldots t_{\lambda_r})^{-1} t_\mu t_{\lambda_1} \ldots t_{\lambda_r} \quad (\lambda_1 \geq \ldots \geq \lambda_r, \lambda_1 > \mu, r \geq 1), \quad (3)$$

as suggested by (2); then no element of K other than those of k can centralize any t_λ:

LEMMA 2.5.3. *The centralizer of any t_ν in the field K just constructed is k.*

Proof. Let G be the subgroup of $F = F_T$ generated by the right-hand sides of (3). Each generator has odd length, so that we can speak of a middle factor. When we form a group element of G, the middle factor of any generator cannot be affected by cancellation, hence any element of G begins with a letter t_λ^{-1} and ends with a letter t_μ even after cancellation; it follows that $t_\nu^n \notin G$ for all $n \neq 0$. Now any $a \in K$ can be written as a series of the form $a = \sum u a_u$, where u runs over G, and conjugation by t_ν maps K into itself:

$$t_\nu^{-1} u_{\mu\lambda_1\ldots\lambda_r} t_\nu = u_{\lambda,\nu}^{-1} \ldots u_{\lambda_i\nu}^{-1} u_{\mu\lambda_1\ldots\lambda_{i-1}\nu} u_{\lambda_i\nu} \ldots u_{\lambda,\nu},$$

where $\lambda_{i-1} \geq \nu > \lambda_i$. Now t_ν commutes only with t_ν^n in F, so conjugation by t_ν fixes 1 and moves all other elements of G in infinite orbits. Each of these orbits is generated from a single element of G by conjugation by a positive power of t_ν. Hence $t_\nu^{-1} a t_\nu = a$ can hold only if $\mathfrak{D}(a) = \{1\}$, thus $a = a_1 \in k$, as claimed. ∎

We have seen that conjugation by t_ν induces an endomorphism of G which we shall denote by α_ν. For any ordering of F or G conjugation is order-preserving, therefore α_ν extends to an endomorphism of K, again denoted by α_ν. Thus for any $a \in K$ we have

$$a t_\nu = t_\nu a^{\alpha_\nu}. \quad (4)$$

Let R be the subring of $E = k((F))$ generated by K and all t_ν $(\nu < \tau)$. By (4) each element of R can be written in the form of a finite sum

$$\sum t_{\lambda_1} \ldots t_{\lambda_r} a_{\lambda_1\ldots\lambda_r}, \quad \text{where } a_{\lambda_1\ldots\lambda_r} \in K. \quad (5)$$

If the λ_i are not already in descending order, then for some $i = 1, 2, \ldots,$

$r - 1$ we have $\lambda_i < \lambda_{i+1} \geqslant \ldots \geqslant \lambda_r$. Now we can pull t_{λ_i} through to the right, using (2). By repeating this process, if necessary, we ensure that $\lambda_1 \geqslant \ldots \geqslant \lambda_r$ in each term of (5). We claim that with this proviso, the expression (5) is unique. To establish this claim, suppose that we have a relation

$$\sum t_{\lambda_1} \ldots t_{\lambda_r} a_{\lambda_1 \ldots \lambda_r} = 0, \text{ where } a_{\lambda_1 \ldots \lambda_r} \in K, \lambda_1 \geqslant \ldots \geqslant \lambda_r. \tag{6}$$

If the highest subscript occurring in (6) is λ, then we can write (6) as $\sum t_\lambda^i c_i = 0$, where each c_i is a polynomial in the t_μ $(\mu < \lambda)$. Conjugating by t_λ we obtain coefficients $c_i^{\alpha_\lambda}$ which lie in K; thus t_λ satisfies an equation over K. We write down the minimal equation:

$$t_\lambda^n + t_\lambda^{n-1} b_1 + \ldots + b_n = 0, \text{ where } b_i \in K. \tag{7}$$

If we conjugate by t_λ we obtain another monic equation of degree n for t_λ and by the uniqueness of the minimal equation it must coincide with (7). Thus $b_i^{\alpha_\lambda} = b_i$ for $i = 1, \ldots, n$, and by the lemma $b_i \in k$, so t_λ is algebraic over k. But this is clearly false, and this contradiction shows that all the coefficients in (6) must vanish. Hence (5) is unique; this means that by adjoining the ts one by one to K we obtain a J-skew polynomial ring at each stage, hence R itself is a J-ring of type τ and we have proved

THEOREM 2.5.4. *For any commutative field k and any ordinal τ there exists a J-ring of type τ which is a k-algebra.* ∎

J-rings of type at least 2 have various unusual properties. In the first place there are elements with arbitrarily long factorizations, as the equation

$$t_2 = t_1^n t_2 u_{12}^{-n} \quad n = 1, 2, \ldots$$

shows: from t_2 we can split off arbitrarily many factors on the left, though not on the right. In fact a J-ring of type τ is right Noetherian, but it has descending chains of length τ, such as $\{t_\lambda R\}$; we note that $\bigcap t_\lambda^n R = t_{\lambda+1} R$. The ideal structure of R is further illuminated by

LEMMA 2.5.5. *In the J-ring of type τ constructed in Th. 5.4, $t_\lambda R$ is a two-sided ideal for any $\lambda < \tau$, and for any $f \in R^\times$, $fR \supseteq t_\lambda R$ for some $\lambda > \mu$ for all t_μ occurring in f.*

Proof. That $t_\lambda R$ is a two-sided ideal follows because

$$t_\mu t_\lambda = \begin{cases} t_\lambda u_{\mu\lambda} & \text{if } \mu < \lambda, \\ t_\lambda^2 & \text{if } \mu = \lambda, \\ t_\lambda t_\mu u_{\lambda\mu}^{-1} t_\lambda & \text{if } \mu > \lambda. \end{cases}$$

Further, if $\lambda > \mu$ for all t_μ occurring in f, we have $ft_\lambda = t_\lambda u$ for a unit u, hence $t_\lambda = ft_\lambda u^{-1}$ and so $t_\lambda R \subseteq fR$. ∎

From this lemma it follows in particular that when τ is a limit ordinal, then in a J-ring of type τ any non-zero right ideal contains an ideal of the form $t_\lambda R$. It follows that R cannot be right primitive, for if $\mathfrak{a}$ is any non-zero right ideal, then $R/\mathfrak{a}$ is not faithful. However, R is left primitive; to show this we need to exhibit a faithful simple left R-module, or equivalently, a maximal left ideal containing no non-zero ideal. From the normal form (5) it is clear that the elements t_λ are left linearly independent over R, hence the same is true of the elements $1 + t_\lambda$. Hence the left ideal $\mathfrak{a} = \sum R(1 + t_\lambda)$ is proper and so is contained in a maximal left ideal $\mathfrak{a}_0$. If $\mathfrak{a}_0$ contained a non-zero ideal, it would contain some t_λ and so also $1 = (1 + t_\lambda) - t_\lambda$, which is absurd. So $\mathfrak{a}_0$ is the required maximal left ideal, and this shows R to be left but not right primitive. Whether such rings exist was a question first raised by Jacobson in 1956 (and first answered by Bergman [64]).

Another question of Jacobson asks whether in a right Noetherian ring the powers of the Jacobson radical always meet in 0. To answer it we take a J-ring of type τ and localize at the set of all polynomials with non-zero constant term. The result is a ring P whose Jacobson radical is $\mathfrak{J} = t_1 P$. If we define the transfinite powers of $\mathfrak{J}$ by the rule $\mathfrak{J}^{\lambda+1} = \mathfrak{J}^\lambda \mathfrak{J}$, $\mathfrak{J}^\lambda = \bigcap_{\mu < \lambda} \mathfrak{J}^\mu$ at a limit ordinal λ, then $\mathfrak{J}^\lambda \supseteq t_\lambda P$ and it follows that $\mathfrak{J}^\lambda \neq 0$ for all $\lambda < \tau$.

Exercises

1. Let R be a J-ring of infinite order type τ. Show that the complement of $t_1 R$ is a right Ore set Σ. Further, show that the localization of R at Σ, $R_\Sigma = L$, is a local ring in which $\{t_\lambda L\}$ forms a well-ordered descending chain of right ideals.

2. By applying the factorial duality to Ex. 1, show that L/Lt_τ is a cyclic left L-module which is Artinian but not Noetherian, with a well-ordered ascending chain of submodules of order type τ.

3. Let k be a commutative field with an endomorphism α and containing an element t such that t is transcendental over k^α. Form the function field $k(y)$ and let K be the subring of all rational functions of the form f/g, where g is not divisible by y. Extend α to K by the rule $y^\alpha = t$ and form the skew power series ring $R = K[\![x; \alpha]\!]$. Show that R is a J-ring of type 2.

4. (H. H. Brungs) Show that a ring in which the set of all principal right ideals is (descending) well-ordered by inclusion is a principal right ideal ring in which every left regular element is right invariant.

5. Show that the ring constructed at the end of this section satisfies the conditions of Ex. 4.

2.6 Fields of fractions for a class of filtered rings

Our aim in this section is to prove that a filtered ring has a field of fractions whenever the associated graded ring is an Ore domain. We begin by briefly recalling the definitions of filtered and graded rings.

By a *filtered ring* we understand a ring R with a series of submodules indexed by $\mathbf{Z}$,

$$\cdots \supseteq R_{-1} \supseteq R_0 \supseteq R_1 \supseteq \cdots, \tag{1}$$

such that

(F.1) $R_i R_j \subseteq R_{i+j}$,
(F.2) $\bigcap R_n = 0$,
(F.3) $\bigcup R_n = R$.

If in (1) $R_0 = R$ and $R_{-1}, R_{-2}, \ldots$ are absent, we speak of a *positive* or *descending filtration*. Similarly, when $R_1 = R_2 = \ldots = 0$, we have a *negative* or *ascending filtration* (in this case it is often convenient to change the sign of the suffix).

On a filtered ring we can define a $\mathbf{Z}$-valued function by putting

$$v(x) = \sup \{ n \mid x \in R_n \}. \tag{2}$$

It is easily checked that v satisfies

(V.1) $v(x) \in \mathbf{Z}$ if $x \neq 0$, $v(0) = \infty$,
(V.2) $v(x - y) \geq \min \{ v(x), v(y) \}$,
(V.3) $v(xy) \geq v(x) + v(y)$.

Such a function is called a *subvaluation*, *pseudo-valuation* or also *filtration* on R. If equality holds in (V.3), v is called a *valuation*, more familiar in field theory (see Ch. 9). For any subvaluation v we have by (V.2), $v(-y) \geq v(y)$, hence $v(-y) = v(y)$. Further, if $v(x) > v(y)$, then $v(x - y) \geq v(y)$. If this inequality were strict, we would have $v(y) = v(x - (x - y)) \geq \min \{ v(x), v(x - y) \}$, which is a contradiction. Hence we have (as for valuations)

(V.4) *If* $v(x) \neq v(y)$, *then* $v(x - y) = \min \{ v(x), v(y) \}$.

Suppose now that conversely, R is a ring with a $\mathbf{Z}$-valued subvaluation; then R may be filtered by the submodules

$$R_n = \{ x \in R \mid v(x) \geq n \},$$

and it is easily checked that (F.1–3) hold.

A ring R is said to be *graded* if it is a direct sum of submodules indexed by $\mathbf{Z}$: $R = \sum A_n$ such that $A_i A_j \subseteq A_{i+j}$. Any element of A_i is said to be *homogeneous* of *degree i*. Such a ring can always be filtered by setting $R_n = \sum_n^\infty A_i$. But more significantly, we can with every filtered ring R associate a graded ring $G(R)$, which may be thought of as the ring of 'leading terms'. Its additive group is the direct sum $\sum(R_n/R_{n+1})$, with multiplication defined as follows: Given $\alpha \in R_i/R_{i+1}$, $\beta \in R_j/R_{j+1}$, take representatives $a \in R_i$ for α and $b \in R_j$ for β and put

$$\alpha\beta \equiv ab \pmod{R_{i+j+1}}.$$

The product lies in R_{i+j}/R_{i+j+1} and depends only on α, β, not on a, b, as can be verified without difficulty. We have a natural mapping from R to $G(R)$ which assigns to $a \in R$ its 'leading term' $\bar{a}$ defined by the rule: if $a = 0$, then $\bar{a} = 0$; if $a \neq 0$ and $v(a) = n$, then $\bar{a} \equiv a \pmod{R_{n+1}}$. We have $\overline{ab} = \bar{a}\bar{b}$, but this is *not* generally a homomorphism, because if $\bar{a} = \bar{b}$ but $a \neq b$, then $\overline{a - b} \neq \bar{a} - \bar{b}$.

We note the following criterion for the associated filtration v to be a valuation:

PROPOSITION 2.6.1. *Let R be a filtered ring with associated filtration v and graded ring $G(R)$. Then $G(R)$ is an integral domain precisely when v is a valuation; when this is so, R itself is an integral domain.*

Proof. Assume that $G(R)$ is an integral domain and take $a, b \in R^\times$; if $v(a) = r$, $v(b) = s$, we have $v(ab) \geq r + s$ and we must show that equality holds here. We have $\bar{a}, \bar{b} \neq 0$, hence $\overline{ab} = \bar{a}\bar{b} \neq 0$ and so $ab \notin R_{r+s+1}$, but this means that $v(ab) < r + s + 1$ and the desired equality follows. The converse is clear; moreover, when v is a valuation, then for any $a, b \neq 0$, $v(a), v(b) < \infty$, hence $v(ab) < \infty$ and so $ab \neq 0$, which shows R to be an integral domain. ■

As an example of a positively filtered ring consider a ring R with an ideal $\mathfrak{a}$ such that

$$\bigcap \mathfrak{a}^n = 0. \tag{3}$$

We can filter R by the powers of $\mathfrak{a}$, writing $R_0 = R$, $R_n = \mathfrak{a}^n$ $(n \geq 1)$, and so obtain the $\mathfrak{a}$-*adic filtration* of R. If $\mathfrak{a}$ is generated by a single central element t, the condition (3) becomes

$$\bigcap t^n R = 0. \tag{4}$$

The associated filtration is then a valuation provided that t is not nilpotent and R/tR is an integral domain. For this condition is clearly necessary by Prop. 6.1, because $R/tR = G_0(R)$. Conversely, when it holds, take any $a, b \in R^\times$ and let $v(a) = r$, $v(b) = s$. Then $a = t^r u$,

$b = t^s v$, where $u, v \notin tR$, hence $uv \notin tR$ and $ab = t^{r+s}uv$, so $v(ab) = r + s$. We shall refer to v as the *t-adic valuation* on R.

With an additional hypothesis we can show that R has a field of fractions. For any central element c of R we shall write (c) for the ideal generated by c when the ring is clear from the context.

THEOREM 2.6.2. *Let R be a ring and t a regular central element of R satisfying (4) and such that $R/(t)$ is a right Ore domain. Then R can be embedded in a field D and if the t-adic valuation of R is extended to D in the natural way, then $R(R^\times)^{-1}$ is dense in D. Moreover, if S is the ring of valuation integers in D, then $S/(t)$ is isomorphic to the field of fractions of $R/(t)$.*

For the proof we shall need a couple of lemmas. Let us fix $n \geqslant 1$ and write $\bar{x}$ for the image of $x \in R$ in $R/(t^n)$, and for $X \subseteq R$ put $\bar{X} = \{\bar{x} | x \in X\}$. Further, we shall put $U = R\backslash(t)$.

LEMMA 2.6.3. *Given $n \geqslant 1$, with the above notation, $\bar{U}$ is a regular right Ore set in $\bar{R}$.*

Proof. We shall use induction on n; for $n = 1$ the conclusion is true by hypothesis. Let $n > 1$ and take $u \in U$, $x \in R$. If $\bar{x}\bar{u} = 0$, this means that $v(xu) \geqslant n$, but $v(u) = 0$, so $v(x) \geqslant n$ and hence $\bar{x} = 0$. The same argument applies if $\bar{u}\bar{x} = 0$, so $\bar{U}$ is regular.

To check the Ore condition, we have by the induction hypothesis $y \in R$ and $v \in U$ such that

$$uy - xv = z \in (t^{n-1}).$$

If $v(z) \geqslant n$, the result follows, so assume that $v(z) < n$, say $z = z_0 t^{n-1}$, where $z_0 \in U$ and

$$uy - xv - z_0 t^{n-1} = 0. \tag{5}$$

Likewise there exist $y_1 \in R$, $v_1 \in U$ such that $uy_1 - z_0 v_1 \in (t)$, hence $uy_1 t^{n-1} - z_0 v_1 t^{n-1} \in (t^n)$, and by (5),

$$uyv_1 - xvv_1 - z_0 t^{n-1} v_1 = 0,$$

hence

$$u(yv_1 - y_1 t^{n-1}) - xvv_1 \in (t^n),$$

and $vv_1 \in U$. This shows $\bar{U}$ to be a right Ore set. ∎

We shall write $R_n = R/(t^n)$ and denote its localization at $\bar{U}$ by S_n. Then

$\bar{t}S_n$ is a nilpotent ideal, $(\bar{t}S_n)^n = 0$, and writing Δ for the field of fractions of $R/(t)$, we have $S_n/(\bar{t}^n) \cong \Delta$. Moreover, we have

$$\bar{t}^{n-1}S_n \cap R_n = \bar{t}^{n-1}R_n, \qquad (6)$$

because the intersection just represents the elements of R_n with value at least $n-1$. It follows that the natural homomorphism from R_n to $R_n/(\bar{t}^{n-1}) \cong R_{n-1}$ extends to a homomorphism from S_n to $S_n/(\bar{t}^{n-1}) \cong S_{n-1}$. We thus have two inverse systems of rings (R_n) and (S_n) with $R_n \subseteq S_n$, such that the diagram

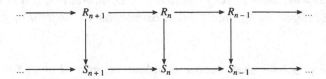

commutes. We put $\tilde{R} = \varprojlim R_n$, $S = \varprojlim S_n$, so that $R \subseteq \tilde{R} \subseteq S$, and write $\varphi_n: S_{n+1} \to S_n$, $\gamma_n: S \to S_n$ for the natural homomorphisms.

LEMMA 2.6.4. *With the above notations S is a local ring and an integral domain, with maximal ideal (t) and residue-class ring $S/(t)$ isomorphic to Δ. Moreover, $tS \cap R = tR$ and $\bigcap t^nS = 0$, and S is complete in the t-adic topology.*

Proof. The homomorphism $\varphi_n: S_{n+1} \to S_n$ maps S_{n+1} to S_n and induces an isomorphism of residue-class rings, $S_{n+1}/(t) \cong S_n/(t) \cong \Delta$, by (6). Hence S contains an ideal tS with residue-class ring $S/tS \cong \Delta$. Any element of $S \backslash tS$ is invertible in S, because its image under the map $\gamma_n: S \to S_n$ is invertible, for all n. Thus S is a local ring with maximal ideal tS and residue-class ring $S/tS \cong \Delta$. We have $\bigcap(t^nS) = 0$, because tS maps to tS_n and $(t\gamma_n)^n = 0$, but $(t\gamma_n)^{n-1} \neq 0$, so t is not nilpotent in S. Since v is a valuation on S, it follows that S is an integral domain.

It remains to show that the t-adic valuation on S reduces to v on R. Given $x \in t^rS \backslash t^{r+1}S$, there exists n_0 such that for all $n \geq n_0$,

$$x\gamma_n \in t^rS_n \backslash t^{r+1}S_n.$$

Since S_n is the localization of R_n at $\bar{U}$ and $x\gamma_n \in R_n$, we see that $x\gamma_n \in t^rR_n \backslash t^{r+1}R_n$ for all $n \geq n_0$, therefore $x \in t^rR \backslash t^{r+1}R$, i.e. $v(x) = r$. ∎

To complete the proof of Th. 6.2, let D be the localization of S at the multiplicative set generated by t. Every element of $D^\times$ has the form $x = t^ru$, where $u \in S \backslash tS$, hence u is a unit and $x^{-1} = t^{-r}u^{-1}$. This shows D to be a field. Finally, choose a set X of representatives for $\Delta^\times$ in S;

then given $x \in D^\times$, say $v(x) = r$, we have $xt^{-r} \in S\backslash tS$, so there exists $a_r \in X$ such that $v(xt^{-r} - a_r) > 0$, or equivalently, $v(x - t^r a_r) > r$. An induction shows that

$$x = \sum_{r}^{\infty} t^i a_i, \quad \text{where } v(x) = r, \, a_i \in X \cup \{0\}. \tag{7}$$

Clearly each a_i is uniquely determined; thus each non-zero element of D has the unique form (7); this also shows $R(R^\times)^{-1}$ to be dense in D. ∎

This proof is due to A. I. Lichtman [a], who has also used the following trick to extend the result.

Let A be any filtered ring whose associated graded ring is an integral domain, and denote the corresponding valuation by v. Let $B = A[t, t^{-1}]$ be the ring of Laurent polynomials in a central indeterminate t over A and extend v to B by the formula

$$v\left(\sum_{k}^{m} t^i a_i\right) = \min_i \{v(a_i) + i\}.$$

It is clear that this defines a filtration on B which extends v. To check that it is in fact a valuation, take $a, b \in B$, and write $a = \sum t^i a_i$, $b = \sum t^i b_i$ where $a_i, b_i \in A$. If r is the least suffix for which $v(a) = v(a_r) + r$ and s is the least suffix for which $v(b) = v(b_s) + s$, and $n = r + s$, then we have

$$v(ab) = \min_k \left\{ v\left(\sum a_i b_{k-i}\right) + k \right\}.$$

For $k = n$ the sum on the right-hand side has a term

$$v(a_r b_s) + n = v(a_r) + r + v(b_s) + s.$$

For any other term in the sum we have

$$v(a_i) + i \geqslant v(a_r) + r, \quad v(b_{n-i}) + n - i \geqslant v(b_s) + s,$$

and at least one of these inequalities is strict, therefore

$$v(a_i b_{n-i}) + n > v(a_r b_s) + n,$$

and it follows that $v(ab) \leqslant v(a_r b_s) + n = v(a) + v(b)$. This proves equality in (V.3), so we have a valuation. Now the ring of valuation integers in B is defined as

$$C = \{x \in B | v(x) \geqslant 0\}.$$

We claim that

$$C/tC \cong G(A). \tag{8}$$

For a proof consider the map $f\colon G(A) \to C/tC$ defined as follows. Given $\alpha \in A_r/A_{r+1}$, take $x \in A_r$ mapping to α and put $\alpha f = xt^{-r}$. If x' is another element mapping to α, then $x - x' \in A_{r+1}$, hence $(x - x')t^{-r} \in tC$, so as map into C/tC, f is well-defined; moreover $xt^{-r} \in tC$ if and only if $x \in A_{r+1}$, so it is injective. Thus we have an injective mapping $A_r/A_{r+1} \to C/tC$, which is easily seen to be additive. Since this holds for all r, we have an additive group homomorphism

$$f\colon G(A) \to C/tC, \tag{9}$$

and it only remains to verify the multiplicative property. But if $v(x) = r$, $v(y) = s$, then $v(xy) = r + s$ and $xt^{-r} \cdot yt^{-s} = xyt^{-r-s}$, while 1 clearly acts as unit-element. The distributive laws are easily checked, so we have an injective ring homomorphism (9). It is an isomorphism since C is spanned by the elements xt^{-r} ($x \in A_r$, $r \in \mathbf{Z}$).

Given $a \in C^\times$, if $v(a) = n$, then $a \notin t^{n+1}C$; thus $\bigcap t^n C = 0$ and of course t lies in the centre of C. Thus all the hypotheses of Th. 6.2 are satisfied and we obtain

THEOREM 2.6.5. *Let R be a filtered ring whose associated graded ring is a right Ore domain. Then R can be embedded in a field D, which is complete in the topology defined by the valuation induced from the filtration on R, and $R(R^\times)^{-1}$ is dense in D.* ∎

As an application we show how to embed the universal associative envelope of any Lie algebra in a field. Let us recall the necessary definitions.

A *Lie algebra L* (over a commutative field k) is a k-space with a bilinear map from $L \times L$ to L, usually called *Lie multiplication* or simply 'multiplication', denoted by $[x, y]$ and satisfying the identities:

$$[x, x] = 0, \tag{10}$$

$$[[x, y], z] + [[y, z], x] + [[z, x], y] = 0 \quad (\textit{Jacobi-identity}). \tag{11}$$

As a consequence of (10), $[x, y] = -[y, x]$, i.e. the multiplication is anticommutative, but it is not generally associative.

An important example of a Lie algebra is derived from an associative algebra. Let A be a ring which is also a k-algebra and on A define a multiplication by the rule

$$[x, y] = xy - yx. \tag{12}$$

The k-space A with the multiplication (12) is easily seen to be a Lie algebra, denoted by A^- and called the Lie algebra *derived from A*. More generally, if V is any subspace of A closed under the multiplication (12), then we can form V^- as a subalgebra of A^-.

If L is any Lie algebra and A an associative algebra, then by a *representation* of L in A one understands a homomorphism of L into A^-. It is a basic result that every Lie algebra has a faithful representation in a suitable associative algebra. This follows from the more precise Birkhoff–Witt theorem (also called Poincaré–Birkhoff–Witt theorem) stated here without proof:

THEOREM 2.A. *For any Lie algebra L there exists an associative algebra $U(L)$ with a representation $L \to U(L)^-$ which is universal for representations of L in associative algebras. This representation is faithful; more precisely, if (u_λ) is a totally ordered basis of L, then a basis of $U(L)$ may be taken in the form of ascending monomials:*

$$u_{\lambda_1} u_{\lambda_2} \ldots u_{\lambda_r}, \quad \lambda_1 \leq \lambda_2 \leq \ldots \leq \lambda_r, \, r = 0, 1, \ldots . \tag{13}$$

The proof of the first sentence follows easily by abstract nonsense. For the proof of the rest (the real content of the theorem) see e.g. Jacobson [62], p. 159 or UA, p. 294.

Let L be a Lie algebra with basis $u_1, u_2, \ldots$ (taken countable for simplicity) and let $U(L)$ be its universal associative envelope. Since $U(L)$ has a basis of elements (13), we can embed L in $U(L)$ by identifying its basis with the elements (13) of degree 1. If the multiplication table for L is given by $[u_i, u_j] = \sum \gamma_{ij}^k u_k$, then the defining relations of $U(L)$ may be written as

$$u_i u_j - u_j u_i = \sum \gamma_{ij}^k u_k. \tag{14}$$

In particular, if $U(L)$ is filtered by the powers of L, we see that L^r/L^{r-1} has a basis consisting of the ascending monomials (13) of degree r; thus $U(L)$ has a negative filtration. Moreover, in the associated graded ring $G(U(L))$ the us commute by (14), hence $G(U(L))$ is a commutative integral domain, in fact it is isomorphic to the polynomial ring $k[u_1, u_2, \ldots]$, as is easily verified. Invoking Th. 6.5, we obtain the following embedding:

THEOREM 2.6.6. *The universal associative envelope of any Lie algebra can be embedded in a skew field.* ∎

This result can be used to provide another way of embedding the free algebra $k\langle X \rangle$ in a field. For let L_0 be the free Lie algebra on X and $U(L_0)$ its universal associative envelope. Then $U(L_0)$ is generated by X

and so is a homomorphic image of $k\langle X \rangle$, while the Lie subalgebra of $k\langle X \rangle^-$ generated by X, L_1 say, is a homomorphic image of L_0. Thus we have the commutative diagram shown, where the map $L_0 \to U(L_0)$ is injective, by Th. 2.A. Hence the homomorphism $L_0 \to L_1$ is injective, and so L_1 is free on X, while $U(L_1) \cong k\langle X \rangle$. Thus the free algebra, as the universal associative envelope of the free Lie algebra, has a field of fractions.

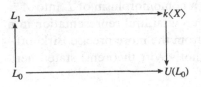

Exercises

1. Let R be a filtered ring and $G(R)$ the associated graded ring. Determine when the 'leading term' map $a \mapsto \bar{a}$ is a homomorphism.

2. Define a filtered ring R to be a BW-algebra (BW = Birkhoff–Witt) if $G(R)$ is a polynomial ring over k in a number of indeterminates. For any Lie algebra L over k with an alternating bilinear form b defined on it, show that the associative algebra $A(L; b)$ generated by L as vector space and 1, with the defining relations

$$xy - yx = [x, y] + b(x, y)1 \quad \text{for all } x, y \in L,$$

is a BW-algebra and that every BW-algebra arises in this way.

3. Give an example of a filtered ring R which is an integral domain such that $G(R)$ is not an integral domain.

4. Define a partial ordering on the set $\mathbf{N}'$ of r-tuples of positive integers by the rule: $(m_1, \dots, m_r) \leqslant (n_1, \dots, n_r)$ if and only if $m_i \leqslant n_i$ for $i = 1, \dots, r$. Show that any infinite subset of $\mathbf{N}'$ contains an infinite ascending sub-sequence. Deduce that the polynomial ring $k[x_1, \dots, x_n]$ is Noetherian (Hilbert basis theorem).

5. Show that a filtered ring R is right Noetherian whenever its associated graded ring $G(R)$ is right Noetherian. By applying Prop. 1.3.6 and Ex. 4, deduce that the universal associative envelope of a finite-dimensional Lie algebra is an Ore domain and hence has a field of fractions.

6. Let $F = k\langle X \rangle$ be the free algebra on X and L the Lie algebra generated by X as subalgebra of F^-. Show that any non-zero element of L is an atom in F. (Hint. If $u \in L$ and $\mathfrak{a}$ is the ideal of F generated by u, verify that $F/\mathfrak{a}$ is the universal associative envelope of $L/(u)$.)

7. Let L be a Lie algebra with an injective endomorphism α. Show that α extends to a unique endomorphism of $U(L)$, and this extends to an endomorphism of its field of fractions, constructed as in the proof of Th. 6.6.

8. Let k be a commutative field of characteristic 0 and on $k\langle X \rangle$, where $X = \{x_{ij} | i, j \in \mathbf{N}\}$, define α, δ by

$$x_{ij}^{\alpha} = x_{i+1\,j}, \quad x_{ij}^{\delta} = x_{i\,j+1}.$$

Show that α extends to an endomorphism of the field of fractions K of $k\langle X \rangle$, formed as in Th. 6.6. Verify that α is not an automorphism and that δ is not inner on im α^n for any n. Deduce that $K[t; \alpha, \delta]$ is a simple principal right ideal domain (see Cozzens and Faith [75], Cohn [77]).

Notes and comments

Skew polynomial rings first arose in the study of linear differential equations. If $f' = \mathrm{d}f/\mathrm{d}x$, then the ring of linear differential operators may be written as $k(x)[D; 1, ']$. In this form it appears in Schlesinger [1897], who proves that it is an integral domain (see also the references in FR, p. 191). The first abstract study was undertaken by Ore [32]; there have been many papers since then and most of 2.1 is folklore (Th. 1.3 is taken from Cohn [61″]). Much of 2.2 follows Lam and Leroy [88], see also Cohn [77]. The observation that power series over $K[t; \alpha, \delta]$ need to be formed in t^{-1} rather than t goes back to Schur [04]. If t is interpreted as differentiation and t^{-1} as integration, it is just an expression of the familiar fact that convergence is improved by integration, but not by differentiation.

As was noted in 2.4, it is still not known whether the group algebra kG is always an integral domain whenever the obvious necessary condition that G be torsion-free is satisfied. Farkas and Snider [76] have shown this to be the case when G is also polycyclic (i.e. soluble with maximum condition on subgroups); since kG is Noetherian in this case it is then embeddable in a field. More generally this has now been established for all torsion-free soluble groups by Kropholler, Linnell and Moody [88]. In another direction, J. Lewin and T. Lewin [78] have shown (using methods of Magnus and some results from Ch. 4 below) that for any torsion-free group given by a presentation with a single defining relation the group algebra can be embedded in a field. Dicks [83] gives another relatively brief proof of this result, based on his theory of HNN-constructions.

The power series ring over an ordered group, $\mathbf{R}((G))$ was introduced (for abelian G) by Hahn [07], who used it to show that every abelian ordered group can be embedded in an ordinal power of $\mathbf{R}$, lexicographically ordered. Th. 4.5 was proved independently by Malcev [48] and Neumann [49]; our proof follows Higman [52] who establishes a version of Lemma 4.3 for general algebras. The fact that the free group can be ordered was proved by Shimbireva [47] and independently by Neumann

[49']. Power series over a free metabelian group were considered by R. Moufang [37] who used them to show that group algebras of free metabelian groups, and hence free algebras, could be embedded in ordered fields (see Ex. 5, 6 of 2.4).

The construction of 2.4 can also be carried out for ordered monoids, and for an ordered cancellation monoid M and an ordered field k, the power series ring $k((M))$ is totally ordered (see 9.6). Dauns [70'] shows that when M is an ordered cancellation monoid not embeddable in a group (see Chehata [53]), then $k((M))$ is an integral domain not embeddable in a field.

The J-rings studied in 2.5 were constructed by Jategaonkar [69], by a method using transfinite induction. The more direct proof given here first appeared in the first edition of FR. Jategaonkar used his construction to give (i) an example of a right but not left primitive ring, (ii) an example of a left Noetherian ring having elements with infinite factorizations, (iii) an example of a left Noetherian ring whose Jacobson radical $\mathfrak{J}$ satisfies $\bigcap \mathfrak{J}^n \neq 0$, (iv) a ring in which the left and right global dimensions differ by a prescribed number (in previous examples this difference had been 1, Kaplansky [58] or 2, Small [66]). This is proved by showing that for a J-skew polynomial ring $R[t; \alpha]$ the left global dimension equals l.gl.dim.$R + 1$ (see Jategaonkar [69] or Rowen [88], 5.1). Points (i)–(iii) are sketched in the text ((ii) was first answered in Cohn [67] by an example using a J-ring of type 2).

Th. 6.5 and its application, Th. 6.6 were proved by Cohn [61'] by constructing the multiplicative group of the field as an inverse limit of monoids and then defining addition with the help of Lemma 1.1.1. The simpler proof given here is due to Lichtman [a]; another proof by valuations is due to Dauns [70]. For a fourth proof, using inverse limits of quotient groups of p-jets, see Wehrfritz [92]. A fifth proof was recently given by A. I. Valitskas [a], using the matrix ideals of Ch. 4. The construction has also been used more recently in micro-localization (see e.g. v.d. Essen [86]).

3

Finite skew field extensions and applications

The beginnings of commutative field theory are to be found in the theory of equations. The analysis of algebraic equations with the help of groups led to Galois theory, but in a modern treatment Galois theory is developed abstractly and equations enter at a relatively late stage. In the non-commutative case it turns out that a Galois theory can be developed which closely parallels the commutative theory, and this is done in 3.3, using the Jacobson–Bourbaki correspondence (3.2) and some basic facts on dimensions in 3.1. By contrast, equations over skew fields are much harder to handle and what little is known is presented in 3.4. In any case, the appropriate tool to use is a matrix; our knowledge of matrix singularities is even more sparse, and an account will have to wait until Ch. 8.

The rest of the chapter deals with various special cases, in which more can be said: quadratic extensions (3.6) and the slightly more general case of extensions generated by a single element with a skew commutation rule, the pseudo-linear extensions (3.5). For outer cyclic Galois extensions 3.7 gives a fairly complete description, due to Amitsur, while the infinite case is briefly dealt with in 3.8.

The last section, 3.9, dealing with the multiplicative structure of fields, forms a separate subject not directly related to the rest. Its location here is determined by the fact that it uses some results from 3.4 but none from later chapters.

3.1 The degree of a field extension

It is a familiar observation that if k is a commutative field and α is algebraic over k, then the field $k(\alpha)$ generated by α over k has finite degree and coincides with the k-algebra generated by α. Moreover, any finite set of algebraic elements over k generates an extension of finite degree. For skew fields there is no corresponding statement; here the extensions of finite degree are much more complicated and there is no simple way of producing them all, as in the commutative case. In fact there could be ambiguity about what is meant by an extension of finite degree.

Let K be a field and E a K-ring. Then E may be regarded as left or right vector space over K and their dimensions provide two numbers (possibly infinite), which will be denoted by

$$[E:K]_L \text{ and } [E:K]_R.$$

When E is itself a field, these numbers are usually called the *left* and *right degree*, and we shall use this terminology generally for K-rings.(*) For many fields these two numbers coincide, but not always, as we shall see in Ch. 5. We shall often call $[E:K]_R$ simply the *degree* of E over K and call the extension E/K *finite* when its degree is finite. As in the commutative case we have the product formula for the degrees (see A.2, pp. 63f.), which follows from the next result:

PROPOSITION 3.1.1. *If $K \subseteq E$ are any fields and V is a right E-module, then the dimensions of V over E and K are related by the formula*

$$[V:K] = [V:E][E:K]_R, \tag{1}$$

whenever either side is finite.

The proof is as in the commutative case, by showing that if $\{u_\lambda\}$ is a right E-basis for V and $\{v_i\}$ a right K-basis for E, then $\{u_\lambda v_i\}$ is a right K-basis for V. ∎

At least one of our difficulties disappears for extensions of finite degree, the difference between zero-divisors and non-units:

PROPOSITION 3.1.2. *Let K be a field and A a K-ring of finite right degree over K. Then every right regular element of A is a unit; hence if A is an integral domain, then it is a field.*

Proof. Let $a \in A$ and suppose that a is right regular. Then the mapping

(*) Care is needed to avoid confusion with the 'degree' of a central simple algebra, which is usually defined as the square root of the dimension. We shall not have occasion to use the term in that sense, but shall avoid confusion by speaking of the 'dimension' in cases of doubt.

$\lambda_a: x \mapsto ax$ is injective, and it is clearly right K-linear on a finite-dimensional K-space, hence it is surjective, and so $ab = 1$ for some $b \in A$. Now b is again right regular: if $bx = 0$, then $x = abx = 0$. Hence there exists $c \in A$ such that $bc = 1$, but now $c = abc = a$, which shows that $ab = ba = 1$, so a is a unit. The rest is clear. ∎

There is one important case where the left and right degrees are the same:

THEOREM 3.1.3. *Let E be a field of finite dimension over its centre. Then for any subfield K of E the left and right degrees of E over K coincide.*

Proof. Let K be a subfield of E and denote the centre of E by C. By hypothesis E is a C-algebra of finite degree, and it is clear that $A = KC = \{\sum x_i y_i | x_i \in K, y_i \in C\}$ is a subalgebra. If we regard A as a K-ring, we can choose a basis of A as left K-space consisting of elements of C; this will also be a right K-basis of A, hence

$$[A:K]_L = [A:K]_R. \tag{2}$$

Now A is a C-algebra of finite degree, and an integral domain, as subalgebra of E, hence A is a field. By (1) we have

$$[E:C] = [E:A]_L[A:C] = [E:A]_R[A:C].$$

Since $[E:C]$ is finite, so is $[A:C]$. If we divide by $[A:C]$ and then multiply by (2), we get, on using (1) again,

$$[E:K]_L = [E:K]_R,$$

as we had to show. ∎

As a consequence the left and right degrees also coincide when the subfield is commutative:

PROPOSITION 3.1.4. *Let E be a field and K a commutative subfield for which $[E:K]_R$ is finite. Then E is of finite degree over its centre and so $[E:K]_L = [E:K]_R$.*

Proof. Denote the centre of E by C and by F the subfield of E generated by C and K. Then F is a commutative subfield containing C and $[E:F]_R$ is again finite. Consider the tensor product $E \otimes_C F$; it is a simple algebra with centre F, by Cor. 7.1.3, p. 260 of A.3, and we have a homomorphism

$$E \otimes_C F \to \mathrm{End}_F(E_F), \tag{3}$$

which maps $\sum u_i \otimes \alpha_i$ $(u_i \in E,\; \alpha_i \in F)$ to the endomorphism $x \mapsto \sum u_i x \alpha_i$. By the simplicity of $E \otimes F$ it is an embedding, and if $[E{:}F]_{\mathrm{R}} = n$, then the right-hand side of (3) is $\mathfrak{M}_n(F)$. Hence

$$[E{:}C] = [E \otimes_C F{:}F] \le [\mathfrak{M}_n(F){:}F] = n^2.$$

Thus E is of finite degree over C and now the conclusion follows by Th. 1.3. ∎

Exercises

1. Let k be a commutative field and $K = k(t)$ the rational function field, with endomorphism $\alpha{:} f(t) \mapsto f(t^2)$. Show that the truncated polynomial ring $K[x; \alpha]/(x^2)$ has left degree 3 and right degree 2 over K. (Hint. Show that $1, x, xt$ is a left K-basis.) Find K-algebras of arbitrary left and right degrees.

2. Let K be a field, E a subfield and F its centralizer in K. Verify that $C = E \cap F$ is the centre of E and give a direct proof that the natural homomorphism of $E \otimes_C F$ into K given by $x \otimes y \mapsto xy$ is injective (this is expressed by saying that E and F are *linearly disjoint* in K over C, see 6.4 below).

3. Give a direct proof that a field cannot be of prime dimension over its centre.

4. (Schofield [85]) Let A, D be k-algebras which are skew fields, A of finite degree over k and D with centre F such that $E \otimes_k F$ is an integral domain for all commutative field extensions E/k (such an extension F/k is called *regular*). Show that $A^\circ \otimes_k D$ is a simple Artinian ring with a unique simple module S of finite dimension over D; verify that A can be embedded in $\mathfrak{M}_n(D)$ if and only if $[S{:}D] \mid n$.

3.2 The Jacobson–Bourbaki correspondence

Let A be any ring and M, N two A-bimodules. We shall write $\mathrm{Hom}(M, N)$, $\mathrm{Hom}_{A-}(M, N)$, $\mathrm{Hom}_{-A}(M, N)$, $\mathrm{Hom}_{A,A}(M, N)$ for the set of all additive, left-A, right-A and A-bimodule homomorphisms from M to N respectively, and we shall use a corresponding notion for $\mathrm{End}(M) = \mathrm{Hom}(M, M)$. Of course $\mathrm{End}(M)$ also has a multiplication which together with the addition gives it a ring structure. In particular, for $M = A$ we have in $\mathrm{End}(A)$ the subring $\rho(A)$ of right multiplications $\rho_a{:}\ x \mapsto xa$; this ring is isomorphic to A, as is well known (see Th. 2.1 below). Similarly the subring $\lambda(A)$ of left multiplications $\lambda_a{:}\ x \mapsto ax$ is anti-isomorphic to A, because (writing maps on the right) we have $\lambda_a \lambda_b = \lambda_{ba}$:

$x \mapsto bax$. These two subrings can be identified with each other's centralizers in $\mathrm{End}\,(A)$; we recall that the set of module endomorphisms of $_AA$ is just the centralizer of $\mathrm{End}_{A-}\,(A)$ in $\mathrm{End}\,(A)$ and the set of module endomorphisms of A_A is the centralizer of $\mathrm{End}_{-A}\,(A)$ in $\mathrm{End}\,(A)$. Further we denote by A° the *opposite ring* of A, i.e. the ring on A^+ as additive group with the multiplication

$$x \circ y = yx.$$

THEOREM 3.2.1. *Let A be a ring and $\rho(A)$, $\lambda(A)$ the rings of all right and left multiplications respectively. Then $\rho(A) \cong A$, $\lambda(A) \cong A^\circ$ and*

$$\rho(A) = \mathrm{End}_{A-}\,(A), \quad \lambda(A) = \mathrm{End}_{-A}\,(A). \tag{1}$$

Proof. Any right multiplication is a left-A homomorphism, by the associative law: $(bx)a = b(xa)$. Hence the map $a \mapsto \rho_a$ is a ring homomorphism from A to $\mathrm{End}_{A-}(A)$. It is injective, because $\rho_a = 0$ means that $a = 1 \cdot \rho_a = 0$; to show that it is surjective, take $\theta \in \mathrm{End}_{A-}(A)$ and let $1\theta = b$. Then for any $x \in A$, $x\theta = x(1\theta) = xb = x\rho_b$, hence $\theta = \rho_b$, as claimed. This proves the first equation (1) and the fact that $\rho(A) \cong A$. The proof for the other side is similar, bearing in mind that the left multiplications define a left A-module structure on A, corresponding to a right A°-module structure. ∎

We shall also need a lemma on centralizers of subrings of $\mathrm{End}\,(A)$:

LEMMA 3.2.2. *Let A be a ring and F a subring of $\mathrm{End}\,(A)$ containing $\rho(A)$. Define a subset of A by the condition*

$$C = \{x \in A \,|\, \lambda_x \text{ centralizes } F\}$$

$$= \{x \in A \,|\, (xy)f = x \cdot yf \text{ for all } y \in A, f \in F\}.$$

Then the centralizer of F in $\mathrm{End}\,(A)$ is $\lambda(C)$, hence C is a subring of A and we have

$$\rho(A) \subseteq F \subseteq \mathrm{End}_{C-}(A). \tag{2}$$

Moreover, C may be defined by the equation

$$C = \{x \in A \,|\, xf = x \cdot 1f \text{ for all } f \in F\}. \tag{3}$$

Proof. Since F contains $\rho(A)$, the centralizer of $\lambda(A)$, it follows that the centralizer of F is contained in $\lambda(A)$; now the definition of C states that $\lambda_x f = f\lambda_x$ for $x \in C$, so the centralizer is just $\lambda(C)$ and it follows that C is a subring of A. Now (2) follows because the elements of F centralize C.

To prove (3), denote the right-hand side by C'; then it is clear that $C \subseteq C'$ and we must show that equality holds. Let $x \in C'$, so that $xf = x \cdot 1f$ for all $f \in F$. It follows that for any $y \in A$,

$$(xy)f = x\rho_y f = x \cdot 1(\rho_y f), \quad \text{because } \rho_y f \in F,$$

$$= x \cdot yf,$$

and so $x \in C$, as claimed. ∎

For the moment let A be any abelian group. In $\text{End}(A)$ there is a natural topology which is induced by regarding it as a subset of A^A, endowed with the product topology (taking A as a discrete space). This is known as the *finite topology* or the *topology of simple convergence*. If $f \in \text{End}(A)$, a typical neighbourhood of f, specified by a finite subset X of A, consists of all $\varphi \in \text{End}(A)$ such that $xf = x\varphi$ for all $x \in X$. It follows that all centralizers are closed:

PROPOSITION 3.2.3. *Let A be an abelian group and $F \subseteq \text{End}(A)$. Then the centralizer of F is closed in the finite toplogy on $\text{End}(A)$.*

Proof. Let $f \in \text{End}(A)$ and $C = \{g \in \text{End}(A) | gf = fg\}$. Given $h \notin C$, we have $hf \neq fh$, so there exists $x \in A$ such that $xhf \neq xfh$. Consider the neighbourhood of h determined by x, xf. It consists of all $\varphi \in \text{End}(A)$ such that $xh = x\varphi$, $xfh = xf\varphi$. Hence

$$x\varphi f = xhf \neq xfh = xf\varphi,$$

and so $\varphi \notin C$. Thus a neighbourhood of h is disjoint from C, hence C is closed. Now for any subset F of $\text{End}(A)$, the centralizer of F is closed, as the intersection of the centralizers of all $f \in F$. ∎

Let us now consider any ring A and a simple right A-module M. By Schur's lemma the endomorphism ring of M_A, i.e. the centralizer of A in $\text{End}(M)$, is a field D, say, and the action of A on M is dense in the centralizer of D in $\text{End}(M)$. This is just the density theorem for simple modules (see e.g. A.3, Cor. 10.2.6, p. 401), which will be applied in the next result.

Let K be a field and consider the ring $E = \text{End}(K)$ of all endomorphisms of K^+. For any subfield D of K define $D' = \text{End}_{D^-}(K)$, the centralizer of $\lambda(D)$ in E; this is a subring containing $\rho(K)$, i.e. a $\rho(K)$-subring of E. If F is any $\rho(K)$-subring of E, we define $F' = \{x \in K | (xy)f = x \cdot yf$ for all $y \in K, f \in F\}$, the centralizer of F, which is a subring of K by Lemma 2.2, in fact a subfield. The mappings

$D \mapsto D'$, $F \mapsto F'$ satisfy the usual rules for a Galois connexion (see A.2, p. 85f.):

(G.1) $D'' \supseteq D$, $F'' \supseteq F$,

(G.2) $D_1 \subseteq D_2 \Rightarrow D_1' \supseteq D_2'$, $F_1 \subseteq F_2 \Rightarrow F_1' \supseteq F_2'$,

(G.3) $D''' = D'$, $F''' = F'$.

Our problem will be to find the sets which correspond under this Galois connexion, i.e. the subfields of K of the form F' and the subrings of E of the form D'. This question is answered by the Jacobson–Bourbaki correspondence, which tells us that the correspondence is between all subfields of K and all closed $\rho(K)$-subrings of E:

THEOREM 3.2.4 (Jacobson–Bourbaki correspondence). *Let K be a field and* $\mathrm{End}\,(K)$ *the endomorphism ring of K^+ with the finite topology. There is an order-reversing bijection between the subfields D of K and the closed $\rho(K)$-subrings F of* $\mathrm{End}\,(K)$, *given by the rules*

$$D \mapsto F = \mathrm{End}_{D-}(K), \qquad F \mapsto D = \{x \in K | (xy)f = x \cdot yf$$

$$\text{for all } y \in K, f \in F\}. \quad (4)$$

Moreover, if D and F correspond in this way, then

$$[K:D]_L = [F:K]_R, \qquad (5)$$

whenever either side is finite.

Proof. Given a closed $\rho(K)$-subring F of $\mathrm{End}\,(K)$, define D as in (4). By Lemma 2.2, $\lambda(D)$ is the centralizer of F in $\mathrm{End}\,(K)$. Now K is a simple right K-module, hence it is simple as right F-module, so the centralizer D of F is a field, by Schur's lemma, and F is dense in the bicentralizer (by the density theorem). But F is closed, so we have equality:

$$F = \mathrm{End}_{D-}(K). \qquad (6)$$

Conversely, given D, put $F = \mathrm{End}_{D-}(K)$ and define D_1 as

$$D_1 = \{x \in K | xf = x \cdot 1f \text{ for all } f \in F\}.$$

Then $D_1 \supseteq D$ and by the first part, $F = \mathrm{End}_{D_1-}(K)$, so $D_1 = D$ by Lemma 2.2. Thus we have the correspondence (4), which is clearly order-reversing.

To establish (5), let us take a finite subset X of K which is left D-independent, and for each $y \in X$ choose $\delta_y \in \mathrm{End}_{D-}(K)$ such that

$$\left(\sum \alpha_x x\right) \delta_y = \alpha_y.$$

This is clearly possible, e.g. by completing X to a left D-basis of K and defining δ_y as 0 on the complement of X. The δ_y are right linearly K-independent, for if $\sum \delta_y a_y = 0$ $(a_y \in K)$, we can apply this relation to $x \in X$ and get $0 = x(\sum \delta_y a_y) = a_x$, so all the a_x vanish and the relation was trivial. This shows that

$$[K:D]_L \leqslant [F:K]_R, \tag{7}$$

and it proves (5) when the left-hand side is infinite. If the left-hand side is finite, we can take X to be a left D-basis of K. Thus every element of K has the form $\sum \alpha_x x$, and for any $f \in F$ we have

$$\left(\sum \alpha_x x \right)\left(\sum \delta_y \cdot yf \right) = \sum \alpha_y yf = \left(\sum \alpha_x x \right)f,$$

hence $f = \sum \delta_y \cdot yf$ and this shows that the δ_y span F as right K-space. Therefore equality holds in (7) and (5) follows. ∎

Exercises

1. Verify that the finite topology on $\mathrm{End}(A)$, for any abelian group A, is Hausdorff.

2. Let K be a field. What is the condition for the finite topology on K^+ to be discrete?

3°. Show that for any abelian group A, a topology on $\mathrm{End}(A)$ may be defined by taking as the closed sets all finite unions of centralizers. Find conditions for this topology to be equal to the finite topology.

4. Show that any $\rho(K)$-subring F of $\mathrm{End}(K)$ satisfying $[F:K]_R < \infty$ is closed in the finite topology.

3.3 Galois theory

Let K be a field and G the group of all its automorphisms. For a subfield E of K define $E^* = \{\sigma \in G | x^\sigma = x \text{ for all } x \in E\}$ and for any subgroup H of G put $H^* = \{x \in K | x^\sigma = x \text{ for all } \sigma \in H\}$. It is clear that E^* is a subgroup of G, H^* is a subfield of K and we again have a Galois connexion:

(G.1) $E^{**} \supseteq E$, $\quad H^{**} \supseteq H$,
(G.2) $E_1 \subseteq E_2 \Rightarrow E_1^* \supseteq E_2^*$, $H_1 \subseteq H_2 \Rightarrow H_1^* \supseteq H_2^*$,
(G.3) $E^{***} = E^*$, $\quad H^{***} = H^*$.

   Given a subgroup H of G, we call H^* the *fixed field* of H. Given a subfield E of K, we call E^* the *group* of K/E; if $E^{**} = E$, we call E^* a *Galois group*, also denoted by $\mathrm{Gal}(K/E)$, and we say that K/E is a *Galois extension*.

Our problem will be to find which fields and groups correspond in this Galois connexion, i.e. which subfields of K are of the form H^* and which subgroups are of the form E^*. We recall that in the case of commutative fields the finite Galois extensions are just the separable normal extensions, while every subgroup of $\mathrm{Gal}(K/E)$ has the form F^* for a suitable field F between K and E. The account which follows is based on Jacobson [56], Ch. VII.

The commutative theory rests on two basic results (see e.g. A.2, Lemma 3.5.1, p. 81, Th. 3.5.5, p. 84):

Dedekind's lemma. *Distinct homomorphisms of a field E into a field F are linearly independent over F.*

Artin's theorem. *If G is a group of automorphisms of a field E and F is the fixed field, then $[E:F] = |G|$ whenever either side is finite.*

Our object is to find analogues in the general case. We begin with Dedekind's lemma; here we have to define what we mean by the linear dependence of homomorphisms over a skew field. Given any fields K, L, we write $H = \mathrm{Hom}(K, L)$ for the set of all field homomorphisms from K to L; H is a subset of the space L^K of all maps from K to L, and we can form HL, the right L-submodule of L^K generated by H. We shall write the elements of H as exponents and also write $\alpha^\beta = \alpha\beta$ for $\alpha, \beta \in K$. We have

$$x^{\alpha s} = (x\alpha)^s = x^s \alpha^s = x^{s\alpha^s} \quad \text{for all } x, \alpha \in K, s \in H;$$

hence we can define HL as left K-module by the rule

$$\alpha s = s\alpha^s \text{ for } \alpha \in K, s \in H = \mathrm{Hom}(K, L). \tag{1}$$

It is easily checked that with this definition HL is a (K, L)-bimodule.

Each $\mu \in L^\times$ defines an inner automorphism of L,

$$I(\mu): \lambda \mapsto \mu\lambda\mu^{-1},$$

and it is clear that for $s \in H$ we have $sI(\mu) \in H$. Two homomorphisms s, t from K to L are called *equivalent*, $s \sim t$, if they differ by an inner automorphism: $t = sI(\mu)$. We note that for each $s \in H$, sL is a (K, L)-submodule of HL which is simple as (K, L)-module, since it is already simple as L-module. The next result shows how HL is made up of these parts.

Lemma 3.3.1. *Let K, L be fields and $H = \mathrm{Hom}(K, L)$. Then HL is a semi-simple (K, L)-bimodule, as sum of the simple modules sL. Two sub-modules sL, tL are isomorphic if and only if s and t are equivalent.*

Proof. We have seen that sL is simple; HL is a sum of these simple modules, and so is semisimple. If $sL \cong tL$, let t correspond to $s\mu$, where $\mu \in L^{\times}$. Then $\alpha \cdot s\mu = s\mu \cdot \alpha^t$ for all $\alpha \in K$, and since $\alpha s = s\alpha^s$, we find that $\alpha^s \mu = \mu\alpha^t$, so $s = tI(\mu)$. Conversely, if $s \sim t$, then $\alpha^s \mu = \mu\alpha^t$ for some $\mu \in L^{\times}$ and retracing our steps, we find that $sL \cong tL$ with t corresponding to $s\mu$ in the isomorphism. ■

We now consider the action of the (K, L)-bimodule HL. Each element of HL defines a mapping $K \to L$ as follows:

$$\sum s_i \lambda_i : \alpha \mapsto \sum \alpha^{s_i} \lambda_i, \quad \text{where } \alpha \in K, s_i \in H, \lambda_i \in L. \tag{2}$$

By (1) the left K-module structure of HL acts on K in the expected way, namely by right multiplication. Let N be the kernel of the mapping (2) from HL to L^K; thus N consists of all sums $\sum s_i \lambda_i$ such that

$$\sum \alpha^{s_i} \lambda_i = 0 \text{ for all } \alpha \in K. \tag{3}$$

The quotient $M = HL/N$ is a (K, L)-bimodule whose elements have the form $\sum s_i \lambda_i$ $(s_i \in H, \lambda_i \in L)$, with $\sum s_i \lambda_i = 0$ if and only if (3) holds. By Lemma 3.1, HL is semisimple, hence so is the quotient M; we recall also that a semisimple module is a direct sum of its homogeneous components or type components, where each component is a direct sum of simple modules of a given type (A.2, 4.2). Now we have the following generalization of Dedekind's lemma:

THEOREM 3.3.2. *Let K, L be any fields, put $H = \text{Hom}(K, L)$ and consider $M = HL/N$, the homomorphic image of HL in L^K as (K, L)-bimodule, as defined above.*

(i) *Given $s, s_1, \ldots, s_n \in H$, if $s = \sum s_i \lambda_i$ in M, then $s = s_i I(\mu)$ for some i and some $\mu \in L^{\times}$.*

(ii) *Given $s \in H$, $\mu_1, \ldots, \mu_r \in L^{\times}$, if the elements $sI(\mu_i)$ are linearly dependent over L, then the μ_i are linearly dependent over $\mathscr{C}_L(K^s)$, the centralizer of K^s in L.*

Proof. (i) If $s = \sum s_i \lambda_i$, then the simple module sL lies in the sum of the $s_i L$ and so it lies in the same type component as some $s_i L$, hence s and s_i generate isomorphic modules and so are equivalent: $s = s_i I(\mu)$ for some $\mu \in L^{\times}$.

(ii) If the $sI(\mu_i)$ are linearly dependent in M, take a relation of shortest length:

$$\sum_1^p sI(\mu_i)\lambda_i = 0, \quad \text{where } \lambda_i \in L^{\times}.$$

Multiplying on the right by a suitable factor, we may assume that $\lambda_1 = \mu_1$. By applying the relation to $1 \in K$, we find that $\sum \lambda_i = 0$, so

$$\sum \mu_i \cdot \mu_i^{-1} \lambda_i = 0, \tag{4}$$

and the result will follow if we can show that $\mu_i^{-1} \lambda_i \in \mathscr{C}_L(K^s)$. Let us apply the relation to a product $\alpha\beta$ in K:

$$0 = \sum (\alpha\beta)^s I(\mu_i) \lambda_i = \sum \mu_i \alpha^s \beta^s \mu_i^{-1} \lambda_i. \tag{5}$$

Next apply the relation to α and multiply by β^s on the right:

$$0 = \sum \alpha^s I(\mu_i) \lambda_i \beta^s = \sum \mu_i \alpha^s \mu_i^{-1} \lambda_i \beta^s.$$

Taking the difference, we get

$$\sum_1^p \mu_i \alpha^s \mu_i^{-1} (\lambda_i \beta^s - \mu_i \beta^s \mu_i^{-1} \lambda_i) = 0.$$

The first coefficient is $\lambda_1 \beta^s - \lambda_1 \beta^s = 0$, hence by the minimality the other coefficients also vanish, so $\mu_i^{-1} \lambda_i \beta^s = \beta^s \mu_i^{-1} \lambda_i$ and hence $\mu_i^{-1} \lambda_i$ centralizes K^s, as we had to show. ∎

Suppose that s is an isomorphism between K and L. Then $K^s = L$ and so the centralizer of K^s in L is just the centre of L. This yields

COROLLARY 3.3.3. *Let* $s_1, \ldots, s_r$ *be pairwise inequivalent isomorphisms between two fields K and L and let $\lambda_1, \ldots, \lambda_t \in L$ be linearly independent over C, the centre of L. Then the isomorphisms $s_i I(\lambda_j)$ are linearly independent over L and any isomorphism which is a linear combination (over L) of these isomorphisms has the form $s = s_i I(\lambda)$, where $\lambda = \sum \lambda_j \beta_j$ ($\beta_j \in C$).*

Proof. If the $s_i I(\lambda_j)$ were linearly dependent, then by (i), for some $s = s_i$ the $s I(\lambda_j)$ would be linearly dependent, so by (ii) the λ_j would then be linearly dependent over C, but this contradicts the hypothesis.

Now assume that the isomorphism s satisfies

$$s = \sum s_i I(\lambda_j) \alpha_{ij}, \quad \text{where } \alpha_{ij} \in L. \tag{6}$$

Then by (i), $s = s_i I(\lambda)$ for some i, say $i = 1$ and $\lambda \in L$. Hence we have $s_1 I(\lambda) = \sum s_i I(\lambda_j) \alpha_{ij}$, and by equating terms of the same type we may omit all terms s_k with $k \neq 1$. Now by what has been shown, $\lambda, \lambda_1, \ldots, \lambda_t$ are linearly dependent over C, but $\lambda_1, \ldots, \lambda_t$ are linearly independent, hence λ depends linearly on the λ_i, so $\lambda = \sum \lambda_j \beta_j$ for some $\beta_j \in C$. ∎

Next we have to translate Artin's theorem. Without using Dedekind's lemma we can state the result as $[K:D] = [G:K]$, where G, or rather GK, is regarded as right K-space. We shall replace G, a group of D-linear automorphisms of K, by D-linear transformations of $_DK$. Every such $s \in \text{End}_{D-}(K)$ satisfies the rule

$$(\alpha x)^s = \alpha x^s \quad \text{for all } x \in K, \alpha \in D,$$

which generalizes the rules $(xy)^s = x^s y^s$, $\alpha^s = \alpha$ satisfied for $s \in G$. For any field K we shall consider the ring $\text{End}(K)$ again as a topological ring, as in 3.2. It contains $\rho(K)$ as a subring and by the Jacobson–Bourbaki correspondence (Th. 2.4) there is an order-reversing bijection between the subfields D of K and the closed $\rho(K)$-subrings F of $\text{End}(K)$ such that

$$[K:D]_L = [F:K]_R$$

whenever either side is finite.

Given a group G of automorphisms of K, we have a right K-space GK, and we have only to show that this is a ring in order to be able to apply the preceding result. Thus we need

LEMMA 3.3.4. *Let K be a field and G a group of automorphisms of K. Then GK is a $\rho(K)$-subring of $\text{End}(K)$ and its closure $\overline{GK}$ in the finite topology is $\text{End}_{D-}(K)$, where D is the subset of K fixed by G.*

Proof. In GK we have the rule $\alpha g = g \alpha^g$ ($\alpha \in K$, $g \in G$); using this rule we have

$$(g_1 \alpha_1)(g_2 \alpha_2) = g_1 g_2 \alpha_1^{g_2} \alpha_2.$$

Since every element of GK is a sum of terms $g\alpha$, it follows that GK is closed under products and $GK \supseteq \rho(K)$, so GK is a $\rho(K)$-ring. By Th. 2.4, $\overline{GK} = \text{End}_{D-}(K)$, where $D = \{x \in K | x^f = x \cdot 1^f \text{ for all } f \in \overline{GK}\}$. Thus if $\alpha \in K$, then

$$\alpha \in D \Leftrightarrow \alpha^f = \alpha \cdot 1^f \text{ for all } f \in \overline{GK},$$

$$\Leftrightarrow \alpha^{g\beta} = \alpha \cdot 1^{g\beta} \text{ for all } g \in G, \beta \in K,$$

$$\Leftrightarrow \alpha^g \beta = \alpha\beta;$$

hence $\alpha \in D$ if and only if $\alpha^g = \alpha$ for all $g \in G$. ∎

By combining this result with Th. 3.2, we obtain a result on extending homomorphisms. By an automorphism of K *over* a subfield D or an automorphism of K/D we shall understand an automorphism of K leaving D elementwise fixed.

PROPOSITION 3.3.5. *Let K be a field, G a group of automorphisms of K and D the fixed field of G. Further, assume that G contains all inner automorphisms of K over D. If E is a D-subring of K of finite left degree over D, then every D-ring homomorphism of E into K is induced by an element of G.*

Proof. Let $s: E \to K$ be a D-ring homomorphism. Regarding E and K as left D-spaces, we see that E is a subspace of K, therefore s can be extended to a D-linear endomorphism of K, i.e. an element of $\text{End}_{D_-}(K)$. By Lemma 3.4, $\text{End}_{D_-}(K) = \overline{GK}$ and so on any finite-dimensional subspace s can be written as $\sum g_i \lambda_i$ ($g_i \in G$, $\lambda_i \in K$). In particular, since $[E:D]_{\mathrm{L}} < \infty$, we have $s = \sum g_i \lambda_i$ on E. Now E is a D-subring of K of finite left degree, hence a field (Prop. 1.2) and by Th. 3.2(i), $s = g_i I(\mu)$ for some i and some $\mu \in K^\times$. Applying this expression to $\alpha \in D$ we find

$$\alpha = \alpha^s = \alpha^{g_i} I(\mu) = \mu \alpha \mu^{-1}.$$

Hence the inner automorphism $I(\mu)$ leaves D fixed, so $I(\mu) \in G$ by hypothesis and so $g_i I(\mu) \in G$ is an automorphism which induces s. ∎

This yields the Skolem–Noether theorem for the special case of fields:

COROLLARY 3.3.6. *Let K be a field with centre C and let D be a C-subalgebra of K finite-dimensional over C. Then any C-algebra homomorphism from D to K can be extended to an inner automorphism of K. In particular, if $[K:C] < \infty$, every automorphism of K over C is inner.*

Proof. In Prop. 3.5 take G to be the group of all inner automorphisms of K. Then C is the fixed field and every inner automorphism belongs to G; now the conclusion follows. ∎

We now return to our initial task of finding which automorphism groups and subfields correspond to each other in the Galois connexion. We first deal with a condition which is obviously satisfied by all Galois groups. Let K be a field with centre C and let D be any subfield of K; then the centralizer of D in K is a subfield D' containing C. Any non-zero element of D' defines an inner automorphism of K over D and so belongs to the group D^*; conversely, an inner automorphism of K belongs to D^* only if it is induced by an element of D'. Thus we see that the $\alpha \in K$ for which $I(\alpha) \in D^*$ together with 0 form a subfield containing C. This then is a necessary condition for an automorphism group to be Galois and it suggests the following

Definition. A group G of automorphisms of a field K is called an *N-group* (after E. Noether) if the set

$$A = \{\alpha \in K | \alpha = 0 \text{ or } I(\alpha) \in G\}$$

is an algebra over the centre C of K; A will be called the *C-algebra associated* with G. Clearly the associated C-algebra is necessarily a field.

We note that the property of G being an N-group is not just a group property, but refers to its action on the field. It is also clear that every Galois group is an N-group: if $G = E^*$ and $E = G^*$, then $I(c) \in G$ if and only if $c \in \mathscr{C}_D(E)$, so the associated algebra is just the centralizer of E in K.

If G is any N-group with associated algebra A, and G_0 is the subgroup of all inner automorphisms $I(\alpha)$ ($\alpha \in A^\times$), then G_0 is normal in G, for if $x \in K$, $\alpha \in A^\times$, $s \in G$, then

$$x^{s^{-1}}I(\alpha)^s = (\alpha x^{s^{-1}} \alpha^{-1})^s = \alpha^s x (\alpha^s)^{-1} = xI(\alpha^s);$$

hence

$$s^{-1}I(\alpha)s = I(\alpha^s),$$

and this shows G_0 to be normal in G. We define the *reduced order* of G as

$$|G|_{\text{red}} = (G:G_0)[A:C]. \tag{7}$$

With this notation we have the following replacement for Artin's theorem:

THEOREM 3.3.7. *Let K be any field, G an N-group of automorphisms of K and put $D = G^*$, the fixed field of G. Then*

$$[K:D]_{\text{L}} = |G|_{\text{red}}, \tag{8}$$

whenever either side is finite, and when this is so, G is a Galois group, i.e. $G = D^* = G^{**}$.

Proof. Suppose first that $[K:D] = m < \infty$; then by the Jacobson–Bourbaki correspondence (Th. 2.4), $[\text{End}_{D_-}(K):K]_{\text{R}} = m$. Let $s_1, \ldots, s_r \in G$ be pairwise incongruent (mod G_0) and $\lambda_1, \ldots, \lambda_t$ any elements of the associated C-algebra A that are linearly independent over C. Then the maps $s_i I(\lambda_j)$ are in $\text{End}_{D_-}(K)$ and by Cor. 3.3 they are linearly independent over K. Hence $rt \leqslant m$ and it follows that $|G|_{\text{red}} \leqslant m < \infty$.

We may now assume that G has finite reduced order. Let $s_1, \ldots, s_r$ be a transversal for G_0 in G and $\lambda_1, \ldots, \lambda_t$ a C-basis for A; then we know that the $s_i I(\lambda_j)$ are linearly independent over K. We shall show that they form a right K-basis for $\mathrm{End}_{D_-}(K)$.

Given $s \in G$, let $s = s_1 I(\lambda)$, say, where $\lambda \in A$ and so $\lambda = \sum \lambda_j \beta_j$ $(\beta_j \in C)$. For any $c \in K$ we have

$$c^s = c^{s_1 I(\lambda)} = \lambda c^{s_1} \lambda^{-1}$$

$$= \sum \lambda_j \beta_j c^{s_1} \lambda^{-1}$$

$$= \sum \lambda_j c^{s_1} \lambda_j^{-1} \cdot \lambda_j \beta_j \lambda^{-1}$$

$$= \sum c^{s_1} I(\lambda_j) \gamma_j,$$

where $\gamma_j = \lambda_j \beta_j \lambda^{-1} \in K$. This shows that the $s_i I(\lambda_j)$ span G, hence also GK. Now GK has finite dimension over K, and so $GK = \overline{GK} = \mathrm{End}_{D_-}(K)$, by Lemma 3.4. It follows that $|G|_{\mathrm{red}} = [\mathrm{End}_{D_-}(K):K]_R = [K:D]_L$.

Next it is clear that $D^* = G^{**} \supseteq G$. Conversely, if $s \in D^*$, then $s \in \mathrm{End}_{D_-}(K)$, hence $s = \sum s_i I(\lambda_j) \alpha_{ij}$, where $\alpha_{ij} \in K$. By Th. 3.2, since s is an automorphism of K, we have $s = s_i I(\lambda)$ for some i, where $\lambda = \sum \lambda_j \beta_j$ $(\beta_j \in C)$, but then $I(\lambda) \in G_0$ and $s_i \in G$, hence $s \in G$, as we had to show. ■

Here is a simple example (noted by G. M. Bergman) to illustrate the need for introducing the reduced order. Let ω be a primitive nth root of 1 (for any odd $n \geqslant 3$) or a primitive $2n$th root of 1 (for any even $n \geqslant 4$), and in the real quaternions $\mathbf{H}$ consider $I(\omega)$. This automorphism has order n, though its fixed field is $\mathbf{C}$ and $[\mathbf{H}:\mathbf{C}] = 2$.

We note some consequences of Th. 3.7. In the first place (8) allows us to assert the equality of left and right degrees in some cases:

COROLLARY 3.3.8. *Let K be a field, G an N-group of automorphisms of K and $D = G^*$. Then*

$$[K:D]_L = [K:D]_R,$$

whenever either side is finite.

This follows from (8) by symmetry. ■

Secondly we have a result first obtained by Brauer [32] in the theory of algebras.

COROLLARY 3.3.9. *Let K be a field with centre C and let A be any C-subalgebra of K. Then the centralizer A' of A in K is again a C-algebra* *and the bicentralizer A'' contains A. Moreover,*

$$[K:A']_L = [A:C], \tag{9}$$

whenever either side is finite, and when this is so, then $A'' = A$.

Proof. Suppose first that $[K:A']_L$ is finite. Clearly A' is a subfield of K; let G be the group of all inner automorphisms of K over A' and let A_1 be the associated algebra. Then $A_1 \supseteq A$ and by Th. 3.7, $|G|_{red} = [A_1:C] = [K:A']_L$, hence A is then of finite degree. Thus we may assume that $[A:C]$ is finite, and hence A is a field. Let G be the group of all inner automorphisms induced by A; then $G^* = A'$ and (9) follows from (8). Moreover, $(A')^* = G$, which means that $A'' = A$. ■

We still need a lemma to identify the precise form taken by subrings of GK.

LEMMA 3.3.10. *If K/D is a Galois extension with group G and $[K:D]_L < \infty$, then any $\rho(K)$-subring B of GK has the form HK, where $H = G \cap B$ is an N-subgroup of G.*

Proof. Put $H = G \cap B$; then clearly $HK \subseteq B$. To prove equality we note that HK and B are both K-bimodules contained in $GK = \mathrm{End}_{D-}(K)$, by Lemma 3.4. Now GK is semisimple and hence so is B; moreover, every simple submodule of B is isomorphic to a simple submodule of GK and so is of the form uK, where $\alpha u = u\alpha^s$ for all $\alpha \in K$ and some $s \in G$. Replacing u by $u\gamma$ ($\gamma \in K$) if necessary, we may suppose that

$$1 \cdot u = u \quad \text{and still } \alpha u = u\alpha^s \quad \text{for some } s \in G.$$

Hence $\alpha \cdot u = 1 \cdot u\alpha^s = \alpha^s$, i.e. u is an automorphism of K, viz. s, and since $u \in \mathrm{End}_{D-}(K)$, s fixes D, i.e. $s \in G$, so $s \in G \cap B = H$. This shows that $B = HK$.

That H is a group follows because B is a centralizer (being of finite degree over K, by Jacobson–Bourbaki). To show that H is an N-group, let $I(\alpha_1), I(\alpha_2) \in H$ and $\alpha = \alpha_1\beta_1 + \alpha_2\beta_2 \neq 0$; we must show that $I(\alpha) \in H$, and clearly it will be enough to show that $I(\alpha) \in B$. We have $I(\alpha) = \lambda(\alpha)\rho(\alpha^{-1})$ and $B \supseteq \rho(K)$, hence $I(\alpha) \in B \Leftrightarrow \lambda(\alpha) \in B$. By hypothesis, $\lambda(\alpha_1), \lambda(\alpha_2) \in B$ and $\lambda(\alpha) = \lambda(\alpha_1)\beta_1 + \lambda(\alpha_2)\beta_2$, hence $\lambda(\alpha) \in B$, and it follows that H is an N-group. ■

Finally we come to the fundamental theorem. In order to describe extensions, we need an analogue of normal subgroups, bearing in mind that we admit only N-groups. We therefore define: a subgroup H of an N-group G is said to be *N-invariant* in G if the N-subgroup generated by the normalizer of H in G is G itself.

THEOREM 3.3.11 (Fundamental theorem of Galois theory for skew fields). *Let K/E be a Galois extension with group G and assume that $[K:E]_L < \infty$.*

(i) *There is a bijection between N-subgroups of G and intermediate fields D, $E \subseteq D \subseteq K$:*

$$H \mapsto H^* = \{x \in K | x^\sigma = x \text{ for all } \sigma \in H\},$$

$$D \mapsto D^* = \{\sigma \in G | x^\sigma = x \text{ for all } x \in D\}.$$

If $H \leftrightarrow D$, then K/D is Galois with group H and $[K:D]_L = |H|_{red}$.

(ii) *If $H \leftrightarrow D$, then the group of automorphisms of D/E is isomorphic to N_H/H, where N_H is the normalizer of H in G. Moreover, D/E is Galois if and only if H is N-invariant in G.*

Proof. (i) By Th. 3.7, $|G|_{red} < \infty$ and so any N-subgroup of G has finite reduced order: $H/H_0 = H/(H \cap G_0) \cong HG_0/G_0 \subseteq G/G_0$, where G_0 is the subgroup of G consisting of inner automorphisms, and similarly for H_0 and H; thus the C-algebra associated with H is contained in that of G. It follows by Th. 3.7 that, given H, we have $H^{**} = H$. Conversely, given D, we put $H = G \cap \text{End}_{D^-}(K)$; then $\text{End}_{D^-}(K) = HK$, by Lemma 3.10. Now H consists of all D-linear transformations in G, i.e. the elements of G fixing D, thus $H = D^*$. Since $\text{End}_{D^-}(K) = HK$, we have $D = H^*$ by Lemma 3.4. Thus $D^{**} = D$ and now $[K:D]_L = |H|_{red}$ by Th. 3.7.

(ii) If $D \leftrightarrow H$, then for any $s \in G$, $D^s \leftrightarrow s^{-1}Hs$, therefore the members of N_H and only these are automorphisms of D/E. Every automorphism of D/E is induced by one of K/E (Prop. 3.5), hence its Galois group, G_1 say, is a homomorphic image of N_H. The kernel consists of those automorphisms of K which fix D, i.e. H, so that $G_1 \cong N_H/H$. So far D was any field between K and E. Now D/E is Galois if and only if E is the fixed field of G_1, i.e. E is the set of elements fixed by N_H. By (i) this holds if and only if the N-subgroup of G generated by N_H corresponds to E, which happens precisely when this group is G. But this just means that H is N-invariant in G. ∎

We note the special case where the Galois group is *inner*, i.e. consists entirely of inner automorphisms. Let K/E be a Galois extension with inner Galois group G and let E' be the centralizer of E in K; then C, the centre of K, is contained in E' and $G \cong E'^{\times}/C^{\times}$. Such extensions are described in Cor. 3.9.

Let us return to the case of a general Galois extension K/E with group G. As we have seen, the subgroup G_0 of inner automorphisms is normal in G; moreover, G/G_0 consists entirely of outer automorphisms (apart from 1). For if the fixed field of G_0 is K_0 and $\bar{\sigma} \in G/G_0$, suppose that $\bar{\sigma} = I(c)$ and take $\sigma \in G$ mapping to $\bar{\sigma}$. Then $\sigma^{-1} I(c)$ fixes K_0 and so lies in G_0, i.e. $\sigma \in G_0$ and hence $\bar{\sigma} = 1$, as claimed.

A Galois group will be called *outer* if the only inner automorphism it contains is 1. E.g. in the commutative theory all Galois groups are outer. It follows from Th. 3.11 that every Galois extension is an outer extension followed by an inner extension:

COROLLARY 3.3.12. *Given any Galois extension K/E, there is a field K_0, $E \subseteq K_0 \subseteq K$, such that K_0/E is Galois with outer Galois group and K/K_0 has inner Galois group. Here K_0 is uniquely determined as the bicentralizer of E in K.*

Proof. Let G be the Galois group of K/E and G_0 the subgroup of inner automorphisms. The automorphisms in G_0 fix E and so are induced by elements of the centralizer E' of E. If K_0 denotes the centralizer of E', i.e. the bicentralizer of E, then $K_0 \supseteq E$ and K_0 is the fixed field of G_0. Thus K/K_0 is Galois with inner Galois group G_0, while K_0/E has Galois group G/G_0 and this is outer, as we have seen. ∎

The 'theorem of the primitive element' has an analogue in the skew case.

THEOREM 3.3.13. *Let K be a field with an infinite subfield E. If there are only finitely many subfields between K and E, then K can be generated by a single element over E, i.e. K/E is a simple extension.*

Proof. Among the simple extension fields of E in K take a maximal one, $E(a)$ say. If $E(a) \neq K$, take $b \notin E(a)$ in K and consider the fields $L_\lambda = E(a + \lambda b)$ $(\lambda \in E)$. Two of these must coincide, because E is infinite. If L_γ contains $a + \gamma b$ and $a + \gamma' b$, where $\gamma \neq \gamma'$, then it also contains $(\gamma - \gamma')b$, hence b and a, so L_γ is a simple extension strictly larger than $E(a)$, which is a contradiction. ∎

The conclusion holds for any Galois extension with finite Galois group. However, a Galois group of finite reduced order may well have infinitely many N-subgroups (see Ex. 3).

Exercises

1. (S. A. Amitsur) Let $F = \mathbf{Q}(u)$, where u is a primitive cube root of 1, with automorphism $\alpha\colon u \mapsto u^2$. Put $R = F[v; \alpha]$ and verify that $v^2 - 2$ is central and irreducible. If $K = R/(v^2 - 2)$, show that K/F is inner Galois, with group generated by $I(u)$, of order 3 but reduced order 2.

2. Show that any field of finite degree over its centre is an inner Galois extension.

3. Let K be a field with centre $C \neq K$ and suppose that K/C is a Galois extension. Show that there are infinitely many fields between K and C, and deduce that the Galois group must be infinite. Give an example where the Galois group has finite reduced order.

4. Show that the set of all automorphisms and antiautomorphisms (reversing multiplication) of a field K forms a group in which the set of automorphisms forms a subgroup of index at most 2. Give an example of an antiautomorphism of a field whose fixed set is not a subfield. (Hint. Try a field of fractions of the free algebra.)

3.4 Equations over skew fields and Wedderburn's theorem

The solution of equations over a commutative field k is closely bound up with the problem of constructing algebraic extensions of k. For skew fields, matrices are needed to treat the general problem, and we shall present what little is known so far in Ch. 8. In this section we shall describe a special situation that allows the use of ordinary polynomials in $K[t]$.

Let K be a field and L an extension of K. An element α of L is said to be *right algebraic* over K if its powers are right linearly dependent over K, i.e. α is a *left root* of an equation

$$f(t) = a_0 + ta_1 + \ldots + t^n a_n = 0, \quad a_i \in K, \text{ not all } 0. \tag{1}$$

We shall also say that α is a *left zero* of the polynomial $f(t) = \sum t^i a_i$. There is a corresponding definition of *left algebraic, right root*.

To find the relation between the left zeros of a polynomial and those of its factors, let $g = \sum t^j b_j$ be another polynomial. We shall put $fg = \sum t^{i+j} a_i b_j$, so formally polynomials may be regarded as members of the polynomial ring $K[t]$; further we write $f(c)$ or f_c for the element obtained by replacing t by c, keeping the coefficients on the right. Then we have, assuming that $f_c \neq 0$,

$$(fg)_c = \sum c^{i+j} a_i b_j = \sum c^j f_c b_j = f_c \sum (f_c^{-1} c^j f_c) b_j,$$

and so

$$(fg)_c = f_c g(f_c^{-1} c f_c), \tag{2}$$

whenever $f_c \neq 0$. If we apply this formula to the usual division algorithm, $f = (t - c)q + r$, putting $t = c$, we find that $r = f_c$ and so

$$f = (t - c)q + f_c \quad \text{for any } f \in K[t] \text{ and } c \in K. \tag{3}$$

The conclusions (3) and (2) may be stated as follows:

PROPOSITION 3.4.1. *Let K be any field, $c \in K$ and $f \in K[t]$. Then $f_c = 0$ if and only if $t - c$ is a left factor of f. Further, for any $f, g \in K[t]$ and $c \in K$,*

$$(fg)_c = \begin{cases} 0 \text{ if } f_c = 0, \\ f_c g(f_c^{-1} c f_c) \text{ if } f_c \neq 0. \end{cases} \tag{4}$$

Hence the left zeros of fg are either left zeros of f or conjugates of left zeros of g. ∎

We shall also need a criterion for the similarity of linear factors in $K[t]$, but this is no harder to prove for the skew polynomial ring, and is sometimes useful in that form.

LEMMA 3.4.2. *Let K be a field with an endomorphism α, an α-derivation δ and put $R = K[t; \alpha, \delta]$. Then $t - a$ is similar to $t - a'$ if and only if*

$$a' = c^{-1} a c^\alpha + c^{-1} c^\delta \quad \text{for some } c \in K^\times. \tag{5}$$

In particular, in $K[t]$, $t - a$ is similar to $t - a'$ if and only if a' is conjugate to a.

Proof. From Prop. 1.5.2 we know that $t - a$ is similar to $t - a'$ if and only if there is a comaximal relation

$$p(t - a') = (t - a)q, \tag{6}$$

where $p, q \in R$. By the division algorithm we can replace p by $p - (t - a)f$, which for suitably chosen f is of degree 0. Thus we may take $p = c \in K^\times$ and then a comparison of degrees shows that $q \in K^\times$. Now (6) can be written as

$$tc^\alpha + c^\delta - ca' = tq - aq,$$

and on comparing coefficients we find that $q = c^\alpha$, $c^\delta - ca' = -aq$; hence $ca' = ac^\alpha + c^\delta$, which yields (5). Conversely, when this holds, then we obtain (6) with $p = c$, $q = c^\alpha$ by reversing the steps of the argument. ∎

Going back to (4), we see that if a is a left zero of f, then it is also a left zero of fg, for any polynomial g, and if a is a left zero of f and g, then it is a left zero of $f - g$. Thus the polynomials having a as left zero form a right ideal in $K[t]$, viz. the kernel of the substitution map $f(t) \mapsto f(a)$. When $a \in K$, this is of course the right ideal generated by $t - a$, while for a right algebraic over K it will be a non-zero principal right ideal. The monic polynomial generating this right ideal is called the *minimal polynomial* of a over K; it is the unique monic polynomial of least degree with a as left zero. In Galois extensions such minimal polynomials have a complete factorization:

THEOREM 3.4.3. *Let L/K be any Galois extension, with group G. If $a \in L$ is right algebraic over K, with minimal polynomial $p(t)$, then*

$$p(t) = (t - a_1) \ldots (t - a_n), \tag{7}$$

where $a_1, \ldots, a_n$ are conjugates of G-transforms of a, and every G-transform of a is conjugate to (at least) one of the a_i.

Proof. The minimal polynomial of a over L is $t - a$, hence we have

$$p(t) = (t - a)f(t) \text{ for some } f \in L[t].$$

If $\sigma \in G$, then $p(t) = (t - a^\sigma)f^\sigma(t)$, hence all the $t - a^\sigma$ have a LCRM $q(t)$ which is a left factor of $p(t)$. But q is invariant under G and so has coefficients in K, therefore $q = p$. Let us write

$$p(t) = (t - a_1)(t - a_2) \ldots (t - a_r)p'' = p'p'',$$

where $a_1 = a$, each a_i is a conjugate of a G-transform of a and r is as large as possible. We claim that $p'' = 1$; if not, then there exists a G-transform a' of a such that $t - a'$ is not a left factor of p'. However, a' is a left zero of p, so $t - a'$ is a left factor of p, while the LCRM of $t - a'$ and p' has the form

$$(t - a')p_1 = p'p_2. \tag{8}$$

This is a right coprime relation, so p_2 is similar to a right factor of $t - a'$, but it is not a unit, so it must have degree 1 and hence can be taken in the form $p_2 = t - a_{r+1}$, where a_{r+1} is similar to a', by Lemma 4.2. Now p is a common right multiple of $(t - a')$ and p', and hence a right multiple of the LCRM, i.e.

$$p = p'(t - a_{r+1})q;$$

but this contradicts the choice of r. This proves (7); now any $t - a^\sigma$ is a left factor of p and by the same argument a^σ is conjugate to one of the a_i. ∎

In particular, since any left zero of p is conjugate to one of $a_1, \ldots, a_n$, we obtain the following generalization of the well known theorem that an equation of degree n cannot have more than n roots:

COROLLARY 3.4.4. *The left zeros of a polynomial of degree n fall into at most n conjugacy classes.* ■

Another special case is obtained by taking G to consist of inner automorphisms:

COROLLARY 3.4.5. *Let K be a field with centre C and let f be an irreducible polynomial over C. Then all left zeros (if any) of f in K are conjugate.*

Proof. If G is the group of all inner automorphisms of K, then the fixed field is just C. Now if α is a left root of $f = 0$ in K, then $t - \alpha$ is a left factor of f, and by irreducibility f is the minimal polynomial for α over C, so we can apply Th. 4.3 to reach the conclusion. ■

This result has another useful consequence:

COROLLARY 3.4.6. *Let K be a field with centre C and let $a \in K \backslash C$ be such that $a^r = 1$. Then there exists $b \in K^\times$ such that $b^{-1}ab = a^m \neq a$.*

Proof. Since $a^r = 1$, a is algebraic over C; moreover if K has finite characteristic p, we may assume that $p \nmid r$, for otherwise $r = pr'$ and $(a^{r'} - 1)^p = a^r - 1 = 0$, hence $a^{r'} = 1$. Let f be the minimal polynomial of a over C; all its zeros are distinct, because they are zeros of $t^r - 1$ and they may be taken as powers of a. Since $a \notin C$, f has degree > 1, so there is a zero $a^m \neq a$ of f, and by Cor. 4.5, $b^{-1}ab = a^m$ for some $b \in K^\times$. ■

It is possible to construct polynomials with prescribed left zeros:

PROPOSITION 3.4.7. *Let K be any field and let $a_1, \ldots, a_n \in K$ be pairwise inconjugate. Then there is a unique monic polynomial of degree n with $a_1, \ldots, a_n$ as left zeros.*

Proof. If f, g are two monic polynomials of degree n with the as as left zeros, then $f - g$ is a polynomial of degree less than n with the as as left zeros, and so must vanish, by Cor. 4.4; so there can be at most one such polynomial. To find it we may, by induction on n, assume that g is monic of degree $n - 1$ with left zeros $a_1, \ldots, a_{n-1}$. Then $g(a_n) \neq 0$, again by

Cor. 4.4, and hence, by Prop. 4.1, $g(t)(t - g(a_n)^{-1}a_n g(a_n))$ is a monic polynomial of degree n with $a_1, \ldots, a_n$ as left zeros. ∎

It is also possible to construct polynomials with conjugate zeros, but then we no longer have uniqueness (see Ex. 3). This still leaves the question whether an equation over a field always has a left root in a suitable extension field; the answer is 'yes', but we shall have to wait until Ch. 8 for the proof.

As a natural consequence of the above results we have Wedderburn's theorem, that every finite field is commutative. For the proof we shall need some facts about finite fields (see A.2, 3.7–8) and finite groups.

In any finite abelian group A of exponent r the equation $x^r = 1$ has at least r solutions, with equality precisely when A is cyclic, as we see from the basis theorem for abelian groups. If A is a subgroup of a field, there are at most r solutions, so A must then be cyclic. Thus every finite abelian subgroup of the multiplicative group of a field is cyclic.

We recall that a finite field F has finite characteristic p; as vector space over $\mathbf{F}_p$, the field of p elements, F has finite degree n, say. Then F has $q = p^n$ elements, and it may be described as the splitting field of the equation $x^q - x = 0$ over $\mathbf{F}_p$; hence the field of p^n elements exists and is unique up to isomorphism. Moreover, $F^\times$ is cyclic.

We shall also need the fact that a finite group cannot be written as a union of a proper subgroup and all its conjugates. This fact is easily proved directly (see Ex. 1); it is also used in some other proofs of Wedderburn's theorem (see e.g. A.3, Th. 7.1.13, p. 265).

With a view to later applications we shall establish the following slightly more general result:

THEOREM 3.4.8. *Every non-commutative field is infinite. More generally, in a non-commutative field, every element is contained in an infinite commutative subfield.*

Proof. Let K be a field with centre C, where $C \neq K$ and $|C| = q < \infty$. Given $a \in K \backslash C$, let f be the minimal polynomial of a over C, of degree r, say. The map $x \mapsto x^q$ is an automorphism of $C(a)$ over C of order r. Hence a^q is conjugate to a, by Cor. 4.5, so there exists $b \in K^\times$ such that

$$bab^{-1} = a^q. \qquad (9)$$

It follows that $b^s a b^{-s} = a^{q^s}$ and $a^{q^r} = a$, so a commutes with b^r, but of course not with b, by (9). Put $E = C(a, b^r)$; this is a commutative field and the right E-module spanned by $1, b, \ldots, b^{r-1}$ is a non-commutative field extension of E of degree at most r. In fact the degree is exactly r, for let the minimal equation of b over E be

$$b^s + b^{s-1}\lambda_1 + \ldots + \lambda_s = 0, \quad \text{where } \lambda_i \in E \text{ and } s \leqslant r. \quad (10)$$

Multiplying on the left by a^{q^s} and subtracting the same equation multiplied on the right by a, we obtain

$$b^{s-1}(a^q - a)\lambda_1 + \ldots + (a^{q^s} - a)\lambda_s = 0.$$

Since this equation has lower degree, all coefficients vanish. If $s < r$, this means that $\lambda_1 = \ldots = \lambda_s = 0$ and (10) reduces to $b^s = 0$, which is a contradiction. Hence $s = r$ and (10) becomes $b^r - \lambda_r = 0$, so $[E(b):E] = r$.

If there are finite skew fields, let K be a smallest one, and again denote its centre by C. Every element of K generates a commutative subfield over C and so is contained in a maximal subfield E of K. By the minimality of K, E is commutative and so has the form $E = C(a)$ for some $a \in K$. If $[E:C] = r$ and b is as before, then $b^r \in E$ by maximality, but $b \notin E$ and so $E(b) = K$, by the maximality of E. It follows that $[K:E] = [E:C] = r$, hence $[K:C] = r^2$ and this shows that r is independent of the choice of E. Thus all maximal subfields of K have the same number of elements, so all are isomorphic and each can be written as $C(a)$, where the minimal equation for a over C is the same in each case. By Cor. 4.5 these maximal subfields are all conjugate in K, so $K^\times$ can be written as the union of all the conjugates of a proper subgroup of $K^\times$. But this contradicts the fact about groups mentioned earlier, so we conclude that any genuinely skew field must be infinite (Wedderburn's theorem).

To complete the proof, we may assume that C is finite and a is algebraic over C, since otherwise the result holds trivially. Let a be algebraic of degree r over C and find $b \in K$ to satisfy (9). Then $[C(a, b):C(a, b^r)] = r$ and $C(a, b^r)$ is commutative, but $C(a, b)$ is not; if $C(a, b^r)$ were finite, this would mean that $C(a, b)$ is also finite, contradicting the first part. Hence $C(a, b^r)$ must be infinite and the conclusion follows. ■

We can now make a more precise statement about the occurrence of conjugate zeros.

THEOREM 3.4.9. *Let K be a field. If a polynomial over K has two distinct left zeros in a conjugacy class of K, then it has infinitely many left zeros in that class.*

Proof. Let $f = \sum t^i c_i$ and suppose that f has a and $b^{-1}ab$ as left zeros. Then $\sum a^i c_i = 0$ and $b^{-1}\sum a^i b c_i = \sum (b^{-1}ab)^i c_i = 0$, hence

$$\sum a^i(1 + \lambda b)c_i = 0$$

for all λ in $\mathscr{C}_K(a)$, the centralizer of a in K. It follows that $(1 + \lambda b)^{-1}a(1 + \lambda b)$ is a left zero of f for all $\lambda \in \mathscr{C}_K(a)$, and these elements are all distinct, for if $(1 + \lambda b)^{-1}a(1 + \lambda b) = (1 + \lambda' b)^{-1}a(1 + \lambda' b)$, then $(1 + \lambda b)(1 + \lambda' b)^{-1} = \mu \in \mathscr{C}_K(a)$. If $\mu \neq 1$, then $1 + \lambda b = \mu + \mu \lambda' b$, and since $\mu \neq 1$, it follows that $\mu \lambda' \neq \lambda$ and so $b = (\mu \lambda' - \lambda)^{-1}(1 - \mu)$, which contradicts the fact that $b \notin \mathscr{C}_K(a)$. Hence $\mu = 1$ and so $\lambda' = \lambda$. Now $\mathscr{C}_K(a)$ is infinite, by Th. 4.8, and this gives infinitely many left zeros conjugate to a. ∎

By combining this result with Cor. 4.4, we obtain

COROLLARY 3.4.10. *Let K be a field and f a polynomial of degree n over K. Then the number of left zeros of f in K is either at most n or infinite.* ∎

Exercises

1. Let G be a finite group, H a proper subgroup of h elements and write $(G:H) = n$. Verify that there are at most n conjugates of H in G; deduce that the union of all the conjugates has at most $n(h - 1) + 1$ elements. Hence show that G cannot be covered by all the conjugates of H.

2. Let K be any field. Given $a \in K$ and a non-zero polynomial f over K, show that there exists an integer $r \geqslant 0$ such that $f = (t - a)^r f_1$, where $f_1(a) \neq 0$. Deduce that $f(t)(t - f_1(a)^{-1}af_1(a))$ has as left zeros all the left zeros of f as well as a.

3. Given a field K and $a_1, \ldots, a_n \in K$, not necessarily inconjugate or even distinct, show that there exists a monic polynomial f of degree n with $a_1, \ldots, a_n$ as left zeros. Is f unique?

4. Show that if a is a left zero of a polynomial f, then $f(t) = (t - a)g(t)$, and the left zeros of g are conjugates of the left zeros of f. Hence obtain another proof of Cor. 4.4.

5. Let K be a skew field with finite centre $C \neq K$. Show that the centralizer of any element of K cannot be algebraic over C.

6. (I. N. Herstein) Show that any finite subgroup of a skew field of prime characteristic is cyclic. (Hint. Recall that a finite subgroup of the multiplicative group of a commutative field is cyclic. See also Prop. 9.4 below.)

7. Show that any finite subring of a skew field is a commutative field.

8. Let K be any field. Show that any element of K not in the centre has infinitely many conjugates.

9. Let D be a finite field with centre C. Put $|C| = q$, $[D:C] = n$ and write down the class equation for the finite group $D^\times$. Deduce that the cyclotomic polynomial $\Phi_n(q)$ divides $q - 1$. By taking a complete factorization of Φ over C and comparing absolute values, obtain a contradiction. (This is essentially the Wedderburn–Witt proof of Wedderburn's theorem.)

10. Assuming Wedderburn's theorem (Ex. 9), give a direct proof of Th. 4.8, using Cor. 4.6 and Prop. 9.4.

11. Let D be a field which is a k-algebra and let $E = k(\alpha)$ be a simple algebraic extension of k, generated by an element α with minimal polynomial f over k. Show that $D \otimes_k E$ is an Artinian ring, simple if and only if f is irreducible over the centre C of D. Moreover, if $f = p_1 \ldots p_r$ is a complete factorization over D, then $D \otimes_k E$ is an $r \times r$ matrix ring over a field, in particular, $D \otimes E$ is itself a field precisely when f is irreducible over D.

12. Let E be a field of degree 4 over its centre k and assume char $k \neq 2$. Show that E may be generated by u, v such that $u^2 = \alpha$, $v^2 = \beta$, where $\alpha, \beta \in k$ and $uv = -vu$. (Such an algebra is called a *quaternion algebra*.)

13. Let K be an infinite field with finite centre C. Use Cor. 4.6 to show that each element of K commutes with an element transcendental over C.

14. Let D be a field with an involution * whose fix-point set is a commutative subfield. Show that every element of $D \backslash k$ is quadratic over k.

15. Show that a commutative field k has an extension with an involution having k as fixed field if and only if k either has a separable quadratic extension or is a separable quadratic extension.

16. (Amitsur [56]) Let E be a field which is a k-algebra. Show that if $s \in E$ is transcendental over k, then the elements $(s - \alpha)^{-1}$, where α ranges over k, are linearly independent over k. Deduce that if E is finitely generated as k-algebra and k is uncountable, then E is algebraic over k.

3.5 Pseudo-linear extensions

It seems hopeless at present to try to describe all skew field extensions of finite degree. We shall therefore single out some special classes that are more manageable.

We begin by looking at *quadratic extensions*, i.e. extensions L/K such that $[L:K]_R = 2$. Then for any $u \in L \backslash K$, the pair 1, u is a right K-basis of L. Since every element of L has a unique expression of the form $ux + y$, where $x, y \in K$, we have in particular,

$$au = ua^{\alpha} + a^{\delta} \quad \text{for all } a \in K, \tag{1}$$

and

$$u^2 + u\lambda + \mu = 0, \quad \text{for certain } \lambda, \mu \in K. \tag{2}$$

Here a^{α}, a^{δ} are uniquely determined by a and a calculation as in 2.1 shows that α is an endomorphism and δ an α-derivation of K. Moreover, the structure of L is completely determined by K and (1), (2).

Conversely, let K be any field with endomorphism α and α-derivation δ. In $R = K[t; \alpha, \delta]$ consider a quadratic polynomial $f = t^2 + t\lambda + \mu$. It is easy to write down conditions on λ, μ for f to be right invariant, so that fR is a two-sided ideal, and when these conditions hold, then $L = R/fR$ is a K-ring of right degree 2. It will be a field precisely if the equation (2) has no solution for u in K. We shall carry out the details in 3.6, but for the moment treat a more general case where we keep (1) but modify (2), so as to allow extensions of higher degree.

Thus we shall define a *pseudo-linear* extension of right degree n, with generator u, as an extension field L of K with right K-basis $1, u, \ldots, u^{n-1}$ such that (1) holds and (in place of (2)):

$$u^n + u^{n-1}\lambda_1 + \ldots + u\lambda_{n-1} + \lambda_n = 0, \quad \text{where } \lambda_i \in K. \tag{3}$$

The remarks made earlier show that every quadratic extension is pseudo-linear, but this does not remain true for extensions of higher degree. Our first concern is to obtain a formula for the left degree of a pseudo-linear extension:

PROPOSITION 3.5.1. *Let L/K be a pseudo-linear extension of right degree n. Then the left degree is given by*

$$[L:K]_{\mathrm{L}} = 1 + [K:K^{\alpha}]_{\mathrm{L}} + [K:K^{\alpha}]_{\mathrm{L}}^2 + \ldots + [K:K^{\alpha}]_{\mathrm{L}}^{n-1}, \tag{4}$$

where α is the associated endomorphism. In particular, we have

$$[L:K]_{\mathrm{L}} \geqslant [L:K]_{\mathrm{R}}, \tag{5}$$

with equality if and only if α is an automorphism of K.

Proof. Let us write $L_0 = K$, $L_i = uL_{i-1} + K$ ($i \geqslant 1$); then by an easy induction, $L_i = K + uK + \ldots + u^i K$, hence we have a chain of right K-modules

$$K = L_0 \subset L_1 \subset \ldots \subset L_{n-1} = L,$$

and (4) will follow if we can show that each L_i is a left K-module and

$$[L_i:L_{i-1}]_L = [K:K^\alpha]_L^i. \tag{6}$$

That L_i is a left K-module is clear by induction, using pseudo-linearity. Now let $\{v_\lambda\}$ be a left K^α-basis for K; we claim that the elements

$$u^i v_{\lambda_{i-1}}^{\alpha^{i-1}} \ldots v_{\lambda_1}^\alpha v_{\lambda_0}, \tag{7}$$

where $\lambda_0, \lambda_1, \ldots, \lambda_{n-1}$ range independently over the index set used for $\{v_\lambda\}$, form a basis of $L_i \pmod{L_{i-1}}$. This will prove (6) and hence (4). For any $c \in K$ we have

$$c = \sum c_{\lambda_0}^\alpha v_{\lambda_0} = \sum c_{\lambda_0 \lambda_1}^{\alpha^2} v_{\lambda_1}^\alpha v_{\lambda_0} = \ldots = \sum c_{\lambda_0 \ldots \lambda_{i-1}}^{\alpha^i} v_{\lambda_{i-1}}^{\alpha^{i-1}} \ldots v_{\lambda_0}.$$

Therefore

$$u^i c \equiv \sum u^i c_{\lambda_0 \ldots \lambda_{i-1}}^{\alpha^i} v_{\lambda_{i-1}}^{\alpha^{i-1}} \ldots v_{\lambda_0}$$

$$\equiv \sum c_{\lambda_0 \ldots \lambda_{i-1}} u^i v_{\lambda_{i-1}}^{\alpha^{i-1}} \ldots v_{\lambda_0} \pmod{L_{i-1}},$$

and this shows that the elements (7) span $L_i \pmod{L_{i-1}}$ over K. To prove their independence, suppose that

$$\sum c_{\lambda_0 \ldots \lambda_{i-1}} u^i v_{\lambda_{i-1}}^{\alpha^{i-1}} \ldots v_{\lambda_0} \equiv 0 \pmod{L_{i-1}},$$

Then on retracing our steps we find that all coefficients

$$c_{\lambda_0 \ldots \lambda_{i-1}} = 0,$$

hence the elements (7) are linearly independent and so form a basis of $L_i \pmod{L_{i-1}}$, so (4) is established. Now the remaining assertion is clear from (4). ∎

As in the case of quadratic extensions, it is clear that every pseudo-linear extension is of the form R/fR, where $R = K[t; \alpha, \delta]$ and f is a right invariant polynomial, which is irreducible over K.

When K is of finite degree over its centre, the pseudo-linear extensions of K can still be simplified by Prop. 2.1.4, which tells us that α and δ are now inner. If we assume that $c^\alpha = ece^{-1}$ and $c^\delta = cd - dc^\alpha$ for all $c \in K$, then the formula $cu = uc^\alpha + c^\delta$ becomes

$$cu = uece^{-1} + cd - dece^{-1};$$

hence on writing $u' = ue - de$, we have

$$cu' = u'c,$$

so that for fields of finite degree over the centre, no generality is lost by taking $\alpha = 1$, $\delta = 0$. However, it is still possible to have an endomorphism or a derivation which is not C-linear (but not both, by Th. 2.1.3).

We recall that an extension L/K is called *central* if $L = K \otimes_C E$, where C is the centre of K and E is a commutative field extension of C. It is clear that a pseudo-linear extension is central precisely if the associated endomorphism is 1, the derivation is 0 and there is a generator satisfying a monic equation with central coefficients.

Secondly we define a *binomial* or *pure* extension as a pseudo-linear extension in which the generator satisfies a binomial equation,

$$u^n - \lambda = 0,$$

and whose associated endomorphism α is such that α^n is the inner automorphism induced by λ. In the case of zero derivations it turns out that every pseudo-linear extension can be built up from these two types:

THEOREM 3.5.2. *Let K be a field with an endomorphism α. Then every pseudo-linear extension of K with endomorphism α and zero α-derivation is obtained by taking a central extension, followed by a binomial extension.*

Proof. As we have seen, every pseudo-linear extension with zero derivation has the form R/fR, where $R = K[t; \alpha]$ and f is a right invariant polynomial in R which is irreducible over K. If α has infinite inner order, then the only such polynomial is t and there are no proper extensions, by Prop. 2.2.8. If α has inner order r, say $\alpha^r = I(e)$, then $u = t^r e$ centralizes K and again by Prop. 2.2.8, any invariant irreducible polynomial is a polynomial in u with central coefficients. Let g be such a polynomial and $F = C[u]/(g)$ the commutative field extension of the centre C of K defined by g. It is indeed a field, since g is *a fortiori* irreducible over C. Then $L = K \otimes_C F$ is a central extension of K and the given extension is a binomial extension of L (to which α has been extended by the rule $u^\alpha = ue^{-1}e^\alpha$), with the defining equation $x^r - ue^{-1} = 0$. ∎

If we are looking among pseudo-linear extensions for examples with different left and right degrees, we can concentrate on binomial extensions, by this result. Although as a matter of fact, we shall need to have non-zero derivations too, this suggests that we begin by looking at binomial extensions and our next aim is a result which provides a supply of them. First a commutation rule, which we shall find useful:

LEMMA 3.5.3. *Let n be a positive integer and ω a primitive n-th root of 1. If u, v are indeterminates over $\mathbf{Z}[\omega]$ such that*

$$vu = \omega uv, \tag{8}$$

then

$$(u + v)^n = u^n + v^n. \qquad (9)$$

This formula holds also in characteristic dividing n with suitable ω.

Proof. Over **Z**, when we expand the left-hand side of (9), we obtain a sum of terms which are products of us and vs. The terms of degree i in u include with any product f, all the terms obtained by cyclic permutation of the factors. By moving the last factor u or v to first place we obtain $\omega^i f$, whether that factor is u or v. Hence the sum s_f of all cyclic permutations of f satisfies $s_f = \omega^i s_f$; it follows that $s_f = 0$ for $0 < i < n$. Since $(u + v)^n$ can be written as a sum of such terms s_f, we see that the only terms to survive are u^n and v^n and (9) follows. If the characteristic is p, write $n = p^r m$ where $p \nmid m$ and take ω to be a primitive mth root of 1. Then $(u + v)^m = u^m + v^m$ follows as before, and $u^m v^m = v^m u^m$, hence we obtain again (9). ∎

We can now describe a particular class of binomial extensions.

THEOREM 3.5.4. *Let n be a positive integer and E a field with an endomorphism α and a primitive n-th root of 1, ω say, in the centre C of E and fixed by α (if the characteristic p divides n, say $n = p^r m$, $p \nmid m$, then ω is understood to be a primitive m-th root of 1). Let δ be an α-derivation of E such that*

$$\delta\alpha = \omega\alpha\delta, \qquad (10)$$

and write $L = E(t; \alpha, \delta)$. Then α may be extended to an endomorphism of L, again written α, by putting

$$t^\alpha = \omega t, \qquad (11)$$

and δ can be extended to an α-derivation of L by writing

$$t^\delta = (1 - \omega)t^2. \qquad (12)$$

With these definitions we have

$$ct = tc^\alpha + c^\delta \text{ for all } c \in L. \qquad (13)$$

Further, $\beta = \alpha^n$ is an endomorphism of L, $\varepsilon = \delta^n$ is a β-derivation, and if K is the subfield of L generated by $u = t^n$ over E, then we have

$$K = E(u; \beta, \varepsilon),$$

and L/K is a binomial extension of degree n.

Proof. We first observe that ω lies in the centre of L. For we have $\omega^n = 1$, hence $0 = (\omega^n)^\delta = n\omega^{n-1}\omega^\delta$ and so $\omega^\delta = 0$, with a similar argument, using $\omega^m = 1$, in characteristic p. By hypothesis, $\omega^\alpha = \omega$, so $t\omega = \omega t$ and the conclusion follows. Now a straightforward computation, using (13), shows that (11) determines an endomorphism of $E[t; \alpha, \delta]$ and this clearly extends to L, since α is injective. We now define δ on t by (12); it extends to a unique α-derivation on $E[t; \alpha, \delta]$ and hence it extends to L. Now $\delta\alpha$ and $\omega\alpha\delta$ agree on E by hypothesis; an easy calculation shows that they agree on t, thus we have two (α, α^2)-derivations agreeing on a generating set of L, hence they are equal.

Let us write α_0 and δ_0 for α and δ acting on $E[t; \alpha, \delta]$ by action on the coefficients only. We can write (13) in operator form as

$$\rho(t) = \lambda(t)\alpha_0 + \delta_0,$$

where $\rho(t)$, $\lambda(t)$ indicate right and left multiplication by t. We have $\lambda(t)\alpha_0 = \alpha_0\lambda(t)$ and $\delta_0 \cdot \lambda(t)\alpha_0 = \omega\lambda(t)\alpha_0 \cdot \delta_0$, hence by Lemma 5.3,

$$\rho(t^n) = \lambda(t^n)\alpha_0^n + \delta_0^n. \tag{14}$$

But by (11), α^n fixes t, so $\alpha_0^n = \alpha^n = \beta$. If we define ε on L by

$$ct^n = t^nc^\beta + c^\varepsilon, \tag{15}$$

then ε will be a β-derivation and will agree with δ_0^n on $E[t; \alpha, \delta]$, by (14), in particular it will equal δ^n on E. Hence the E-subring of L generated by t^n is of the form $E[u; \beta, \varepsilon]$, where $u = t^n$, and so the subfield of L generated by t^n over E has the form $K = E(u; \beta, \varepsilon)$. We claim that $1, t$, $\ldots, t^{n-1}$ are right linearly independent over K; for if we had a relation $\sum_0^{n-1} t^i a_i = 0$ ($a_i \in K$), then on multiplying by a common denominator we could take the a_i to be polynomials in u and now a comparison of degrees shows that $a_i = 0$ for $i = 0, 1, \ldots, n - 1$. To show that L/K is pseudo-linear we shall verify that K is mapped into itself by β and ε. Clearly E admits β and ε, while $u^\beta = u$ by (11), with $u = t^n$, and so $u^\varepsilon = 0$, by (15). Finally L/K is binomial because $t^n = u$. ∎

Later, in Ch. 5, we shall use this result to construct extensions of finite right and infinite left degrees. The problem will be to choose E, α and δ so that $[K:K^\alpha]_L = \infty$.

Pseudo-linear extensions may be described as simple extensions with a commutation rule defined by an automorphism. A more general class of finite extensions with a commutation rule defined by a group of automorphisms is formed by the crossed product construction. This can be defined for general skew fields and is sometimes useful; to describe it we first recall the usual crossed product construction.

Given a finite Galois extension of commutative fields F/k, put $\Gamma = \mathrm{Gal}\,(F/k)$; then each element c of the cohomology group $H^2(\Gamma, F^\times)$ defines a crossed product as follows. Let $\{u_\sigma\}$ be a family of symbols indexed by Γ and let A be the right F-space on the u_σ as basis; A may be defined as an F-bimodule by the equations

$$au_\sigma = u_\sigma a^\sigma \quad (a \in F, \sigma \in \Gamma). \tag{16}$$

Now a ring structure may be defined on A by the equations

$$u_\sigma u_\tau = u_{\sigma\tau} c_{\sigma,\tau}, \tag{17}$$

where $c_{\sigma,\tau}$ is the cocycle (factor set) representing $c \in H^2(\Gamma, F^\times)$. Then A is a central simple k-algebra, and this is what one usually understands by a crossed product. Conversely, whenever a central simple k-algebra A has a maximal commutative subfield F which is a splitting field of A and is Galois over k, then A is a crossed product (see e.g. A.3, 7.5). It turns out that it is not necessary to assume F maximal. If we drop this assumption (but restrict to the case of central division algebras for simplicity), we obtain

THEOREM 3.5.5. *Let D be a field with centre k and let F be a commutative subfield containing k, such that F/k is a finite Galois extension, then D is a crossed product over F', the centralizer of F in D, with group $\mathrm{Gal}\,(F/k)$.*

Proof. We observe that F' is a field and by Cor. 3.9, $F'' = F$, hence F is the centre of F'. Let U be the normalizer of $F^\times$ in $D^\times$. Any element u of U induces an automorphism of F/k, hence we have a mapping σ: $U \to \Gamma = \mathrm{Gal}\,(F/k)$, which is clearly a homomorphism. By the Skolem–Noether theorem it is surjective and its kernel is the centralizer of $F^\times$ in $D^\times$, i.e. $F'^\times$. Thus we have a short exact sequence

$$1 \to F'^\times \to U \to \Gamma \to 1.$$

Choose a transversal $\{u_\sigma\}$ of Γ in U and let D_1 be the right F'-space spanned by the u_σ. It is easily checked that $c_{\sigma,\tau} = u_{\sigma\tau}^{-1} u_\sigma u_\tau \in F'$, hence D_1 is a crossed product over F' with group Γ. The linear independence of the u_σ over F' follows by the familiar argument: if

$$\sum u_\sigma a_\sigma = 0, \quad \text{where } a_\sigma \in F',$$

is a shortest non-trivial relation, then $a_1 \neq 0$ and for any $b \in F$,

$$0 = b \cdot \sum u_\sigma a_\sigma - \sum u_\sigma a_\sigma \cdot b = \sum u_\sigma (b^\sigma - b) a_\sigma$$

is a shorter relation and for suitable b is non-trivial, a contradiction.

Now $[D:F']_\mathrm{R} = [F:K]$ by Cor. 3.9, hence $[D:F'] = [D_1:F']$ and so $D_1 = D$; thus D has been expressed as a crossed product over F' with group $\Gamma = \mathrm{Gal}\,(F/k)$. ∎

Exercises

1. Let E be a field with an endomorphism α. Show that α can be extended to $K = E(t; \alpha)$ by defining $t^\alpha = t\lambda$ for any λ in the centre of E such that $\lambda^\alpha = \lambda \neq 0$ and that $K^\alpha = K$.

2. Let F/k be any commutative field extension and let D be a field with k as central subfield. Show that $D \otimes_k F$ is an Ore domain whenever it is an integral domain, and that its field of fractions is then a central extension of D.

3. Let L be a central extension of K with centre F. Show that L is the field of fractions of the Ore domain $K \otimes_C F$, where C is the centre of K.

4. Let D be a field with central subfield k. Show that the following conditions are equivalent: (a) every irreducible polynomial over k remains irreducible over D, (b) for every simple algebraic extension F/k, $D \otimes_k F$ is a field, (c) every monic left factor in $D[t]$ of an element of $k[t]$ lies in $k[t]$. (When (a)–(c) hold, k is said to be *totally algebraically closed* in D, see Cohn and Dicks [80]).

5. Let K be a field and G a finite group of outer automorphisms of K. Use the construction of the text to define for any factor set of G in K a crossed product field. What goes wrong if G contains non-trivial inner automorphisms?

6. Let K be a field and G a group of outer automorphisms with a finite normal subgroup N such that G/N is torsion-free abelian. Define the crossed product algebra of G over K, for a given factor set and verify that it is an Ore domain.

7. (Ikeda [63]) Let K be a field and G a group which is the union of a well-ordered ascending chain of groups G_α ($\alpha < \tau$) of outer automorphisms of K, such that $G_0 = 1$, G_α for each non-limit ordinal $\alpha < \tau$ is an extension of $G_{\alpha-1}$ by a torsion-free abelian group, while at a limit ordinal α, $G_\alpha = \bigcup \{G_\beta | \beta < \alpha\}$. Define a crossed product field D for any factor set of G in K as the limit of an ascending chain $\{D_\alpha\}$, where D_α is the field of fractions of the crossed product algebra of G_α over K. Verify that the centralizer of K in D is the subfield of the centre of K left fixed by G.

8. Let A be a central simple algebra over a field F. If k is a subfield of F such that F/k is a finite Galois extension with group Γ, show that for any factor set $\{c_{\sigma,\tau}\}$ representing $c \in H^2(\Gamma, F^\times)$ there is a central simple algebra B over k which is a crossed product (in the sense of 5.5) over A with factor set $\{c_{\sigma,\tau}\}$.

3.6 Quadratic extensions

Let us now consider quadratic field extensions more closely. As we saw in 3.5, a quadratic extension L/K has a basis 1, u for any $u \in L\backslash K$, and the defining equations are

$$au = ua^\alpha + a^\delta \quad \text{for all } a \in K, \tag{1}$$

$$u^2 + u\lambda + \mu = 0, \tag{2}$$

where $\lambda, \mu \in K$, α is an endomorphism and δ an α-derivation of K.

To construct such an extension, let the field K be given, with an endomorphism α and an α-derivation δ, and form the skew polynomial ring $R = K[t; \alpha, \delta]$. Given $\lambda, \mu \in K$, the quadratic polynomial

$$f = t^2 + t\lambda + \mu \tag{3}$$

defines a quadratic extension of K if and only if (i) the right ideal fR of R is two-sided, and (ii) the residue-class ring R/fR has no zero-divisors. Condition (i) holds precisely when f is right invariant; thus $af \in fR$ for all $a \in K$ and $tf \in fR$. Conditions for this to hold were given in Prop. 2.2.3, but in this case it is just as quick to work out the result directly, using the fact that every polynomial in t is congruent (mod fR) to a unique linear polynomial. Thus we have

$$af = at^2 + at\lambda + a\mu$$

$$= (ta^\alpha + a^\delta)(t + \lambda) + a\mu$$

$$= t^2 a^{\alpha^2} + ta^{\alpha\delta} + ta^{\delta\alpha} + a^{\delta^2} + ta^\alpha\lambda + a^\delta\lambda + a\mu$$

$$\equiv t(a^{\alpha\delta} + a^{\delta\alpha} + a^\alpha\lambda - \lambda a^{\alpha^2}) + a^{\delta^2} + a^\delta\lambda + a\mu - \mu a^{\alpha^2} \pmod{fR}.$$

If $af \in fR$, the coefficients on the right must vanish and we obtain

$$a^{\alpha\delta} + a^{\delta\alpha} = \lambda a^{\alpha^2} - a^\alpha\lambda, \tag{4}$$

$$a^{\delta^2} + a^\delta\lambda = \mu a^{\alpha^2} - a\mu. \tag{5}$$

We note that (4) may be expressed by saying that $\alpha\delta + \delta\alpha$, regarded as operator, is the inner (α, α^2)-derivation induced by λ; similarly (5) expresses the fact that $\delta^2 + \delta\lambda$ is the inner α^2-derivation induced by μ. Next we consider $tf = t^3 + t^2\lambda + t\mu$. We have

$$ft = t^3 + t^2\lambda^\alpha + t(\mu^\alpha + \lambda^\delta) + \mu^\delta;$$

hence

$$tf \equiv t^2(\lambda - \lambda^\alpha) + t(\mu - \mu^\alpha - \lambda^\delta) - \mu^\delta \pmod{fR},$$

$$\equiv t(\mu - \mu^\alpha - \lambda^\delta - \lambda(\lambda - \lambda^\alpha)) - \mu^\delta - \mu(\lambda - \lambda^\alpha) \pmod{fR},$$

and this lies in fR if both coefficients on the right vanish:

$$\lambda^\delta = \mu - \mu^\alpha - \lambda(\lambda - \lambda^\alpha), \tag{6}$$

$$\mu^\delta = \mu(\lambda - \lambda^\alpha). \tag{7}$$

Thus f is invariant precisely when (4)–(7) hold.

We now assume that these equations are satisfied and ask when R/fR is an integral domain (and hence a field). The ring R/fR has zero-divisors if and only if there exist $a, b, c, d \in K$ such that neither a, b nor c, d are both zero and

$$(ta + b)(tc + d) \equiv 0 \quad (\mathrm{mod}\, fR).$$

Such a congruence is possible only if the left-hand side is of degree 2 in t, i.e. a, c are both non-zero. We may therefore replace b by $-ba$ and d by $-dc$. Dividing by c on the right, we thus obtain

$$(t - b)a(t - d) \equiv 0 \quad (\mathrm{mod}\, fR).$$

Now $(\mathrm{mod}\, fR)$ we have

$$(t - b)a(t - d) = tat - bat - tad + bad$$
$$= t^2 a^\alpha + ta^\delta - t(ba)^\alpha - (ba)^\delta - tad + bad$$
$$= t(a^\delta - (ba)^\alpha - ad - \lambda a^\alpha) + bad - (ba)^\delta - \mu a^\alpha.$$

Equating coefficients, we find

$$ad = a^\delta - (ba)^\alpha - \lambda a^\alpha, \tag{8}$$

$$bad = (ba)^\delta + \mu a^\alpha. \tag{9}$$

Here we can substitute for ad from (8) and obtain

$$b(a^\delta - (ba)^\alpha - \lambda a^\alpha) = (ba)^\delta + \mu a^\alpha.$$

Recalling that $(ba)^\delta = b^\delta a^\alpha + ba^\delta$, we can simplify this to

$$b^\delta a^\alpha + \mu a^\alpha + bb^\alpha a^\alpha + b\lambda a^\alpha = 0.$$

Since $a \neq 0$, we have $a^\alpha \neq 0$, and dividing by a^α we find

$$bb^\alpha + b\lambda + b^\delta + \mu = 0. \tag{10}$$

Thus if R/fR has zero-divisors, then (10) has a solution in K. Conversely, if (10) has a solution b in K, we can put $a = 1$ and define d by (8). Then (10) ensures that (9) also holds and by retracing our steps we obtain zero-divisors in R/fR. Hence the solubility of (10) in K is necessary and sufficient for the existence of zero-divisors.

If L/K is generated by u satisfying (1) and (2), then it is also generated by $v = ua + b$, where $a, b \in K$ and $a \neq 0$. The commutation rule for v is obtained by substituting in (1):

$$cv = c(ua + b) = uc^\alpha a + c^\delta a + cb$$

$$= (v - b)a^{-1}c^\alpha a + c^\delta a + cb$$

$$= va^{-1}c^\alpha a + c^\delta a - ba^{-1}c^\alpha a + cb.$$

Thus α is changed by an inner automorphism. If α is kept fixed, then we may take $a = 1$ and in place of c^δ we now have $c^\delta + cb - bc^\alpha$, so δ is modified by an inner α-derivation. The results may be summed up as

THEOREM 3.6.1. (i) *Let L/K be a quadratic extension. Then for any $u \in L\backslash K$, 1, u form a right K-basis of L and there is an endomorphism α of K with an α-derivation δ of K such that*

$$au = ua^\alpha + a^\delta \text{ for all } a \in K, \tag{1}$$

and there exist $\lambda, \mu \in K$ such that

$$u^2 + u\lambda + \mu = 0. \tag{2}$$

Here the endomorphism α is determined up to an inner automorphism of K, and for fixed α, δ is determined up to an inner α-derivation of K.

(ii) *Given a field K with an endomorphism α and an α-derivation δ, there is a quadratic extension of K with right K-basis 1, u and defining equations (1), (2) if and only if $\alpha\delta + \delta\alpha$ is the inner (α, α^2)-derivation induced by λ, $\delta^2 + \delta\lambda$ is the inner α^2-derivation induced by μ, λ^δ and μ^δ are given by (6), (7) respectively and K contains no element b satisfying (10).* ∎

As in the case of the polynomial ring the derivation becomes inner in the extension:

COROLLARY 3.6.2. *If L/K is a quadratic extension defined by (1), (2), then the endomorphism α of K may be extended to an endomorphism $\bar{\alpha}$ of L by putting*

$$u^{\bar{\alpha}} = -\lambda - u,$$

and δ is then the inner $\bar{\alpha}$-derivation induced by u. Further, the endomorphism $\bar{\alpha}$ is an automorphism of L if and only if α is an automorphism.

Proof. For the first point we need only verify that $\bar{\alpha}$ preserves the defining relations (1), (2) of L, i.e. we have to show that

$$a^\alpha(-\lambda - u) = (-\lambda - u)a^{\alpha^2} + a^{\delta\alpha},$$

$$(-\lambda - u)^2 + (-\lambda - u)\lambda^\alpha + \mu^\alpha = 0.$$

The first equation follows by (1) and (4), while the second is a consequence of (2) and (6). Now (1) shows δ to be inner and finally, if α is an automorphism with inverse β, then $\bar\beta$, defined by $u^{\bar\beta} = -\lambda^\beta - u$, is an endomorphism extending β and is inverse to $\bar\alpha$. Conversely, if $\bar\alpha$ is an automorphism of L with inverse $\bar\beta$, then for any $a \in K$, $a^{\bar\beta} \in L$, say

$$a^{\bar\beta} = ua_1 + a_2, \quad \text{where } a_1, a_2 \in K.$$

Hence

$$a = -(\lambda + u)a_1^\alpha + a_2^\alpha;$$

comparing coefficients of u, we find that $a_1^\alpha = 0$, hence $a_1 = 0$ and $a = a_2^\alpha$, which shows α to be an automorphism of K. ∎

In the commutative case a quadratic equation can be simplified by completing the square, at least in characteristic not 2. A similar reduction applies in general and there is now no restriction on the characteristic.

PROPOSITION 3.6.3. *Let L/K be a quadratic extension defined by (1) and (2). If the equation*

$$x + x^\alpha = \lambda \tag{11}$$

has a solution in K, then L/K may be defined by an element v of L satisfying the equation

$$v^2 + v = 0. \tag{12}$$

In particular this holds if $\lambda^\alpha = \lambda$ and char $K \neq 2$.

Proof. Let $x = c$ be a solution of (11) in K. Replacing u by $u + c$, we have $(u + c)^2 = u^2 + uc + cu + c^2 = -u\lambda - \mu + uc + uc^\alpha + c^\delta + c^2$, hence

$$(u + c)^2 = c^2 + c^\delta - \mu.$$

Thus we obtain (12) on putting $v = u + c$, $v = \mu - c^2 - c^\delta$. ∎

We now ask for conditions for a quadratic extension to be Galois. Let L/K again be a quadratic extension generated by $u \in L$ subject to (1) and (2) and assume that it is a Galois extension. This means that there is an automorphism of L over K other than the identity. Suppose that in this automorphism

$$u \mapsto uc + d, \text{ where } c, d \in K, c \neq 0. \tag{13}$$

Then by (1), for any $a \in K$,

$$a(uc + d) = (uc + d)a^\alpha + a^\delta.$$

Since $a(uc + d) = ua^\alpha c + a^\delta c + ad$, we obtain on equating the coefficients of u and of 1,

$$ca^\alpha = a^\alpha c, \tag{14}$$

$$a^\delta(c - 1) = da^\alpha - ad. \tag{15}$$

If $c \neq 1$, then on putting $e = d(c - 1)^{-1}$ and using (14), we can reduce (15) to the form

$$a^\delta = ea^\alpha - ae;$$

thus δ is then inner. If $c = 1$, then (15) states

$$da^\alpha = ad. \tag{16}$$

Now by hypothesis the automorphism defined by (13) is not the identity, so $d \neq 0$ and by (16) α is then inner. Thus for a quadratic Galois extension either α or δ must be inner; hence by a suitable choice of generator we may assume that either $\alpha = 1$ or $\delta = 0$. The precise conditions for L/K to be Galois can now be stated as follows.

THEOREM 3.6.4. *Let L/K be a quadratic field extension with endomorphism α and α-derivation δ, and denote the centre of K by C.*
   (i) *If char $K \neq 2$, then L/K is Galois if and only if δ is inner;*
   (ii) *if char $K = 2$, then L/K is Galois if and only if*
(ii.a) *α but not δ is inner, say $\alpha = 1$, and either the coefficient λ in (1) is not zero or there exists $c \in C^\times$ such that $c^\delta = c^2$, or*
(ii.b) *δ but not α is inner, say $\delta = 0$, and either $\lambda \neq 0$ or there exists $c \in C, c \neq 1$ such that $cc^\alpha = 1$, or*
(ii.c) *both α and δ are inner, say $\alpha = 1$, $\delta = 0$ and $\lambda \neq 0$.*

Proof. (i) If L/K is Galois and char $K \neq 2$, then by the earlier remark, α or δ is inner. But if α is inner, we may take $\alpha = 1$; then by (4), 2δ is the inner derivation induced by λ, so δ is inner in any case. Now when δ is inner, then taking $\delta = 0$, we see from (6), (7), that $\lambda^\alpha = \lambda$, $\mu^\alpha = \mu$ and (5), (4) reduce to

$$\mu a^{\alpha^2} = a\mu, \quad \lambda a^{\alpha^2} = a^\alpha \lambda. \tag{17}$$

If we change the generator of the extension to $v = 2u + \lambda$, then

$$v^2 = 4u^2 + 4u\lambda + \lambda^2 = \lambda^2 - 4\mu.$$

Now $v \mapsto -v$ defines an automorphism of L/K of order 2.

(ii) Suppose now that L/K is Galois and char $K = 2$. We already know that α or δ must be inner and so we may assume that $\alpha = 1$ or $\delta = 0$. In case $\delta = 0$ we see from (17) that α is an automorphism; thus α is an automorphism in any case and by (14), c lies in the centre of K. If we assume further that $\lambda = 0$, then $u^2 + \mu = 0$ and hence

$$
\begin{aligned}
0 = (uc + d)^2 + \mu &= ucuc + ucd + duc + d^2 + \mu \\
&= u^2 c^\alpha c + uc^\delta c + ucd + ud^\alpha c + d^\delta c + d^2 + \mu \\
&= u(c^\delta c + cd + d^\alpha c) + \mu(1 + c^\alpha c) + d^\delta c + d^2.
\end{aligned}
$$

Equating coefficients (and remembering that c is a non-zero element of the centre of K), we find

$$c^\delta + d + d^\alpha = 0, \tag{18}$$

$$\mu(1 + cc^\alpha) + d^\delta c + d^2 = 0. \tag{19}$$

We now consider the cases $\alpha = 1$ and $\delta = 0$ separately.

(ii.a) $\alpha = 1$ but δ is outer. By (15), $c = 1$ and now (18) holds identically, while (19) becomes

$$d^\delta = d^2. \tag{20}$$

Thus there must be a non-zero element d satisfying this equation and by (16) $d \in C$. Conversely, if $\lambda = 0$ and there exists $d \in C^\times$ such that (20) holds, then the map $u \mapsto u + d$ defines an automorphism of order 2 of L/K, because (14), (15), (18), (19) then hold. On the other hand, if $\lambda \neq 0$, then $u \mapsto u + \lambda$ defines the required automorphism, since $\lambda^\delta = 0$ by (6), bearing in mind that $\alpha = 1$.

(ii.b) $\delta = 0$ but α is outer. By (15), $d = 0$, hence $c \neq 1$. Again (18) is satisfied, while (19) becomes

$$cc^\alpha = 1. \tag{21}$$

Therefore this equation must be satisfied by an element of C other than 1. Conversely, when (21) holds for an element $c \neq 1$ in C, then $u \mapsto uc$ defines an automorphism of order 2 of L/K, while for $\lambda \neq 0$ we can again take $u \mapsto u + \lambda$ as our automorphism.

(ii.c) $\alpha = 1$ and $\delta = 0$. Now c and d lie in C, (18) holds identically and (19) becomes

$$\mu(1 + c)^2 + d^2 = 0.$$

Since either $c \neq 1$ or $d \neq 0$, this means that μ must be the square of an element of C, which contradicts the fact that $u \notin K$. Thus we must have $\lambda \neq 0$ in this case. When this is so, we can again take $u \mapsto u + \lambda$ to obtain an automorphism of order 2 of L/K. ■

As a consequence we obtain a result of Jacobson [55] on quadratic Galois extensions:

COROLLARY 3.6.5. *If K is a field of finite degree over its centre and of characteristic not 2, then any quadratic extension of K is Galois.*

For in this case α is inner and now the argument under (i) of Th. 6.4 shows δ to be inner, hence the extension is then Galois. ■

Exercises

1. Let L/K be a quadratic extension defined by (1), (2). Show that λ and μ are fixed under α if and only if they are constant under δ, and when this is so, then $\lambda\mu = \mu\lambda$. In this case show further that when char $K \neq 2$, then L/K may be generated by $v \in L$ such that $v^2 + v = 0$, where $v^\alpha = v$, $v^\delta = 0$; the same holds in characteristic 2 if $\lambda \neq 0$, while for char $K = 2$ and $\lambda = 0$, L/K may be generated by $v \in L$ such that $v^2 + v + v = 0$, where $v^\alpha = v$, $v^\delta = 0$.

2. Let L/K be a quadratic extension defined by (1), (2). Show that when α is an inner automorphism, then δ is inner and the conclusion of Ex. 1 holds.

3. Show that a quadratic extension defined by (1), (2) is central if and only if α and δ are inner.

4. Let L/K be a quadratic Galois extension. Verify that the generator v can be chosen so that the automorphism is $v \mapsto -v$ in characteristic $\neq 2$, and $v \mapsto v + 1$ or $v \mapsto vc$, where $cc^\alpha = 1$ in characteristic 2.

5. Let char $k = 0$ and put $F = k(x)$ with automorphism $\alpha: f(x) \mapsto f(2x)$; if $L = F(t; \alpha)$ and K is the subfield generated by t^2 over F, show that L/K is Galois with outer Galois group.

6. Let char $k = 2$ and put $F = k(x)$ with derivation $\delta: f(x) \mapsto \mathrm{d}f/\mathrm{d}x$. If $L = F(t; 1, \delta)$ and K is the subfield generated over F by $t^2 + t$, show that L/K is Galois with outer Galois group.

7. Let E be a commutative field. Find all quadratic extensions F/E such that F is non-commutative and has an anti-automorphism fixing E.

3.7 Outer cyclic Galois extensions

To illustrate some of the results of this chapter, we shall now determine outer Galois extensions with cyclic Galois group, briefly *outer cyclic extensions*, following Amitsur [48, 54].

Let K be a field with an endomorphism α and an α-derivation δ. We shall often write a' for a^δ and $a^{(n)}$ for a^{δ^n}. The set of constants

$$C = \{a \in K | a' = 0\} \tag{1}$$

forms a subfield of K. We consider the 'differential equation'

$$p(z) = z^{(n)}a_0 + z^{(n-1)}a_1 + \ldots + za_n = 0, \quad \text{where } a_i \in K, a_0 \neq 0, \tag{2}$$

and show that the set of elements of K satisfying (2) form a finite-dimensional C-subspace of K.

THEOREM 3.7.1. *In a field K with endomorphism α, α-derivation δ and with field of constants C defined by (1), the solutions of (2) in K form a left C-space of dimension at most n.*

Proof. The linearity of the solution set is clear from the relation $p(cz) = cp(z)$ for $c \in C$. We shall put $p = \sum \delta^i a_{n-i}$, so that (2) can be written $zp = 0$. The proof proceeds by induction on n, the case $n = 0$ being trivial. Let $n > 0$ and suppose first that $a_n = 0$; then $p = \delta q$, where q has degree $n - 1$ in δ, and by induction, $U = \ker q$ has dimension $\leq n - 1$. Let $u_1, \ldots, u_r$ be left C-independent solutions of (2), where we may take $u_1 = 1$ without loss of generality, because $a_n = 0$. Then $u_2', \ldots,$ u_r' satisfy $zq = 0$ and they are left C-independent, for if $\sum_2^r c_i u_i' = 0$, then $v = \sum c_i u_i \in C$. By the independence of $u_1 = 1, u_2, \ldots, u_r$, we conclude that $v = c_2 = \ldots = c_r = 0$. Hence $r - 1 \leq n - 1$ and so $r \leq n$, as claimed.

In the general case, when $a_n \neq 0$, let u be a solution of (2). If the only solution is 0, then there is nothing to prove, so take $u \neq 0$, and consider the equation $zup = 0$. The coefficient of z is $up = 0$, so its solution space U_0 has dimension $\leq n$. Now $\ker p = \{zu | z \in U_0\} = U_0 u$, and this has the same dimension as U_0, so the solution space of (2) has dimension at most n. ∎

If L/K is a cyclic extension of degree n, then

$$n = [L:K]_L = [L:K]_R \leq |\text{Gal}(L/K)|,$$

by Th. 3.7, with equality when the Galois group is outer. We shall here confine ourselves to outer cyclic extensions. Let σ be a generator of the Galois group and write $\delta = \sigma - 1$; then δ is a σ-derivation, for we have

$$(ab)^\delta = (ab)^\sigma - ab = a^\sigma b^\sigma - ab = (a^\sigma - a)b^\sigma + a(b^\sigma - b)$$
$$= a^\delta b^\sigma + ab^\delta.$$

We also note that the field of δ-constants is just the fixed field under σ. By hypothesis $\sigma^n = 1$, hence

$$h(\delta) = \sum_1^n \binom{n}{v}\delta^v = (\delta + 1)^n - 1 = 0. \tag{3}$$

We first construct a basis relative to which σ is represented by a diagonal matrix:

PROPOSITION 3.7.2. *Let L/K be an outer cyclic extension of degree n and assume that K contains a primitive n-th root of 1, ω say. For any generator σ of the Galois group, L has a right K-basis $a_1, \ldots, a_n$ such that*

$$a_v^\sigma = \omega^v a_v, \quad v = 1, \ldots, n. \tag{4}$$

Proof. Let k be a commutative subfield of K containing ω and consider L as left k-space. Then L is a direct sum of eigenspaces L_v on which σ acts by multiplication by ω^v, and $K = L_0$ by definition. For any non-zero elements $a, b \in L_v$ we have $ab^{-1} \in L_0 = K$, hence $L_v = a_v K$, for some $a_v \in L_v$. Thus each L_v is at most 1-dimensional over K, and it is non-zero because $[L:K]_R = n$. ∎

The hypothesis on ω can be satisfied whenever n is prime to the characteristic of K. In constructing outer cyclic extensions we shall for simplicity assume that ω is central, but this is not essential.

THEOREM 3.7.3. *Let K be a skew field with a central primitive n-th root of 1, ω say. Then there is an outer cyclic extension L/K of degree n and containing ω in its centre if and only if there exists an automorphism α of K and $a \in K$ such that* (i) $\alpha^n = I(a)$, $a^\alpha = a$, $\omega^\alpha = \omega$ *and no lower power of α is inner,* (ii) $t^n a - 1$ *is irreducible in $R = K[t; \alpha]$. When this is so, then $t^n a - 1$ is invariant in R and $L = R/(t^n a - 1)R$, with generating automorphism*

$$\sigma: \sum t^v c_v \mapsto \sum (\omega t)^v c_v.$$

Proof. (i), (ii) just amount to saying that $t^n a - 1$ is central and irreducible (observe that every right invariant element of R is associated to a central element, by Prop. 2.2.2). So if (i), (ii) hold, we have an outer cyclic extension. Conversely, given an outer cyclic extension L/K, we can by Prop. 7.2 find $u \in L$ such that $u^\sigma = \omega u \neq 0$. Then any $c \in K$ satisfies

$$(u^{-1}cu)^\sigma = u^{-1}\omega^{-1}c\omega u = u^{-1}cu,$$

hence $u^{-1}cu \in K$, so $\alpha: c \mapsto u^{-1}cu$ is an automorphism of K, $\omega^\alpha = \omega$ and we have a homomorphism

$$K[t; \alpha] \to L, \quad \text{given by } t \mapsto u,$$

and the generator of the kernel has the form $t^n a - 1$, where a is such that (i), (ii) hold. ∎

If we drop the condition that $\omega^\alpha = \omega$, then $\omega^\alpha = \omega^r$ for some integer r and it is not hard to write down conditions on r for an extension to exist (see Ex. 2). One can also give conditions for an outer cyclic extension of degree n if K merely contains a primitive dth root of 1, where d is a proper factor of n.

As a consequence we have a form of Hilbert's theorem 90; we shall define the norm of $c \in L$ in our cyclic extension L/K by

$$N(c) = cc^\sigma \ldots c^{\sigma^{n-1}}. \tag{5}$$

COROLLARY 3.7.4. *If L/K is an outer cyclic extension of degree n with generating automorphism σ, then for any $c \in L$, the equation*

$$cx^\sigma = x \tag{6}$$

has a non-zero solution in L if and only if

$$N(c) = 1. \tag{7}$$

Proof. It is clear that $c = a(a^\sigma)^{-1}$ satisfies (7) for any $a \in L^\times$. Conversely, if (7) holds, we have $(\sigma\lambda(c))^n = 1$, where $\lambda(c)$ denotes left multiplication by c, for the left-hand side maps x successively to x, cx^σ, $c^\sigma x^{\sigma^2}, \ldots, cc^\sigma \ldots c^{\sigma^{n-1}} x^{\sigma^n} = x$. Thus we have

$$x[(\sigma\lambda(c))^n - 1] = 0.$$

This has the form $xp(\sigma)[\sigma\lambda(c) - 1] = 0$ for some polynomial $p(\sigma)$ of degree $n - 1$. Now $xp(\sigma) = 0$ can be considered as a differential equation (for $\delta = \sigma - 1$) of order $n - 1$, so its solution space has dimension $\leqslant n - 1$, hence there exists $a \in L$ such that $ap(\sigma) = b \neq 0$, and $b(\sigma\lambda(c) - 1) = 0$, i.e. $cb^\sigma = b$, so (6) holds for $x = b$. ∎

In a similar way one can show that the equation $x^\sigma c = x$ has a non-zero solution if and only if $c^{\sigma^{n-1}} \ldots c^\sigma c = 1$.

We also note the following criterion for reducibility:

PROPOSITION 3.7.5. *Let K be a field with an automorphism σ of order n and let ω be a primitive n-th root of 1 in its centre. Then for any $a \in K$, either $t^n - a$ is irreducible in $K[t; \sigma]$ or it splits into factors of the same degree. In particular, if n is prime, then $t^n - a$ is a product of linear factors or irreducible according as*

$$N(x) = xx^{\sigma} \ldots x^{\sigma^{n-1}} = a$$

has a solution or not.

Proof. Let $p = p(t)$ be an irreducible left factor of $t^n - a$; then so is $p(\omega^\nu t)$, for $\nu = 1, \ldots, n - 1$, therefore we can write

$$t^n - a = p_1 \ldots p_r q,$$

where $p_1 = p(t)$ and each p_i is similar to $p(\omega^{\nu_i} t)$ for some ν_i. If r is chosen as large as possible, then each $p(\omega^\nu t)$ is a factor of $p_1 \ldots p_r$, in fact this is their least common right multiple and so is unchanged by the substitution $t \mapsto \omega t$. This means that it is a polynomial in t^n, of positive degree, and a factor of $t^n - a$. Hence it must be $t^n - a$, so $q = 1$ and we have proved the first part.

Now if p has degree d, then $d \mid n$, hence when n is prime, $d = 1$ or n and now the last part follows from the identity

$$t^n - a = (t - b)(t^{n-1} + t^{n-2}b^{\sigma^{n-1}} + \ldots + b^{\sigma}b^{\sigma^2} \ldots b^{\sigma^{n-1}})$$

$$+ bb^{\sigma} \ldots b^{\sigma^{n-1}} - a. \quad \blacksquare$$

We now turn to the case where n is a power of the field characteristic, $n = p^e$, $p = \text{char } K$. We note that in this case (3) reduces to $\delta^n = 0$.

PROPOSITION 3.7.6. *Let L/K be an outer cyclic extension of degree $n = p^e$, where $p = \text{char } K$, with automorphism $\sigma = \delta + 1$ and write $L_\nu = \ker \delta^\nu = \{c \in L \mid c^{(\nu)} = 0\}$. Then each L_ν is a right K-space of dimension ν and*

$$K = L_1 \subset L_2 \subset \ldots \subset L_n = L, \ L_\nu = L'_{\nu+1}(\nu = 1, \ldots, n - 1).$$

Proof. By Th. 7.1, $[L_\nu : K]_R \leqslant \nu$ and for $\nu = n$ we have equality. We shall use induction on $n - \nu$, thus we assume that $L_{\nu+1}$ has a right K-basis $a_0 = 1, a_1, \ldots, a_\nu$, and we claim that $a'_1, \ldots, a'_\nu$ forms a right K-basis for L_ν. If $\sum a'_i c_i = 0$, then $\sum a_i c_i = c \in K$ and by the linear independence of $a_0, \ldots, a_\nu$ we have $c_1 = \ldots = c_\nu = 0$, hence $a'_1, \ldots, a'_\nu$ are linearly independent; they belong to L_ν, so this shows that $L_\nu = L'_{\nu+1}$ and $[L_\nu : K]_R = \nu$. $\quad \blacksquare$

Since $L' = L_{n-1}$, we have the following solubility criterion:

COROLLARY 3.7.7. *In the situation of Prop. 7.6, the equation $x' = a$ ($a \in L$) has a solution in L if and only if $a^{(n-1)} = 0$.* ∎

When v is a power of p, L_v can also be described as the fixed set of an automorphism:

COROLLARY 3.7.8. *In Prop. 7.6, for $v = p^i$ ($i = 0, 1, \ldots, e$) the subspace L_v is the subfield of L fixed by σ^v.*

This follows because $\delta^v = (\sigma - 1)^v = \sigma^v - 1$, when $v = p^i$. ∎

In our outer cyclic extension L/K let us define the *trace* of $a \in L$ as

$$\operatorname{tr} a = \sum_0^{n-1} a^{\sigma^v}.$$

Since

$$\delta^{n-1} = (\sigma - 1)^{n-1} = \frac{(\sigma - 1)^n}{(\sigma - 1)} = \frac{\sigma^n - 1}{\sigma - 1} = \sum_0^{n-1} \sigma^v,$$

it follows that

$$\operatorname{tr} a = a^{(n-1)} \quad \text{for any } a \in L. \tag{8}$$

This formula enables us to prove a normal basis theorem. We recall that a basis of a Galois extension L/K is said to be *normal* if it consists of the conjugates of a suitably chosen element; such an element is said to be *primitive*.

THEOREM 3.7.9 (Normal basis theorem). *Any outer cyclic extension L/K of degree $n = p^e$, where $p = \operatorname{char} K$, has a normal basis, and $a \in L$ is primitive if and only if $\operatorname{tr} a \neq 0$.*

For by (8), $\operatorname{tr} a \neq 0$ if and only if $a \notin L_{n-1}$. Thus for any $a \notin L_{n-1}$ we have $a^{(n-v)} \in L_v \backslash L_{v-1}$ and so $a, a', \ldots, a^{(n-1)}$ form a basis of $L_n = L$. ∎

Let us now determine the extensions of degree p (the case p^e, $e > 1$, follows by repetition, see Amitsur [54] and Ex. 5). We shall write $x^* = x^p - x$, and we also recall the Jacobson–Zassenhaus formula (Jacobson [62], p. 187, (63)) in characteristic p:

$$(x + y)^p = x^p + y^p + \Lambda(x, y), \tag{9}$$

where Λ is a sum of commutators in x and y. It follows that the expression $V(x)$ defined by

$$V(x) = (t + x)^p - t^p = (t + x)^p - t^p - x, \qquad (10)$$

when evaluated in $K[t; 1, \delta]$ is a polynomial in $x, x', \ldots, x^{(p-1)}$, since e.g. $[x, t] = xt - tx = x'$. We first prove an analogue of the reducibility criterion of Prop. 7.5:

PROPOSITION 3.7.10. *Let K be a field of characteristic p with a derivation δ such that $\delta^p = 0$. For any $a \in K$, the polynomial $t^p - a$ is either a product of commuting linear factors in $K[t; 1, \delta]$ or irreducible according as the equation*

$$V(x) + a = 0,$$

where V is as in (10), has a solution in K or not.

Proof. Let h be a monic irreducible factor of $t^p - a$, of degree d say; then the polynomials $h(t + v)$ $(v = 0, 1, \ldots, p - 1)$ are similar to factors of $t^p - a$. Their least common right multiple is invariant under the map $t \mapsto t + 1$, and hence is a polynomial in t^p, so it has degree at least p, but it is a factor of $t^p - a$, hence it must be equal to $t^p - a$. All the $h(t + v)$ are irreducible of the same degree d, hence $d \mid p$ and so either $d = p$ or $d = 1$. If $V(b) + a = 0$, then $(t + b)^p - t^p = V(b) = -a$, hence

$$t^p - a = (t + b)^p = (t + b)((t + b)^{p-1} - 1),$$

and so $t^p - a$ splits into linear factors $t + b + v$ $(v = 0, 1, \ldots, p - 1)$, which clearly commute with each other. Conversely, if $t^p - a$ has a linear factor $t + b$, then

$$(t + b)^p - V(b) - a = t^p - a = (t + b)h(t),$$

hence $V(b) + a$ has $t + b$ as a factor, but it is of degree 0 in t, so $V(b) + a = 0$. ∎

We can now prove an analogue of Th. 7.3, giving outer cyclic extensions in characteristic p:

THEOREM 3.7.11. *A field K of characteristic p has an outer cyclic extension of degree p if and only if there is a derivation δ in K such that (i) δ^p is inner, induced by $a \in K$ with $a^\delta = 0$, but δ itself is outer, and (ii) $V(x) + a = 0$ has no solution in K.*

When this holds, $t^p - a$ is invariant irreducible in $R = K[t; 1, \delta]$ and $L = R/(t^p - a)R$, with generating automorphism

$$\sigma: \sum t^v c_v \mapsto \sum (t + 1)^v c_v.$$

Proof. Again (i), (ii) ensure that $t^p - a$ is central and irreducible, so when these conditions are satisfied, we have an extension.

Conversely, let L/K be an outer cyclic extension of degree p with generating automorphism σ. Then by Prop. 7.6, L has an element y such that $y^\sigma = y + 1$, hence the map $c \mapsto c^\delta = cy - yc$ induces a derivation on K and we have a homomorphism

$$K[t; 1, \delta] \to L \quad \text{with } t \mapsto y.$$

Here $y^p = a \in K$, so δ^p is inner, induced by a, and $a^\delta = 0$, while $V(x) + a = 0$ has no solution in K, by the irreducibility of $y^p - a$ over K. ∎

Exercises

1. Let L/K be an outer cyclic extension with generating automorphism σ. Show that $a, b \in L$ satisfy $c^{-1}ac = b$ for some $c \in L$ if and only if $N(b) = c^{-1}N(a)c$.

2. Let K be a field with an automorphism α and let ω be a primitive nth root of 1 in K such that $\omega^\alpha = \omega^d$. Show that d is a primitive root of 1 (mod n) and find the conditions for an outer cyclic extension of K of degree n to exist.

3. Let **H** be the field of real quaternions and ω a complex primitive nth root of 1 ($n > 2$). Show that the inner automorphism $\sigma = I(\omega)$ satisfies $\sigma^n = 1$, but there are only two non-zero eigenspaces **C** and j**C** (Bergman).

4. Let K be a field of characteristic p with a derivation. Show that if $c \in K$ commutes with all its derivatives, then $V(c) = c^p + c^{(p-1)}$.

5. (Amitsur [54]) Let K/E be an outer cyclic extension of degree $n = p^e > 1$, with generating automorphism σ. Show that K/E can be embedded in an outer cyclic extension L/K of degree p^{e+1} if and only if there is a derivation δ in K and elements $a, b \in K$ such that $\operatorname{tr} a \ne 0$ and $\sigma^{-1}\delta\sigma - \delta$ is the inner derivation induced by a, $b^\delta = 0$ and δ^p is the inner derivation induced by b, $b^\sigma - b = V(a)$ and $V(x) + b = 0$ has no solution in K.

When these conditions hold, verify that $t^p - a$ is an invariant irreducible element in $R = K[t; 1, \delta]$ and $L = R/(t^p - a)R$, with generating automorphism over K given by $t \mapsto t + a$.

6. Show that all cyclic extensions of degree p^e of a perfect commutative field of characteristic p are commutative. (Hint. Use the fact that a commutative perfect field has no non-zero derivation.)

3.8 Infinite outer Galois extensions

For commutative fields there is a theory of infinite algebraic extensions and profinite Galois groups, first developed by Krull [28]; a non-commutative analogue was given by Jacobson [56], whose exposition we follow here.

We shall consider *algebraic* field extensions L/K, i.e. extensions where every element of L is right algebraic over K. Further we shall confine our attention to outer automorphism groups. Clearly every outer automorphism group is an N-group and the associated algebra reduces to the centre. Thus a Galois extension L/K is outer precisely when the centralizer of K in L is just the centre of L. In particular, every commutative Galois extension is outer.

For infinite Galois extensions the Galois group carries a natural topology, which plays an important role. We recall that a topological group G is called *profinite* if it is an inverse limit of finite groups (see A.3, 3.2, p. 84). As a closed subgroup of a direct product of finite groups it is a compact group in the finite topology and it is totally disconnected, i.e. the connected components are single points. Conversely, it can be shown that every compact totally disconnected group is profinite (see e.g. Gruenberg [67]), but we shall not need these details in what follows.

We begin by establishing some basic properties of outer Galois extensions.

PROPOSITION 3.8.1. *Let L/K be an algebraic field extension which is Galois, with Galois group G which is outer. Then G is profinite; further, for any intermediate field E, $K \subseteq E \subseteq L$, the following hold:*

(i) *Distinct homomorphisms of E into L over K are right linearly independent over L and if $[E:K]_L = m$ is finite, then there are precisely m such homomorphisms.*

(ii) *Any finite subset of L is contained in a subfield E which is Galois over K, admits G and is of finite degree*

$$[E:K]_L = [E:K]_R. \tag{1}$$

In particular, (1) holds for any subfield E of L whenever either side is finite.

Proof. We begin by proving (i) and (ii). (i) By Lemma 3.1, if the homomorphisms $s_1, \ldots, s_r$ are linearly dependent over L, then two are equivalent, say $s_1 = s_2 I(c)$ for some $c \in L^\times$. Hence $I(c) \in G$ and since G is outer, $I(c) = 1$ and $s_1 = s_2$. This shows that distinct homomorphisms are right linearly independent over L. Now let $[E:K]_L = m$; then $[\mathrm{Hom}_K(E, L):L]_R = m$ by Th. 2.4. Further, every homomorphism from

E to L over K is induced by the restriction to E of an element of G, by Prop. 3.5, so there are at least m distinct such homomorphisms, $s_1, \dots,$ s_m say. Since they are linearly independent, they form a basis over L and so every homomorphism s is a linear combination: $s = \sum s_i \lambda_i$ ($\lambda_i \in L$). By Lemma 3.1 and because G is outer, s must be equal to some s_i, so there are exactly m such homomorphisms and (i) is proved.

(ii) For any $c \in L$, $K(c)$ has finite right degree over K, so there are only finitely many homomorphisms $K(c) \to L$ and only finitely many conjugates c^s. Hence for any finite family $c_1, \dots, c_r$, the set $\{c_i^s\}$, where s ranges over G, is finite. Let E be the field generated over K by the c_i^s; then every $s \in G$ maps E into itself and is an automorphism, because $s^{-1} \in G$. The set of induced automorphisms of E forms a group H, which like G is outer. It is a finite group, since it is defined by its effect on the c_i^s. Hence E/K is Galois and (1) follows by Cor. 3.8. Now if E is any field between K and L for which one side of (1) is finite, then E is finitely generated over K and we can embed it in a finite Galois extension D such that $[D:K]_L = [D:K]_R$. We also have $[D:E]_L = [D:E]_R$, hence (1) follows by division.

It remains to show that G is profinite. For any finite subset X of L there is a finite Galois extension E_X of K containing X, and its group G_X is a finite homomorphic image of G. For $Y \supseteq X$ we have a homomorphism $G_Y \to G_X$, so the G_X form an inverse system. Its inverse limit is G, because the action of any element of G on a finite set X is represented by an element of G_X. ∎

Later we shall need the fact that homomorphisms from any intermediate field can be extended to automorphisms:

PROPOSITION 3.8.2. *Let L/K be an algebraic Galois extension, with group G which is outer. If E is an intermediate field, $K \subseteq E \subseteq L$, then any homomorphism from E to L over K can be extended to an automorphism, i.e. an element of G.*

Proof. Let $s: E \to L$ be a homomorphism over K and denote by Φ the set of all fields F between E and K such that $[F:K]_L < \infty$. For any $F \in \Phi$ write s_F for the restriction of s to F; by Prop. 3.5 there exists $t \in G$ such that $t_F = s_F$. We now define a subset of G by the equation

$$G_F = \{t \in G \mid t_F = s_F\}.$$

Each G_F is a non-empty closed subset of G; moreover, if $F_1, \dots, F_r \in \Phi$, then there exists $F \in \Phi$ such that $F_i \subseteq F$ for $i = 1, \dots, r$, and clearly

$$G_{F_1} \cap \dots \cap G_{F_r} \supseteq G_F \neq \varnothing.$$

Thus the family $\{G_F\}$ has the finite intersection property, hence by the compactness of G, we have $\bigcap G_F \neq \varnothing$. This means that there exists $t \in G$ such that $s_F = t_F$ for each $F \in \Phi$; but every element of E can be embedded in a finite Galois extension, by Prop. 8.1, hence $s = t_E$, as we wished to show.

To establish the Galois connexion we need to prove a special case first:

LEMMA 3.8.3. *Let L/K be an algebraic Galois extension with group G which is outer. If H is a subgroup of G with K as fixed field, then H is dense in G. In particular, if H is closed, then $H = G$.*

Proof. Let E be an intermediate field of finite degree admitting G and let G_E be the finite group of automorphisms induced by G in E, and write H_E for the group induced by H. Since G_E, H_E are N-groups, we have $H_E^* = G_E^*$ by the finite theory, and this holds for all E, hence $H^* = G^*$. Now any finite set can be embedded in such a field E, hence $\bar{H} = \bar{G} = G$, i.e. H is dense in G, and when H is closed, it must equal G. ∎

Now the main result, giving the Galois connexion, can be stated:

THEOREM 3.8.4. *Let L/K be an algebraic Galois extension with group G which is outer. Then there is a bijection between the closed subgroups H of G and intermediate fields E, $K \subseteq E \subseteq L$:*

$$H \mapsto H^* = \{x \in L \,|\, x^s = x \text{ for all } s \in H\},$$

$$E \mapsto E^* = \{s \in G \,|\, x^s = x \text{ for all } x \in E\}.$$

If $H \leftrightarrow E$ in this correspondence, then L/E is Galois with group H; E/K is Galois if and only if H is normal in G and in that case the group of E/K is G/H.

Proof. The proof is similar to the finite case. Clearly we have $E \subseteq E^{**}$; if this inclusion is proper, take $c \in E^{**} \backslash E$ and let F be a finite Galois extension of K containing c. Since $c \notin F \cap E$ and F/K is Galois, there is an automorphism s of F fixing $F \cap E$ such that $c^s \neq c$, but the number of conjugates c^s of c is finite, say $c_0 = c, c_1, \ldots, c_r$. We define

$$G_i = \{s \in G \,|\, x^s = x \text{ for all } x \in E \cap F, \, c^s = c_i\}, \, i = 1, \ldots, r.$$

Since every automorphism of $F/F \cap E$ arises by restriction from an element of G, it follows that G_i is not empty, and neither is $G_F = G_1 \cap \ldots \cap G_r$; moreover, this set is again closed. For each finite Galois extension F/K such that $c \in F$ we obtain such a closed set G_F and as before we see that the family $\{G_F\}$ has the finite intersection property. By compactness we have $\bigcap G_F \neq \emptyset$ and any element s of $\bigcap G_F$ fixes any $x \in E \cap F$ for any F, while $c^s \neq c$. Since every element of E is contained in some finite Galois extension F, it follows that s fixes E, thus $s \in E^*$ and so s fixes E^{**}, but this is a contradiction, because $c \in E^{**}$, and it follows that $E^{**} = E$.

Next take a closed subgroup H of G and put $E = H^*$. Then L/E is Galois and its group E^* is clearly outer. Since L/K is algebraic, the same holds of L/E and by Lemma 8.3, $H = E^* = H^{**}$. Thus $H^{**} = H$ and L/E is Galois with group H.

For any $s \in G$, $s^{-1}Hs$ fixes E^s, hence $s^{-1}Hs = H$ if and only if $E^s = E$, so E admits all elements of G precisely when H is normal in G. If s_E denotes the restriction of s to E, then the map $s \mapsto s_E$ is a homomorphism from G to $\mathrm{Gal}\,(E/K)$, which is surjective by Prop. 8.1. This homomorphism is easily verified to be open and continuous and its kernel is H, hence $\mathrm{Gal}\,(E/K) \cong G/H$. ∎

Exercises

1. Show that an open subgroup of a profinite group G has finite index and that the intersection of all open subgroups is 1. Deduce that the topology on G is Hausdorff.

2. Show that any profinite group G can be expressed as $G = \varprojlim (G/N)$, where N runs over all open normal subgroups of G.

$3°$. Examine the problems encountered in extending the theory of this section to mixed (not purely outer) Galois groups.

3.9 The multiplicative group of a skew field

So far very little is known about the general structure of skew fields, but there are some results about their multiplicative groups, mainly asserting that quite weak commutativity conditions imply commutativity. The proofs are usually direct calculations with conjugates or commutators, independent of the general theory.

We adopt the usual notation for group commutators:

$$(x, y) = x^{-1}y^{-1}xy.$$

Let K be any field, let $a, b \in K^\times$ and suppose also that $a \neq 1$. We have

$$b^{-1}(a - 1)b = b^{-1}ab - 1.$$

Since $b^{-1}ab = a(a, b)$, we can rewrite this relation as

$$(a - 1)(a - 1, b) = a(a, b) - 1,$$

or also

$$a((a, b) - (a - 1, b)) = 1 - (a - 1, b). \tag{1}$$

Moreover, the two sides of (1) do not vanish when $ab \neq ba$. This formula shows that any element not in the centre lies in the subfield generated by all the commutators. More precisely, suppose that K is not commutative and let b be a non-central element, say $ab \neq ba$. Then by (1), b cannot commute with both (a, b) and $(a - 1, b)$, for if it did, it would also commute with a. So the derived group cannot be contained in the centre and we obtain

THEOREM 3.9.1. *Let K be a field such that the derived group $(K^\times, K^\times)$ is contained in the centre of K. Then K is commutative.* ∎

Another property of fields which holds quite generally was discovered surprisingly late:

THEOREM 3.9.2 (Cartan–Brauer–Hua theorem). *Let D be a field with centre C. If K is a subfield of D admitting all inner automorphisms of D, then either $K \subseteq C$ or $K = D$.*

Proof. Suppose that K is a subfield of D admitting all inner automorphisms but $K \not\subseteq C$, say $b \in K \backslash C$. Take $a \in D$ such that $ab \neq ba$; then $(a, b) = a^{-1}b^{-1}a \cdot b \in K$ and $(a, b) \neq 1$, and similarly, $(a - 1, b) \neq 1$, so both sides of (1) are non-zero and we can solve (1) for a, hence $a \in K$. This shows that any element not commuting with b lies in K. If c commutes with b but a does not, then $a, a + c \in K$ and so $c \in K$. Hence $K = D$, as we had to show. ∎

We next show that the multiplicative group of a skew field cannot be nilpotent. We recall that a group G is said to be *nilpotent* if for some n, $(\ldots ((x_0, x_1), x_2), \ldots, x_n) = 1$ for all $x_0, \ldots, x_n \in G$; the least n is the *nilpotence class* of G. A group G is nilpotent of class at most n if and only if there is a chain of normal subgroups

$$G = G_1 \supset G_2 \supset \ldots \supset G_{n+1} = 1,$$

such that $(x, y) \in G_{i+1}$ for all $x \in G_i$, $y \in G$ (see e.g. A.1, 9.8).

THEOREM 3.9.3 (Hua). *Let K be a field with centre C. Then $K^\times/C^\times$ has trivial centre, hence $K^\times$ is not nilpotent unless K is commutative.*

Proof. Let $b \in K$ and suppose that b corresponds to a non-trivial element in the centre of $K^\times/C^\times$; thus $b \notin C$ but $(a, b) \in C$ for all $a \in K^\times$. So for any $a \neq 0, 1$ in K, (a, b), $(a - 1, b) \in C$ and hence by (1), $a \in C$. It follows that $K = C$, so if K is not commutative, then $K^\times/C^\times$ has trivial centre; in particular, if $K^\times$ is nilpotent, $K^\times/C^\times$ has a non-trivial centre, so K must then be commutative. ∎

We next generalize Wedderburn's theorem by showing that a field K is commutative if every element is of finite multiplicative order. We recall that a group is called *periodic* if all its elements have finite orders. We also recall the following generalization of the well known fact that every finite subgroup of a commutative field is cyclic (see 3.4 above):

PROPOSITION 3.9.4. *Any finite subgroup of a field K of finite characteristic is cyclic.*

Proof. Let G be a finite subgroup and let P be the prime subfield of K. Then the P-algebra generated by G is finite-dimensional and hence is a finite subfield F. By Wedderburn's theorem (see Th. 4.8) F is commutative, and as subgroup of a commutative field, G is cyclic. ∎

THEOREM 3.9.5 (Jacobson). *Let K be a field such that $K^\times$ is periodic. Then K is commutative.*

Proof. Since $2^r = 2$ for some $r > 1$, K is of finite characteristic p. Let C be the centre of K and assume that $C \neq K$. Given $a \in K \backslash C$, since a has finite order, by Cor. 4.6, there exists $b \in K^\times$ such that $b^{-1}ab = a^m \neq a$, hence $ab \neq ba$. The subgroup H generated by a, b contains $\langle a \rangle$ as a normal subgroup and so is an extension of $\langle a \rangle$ by $\langle b \rangle$, hence it is finite and so is cyclic, contradicting the fact that $ab \neq ba$. This contradiction shows that $K = C$, so K is commutative. ∎

The problem of characterizing the multiplicative group of a field has not yet been solved even in the commutative case, but there are some results on the subgroups of fields. We begin by looking at the commutative case and first recall some properties of abelian groups.

Since our groups occur as subgroups of fields, we shall write them multiplicatively. Every abelian group G has a unique subgroup $T(G)$ consisting of all elements of finite order, the *torsion subgroup* of G. If $T(G) = G$, G is a *torsion group*; such a group can be expressed in just one way as a direct product $G = \prod G_p$, where p runs over all primes and G_p, the *p-primary component* of G, consists of all elements of p-power order. If the torsion group G is *locally cyclic*, i.e. all its finitely generated subgroups are cyclic, then its p-primary component G_p is either cyclic of order p^r, where r is a non-negative integer, or of type Z_{p^∞} (the group of all p^rth roots of 1 in $\mathbf{C}$, for all r). Both cases can be described by saying that G_p has order p^α, where $\alpha = r$ or $\alpha = \infty$. Since $G = \prod G_p$, it follows that any locally cyclic torsion group is completely described by the formal product of the orders of its components:

$$N = \prod p_i^{\alpha_i} \ (\alpha_i = 0, 1, \ldots, \text{ or } \infty). \tag{2}$$

Such a product is called a *supernatural number* or *Steinitz number*. Conversely, to every supernatural number there corresponds exactly one locally cyclic group up to isomorphism. We shall denote the group corresponding to N by Z_N and also say that Z_N is of *type N*. If

$$M = \prod p_i^{\beta_i}$$

is a second supernatural number, we say that M divides N and write $M|N$ if $\beta_i \leqslant \alpha_i$ for all i. Clearly Z_M can be embedded in Z_N if and only if $M|N$. We also note that every supernatural number is completely determined by its natural divisors, so symbolically we have

$$N = \lim_{n|N} n.$$

The torsion group of a commutative field can be described as follows:

THEOREM 3.9.6. *Let k be an arbitrary commutative field. Then $T(k^\times)$, the torsion subgroup of the multiplicative group of k, is locally cyclic of type N, where* (i) $2|N$ *if* char $k = 0$, (ii) $N = \lim \{p^m - 1 | m | M\}$ *for some supernatural number M if* char $k = p$.

Proof. Since every finite subgroup of $k^\times$ is cyclic (see 3.4), $T(k^\times)$ is locally cyclic. Now (i) follows because -1 has order 2. To prove (ii), let k be of prime characteristic p. Then the roots of 1 in k together with 0 form a subfield, namely the relative algebraic closure in k of its prime subfield Π, so we may without loss of generality take k to be algebraic over Π. Let us denote by Π_n the extension of degree n of Π; then Π_m is contained in Π_n if and only if $m|n$. We now put

$$M = \lim \{m | \Pi_m \text{ is embedded in } k\},$$

and note that k is completely determined by the supernatural number M. Conversely, every supernatural number M defines an algebraic extension of Π, unique up to isomorphism. Clearly $k^\times$ is then a torsion group and its type N is determined by its finite subgroups, hence

$$N = \lim \{p^m - 1 | m | M\},$$

as we had to show. ∎

The necessary conditions in this theorem are actually also sufficient. For if $\bar{F}_p$ is the algebraic closure of the field F_p of p elements, then $T(\bar{F}_p^\times)$ is of type $\prod \{p_i^\infty | p_i \neq p\}$, because $\bar{F}_p$ contains a primitive l^nth root of 1 for all n and all primes $l \neq p$. In characteristic 0 the type of $T(\mathbf{C}^\times)$ is clearly $\prod p_i^\infty$. Thus we have

COROLLARY 3.9.7. *An abelian torsion group G can be embedded in the multiplicative group of a commutative field if and only if it is locally cyclic. More precisely, every locally cyclic group can be embedded in a commutative field of characteristic 0, while G can be embedded in a field of prime characteristic p if and only if G is locally cyclic of type prime to p.* ∎

We now turn to groups that are not necessarily torsion-free and show how to extend an embedding of their torsion subgroup to one of the whole group.

PROPOSITION 3.9.8. *Let G be an abelian group whose torsion subgroup is contained as a subgroup in $F^\times$, where F is a commutative field. Then there is a commutative extension field E of F whose multiplicative group contains G as a subgroup and such that F is relatively algebraically closed in E; in particular, $E^\times/F^\times$ is torsion-free.*

Proof. Put $T = T(G)$ and take a transversal $\{g_\alpha\}$, where $\alpha \in G/T$, of T in G, with $g_1 = 1$ and with factor set $\{m_{\alpha,\beta}\}$ in $F^\times$, so that

$$g_\alpha g_\beta = g_{\alpha\beta} m_{\alpha,\beta}. \tag{3}$$

Let A be the F-algebra with basis g_α ($\alpha \in G/T$) and multiplication table (3). Since G is abelian, A is associative and commutative, and G/T may be ordered because it is torsion-free. Now consider the ring B of formal series over G/T with coefficients in F, with well-ordered support, using the multiplication (3). This is a field, by Th. 2.4.5, adapted to take account of the multiplication rule (see Ex. 7 of 2.4). Clearly B contains A as a subring; we claim that the subfield E of B generated by A satisfies the conditions of the proposition. By construction $E^\times$ contains G as a subgroup, so it only remains to prove that F is relatively algebraically closed in E. Let $u \in E$ be algebraic over F, with minimal equation

$$u^n + c_1 u^{n-1} + \ldots + c_n = 0 \quad (c_i \in F). \tag{4}$$

We can write u as a power series:

$$u = g_\alpha \lambda_\alpha + \ldots \quad (\lambda_\alpha \in F^\times),$$

where dots denote terms in g_β with $\beta > \alpha$. Suppose that $\alpha < 1$; then the lowest term in (4) is g_{α^n} and this is not cancelled by any other term, a contradiction. Hence $\alpha \geq 1$; if $\alpha > 1$, we again reach a contradiction, by applying the same argument to u^{-1}, therefore $\alpha = 1$. Now $u - \lambda_1$ is again algebraic over F, with lowest term > 1, and this is possible only if $u - \lambda_1 = 0$, hence $u \in F$, as we wished to show. ∎

If we combine this result with Cor. 9.7, we obtain

COROLLARY 3.9.9. *An abelian group can be embedded in the multiplicative group of a commutative field precisely when its torsion subgroup is locally cyclic.* ∎

In a general skew field the possible finite subgroups have been determined by Amitsur [55]; we shall describe the result without giving a proof. By Prop. 9.4 we can limit ourselves to the case of characteristic zero. We recall that a finite group is called *metacyclic* if it consists of an extension of one cyclic group by another. Further, if P is a class of groups, then a *binary P-group* is a central extension of the two-element group by a P-group.

THEOREM 3.A (Amitsur [55]). *A finite group G can be embedded as a subgroup in the multiplicative group of a field if and only if G is*
   (i) *a cyclic group, or*
  (ii) *a certain form of metacyclic group, or*
 (iii) *a certain form of soluble group with a quaternion subgroup, or*
 (iv) *the binary icosahedral group $\mathrm{SL}_2(\mathbf{F}_5)$ of order* 120.

Here the soluble groups under (iii) include the binary octahedral group (of order 48), extensions of a cyclic group of odd order by a generalized quaternion group and the direct product of a quaternion group (order 8) by a metacyclic group occurring in (ii), of odd order m, where 2 has odd order (mod m). The metacyclic groups in (ii) have the property that all their Sylow subgroups are cyclic. A precise description is somewhat complicated, and can be found in the reference quoted or also in Shirvani and Wehrfritz [86].

Exercises

1. By expressing $x(xy - yx)$ as an additive commutator show that any skew field is generated by its additive commutators.

2. Show that Th. 9.1 can be restated as follows: If in a field K, $((a, b), c) = 1$ for all $a, b, c \in K$ such that the commutator is defined, then K is commutative.

3. Verify that a metacyclic extension of C_m by C_n has the presentation

$$\langle a, b; a^m = 1, b^n = a^r, ab = ba^s \rangle,$$

where $s^n \equiv 1 \pmod{m}$, $r(s - 1) \equiv 0 \pmod{n}$, and show that for any m, n, r, s satisfying these conditions a metacyclic group of order mn with this presentation exists.

If the resulting group is G, show that G can be embedded in a field if and only if, writing ω for a primitive mth root of 1 over $\mathbf{Q}$ and α for the automorphism $\omega \mapsto \omega^s$ of $F = \mathbf{Q}(\omega)$, the polynomial $t^n - \omega^r$ is irreducible in $F[t; \alpha]$.

4. Show that a finite ring which is reduced is a direct product of fields. (Hint. Use induction and the fact that every central idempotent $\neq 0, 1$ leads to a direct product decomposition.)

5. Let D be a field with an involution * whose fix-point set is a commutative subfield k. Show that $D^\times$ normalizes k and so k is central in D.

6. Let $K \subseteq L$ be any pair of fields and define N as the normalizer of K in L: $N = \{x \in L^\times | x^{-1}Kx = K\}$. Show that the elements of $N/K^\times$, i.e. the automorphisms of K induced by N modulo inner automorphisms, are linearly independent over K.

7. A field is said to be *critically skew* if it is infinite-dimensional over its centre and every proper subfield is commutative. Show that in a critically skew field the centre is relatively algebraically closed. (Hint. Use the fact that a subfield of finite degree is its own bicentralizer.)

8. Show that for a critically skew field D with centre C the commutation relation is transitive on $D \backslash C$. (Hint. Show that the complements of C in centralizers form a partition of $D \backslash C$.)

9. Show that in a critically skew field D with centre C there exists a pair of conjugate elements generating D over C. (Hint. Observe that any two non-commuting elements generate D and apply Kaplansky's PI-theorem.)

10°. Give an example of a critically skew field.

11°. Let K be a field which is algebraic over its centre C; is K commutative? Show, using Th. 9.5, that this is so when C is finite.

Notes and comments

The Jacobson–Bourbaki correspondence, as a basis for Galois theory in the non-commutative case, was developed by Jacobson [40, 47] and N. Bourbaki (cf. H. Cartan [47]), but a form of Galois theory for simple Artinian rings already occurred in Noether [33], where the notion (but not the name) of N-group is introduced. The theory was further generalized to prime rings by Kharchenko [91].

There are many papers on algebraic equations over skew fields, usually with a substitution rule keeping all the coefficients on one side (as in the text). Cor. 4.4 was first proved by Richardson [27], while Herstein [56] proved Cor. 4.10 and Gordon and Motzkin [65] proved Th. 4.9 (using Th. 4.8 as in the text). The commutativity of finite fields was established by Wedderburn [05]; here it forms part of Th. 4.8, whose proof is modelled on that of Artin [28]. Cor. 4.5 was first obtained by Wedderburn [21], and Prop. 4.7 appears in Bray and Whaples [83]. For a general survey of equations over skew fields see Lawrence and Simons [89].

Pseudo-linear extensions were introduced in Cohn [61''] and Prop. 5.1 was established in Cohn [66'], where Th. 5.4 is also proved. 3.6 essentially follows Cohn [61'']; the actual description of quadratic Galois extensions already occurs in Dieudonné [52], while Cor. 6.5 was proved by Jacobson [55]. 3.7 is based on Amitsur [48, 54], while 3.8 essentially follows Jacobson [56], VII. 6, generalizing the work of Krull [28].

H. Cartan [47] proved Th. 9.2 for finite-dimensional division algebras as an application of the non-commutative Galois theory. A little later Brauer [49] and Hua [49] independently gave a direct proof of the general case. Th. 9.1 and 9.3 are due to Hua [50], who actually showed that the multiplicative group of a skew field cannot be soluble (see 6.4, Ex. 6). Prop. 9.4 is due to Herstein [53] and Th. 9.5 is a special case of the theorem of Jacobson [45], that a ring satisfying $x^n = x$ where $n = n(x) > 1$ is commutative. More generally, Kaplansky [51] showed that for a field K with centre C, if $K^\times/C^\times$ is a torsion group, then $K = C$.

Th. 9.6 is taken from Cohn [62], where Prop. 9.8 is also proved, though in a slightly weaker form ($E^\times/F^\times$ is torsion-free); the strengthened form was obtained by Schenkman [64]. For further properties of multiplicative groups of commutative fields see W. May [72].

Amitsur's theorem 3.A is proved using results on groups with fix-point-free action and the determination of groups with cyclic Sylow subgroups by Zassenhaus [36]. A proof of Th. 3.A along similar lines was obtained independently at about the same time by J. A. Green (unpublished).

Many individual results on subgroups of skew fields have been established. Thus W. R. Scott [57] has generalized Th. 9.3 by showing that $K^\times/C^\times$ has no non-trivial abelian normal subgroups and M. S. Huzurbazar [60] has shown that $K^\times/C^\times$ has no non-trivial locally nilpotent subgroup. G. R. Greenfield [81] has conjectured: In any field K a non-central subnormal subgroup of $K^\times$ contains a non-central normal subgroup. He proves this conjecture when K is finite-dimensional over a commutative local field with finite residue-class field of characteristic $\neq 2$.

At the other extreme J. Tits [72] proved that a full matrix group over a field either is soluble-by-locally-finite or contains a non-cyclic free subgroup, while A. I. Lichtman [78] has shown that if $K^\times$ has a normal subgroup H containing a non-abelian nilpotent subgroup, then H also contains a non-cyclic free subgroup.

Zalesskii [67] has conjectured that a finite group of $s \times s$ matrices over a skew field K of characteristic zero has a metabelian subgroup of index bounded in terms of s alone (generalizing the classical theorem of Jordan and Schur which asserts this for commutative fields and an abelian subgroup, see e.g. Wehrfritz [73]), and he has proved it when G is soluble. The general case has now been proved by Hartley and Shahabi Shojaei [82].

4

Localization

This chapter deals with the formation of fractions in general rings. In the commutative case a necessary and sufficient condition for the existence of a field of fractions is the absence of zero-divisors (and the condition $1 \neq 0$), and the construction as fractions ab^{-1} is well known. As we saw in 1.3, the same method of construction still applies in Ore domains, though the verification is a little more involved. In the general case the difficulties are both theoretical – the criterion for embeddability is quite complicated and cannot be stated as an elementary sentence – and practical – a sum of fractions cannot generally be brought to a common denominator. The practical problem is overcome by inverting matrices rather than elements. After some general remarks on epimorphisms and localizations in 4.1, we go on to show in 4.2 that all elements of the field of fractions (if one exists) can be found by solving matrix equations, and something like a normal form (in the case of firs) is presented in 4.7. On the theoretical side we shall meet a criterion for a ring to possess a field of fractions in Th. 4.5, but what turns out to be more useful is a sufficient condition for a ring to have a universal field of fractions (Th. 5.3); the latter, when it exists, is unique up to isomorphism, unlike a field of fractions, of which there may be many, e.g. for a free algebra.

The main step is the construction of a field from a ring by inverting certain matrices, and this occupies 4.3, while 4.4 examines the sets of matrices inverted in forming such fields. The result, giving a correspondence of epic R-fields and prime matrix ideals (Th. 4.1), is remarkably similar in appearance to the theorem in the commutative case describing R-fields in terms of the prime spectrum of R. In 4.5 we examine conditions ensuring the existence of a universal field of fractions. Here the main result, which will be much used in later chapters, is that every semifir has a universal field of fractions (Cor. 5.9).

In the construction of localizations the inner rank of a matrix plays a crucial role, and 4.6 is devoted to a general study of rank functions on rings, and the connexion with fields of fractions.

4.1 The category of epic R-fields and specializations

Let R be any ring and consider the category of R-rings. We recall that in any category an *epimorphism* is a map f such that $fg = fg'$ implies $g = g'$. Whereas in the category of groups the epimorphisms are just the surjective homomorphisms, for rings this is no longer so, for example the embedding $\mathbf{Z} \subseteq \mathbf{Q}$ is easily seen to be an epimorphism. The following equivalent ways of expressing the condition are often useful (cf. Knight [70]).

PROPOSITION 4.1.1. *Let* $f: R \to S$ *be any homomorphism of rings. Then the following conditions are equivalent:*

(a) $f: R \to S$ *is an epimorphism,*

(b) *in the S-bimodule $S \otimes_R S$ we have $x \otimes 1 = 1 \otimes x$ for all $x \in S$,*

(c) *the multiplication map $x \otimes y \mapsto xy$ in $S \otimes_R S \to S$ is an isomorphism,*

(d) $S \underset{R}{*} S \cong S$ *under a natural isomorphism.*

Proof. (a) $\Rightarrow$ (b). We form the split null extension $S \oplus (S \otimes_R S)$ with the multiplication

$$(x, u)(y, v) = (xy, xv + uy),$$

and consider the maps from S to $S \oplus (S \otimes_R S)$ given by

$$x \mapsto (x, 1 \otimes x), x \mapsto (x, x \otimes 1), \quad x \in S.$$

They are easily verified to be ring homomorphisms and their restrictions to R agree, hence they are equal, and so $x \otimes 1 = 1 \otimes x$.

(b) $\Rightarrow$ (c). The multiplication homomorphism maps $\sum x_i \otimes y_i$ to $\sum x_i y_i$, but when (b) holds, we have $\sum x_i \otimes y_i = \sum x_i y_i \otimes 1 = 1 \otimes \sum x_i y_i$, hence (c).

(c) $\Rightarrow$ (d). $S \underset{R}{*} S$ has a filtration (S_n), where $S_1 = S$, $S_{n+1} = S_n \otimes_R S$; when (c) holds, we see by induction on n that $S_n \cong S$, whence $S * S \cong S$.

(d) $\Rightarrow$ (a). By definition, $S \underset{R}{*} S$ is the pushout of the map $f: R \to S$ with itself. If $g_i: S \to T$ $(i = 1, 2)$ are two homomorphisms such that $fg_1 = fg_2$, then by the definition of the pushout there is a homomorphism $h: S * S \to T$ such that $g_i = \alpha_i h$, where α_i is the pushout map $S \to S * S$ mapping S on the ith factor. Since $S * S \cong S$, we have $\alpha_1 = \alpha_2 = 1_S$, hence $g_1 = g_2$ and (a) follows. ∎

We shall be interested in R-rings that are fields, R-*fields* for short. If K is an R-field which is generated, as a field, by the image of R, we call R an *epic R-field*. This terminology is justified by the fact that the canonical map is then an epimorphism, and only then:

PROPOSITION 4.1.2. *A homomorphism f from a ring R to a field K is an epimorphism if and only if K is the field generated by* im f.

Proof. Let K be the field generated by im f. Then $K' = \{a \in K | a \otimes 1 = 1 \otimes a\}$ clearly contains im f and if $a \in K'$ and $a \neq 0$, then $a^{-1} \otimes 1 = a^{-1} \otimes aa^{-1} = a^{-1}a \otimes a^{-1} = 1 \otimes a^{-1}$, so $K' = K$, (b) holds and hence f is an epimorphism. Conversely, if the subfield generated by im f is a proper subfield E say, and $u_1 = 1$, u_i is a right E-basis of K, then $K \otimes_E K = \sum u_i \otimes K = K \oplus \sum_{i \neq 1} u_i \otimes K$ and this is not isomorphic to K, so f is not an epimorphism. ∎

Our object is to make the epic R-fields, for a given ring R, into a category, and we must find the morphisms. To take R-ring homomorphisms would be too restrictive, as all maps would then be isomorphisms. For if $f: K \to L$ is such a map between epic R-fields, then f is injective, because its kernel is a proper ideal in a field, and im f is a subfield of L containing the image of R, hence im $f = L$, because L was epic, so f is an isomorphism.

To obtain a workable notion of morphism let us define a *local homomorphism* between any rings A, B as a homomorphism $f: A_0 \to B$, whose domain A_0 is a subring of A, which maps non-units to non-units. If B is a field, this means that the non-units in A_0 form an ideal, viz. ker f, so that A_0 is then a local ring. We recall that a *local ring* is a ring A_0 in which the non-units form an ideal $\mathfrak{m}$; the quotient ring $A_0/\mathfrak{m}$ is then a field, called the *residue-class field* of A_0. Of course when we are dealing with R-rings, a local homomorphism is understood to have a domain which includes the image of R.

Let f be a local homomorphism between R-fields K, L. If its domain is K_0, then by what has been said, K_0 is a local ring with residue-class field $K_0/\text{ker } f$; this is isomorphic to a subfield of L containing the image of R, so if L is an epic R-field, we have

$$K_0/\text{ker } f \cong L. \tag{1}$$

Two local homomorphisms between A and B are said to be *equivalent* if there is a subring of A on which both are defined, and on which they agree and again define a local homomorphism. This is easily seen to be an equivalence; an equivalence class of local homomorphisms between epic R-fields is called a *specialization*. It can easily be checked that the composition of two specializations is again a specialization, i.e. the composition of mappings, when defined, is compatible with the equivalence defined earlier. In this way we obtain for each ring R, a category $\mathcal{F}_R$ of epic R-fields and specializations. It is clear that the number of epic R-fields is bounded by $\max \{|R|, \aleph_0\}$, so $\mathcal{F}_R$ is a small category, and as we shall see later, there is at most one specialization between two epic R-fields, so that $\mathcal{F}_R$ can be represented as a partially ordered set.

At first sight it seems as if there may be several specializations between a given pair of epic R-fields. Thus let $R = k[x, y]$ be the commutative polynomial ring in x and y over a field, take $F = k(x, y)$, its field of fractions with the natural embedding and $E = k$, with the homomorphism $R \to E$ leaving k fixed and mapping x, y to 0. We obtain a specialization from F to E by defining a homomorphism $\alpha: k[x, y] \to E$ in which $x\alpha = y\alpha = 0$. Let F_0 be the localization of $k[x, y]$ at the maximal ideal (x, y); then α can be extended in a natural way to F_0. Now there are local homomorphisms from F to E that are defined on larger subrings than F_0, for we can 'specialize' rational functions $\phi(x, y)$ so that x/y takes on a specified value in k. In this way we obtain many different local homomorphisms from F to E; however, they all agree on F_0, so that there is just one specialization from F to E.

Of course for some rings R there will be no R-fields at all. For example, when $R = 0$, or for a less trivial example, take R to be any simple ring with zero-divisors, say a matrix ring over a field. For then any homomorphism $R \to K$ must be injective and this is impossible when K is a field. Even integral domains R without R-fields exist, e.g. if R is a ring without invariant basis number (see 1.4) and an integral domain. Then any R-ring is again without IBN and so cannot be a field.

What can we say about R-fields in the commutative case? Let R be a commutative ring and K an epic R-field; then K is of course also commutative, being generated by a homomorphic image of R. The kernel $\mathfrak{p}$ of the natural mapping $R \to K$ is a prime ideal and K can be reconstructed in two ways from R and $\mathfrak{p}$. Firstly, we can form $R/\mathfrak{p}$, an integral domain, because $\mathfrak{p}$ is prime, and now K is obtained as the field of fractions of $R/\mathfrak{p}$. Secondly, instead of putting the elements *in* $\mathfrak{p}$ equal to 0, we can make the elements *outside* $\mathfrak{p}$ invertible, by forming the localization $R_\mathfrak{p}$. This is a local ring and its residue-class field is isomorphic to K. The situation can be illustrated by the accompanying commutative diagram.

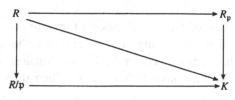

The two triangles correspond to the two methods of constructing K. The route via the lower triangle is perhaps more familiar, but unfortunately it does not seem to generalize to the non-commutative case. We therefore turn to the upper triangle. Even this cannot be used as it stands, for as we have seen in 2.1, the field of fractions need not be unique, which means that in general an epic R-field will not be determined by its kernel alone.

To describe an epic R-field we need more than the elements which map to zero, we need the matrices which become singular; over a field that term has an unambiguous meaning (see 1.4). Given any R-field K, with natural map $\lambda: R \to K$, by the *singular kernel* of K (or of λ), written Ker λ, we understand the collection of all square matrices over R, of all orders, which map to singular matrices over K. If $\mathcal{P}$ is the set of all such matrices, then we can define a localization $R_{\mathcal{P}}$, analogous to $R_\mathfrak{p}$ in the commutative case, as follows. In 1.3 we met the notion of a universal S-inverting ring; we shall need the corresponding construction when S is replaced by a set of matrices over R.

Let R be a ring and Σ a set of matrices over R. A homomorphism $f: R \to R'$ is called Σ-*inverting* if for each $A = (a_{i\alpha}) \in \Sigma$, the image $Af = (a_{i\alpha}f)$ is an invertible matrix over R'. Here we need not limit ourselves to square matrices; A can be of arbitrary shape, although for homomorphisms to fields only square matrices will play a role. Now we have the following analogue of Prop. 1.3.1:

THEOREM 4.1.3. *Given a ring R and a set Σ of matrices over R, there exists a ring R_Σ and a homomorphism $\lambda: R \to R_\Sigma$ which is universal Σ-inverting, in the sense that the images under λ of the members of Σ are invertible over R' and every Σ-inverting homomorphism from R to another ring can be factored uniquely by λ.*

Proof. For every $m \times n$ matrix $A = (a_{i\alpha})$ in Σ we choose mn symbols $a'_{\alpha i}$ which we adjoin to R, with defining relations, in matrix form, writing $A' = (a'_{\alpha i})$:

$$AA' = I_m, \quad A'A = I_n. \tag{2}$$

The resulting ring is denoted by R_Σ and is called the *universal Σ-inverting ring* or the *localization at Σ* of R. Clearly the natural homomorphism $\lambda: R \to R_\Sigma$ is Σ-inverting and as in Prop. 1.3.1 we can show that for every Σ-inverting homomorphism $f: R \to R'$ there is a unique homomorphism $f': R_\Sigma \to R'$ such that $f = \lambda f'$. ∎

We can now describe the construction of an epic R-field in terms of its singular kernel. Let K be an epic R-field, $\mathcal{P}$ its singular kernel and Σ the complement of $\mathcal{P}$ in the set of all square matrices over R. Thus Σ consists of all square matrices over R which become invertible over K. Then the localization R_Σ is a local ring, with residue-class field K. We shall soon see a proof of this fact (Th. 3.5), but we note that it does not solve our problem yet. For we would like to know when a collection of matrices is a singular kernel, just as we can tell when a collection of elements of R is a prime ideal. In fact we shall be able to characterize singular kernels in much the same way in which kernels of R-fields in the commutative case are characterized as prime ideals.

Exercises

1. Carry out the details of the proof that for any ring R the epic R-fields and specializations form a category.

2. Show that if R is a commutative ring and $R \to S$ is an epimorphism, then S is also commutative.

3. Let $f: R \to S$ be a local homomorphism between two rings. Show that the domain of f is a local ring if and only if S is a local ring.

4. Let R be a ring with R-fields K, L and M, and let f, g be specializations from K to L and from L to M, with domains K_0 and L_0 respectively, and define $K_1 = L_0 f^{-1} \cap K_0$, $f_1 = f | K_1$. Verify that $f_1 g$ is a local homomorphism from K_1 to M and so defines a specialization from K to M.

5. Let $A = \prod A_\lambda$ be a direct product of a family of rings. Show that each specialization f to an A-field K is a product of specializations $f_\lambda: A_\lambda \to K$.

6. Let $f: A \to B$ be a ring homomorphism. If A is a local ring, show that f is local if and only if the induced homomorphism of matrix rings $\mathfrak{M}_n(A) \to \mathfrak{M}_n(B)$ (for any n) is local.

4.2 The matrix representation of fractions

To obtain a description of the elements of the localization R_S in the commutative case it is usual to assume S to be multiplicative. In the same way we need to restrict Σ to find a convenient means of describing the elements of R_Σ. A set Σ of matrices over R is called *upper multiplicative* if $1 \in \Sigma$ and with $A, B \in \Sigma$ the matrix $\begin{pmatrix} A & C \\ 0 & B \end{pmatrix}$ is in Σ, for any C of appropriate size. *Lower multiplicative* sets are defined similarly, using $\begin{pmatrix} A & 0 \\ D & B \end{pmatrix}$. Given a Σ-inverting homomorphism $f: R \to R'$, we define the *Σ-rational closure* of f as the set of all entries of all matrices $(Af)^{-1}$ for

$A \in \Sigma$. With these definitions we have the following description of the Σ-rational closure:

THEOREM 4.2.1. *Let R be any ring, Σ an upper multiplicative set of matrices over R and $f: R \to R'$ a Σ-inverting homomorphism. Then for any $x \in R'$ the following conditions are equivalent:*

(a) *x lies in the Σ-rational closure of R in R',*

(b) *x is a component of the solution u of an equation*

$$Au - e_j = 0, \quad A \in \Sigma^f, \tag{1}$$

(c) *x is a component of the solution u of*

$$Au - a = 0, \quad A \in \Sigma^f, \tag{2}$$

where a is a column over im f,

(d) *$x = bA^{-1}c$, where $A \in \Sigma^f$, b is a row and c a column over im f.*

Moreover, the Σ-rational closure of R in R' is a subring of R' containing im f.

Proof. If (a) holds, say x is the (i, j)-entry of A^{-1}, then x is the ith component of the solution of (1), hence (b) holds. Now (c) is a special case of (b), and if (c) holds, then $u_i = e_i^T A^{-1} a$, which establishes (d). Finally, when (d) holds, then

$$\begin{pmatrix} 1 & b & 0 \\ 0 & A & c \\ 0 & 0 & 1 \end{pmatrix}^{-1} = \begin{pmatrix} 1 & -bA^{-1} & bA^{-1}c \\ 0 & A^{-1} & -A^{-1}c \\ 0 & 0 & 1 \end{pmatrix},$$

where the matrix whose inverse is taken is again in Σ, because Σ is upper multiplicative. This shows (a)–(d) to be equivalent.

Let $R_\Sigma(R')$ be the Σ-rational closure of R in R'; it contains im f, because any $c \in \text{im } f$ satisfies the equation $1 \cdot u - c = 0$, which is of the form (2). To prove the ring property, suppose that u_i is the ith component of the solution of (2) and v_j is the jth component of the solution of $Bv - b = 0$, then $u_i - v_j$ is the ith component of the solution of

$$\begin{pmatrix} A & C \\ 0 & B \end{pmatrix} w - \begin{pmatrix} a \\ b \end{pmatrix} = 0,$$

where C has for its jth column the ith column of A and the rest 0. Next $u_i v_j$ is the ith component of the solution of

$$\begin{pmatrix} A & C \\ 0 & B \end{pmatrix} w - \begin{pmatrix} 0 \\ b \end{pmatrix} = 0,$$

where C has as its jth column $-a$ and the rest 0. This shows that $R_\Sigma(R')$ is a ring containing im f, as claimed. ∎

Let R be any ring and K an epic R-field. As we saw in Th. 2.1, any element $p \in K$ can be expressed in the form

$$p = c - uA^{-1}v, \text{ where } c \in R, u \in R^n, v \in {}^nR, A \in R_n, \tag{3}$$

and A is not in the singular kernel of K. Thus p is completely determined by the block

$$\alpha = \begin{pmatrix} u & c \\ A & v \end{pmatrix}. \tag{4}$$

Such a block will be called *admissible* for K or *K-admissible*. Thus the elements of K may be described either in block form, as in (4), or in equational form, as in (2).

Sometimes a different notation is convenient for the system (2), where we now take $(-a, A)$ as our basic matrix. Omitting the reference to f, for simplicity, we shall write our system with a matrix A of index 1 in the form

$$Au = 0, \quad A \in {}^mR^{m+1}. \tag{5}$$

The columns of A are indicated by subscripts:

$$A = (A_0 \quad A_1 \quad \dots \quad A_m),$$

and we also write A_∞ for A_m and $A_* = (A_1 \quad \dots \quad A_{m-1})$. We shall call (5) an *admissible system* and A an *admissible matrix* of order m for the element p of R' if (5) has a unique solution $u \in {}^{m+1}R'$ normalized by the condition $u_0 = 1$, and $u_\infty = p$. The $m \times m$ matrix formed by the last m columns of A, $(A_* \quad A_\infty)$, is called the *denominator*, the matrix of the first m columns, $(A_0 \quad A_*)$ the *numerator* and A_* the *core* of p in the representation (5) and we write $u = (1 \quad u_* \quad p)^T$. These matrices depend not merely on p but on the representation (5). Thus a system (5) is admissible precisely when its denominator is invertible over R'.

For reference we note that if A, B are admissible matrices for p, q respectively, then we have as admissible matrices for $p - q$, pq

$$\begin{pmatrix} B_0 & B_* & B_\infty & 0 & 0 \\ A_0 & 0 & A_\infty & A_* & A_\infty \end{pmatrix} \text{ and } \begin{pmatrix} B_0 & B_* & B_\infty & 0 & 0 \\ 0 & 0 & A_0 & A_* & A_\infty \end{pmatrix}$$

$$\tag{6}$$

respectively. As (6) shows, we need to assume Σ lower multiplicative with the present convention.

We note that a result similar to Th. 2.1 holds for matrices:

PROPOSITION 4.2.2. *Let R be a ring, Σ a lower multiplicative set of matrices over R and $f: R \to R'$ a Σ-inverting homomorphism. Then for any $m \times n$ matrix P over $R_\Sigma(R')$ there exist $r \geq 0$ and an $(r + m) \times (n + r + m)$ matrix $A = (A_0 \quad A_* \quad A_\infty)$ over $\operatorname{im} f$, $u = (I \quad U \quad P)^T$ over R' such that*

$$Au = 0, \quad (A_* \quad A_\infty) \in \Sigma^f. \tag{7}$$

Here A_0, A_*, A_∞ all have $r + m$ rows and n, r, m columns respectively.

Proof. Consider an $m \times n$ matrix P with a single non-zero entry, say $P = p \oplus 0$. If $Cu = 0$ is an admissible system for p, then we have as an admissible system for the matrix P

$$\begin{pmatrix} C_0 & 0 & C_* & C_\infty & 0 \\ 0 & 0 & 0 & 0 & I_{m-1} \end{pmatrix} \begin{pmatrix} 1 & 0 \\ 0 & I_{n-1} \\ u_* & 0 \\ p & 0 \\ 0 & 0 \end{pmatrix} = 0.$$

We complete the proof by showing that the set of matrices determined by a system (7) is closed under addition. If P', P'' are determined by the matrices A', A'' respectively, then $P = P' + P''$ is determined by the system

$$\begin{pmatrix} A_0' & A_*' & A_\infty' & 0 & 0 \\ A_0'' & 0 & -A_\infty'' & A_*'' & A_\infty'' \end{pmatrix} \begin{pmatrix} I \\ U' \\ P' \\ U'' \\ P \end{pmatrix} = 0,$$

hence every matrix over $R_\Sigma(R')$ is so determined, by induction. ∎

In the commutative theory we have Cramer's rule, giving an explicit formula for the solution of the matrix equation (2). If we write $A^{(i)}$ for the matrix obtained from A in (2) by replacing its ith column by a, then *Cramer's rule* states that the ith component u_i of u is given by the formula

$$u_i = \frac{\det A^{(i)}}{\det A}.$$

In the general case we no longer have determinants, but there still is a form of Cramer's rule, using now the form (5):

PROPOSITION 4.2.3 (Cramer's rule). *Given an admissible system*

$$Au = 0, \tag{8}$$

for an element $p = u_\infty$, p is stably associated to its numerator; hence p is (left/right) (regular/invertible) if and only if the corresponding property holds for its numerator.

Proof. The equation (8) can be written $A_0 + A_* u_* + A_\infty p = 0$, hence

$$(A_* \quad -A_0) = (A_* \quad A_\infty) \begin{pmatrix} I & 0 \\ 0 & p \end{pmatrix} \begin{pmatrix} I & u_* \\ 0 & 1 \end{pmatrix}. \tag{9}$$

Now the denominator $(A_* \quad A_\infty)$ is invertible, hence p is stably associated to $(A_* \quad -A_0)$, which, except for a column permutation, is the numerator. ∎

The following form of this rule is often useful:

PROPOSITION 4.2.4. *Let R be a ring, Σ a lower multiplicative system of matrices and R_Σ the universal localization. If R_Σ has unbounded generating number, then every invertible matrix P over R_Σ is stably associated to a full matrix over R.*

Proof: By Cramer's rule in matrix form, P is stably associated to $(A_* \quad -A_0) = A^0$ say. Since P is invertible over R_Σ, so is A^0, hence it is full over R_Σ, by UGN (see Prop. 1.4.3), and *a fortiori* it is full over R. ∎

Exercises

1. Find admissible matrices for xy^{-1}, $x^{-1}y$, $xy^{-1}z$, $x^{-1}yz^{-1}$, $xy^{-1} + zt^{-1}$ over the free algebra $k\langle x, y, z, t \rangle$.

2. Show that over a left Ore domain every element of the field of fractions is given by an admissible system of order 1.

3. Given a ring homomorphism $R \to R'$, let $A \in {}^nR^{n+1}$ be an admissible matrix for $p \in R'$. Show that for any $n \times n$ matrix P over R which is right regular over R', PA is again admissible for p. Similarly for AQ^*, where $Q^* = 1 \oplus Q \oplus 1$ and Q is left regular over R'.

4. (N. G. Greenwood) Show that for any Σ-localization $R \to R_\Sigma$ and any subset I of R the set of solutions of admissible equations with matrices $A = (A_0 \quad A_* \quad A_\infty)$, where the entries of A_0 are in I form a left ideal of R_Σ.

5. Let R be a semifir with a Φ-inverting homomorphism to a field U, where Φ is the set of all full matrices (see 4.5). Given $p, q \in U$ with matrices A, B which are left prime, with cores that are right prime, if $A \cdot B$ is the matrix for pq (as in (6)),

show that $A \cdot B$ is left prime unless the denominator of A and the numerator of B have non-unit left factors that are similar (an $m \times n$ matrix is *left prime* if it has no non-unit left $m \times m$ factor; similarly for right prime).

4.3 The construction of the localization

The localization of a ring R at a set Σ of matrices was defined in 4.1 in terms of a presentation, but we shall want to have a more concrete construction that is easier to handle. The main problems that arise are the following:

(i) *Find the kernel of the natural map $R \to R_\Sigma$, or at least find conditions for this map to be injective,*

(ii) *find the singular kernel of the natural map $R \to R_\Sigma$ and its relation to Σ.*

Since our main interest is in R-fields, there is a third problem:

(iii) *Find conditions for R_Σ to be a local ring, and when they hold, find conditions for the natural map from R to the residue-class field of R_Σ to be injective.*

We shall construct the elements of R_Σ in the form $p - bA^{-1}c$, where p is an element, b a row, c a column and A a matrix in Σ. Since A is invertible over R_Σ, we can regard $p - bA^{-1}c$ as an element of R_Σ, but we shall use a stably associated form to avoid inversion:

$$\begin{pmatrix} p - bA^{-1}c & 0 \\ 0 & I \end{pmatrix} \to \begin{pmatrix} p - bA^{-1}c & b \\ 0 & A \end{pmatrix} \to \begin{pmatrix} p & b \\ c & A \end{pmatrix} \to \begin{pmatrix} b & p \\ A & c \end{pmatrix}.$$

Each term is associated to the next over R_Σ and the last expression has the advantage of being defined as a matrix over R itself. Thus we consider, for any set Σ of square matrices over R, the set $M(\Sigma)$ of all matrix blocks

$$a = \begin{pmatrix} a' & \tilde{a} \\ a^0 & 'a \end{pmatrix}, \tag{1}$$

where $\tilde{a} \in R$, $a' \in R^n$, $'a \in {}^nR$, $a^0 \in \Sigma_n$, an $n \times n$ matrix in Σ. Here n may be any positive integer, or it may be 0, in which case $a = (\tilde{a})$. A matrix block (1) is said to be *pure* if $\tilde{a} = 0$; it is possible to operate with pure blocks only, but we shall use the general form (1), which is no harder to handle; a^0 will be called its *denominator*.

Our aim will be to construct R_Σ as a set of equivalence classes of elements of $M(\Sigma)$, for any ring R and upper multiplicative set Σ of square matrices over R. If $\lambda: R \to R_\Sigma$ is the natural homomorphism, then the matrix block given by (1) is to represent the element

$$\lambda(\tilde{a}) - \lambda(a')\lambda(a^0)^{-1}\lambda('a). \tag{2}$$

On each matrix (1) we can perform certain elementary operations which do not change the element represented.

(E.1) *To a given row (column) add a left multiple of a later row (a right multiple of an earlier column).*

This amounts to left (right) multiplication by an upper unitriangular matrix; thus if a is changed to

$$\begin{pmatrix} 1 & p \\ 0 & P \end{pmatrix} \begin{pmatrix} a' & \tilde{a} \\ a^0 & 'a \end{pmatrix} = \begin{pmatrix} a' + pa^0 & \tilde{a} + p('a) \\ Pa^0 & P('a) \end{pmatrix},$$

then the corresponding element of R_Σ is

$$\lambda(\tilde{a} + p('a)) - \lambda(a' + pa^0)\lambda(a^0)^{-1}\lambda(P)^{-1}\lambda(P('a)),$$

and this simplifies to $\lambda(\tilde{a}) - \lambda(a')\lambda(a^0)^{-1}\lambda('a)$, if we bear in mind that λ is a homomorphism. Similarly for right multiplication by an upper unitriangular matrix.

To describe the second type of operation, let us call a square block T on the main diagonal of a^0 in (1) *superfluous* if either the row block or the column block of T is zero apart from T itself, and any entry below and to the left of T is zero. Now we have a second elementary transformation:

(E.2) *To insert or remove a superfluous block, and the row and column block containing it.*

We shall write $a \to b$ to indicate that b is obtained from a by an elementary transformation (E.1) or (E.2), and $a \sim b$ means that we can pass from a to b by a series of elementary transformations. This is clearly an equivalence relation; we shall call a and b *equivalent* if $a \sim b$. For example, any block is equivalent to a pure block:

$$\begin{pmatrix} a' & \tilde{a} \\ a^0 & 'a \end{pmatrix} \to \begin{pmatrix} a' & \cdot & \tilde{a} \\ a^0 & \cdot & 'a \\ \cdot & 1 & \tilde{a} \end{pmatrix} \to \begin{pmatrix} a' & -1 & \cdot \\ a^0 & \cdot & 'a \\ \cdot & 1 & \tilde{a} \end{pmatrix},$$

where dots stand for zeros. The equivalence class containing $a \in M(\Sigma)$ is written $[a]$ and the set of all such classes is denoted by R_Σ. Our object will be to define a ring structure on R_Σ, but first we shall need to define an operation on matrices, the determinantal sum. Let A, B be two $n \times n$ matrices which differ only in the first column, say $A = (A_1, A_2, \ldots, A_n)$, $B = (B_1, A_2, \ldots, A_n)$. Then the *determinantal sum* of A and B with respect to the first column is defined as the matrix

$$C = (A_1 + B_1, A_2, \ldots, A_n).$$

The determinantal sum with respect to other columns (for suitable matrices) is defined similarly. We shall usually denote the determinantal sum of A and B by $A \nabla B$, without specifying the column to be added, since this is usually clear from the context. For matrices over a commutative ring it is clear that

$$\det(A \; \nabla \; B) = \det A + \det B.$$

In general we have no such relation, because determinants are then not defined, but over a skew field it is easily verified that if A and B are singular, then so is $A \; \nabla \; B$, whenever the latter is defined. We also note the distributive law:

$$(A \; \nabla \; B) \oplus P = (A \oplus P) \; \nabla \; (B \oplus P), \tag{3}$$

whenever $A \; \nabla \; B$ is defined. There is a corresponding definition of determinantal sum with respect to rows, but this will not be needed.

We now define the following operations on $M(\Sigma)$, where dots indicate blocks of zeros:

$$a \oplus b = \begin{pmatrix} a' & b' & \tilde{a} + \tilde{b} \\ a^0 & \cdot & 'a \\ \cdot & b^0 & 'b \end{pmatrix} = \begin{pmatrix} a' & b' & \tilde{a} \\ a^0 & \cdot & 'a \\ \cdot & b^0 & \cdot \end{pmatrix} \nabla \begin{pmatrix} a' & b' & \tilde{b} \\ a^0 & \cdot & \cdot \\ \cdot & b^0 & 'b \end{pmatrix}, \tag{4}$$

$$a \odot b = \begin{pmatrix} a' & \tilde{a}b' & \tilde{a}\tilde{b} \\ a^0 & 'ab' & 'a\tilde{b} \\ \cdot & b^0 & 'b \end{pmatrix} = \begin{pmatrix} a' & \tilde{a} & \cdot \\ a^0 & 'a & \cdot \\ \cdot & \cdot & 1 \end{pmatrix} \begin{pmatrix} 1 & \cdot & \cdot \\ \cdot & b' & \tilde{b} \\ \cdot & b^0 & 'b \end{pmatrix}. \tag{5}$$

Our object will be to show that R_Σ is a ring relative to these operations. In the first place we observe that $\oplus$, $\odot$ are well-defined on the classes; we have $a_1 \sim a_2, b_1 \sim b_2 \Rightarrow a_1 \oplus b_1 \sim a_2 \oplus b_2$, $a_1 \odot b_1 \sim a_2 \odot b_2$, because any elementary operation carried out on a, b can also be carried out on $a \oplus b$, $a \odot b$. The associative laws follow because the sum of a, b and c in $M(\Sigma)$ with either bracketing has the matrix

$$\begin{pmatrix} a' & b' & c' & s \\ a^0 & \cdot & \cdot & 'a \\ \cdot & b^0 & \cdot & 'b \\ \cdot & \cdot & c^0 & 'c \end{pmatrix}$$

where $s = \tilde{a} + \tilde{b} + \tilde{c}$; the matrix for the product of a, b and c is

$$\begin{pmatrix} a' & \tilde{a} & \cdot & \cdot \\ a^0 & 'a & \cdot & \cdot \\ \cdot & \cdot & 1 & \cdot \\ \cdot & \cdot & \cdot & 1 \end{pmatrix} \begin{pmatrix} 1 & \cdot & \cdot & \cdot \\ \cdot & b' & \tilde{b} & \cdot \\ \cdot & b^0 & 'b & \cdot \\ \cdot & \cdot & \cdot & 1 \end{pmatrix} \begin{pmatrix} 1 & \cdot & \cdot & \cdot \\ \cdot & 1 & \cdot & \cdot \\ \cdot & \cdot & c' & \tilde{c} \\ \cdot & \cdot & c^0 & 'c \end{pmatrix}.$$

Next we note the formulae for the sum and product of an element $r \in R$ by a block matrix a:

$$a \oplus (r) = (r) \oplus a = \begin{pmatrix} a' & \tilde{a} + r \\ a^0 & {}'a \end{pmatrix},$$

$$(r) \odot a = \begin{pmatrix} ra' & r\tilde{a} \\ a^0 & {}'a \end{pmatrix}, \qquad a \odot (r) = \begin{pmatrix} a' & \tilde{a}r \\ a^0 & {}'ar \end{pmatrix}.$$

Taking $r = 0, 1$ in turn, we find that $(0) \oplus a \sim a \oplus (0) \sim a$,

$$(0) \odot a \sim \begin{pmatrix} \cdot & \cdot \\ a^0 & {}'a \end{pmatrix} \sim 0, \qquad a \odot (0) \sim \begin{pmatrix} a' & \cdot \\ a^0 & \cdot \end{pmatrix} \sim 0,$$

by using (E.2), and $(1) \odot a \sim a \odot (1) \sim a$. Writing $(-)a = (-1) \odot a$, we have $(-) ((-)a) = a$, and

$$a \oplus (-)a = \begin{pmatrix} a' & -a' & \cdot \\ a^0 & \cdot & {}'a \\ \cdot & a^0 & {}'a \end{pmatrix} \to \begin{pmatrix} a' & \cdot & \cdot \\ a^0 & a^0 & {}'a \\ \cdot & a^0 & {}'a \end{pmatrix} \to \begin{pmatrix} a' & \cdot & \cdot \\ a^0 & \cdot & \cdot \\ \cdot & a^0 & {}'a \end{pmatrix}$$

$$\to \begin{pmatrix} a' & \cdot \\ a^0 & \cdot \end{pmatrix} \to (0).$$

Hence $a \oplus (-)a \sim (0)$ and similarly $((-)a) \oplus a \sim (0)$. There remain the distributive laws. Consider first

$$(a \odot c) \oplus (b \odot c) \sim (a \oplus b) \odot c. \tag{6}$$

We write down the matrix block for the left-hand side of (6), putting $t = \tilde{a}\tilde{c} + \tilde{b}\tilde{c}$ for short, and apply elementary transformations:

$$\begin{pmatrix} a' & \tilde{a}c' & b' & \tilde{b}c' & t \\ a^0 & {}'ac' & \cdot & \cdot & {}'a\tilde{c} \\ \cdot & c^0 & \cdot & \cdot & {}'c \\ \cdot & \cdot & b^0 & {}'bc' & {}'b\tilde{c} \\ \cdot & \cdot & \cdot & c^0 & {}'c \end{pmatrix} \to \begin{pmatrix} a' & \tilde{a}c' & b' & \tilde{b}c' & t \\ a^0 & {}'ac' & \cdot & \cdot & {}'a\tilde{c} \\ \cdot & c^0 & \cdot & -c^0 & \cdot \\ \cdot & \cdot & b^0 & {}'bc' & {}'b\tilde{c} \\ \cdot & \cdot & \cdot & c^0 & {}'c \end{pmatrix}$$

$$\to \begin{pmatrix} a' & \tilde{a}c' & b' & (\tilde{a} + \tilde{b})c' & t \\ a^0 & {}'ac' & \cdot & {}'ac' & {}'a\tilde{c} \\ \cdot & c^0 & \cdot & \cdot & \cdot \\ \cdot & \cdot & b^0 & {}'bc' & {}'b\tilde{c} \\ \cdot & \cdot & \cdot & c^0 & {}'c \end{pmatrix}$$

$$\to \begin{pmatrix} a' & b' & (\tilde{a} + \tilde{b})c' & t \\ a^0 & \cdot & {}'ac' & {}'a\tilde{c} \\ \cdot & b^0 & {}'bc' & {}'b\tilde{c} \\ \cdot & \cdot & c^0 & {}'c \end{pmatrix},$$

where we have used (E.2) to remove c^0 at the last step; the result is the matrix block for the right-hand side of (6). Now the other distributive law follows by symmetry. By the observation in 1.1 the addition is commutative and so we have a ring R_Σ and it is clear that the mapping

$$\lambda: r \mapsto [(r)] \quad (r \in R)$$

is a ring homomorphism from R to R_Σ. We claim that λ is universal Σ-inverting. To establish this fact, we first prove the rule

$$\begin{pmatrix} a' & \cdot \\ a^0 & b \end{pmatrix} \oplus \begin{pmatrix} a' & \cdot \\ a^0 & c \end{pmatrix} \sim \begin{pmatrix} a' & \cdot \\ a^0 & b+c \end{pmatrix}. \tag{7}$$

Writing down the matrix block for the left-hand side and applying elementary transformations, we have

$$\begin{pmatrix} a' & a' & \cdot \\ a^0 & \cdot & b \\ \cdot & a^0 & c \end{pmatrix} \rightarrow \begin{pmatrix} a' & \cdot & \cdot \\ a^0 & -a^0 & b \\ \cdot & a^0 & c \end{pmatrix} \rightarrow \begin{pmatrix} a' & \cdot & \cdot \\ a^0 & \cdot & b+c \\ \cdot & a^0 & c \end{pmatrix}$$

$$\rightarrow \begin{pmatrix} a' & \cdot \\ a^0 & b+c \end{pmatrix}.$$

Thus (7) is proved. Now if e_j denotes the jth column of the unit matrix and e_i^T the i-th row, then the (ij)-entry of $\lambda(A)^{-1}$ is represented by the matrix block

$$u_{ij} = \begin{pmatrix} -e_i^T & 0 \\ A & e_j \end{pmatrix},$$

for on denoting the k-th column of A by A_k, we have

$$\sum_j \begin{pmatrix} -e_i^T & \cdot \\ A & e_j \end{pmatrix} \odot (a_{jk}) = \sum \begin{pmatrix} -e_i^T & \cdot \\ A & e_j a_{jk} \end{pmatrix} \rightarrow \begin{pmatrix} -e_i^T & \cdot \\ A & \sum e_j a_{jk} \end{pmatrix}$$

$$= \begin{pmatrix} -e_i^T & \cdot \\ A & A_k \end{pmatrix} \rightarrow \begin{pmatrix} -e_i^T & \delta_{ik} \\ A & \cdot \end{pmatrix} \rightarrow (\delta_{ik})$$

where we first use (7) with n summands, then (E.1) and finally (E.2). Similarly we have $\sum (a_{ij}) \odot u_{jk} \sim (\delta_{ik})$, therefore $[u_{ij}]$ is the (ij)-entry of $\lambda(A)^{-1}$, and so $\lambda(A)$ is invertible, as claimed. To prove the universal property, let $f: R \rightarrow R'$ be a Σ-inverting homomorphism and for any element $[a]$ of R_Σ define

$$f_1([a]) = f(\tilde{a}) - f(a')f(a^0)^{-1}f('a). \tag{8}$$

As for λ, we can verify that the right-hand side is unaffected by elementary transformations (E.1, 2). Hence f_1 is well-defined on R_Σ by (8), and is uniquely determined by f. Moreover, for any $r \in R$, we have

$$f(r) = f_1([(r)]) = f_1\lambda(r),$$

so f has been factored by f_1 and this shows λ to be universal Σ-inverting. Summing up, we have the following explicit description of R_Σ:

THEOREM 4.3.1. *Let R be any ring and Σ an upper multiplicative set of square matrices over R. Then the set $M(\Sigma)$ of all matrix blocks*

$$a = \begin{pmatrix} a' & \tilde{a} \\ a^0 & {}'a \end{pmatrix}, \quad a^0 \in \Sigma, \ a' \ row, \ 'a \ column \ over \ R, \ \tilde{a} \in R, \quad (9)$$

taken modulo the equivalence $\sim$ defined by the elementary operations (E.1, 2) and with the operations $\oplus$, $\odot$ defined by (4) and (5) forms a ring R_Σ which is the localization of R at Σ, with the natural homomorphism $\lambda: R \to R_\Sigma$ given by $r \mapsto [(r)]$. ∎

To apply the result we need to be able to recognize when a matrix block represents zero in R_Σ (problem (i)). This is accomplished by *Malcolmson's criterion*:

PROPOSITION 4.3.2. *Let R, Σ, R_Σ be as in Th. 3.1 and let a be a matrix block as in (9). Then $[a] = 0$ in R_Σ if and only if there exist F, G, P, $Q \in \Sigma$, a matrix H over R, rows f, u and columns g, v over R such that*

$$\begin{pmatrix} f & \cdot & a' & \tilde{a} \\ H & \cdot & a^0 & {}'a \\ F & \cdot & \cdot & \cdot \\ \cdot & G & \cdot & g \end{pmatrix} = \begin{pmatrix} u \\ P \end{pmatrix} (Q \quad v). \quad (10)$$

Proof. Let us write $a \underset{M}{\sim} 0$ if a satisfies (10). We remark that the left-hand side of (10) can be factorized as

$$\begin{pmatrix} 1 & \cdot & f & \cdot \\ \cdot & 1 & H & \cdot \\ \cdot & \cdot & F & \cdot \\ \cdot & \cdot & \cdot & 1 \end{pmatrix} \begin{pmatrix} a' & \tilde{a} & \cdot & \cdot \\ a^0 & {}'a & \cdot & \cdot \\ \cdot & \cdot & 1 & \cdot \\ \cdot & \cdot & \cdot & 1 \end{pmatrix} \begin{pmatrix} \cdot & \cdot & 1 & \cdot \\ \cdot & \cdot & \cdot & 1 \\ 1 & \cdot & \cdot & \cdot \\ \cdot & G & \cdot & g \end{pmatrix}. \quad (11)$$

This makes it clear that if $a \underset{M}{\sim} 0$, then $a \sim b$, where b is the left-hand side of (10), and we then have

$$[a] = [b] = \lambda(uv) - \lambda(uQ)\lambda(PQ)^{-1}\lambda(Pv) = \lambda(uv) - \lambda(uv) = 0,$$

so $[a] = 0$. Conversely, if $a \sim b$ and $a \underset{M}{\sim} 0$, then we have $b \underset{M}{\sim} 0$, for any elementary operation (E.1) carried out on (10) will change P, Q into other members of Σ, while the effect of (E.2) is just to insert or remove further terms $F \cdot G$. Thus if e.g.

$$\begin{pmatrix} a' & \tilde{a} \\ a^0 & 'a \end{pmatrix} = \begin{pmatrix} u \\ P \end{pmatrix} (Q \quad v),$$

then

$$\begin{pmatrix} \cdot & a' & \tilde{a} \\ \cdot & a^0 & 'a \\ G & \cdot & g \end{pmatrix} = \begin{pmatrix} u & \cdot \\ P & \cdot \\ \cdot & I \end{pmatrix} \begin{pmatrix} \cdot & Q & v \\ G & \cdot & g \end{pmatrix}.$$

If the superfluous terms in (10) appear in the opposite order, we can reduce them to the form (10) by the following steps:

$$\begin{pmatrix} \cdot & f & a' & \tilde{a} \\ \cdot & H & a^0 & 'a \\ G & K & \cdot & g \\ \cdot & F & \cdot & \cdot \end{pmatrix} \rightarrow \begin{pmatrix} \cdot & f & \cdot & a' & \tilde{a} \\ \cdot & H & \cdot & a^0 & 'a \\ G & K & \cdot & \cdot & g \\ \cdot & F & \cdot & \cdot & \cdot \\ \cdot & \cdot & G & \cdot & g \end{pmatrix}$$

$$\rightarrow \begin{pmatrix} \cdot & f & \cdot & a' & \tilde{a} \\ \cdot & H & \cdot & a^0 & 'a \\ G & K & -G & \cdot & \cdot \\ \cdot & F & \cdot & \cdot & \cdot \\ \cdot & \cdot & G & \cdot & g \end{pmatrix}$$

$$\rightarrow \begin{pmatrix} \cdot & f & \cdot & a' & \tilde{a} \\ \cdot & H & \cdot & a^0 & 'a \\ G & K & \cdot & \cdot & \cdot \\ \cdot & F & \cdot & \cdot & \cdot \\ \cdot & \cdot & G & \cdot & g \end{pmatrix}$$

In particular it follows that if $a \sim 0$, then $a \underset{M}{\sim} 0$. ∎

The criterion of Prop. 3.2 is quite general, but not easy to apply directly; in particular it may be quite difficult even to decide when R_Σ is trivial. We shall therefore derive some consequences that are easier to use. We recall from 1.4 that a matrix A is called *full* if it is square, say $n \times n$, and cannot be written in the form $A = PQ$, where Q has fewer than n rows. With the help of this notion we have the following sufficient condition for the natural map λ to be injective:

COROLLARY 4.3.3. *Let R be any ring and Σ an upper multiplicative set of full matrices, which is closed under permutations of rows or of columns. Then in R_Σ, $[a] \neq 0$ for any $a \in \Sigma$.*

Proof. Suppose that $a \in \Sigma$ and $[a] = 0$. Then (10) holds; further,

$$B = \begin{pmatrix} a & \begin{matrix} f \\ H \\ \end{matrix} \\ 0 & F \end{pmatrix} \in \Sigma \quad \text{and} \quad C = \begin{pmatrix} G & 0 & g & 0 \\ 0 & & & B \end{pmatrix} \in \Sigma,$$

and by a permutation of rows and of columns we find that the left-hand side of (10) lies in Σ and so is full. But this contradicts (10), so we conclude that $[a] \neq 0$. ∎

We next derive conditions for the localization to be a local ring. For this purpose we remark that a ring R is local if and only if $1 \neq 0$ and for every $x \in R$ either x has a left inverse or $1 - x$ has a right inverse. For a local ring these conditions are clearly satisfied. Conversely, suppose that they hold in R. An idempotent $e \neq 1$ has no left or right inverse, for if $eu = 1$, say, then $e = e \cdot eu = eu = 1$. Hence, for any idempotent e in R, either $e = 1$ or $1 - e = 1$, so 0, 1 are the only idempotents in R. Now if $xy = 1$, then yx is an idempotent $\neq 0$, hence $yx = 1$, so every inverse in R is two-sided, in other words, R is weakly 1-finite (see 1.4). Thus for any $x \in R$, either x or $1 - x$ has an inverse, hence the non-invertible elements of R form a set closed under addition and so form an ideal; this shows R to be a local ring.

We shall now see that the conditions of Cor. 3.3, together with one other condition, are enough to ensure that R_Σ is a local ring.

THEOREM 4.3.4. *Let R be a ring and Σ a set of full matrices such that Σ is upper multiplicative, closed under permutations of rows or of columns, while the complement of Σ admits determinantal sums of columns. Then R_Σ is a local ring.*

Proof. R_Σ is non-trivial, because $(1) \in \Sigma$, so $[(1)] \neq 0$, by Cor. 3.3. Given $x \in R_\Sigma$, let $x = u_1$ be the first component of the solution of

$$Au - e_1 = 0, \tag{12}$$

and suppose that x has no left inverse in R_Σ. By Cramer's rule the numerator of u_1 has no left inverse in R_Σ. Let us write $A = (A_1 \ A')$, where A_1 is the first column of A; then this numerator is $(e_1 \ A')$. We assert that $(A_1 - e_1 \ A') \in \Sigma$, for if not, then $(e_1 \ A')$ and $(A_1 - e_1 \ A')$ both lie in the complement of Σ, hence so does $A = (A_1 - e_1 \ A') \triangledown$

$(e_1 \quad A')$. But this is a contradiction; hence $(A_1 - e_1 \quad A') \in \Sigma$, and we can form the system

$$(A_1 - e_1 \quad A')v - e_1 = 0. \tag{13}$$

It may be rewritten as

$$Av - e_1(1 + v_1) = 0. \tag{14}$$

By (12) and (14), $Av = e_1(1 + v_1) = Au(1 + v_1)$, so $v = u(1 + v_1)$. Equating the first components, we find that $v_1 = u_1(1 + v_1)$, and so

$$(1 - u_1)(1 + v_1) = 1.$$

This shows that $1 - x$ has a right inverse, and now the remark made earlier shows that R_Σ is a local ring. ∎

As an application let us show that any epic R-field is determined by its singular kernel. Let K be an epic R-field, $\mu: R \to K$ the canonical map and $\mathscr{P} = \text{Ker } \mu$ its singular kernel. If $\mathscr{P}'$ is its complement in $\mathfrak{M}(R)$, then the localization at $\mathscr{P}'$, i.e. $R_{\mathscr{P}'}$ is usually written $R_\mathscr{P}$ (just as we write R_p in the commutative case). From the definition of $\mathscr{P}$ it follows that μ can be factored uniquely by $\lambda: R \to R_\mathscr{P}$ to give a map $\alpha: R_\mathscr{P} \to K$ such that $\mu = \lambda\alpha$. Now it is clear that $\mathscr{P}$ admits determinantal sums of columns and contains all non-full matrices, while $\mathscr{P}'$ contains only full matrices, is upper multiplicative and is closed under permutations of rows or of columns. Hence by Th. 3.4, $R_\mathscr{P}$ is a local ring and K, as homomorphic image of $R_\mathscr{P}$ is just its residue-class field. This establishes

THEOREM 4.3.5. *Let R be any ring and K an epic R-field with singular kernel $\mathscr{P}$. Then the localization $R_\mathscr{P}$ is a local ring with residue-class field K.* ∎

Exercises

1. If a is a matrix block as in (1), find a matrix block for $[a]^{-1}$.

2. Given a ring R and a set Σ of square matrices, consider blocks (1) where $\tilde{a}$ is $r \times r$, $a^0 \in \Sigma$ is $s \times s$, a' is $r \times s$ and $'a$ is $s \times r$. Show that in R_Σ, $[a]$ represents the $r \times r$ matrix (2) and obtain conditions for $[a]$ to be non-full. What are the conditions for $[a]$ to be zero?

3. Let R be any ring and let Σ be the set of all upper triangular matrices over R with non-zero elements along the main diagonal. Show that R has an embedding in an $R^\times$-inverting ring if and only if any matrix equation

$$\begin{pmatrix} f & \cdot & a \\ F & \cdot & \cdot \\ \cdot & G & g \end{pmatrix} = \begin{pmatrix} u \\ P \end{pmatrix} (Q \quad v)$$

where $F, G, P, Q \in \Sigma$, f, u are rows, g, v columns over R and $a \in R$, implies $a = 0$.

4. (Malcolmson [93]) Let R be any ring and Σ a multiplicatively closed set of square matrices. Show that a matrix A over R has a right inverse over R_Σ if and only if there exist $P, Q \in \Sigma$, a matrix S over R and invertible square matrices, E, F over R such that $Q = EPF(I \oplus A)S$.

5. (Malcolmson [93]) Let R be a ring and Σ a multiplicative set of square matrices. Show that R_Σ has UGN if and only if Σ contains only full matrices; R_Σ is weakly finite if and only if $AB \in \Sigma$ for square matrices A, B implies that A, B are invertible over R_Σ.

4.4 Matrix ideals

In the last section we saw in Th. 3.5 how to construct an epic R-field from its kernel. What we need now is a simple way of recognizing singular kernels – just as in the commutative case the kernels of epic R-fields are precisely the prime ideals of R. For this purpose we need to develop a form of ideal theory in which the place of ideals is taken by certain sets of matrices. Instead of addition and multiplication we have determinantal addition and diagonal sums, while the non-full matrices play the role of zero. The use of these notions can be motivated by observing that in the commutative case they reduce to the corresponding operations on determinants, thus diagonal sums correspond to taking the product of determinants, while any non-full matrix A over a commutative ring has zero determinant, for we can write $A = PQ$, where P, Q are square matrices with a zero column and row, respectively. However, this analogy should not be pushed too far: it is quite possible for a full matrix to have zero determinant, for example in the polynomial ring $k[x, y, z]$ the matrix

$$T = \begin{pmatrix} 0 & z & -y \\ -z & 0 & x \\ y & -x & 0 \end{pmatrix} \tag{1}$$

can be shown to be full (see Cohn [89] or Ex. 2 below), though its determinant is zero.

We shall need analogues of ideals and prime ideals. In any ring R a *matrix ideal* is a collection $\mathscr{A}$ of square matrices satisfying the following four conditions:

(MI.1) $\mathscr{A}$ *contains all non-full matrices,*

(MI.2) *If $A \in \mathscr{A}$, then $A \oplus B \in \mathscr{A}$ for all $B \in \mathfrak{M}(R)$,*

(MI.3) *If $A, B \in \mathscr{A}$ and $C = A \triangledown B$ is defined with respect to some column, then $C \in \mathscr{A}$,*

(MI.4) *If $A \oplus 1 \in \mathscr{A}$, then $A \in \mathscr{A}$.*

If, moreover,

(MI.5) $\mathscr{A}$ *is proper, i.e. $\mathscr{A} \neq \mathfrak{M}(R)$,*

(MI.6) $A, B \notin \mathscr{A} \Rightarrow A \oplus B \notin \mathscr{A}$,

then $\mathscr{A}$ is called a *prime matrix ideal*. The analogy with prime ideals is clear, at least at the formal level. It is easily verified that for any epic R-field, its singular kernel is a prime matrix ideal, and conversely, from Th. 3.4–5 it is clear that for any prime matrix ideal $\mathscr{P}$ the ring $R_{\mathscr{P}}$ is a local ring, whose residue-class field is an epic R-field with singular kernel $\mathscr{P}$. This field will be denoted by $R/\mathscr{P}$. Moreover, a specialization between epic R-fields can be characterized in terms of the corresponding singular kernels:

THEOREM 4.4.1. *Let R be a ring and K_1, K_2 epic R-fields with singular kernels $\mathscr{P}_1$, $\mathscr{P}_2$ and corresponding localizations R_1, R_2. Then the following conditions are equivalent:*

 (a) *there is a specialization $\alpha\colon K_1 \to K_2$,*

 (b) $\mathscr{P}_1 \subseteq \mathscr{P}_2$,

 (c) *there is an R-ring homomorphism $R_2 \to R_1$.*

Thus the category of epic R-fields $\mathscr{F}_R$, as a partially ordered set, is order-isomorphic to the set of prime matrix ideals over R, partially ordered by inclusion.

 Further, if there are specializations $K_1 \to K_2$ and $K_2 \to K_1$, then K_1 and K_2 are isomorphic as R-fields.

We note the reversal of direction in (c) compared with (a).

Proof. Write Σ_i for the complement of $\mathscr{P}_i$ and $\mu_i\colon R \to K_i$ for the canonical homomorphism. To prove (a) $\Rightarrow$ (b), assume a specialization $\alpha\colon K_1 \to K_2$ and take $A \in \Sigma_2$; then $A\mu_2$ has an inverse which is the image of a matrix B over K_1: $(A\mu_2)(B\alpha) = I$, hence $(A\mu_1)B = I + C$, where $C\alpha = 0$. It follows that $I + C$ has an inverse over R_1 and so does $A\mu_1$, therefore $A \in \Sigma_1$. This shows that $\mathscr{P}_1 \subseteq \mathscr{P}_2$, i.e. (b) holds.

 Now (b) $\Rightarrow$ (c) is clear, for when (b) holds, then $\lambda_1\colon R \to R_1$ is Σ_2-inverting and so may be factored by λ_2 to give a homomorphism

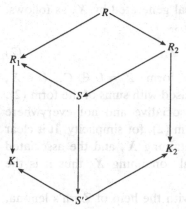

$R_2 \to R_1$. Finally, to show that (c) $\Rightarrow$ (a), given a homomorphism from R_2 to R_1, let S be the image of R_2 in R_1 and write $\mathfrak{m}_1$ for the maximal ideal of R_1. The natural homomorphism $R_1 \to K_1$ maps S to $S' = S/(S \cap \mathfrak{m}_1)$. Now S is a local ring (as homomorphic image of R_2) and $S \cap \mathfrak{m}_1$ is a proper ideal, so the natural homomorphism $R_2 \to K_2$ can be taken via S', giving a homomorphism from S' to K_2, and this is the required specialization.

Now the last remark follows by using (b). ∎

As an immediate consequence we have

COROLLARY 4.4.2. *Let R, K_1, K_2 be as in Th. 4.1, with a specialization $\alpha: K_1 \to K_2$. Then any K_2-admissible block C is also K_1-admissible, and if C defines p in K_1, then p lies in the domain of α and the element of K_2 defined by C is $p\alpha$.*

Proof. Let $\mathcal{P}_i$ be the singular kernel of K_i and let A be the denominator in C. Since K_2 is a specialization of K_1 we have $\mathcal{P}_1 \subseteq \mathcal{P}_2$ by Th. 4.1, and by hypothesis, $A \notin \mathcal{P}_2$, hence $A \notin \mathcal{P}_1$ and so C is K_1-admissible. If $C = \begin{pmatrix} u & a \\ A & v \end{pmatrix}$, then p is given in K_1 by

$$p = a - uA^{-1}v,$$

hence p lies in the domain of α, and $p\alpha$ is given by the same formula over K_2. ∎

The partially ordered set of prime matrix ideals of R is sometimes called the *field spectrum* and is denoted by $\mathrm{Spec}(R)$. When R is commutative it reduces to the familiar prime spectrum of R.

There still remains the problem of constructing prime matrix ideals. Guided by the analogy of the commutative case, we first consider matrix ideals. It will be convenient to begin with sets satisfying only the axioms (MI.1–3); such a set will be called a *matrix preideal*. From a preideal $\mathcal{A}$ we get a matrix ideal $\bar{\mathcal{A}}$ as the set of all matrices A such that $A \oplus I \in \mathcal{A}$, for a unit matrix of suitable size. Given any set X of square matrices over a ring R, we can form (X), the matrix ideal generated by X, as follows. Let $(X)_0$ be the set of all determinantal sums

$$Z_1 \nabla Z_2 \nabla \ldots \nabla Z_r, \tag{2}$$

where Z_i is either non-full or of the form $Z_i = U \oplus C$, $U \in X$, $C \in \mathfrak{M}(R)$. Here some care has to be exercised with sums of the form (2), because determinantal addition is non-associative and not everywhere defined, but we have omitted the brackets in (2), for simplicity. It is clear that $(X)_0$ is the least matrix preideal containing X, and the associated matrix ideal $\overline{(X)_0}$ is the least matrix ideal containing X; thus it is the matrix ideal generated by X.

We can now form prime matrix ideals with the help of Zorn's lemma, as in the commutative case.

THEOREM 4.4.3. *Let R be any ring, $\mathcal{A}$ a matrix ideal and Σ a set of square matrices, including 1 and closed under diagonal sums and disjoint from $\mathcal{A}$. Then* (i) *there exist maximal matrix ideals containing $\mathcal{A}$ and disjoint from Σ, and* (ii) *every such matrix ideal is prime.*

Proof. It is easily checked that the collection of matrix ideals containing $\mathcal{A}$ and disjoint from Σ is inductive, hence by Zorn's lemma there are maximal matrix ideals satisfying these conditions. Let $\mathcal{P}$ be such a matrix ideal; since $1 \in \Sigma$, $\mathcal{P}$ is proper. Now suppose that there exist $A_1, A_2 \notin \mathcal{P}$ such that $A_1 \oplus A_2 \in \mathcal{P}$. If $\mathcal{A}_i$ denotes the matrix ideal generated by $\mathcal{P}$ and A_i, then $\mathcal{A}_i \supset \mathcal{P}$, hence $\mathcal{A}_i \cap \Sigma \neq \varnothing$, by the maximality of $\mathcal{P}$, say $B_i \in \mathcal{A}_i \cap \Sigma$. Now $B_1 \oplus B_2$ is a determinantal sum of terms in $\mathcal{P}$ and terms of the form $A_1 \oplus A_2 \oplus C$, and hence it lies in $\mathcal{P}$. But $B_1 \oplus B_2$ is also in Σ, which contradicts the fact that $\mathcal{P} \cap \Sigma = \varnothing$. This contradiction shows that $A_1 \oplus A_2 \notin \mathcal{P}$, so $\mathcal{P}$ is indeed prime. ∎

For any matrix ideal $\mathcal{A}$ we can define its *radical* as

$$\sqrt{\mathcal{A}} = \{A \mid A \oplus \ldots \oplus A \ (r \text{ terms}) \in \mathcal{A}, \text{ for some } r\};$$

it is easily verified that $\sqrt{\mathcal{A}}$ is a matrix ideal containing $\mathcal{A}$. With the help of Th. 4.3 one can now prove the following analogue of the commutative case:

$$\sqrt{\mathcal{A}} = \bigcap \{\mathcal{P} \mid \mathcal{P} \text{ prime} \supseteq \mathcal{A}\}. \tag{3}$$

To illustrate how these results may be used to answer questions about R-fields, consider the following general situation. We have a ring R and want to find an epic R-field making a given set X of square matrices invertible and a second set Y singular. Let Σ be the set of all diagonal sums of 1 and matrices from X, and take (Y), the matrix ideal generated by Y. Our problem has a solution precisely when

$$\Sigma \cap (Y) = \varnothing. \tag{4}$$

For when such an epic R-field exists, then its singular kernel $\mathcal{P}$ must contain (Y) and be disjoint from X, so (4) holds. Conversely, when (4) holds, then we can by Th. 4.3 find a prime matrix ideal containing Y and disjoint from X. An equivalent condition is $\Sigma \cap \sqrt{(Y)} = \varnothing$.

To find if R has any epic R-field at all, we form the least matrix ideal $\mathcal{N}$, which is the set of all matrices A such that $A \oplus I$ is a determinantal sum of non-full matrices, for some unit matrix I. If $\mathcal{N}$ contains no unit matrix, then $\mathcal{N}$ can be enlarged to a prime matrix ideal, by Th. 4.3, and hence we can find an epic R-field. This sufficient condition is clearly also necessary, so we obtain

THEOREM 4.4.4. *A ring R has a homomorphism into a field if and only if no unit matrix I (of any size) can be written as a determinantal sum of non-full matrices.* ∎

More explicitly, if R has no R-field, then there must be an equation

$$I = C_1 \nabla C_2 \nabla \ldots \nabla C_r, \tag{5}$$

where the C_i are non-full and the right-hand side is bracketed in some way so as to make sense. This is a very explicit condition, but if we are given a ring R with no R-fields, it may be far from easy to construct an equation (5). Likewise, if there is no equation (5), the construction of an R-field may be difficult. The proof will be of no help since it was non-constructive (Zorn's lemma was used).

The situation may be illustrated by a corresponding problem for commutative rings. If R is a commutative integral domain, then any $n \times n$ matrix A, which is nilpotent satisfies the equation $A^n = 0$, as we see by transforming A to triangular form over the field of fractions of R. It follows that for any commutative ring R, if A is a nilpotent $n \times n$ matrix, then the entries of A^n must lie in every prime ideal of R and hence be nilpotent. In particular, let A be a 2×2 matrix over a commutative ring R and denote by $\mathfrak{J}$ the ideal generated by the entries of A^3; then the entries of A^2 lie in $\sqrt{\mathfrak{J}}$. The explicit verification shows that such a problem is not always trivial (see Ex. 5).

Returning to Th. 4.4, to give an example, consider a ring R without UGN. Such a ring has an $n \times r$ matrix A and an $r \times n$ matrix B, where $r < n$, such that $AB = I$, thus I is not full. It is clear that such a ring has no R-fields; what Th. 4.4 shows is that a slightly more general condition of this type is sufficient as well as necessary for the existence of R-fields.

We next ask when there is a field of fractions. For this to exist we need an injective homomorphism to a field, thus the singular kernel must not contain any non-zero elements of R. Let Δ be the set of all diagonal matrices with non-zero entries on the main diagonal. We need a prime matrix ideal disjoint from Δ and by Th. 4.3 this can be found precisely when the least matrix ideal does not meet Δ. Thus we have

THEOREM 4.4.5. *A ring R has a field of fractions if and only if $R \neq 0$ and no diagonal matrix with non-zero diagonal elements can be expressed as a determinantal sum of non-full matrices.* ■

Although this solves our problem, we now see that we would like to know more: Is there more than one field of fractions, and if so, is there one that is universal in some sense? Here a *universal R-field* is understood to be an epic R-field, which has every other epic R-field as specialization. Using the correspondence in Th. 4.1, we see that there is a universal R-field precisely if there is a least prime matrix ideal. Evidently (bearing (3) in mind) this is the case if and only if $\sqrt{\mathcal{N}}$ is prime, where $\mathcal{N}$ is the least matrix ideal.

Exercises

1. Prove the relation (3) for any matrix ideal $\mathcal{A}$.

2. Verify that T, given by (1), is full over $k[x, y, z]$, but that $(x \oplus 1 \oplus 1)T$ is not full. Deduce that T becomes non-full if x^{-1} is adjoined. Do the same for y, z.

3. Writing $T(x, y, z)$ for the matrix (1), show that $T(x, y, z) \oplus T(x - 1, y, -z)$ is not full.

4°. Find a commutative ring in which a determinantal sum of non-full matrices is full.

5. Let A be a 2×2 matrix over a commutative ring R and denote by $\mathfrak{J}$ the ideal generated by the entries of A^3. Find powers of the entries of A^2 which lie in $\mathfrak{J}$. Do the same with $\mathfrak{J}$ replaced by a matrix ideal.

6. Let R be a ring with a universal field of fractions U. Show that every automorphism of R extends to U. Give an example of a field of fractions for which this fails.

7. Show that every ring R which has an epic R-field also has one that is *epi-final*, i.e. one without specializations to other epic R-fields. Show also that R has *epi-initial* R-fields, i.e. epic R-fields to which there are no specializations.

8. Let R be a commutative ring. Show that for any $A \in \mathfrak{M}(R)$, $\det A = 0$ if and only if A becomes singular in every R-field. Deduce that $\det A = 0$ if and only if $\bigoplus^r A$ can be written as a determinantal sum of non-full matrices, for some $r \geq 1$.

9. Verify that a ring can be embedded in a field if and only if $\Delta \cap \sqrt{\mathcal{N}} = \varnothing$, where Δ is the set of all diagonal matrices with non-zero diagonal elements and $\mathcal{N}$ is the least matrix ideal. Deduce another proof of Cor. 1.2.5 by applying (3) to $\mathcal{N}$.

10. Given square matrices A, B of the same size, show that $A \oplus B$ is associated to $AB \oplus I$. Deduce that the complement of a prime matrix ideal is closed under products.

4.5 Universal fields of fractions

In 4.4 we met precise conditions for the existence of a universal field of fractions, but we are also interested in having more manageable sufficient conditions for the existence. In any R-field the only matrices over R that can be inverted are the full matrices; a ring homomorphism is said to be *fully inverting* if every full matrix has an image which is invertible. In particular, if the canonical map from a ring R to an epic R-field K is fully inverting, then the singular kernel is the set of all non-full matrices over R, and K must then be the universal field of fractions of R. It is indeed a field of fractions, since every non-zero element is full as 1×1 matrix and so is inverted over K. Our aim in this section will be a characterization of rings R with a fully inverting homomorphism to an R-field. The corresponding localization is called a *universal localization*.

A ring homomorphism $f: R \to S$ is said to be *honest* if it maps full matrices to full matrices; for example, a homomorphism to an R-field is honest if and only if it is fully inverting. Every honest homomorphism is injective; we also have the following useful condition for honesty:

PROPOSITION 4.5.1. *Let R be any ring and S a retract of R; then the inclusion homomorphism $S \to R$ is honest.*

Proof. Let A be a full matrix over S and denote the retraction $R \to S$ by p. If A is not full over R, say $A = PQ$, where Q has fewer rows than A, then we have $A = Pp \cdot Qp$ over S and this contradicts the fact that A was full over S. ∎

To describe fully inverting maps we first give conditions for an epic R-field to coincide with its localization.

LEMMA 4.5.2. *Let R be any ring and $\mathcal{P}$ a prime matrix ideal of R with universal localization $R_{\mathcal{P}}$ and residue-class field K. Suppose that either* (i) *the map $R \to K$ is fully inverting (and so $\mathcal{P}$ consists of all non-full matrices), or* (ii) *every matrix over R which is full over $R_{\mathcal{P}}$ is invertible over K. Then $R_{\mathcal{P}} = K$ and $\mathcal{P}$ is a minimal prime matrix ideal.*

Proof. By Th. 3.4, $R_{\mathcal{P}}$ is a local ring; we have to show that under the given conditions every non-unit of $R_{\mathcal{P}}$ is 0, for then it will coincide with the residue-class field. So let $p \in R_{\mathcal{P}}$ and take an admissible system for p:

$$Au = 0. \tag{1}$$

Since p is a non-unit, its numerator $(A_0 \quad A_*)$ is not invertible over $R_{\mathcal{P}}$ hence it is not invertible over K. In case (i) $(A_0 \quad A_*)$ is non-full over R, while in case (ii) it is not full over $R_{\mathcal{P}}$; thus in either case we can over $R_{\mathcal{P}}$ write $(A_0 \quad A_*) = PQ$, where P is $n \times (n-1)$ and Q is $(n-1) \times n$. Hence

$$A = (A_0 \quad A_* \quad A_n) = (PQ \quad A_n) = (P \quad A_n)\begin{pmatrix} Q & 0 \\ 0 & 1 \end{pmatrix},$$

so $(P \quad A_n)$ is a factor of the denominator $(A_* \quad A_n)$ over $R_{\mathcal{P}}$; the latter is a unit and the local ring $R_{\mathcal{P}}$ is weakly finite, therefore $(P \quad A_n)$ is also invertible over $R_{\mathcal{P}}$. Cancelling this factor in (1), we obtain

$$\begin{pmatrix} Q & 0 \\ 0 & 1 \end{pmatrix} u = 0,$$

and this has the solution $p = 0$. Hence $R_{\mathcal{P}}$ is a field, so $R_{\mathcal{P}} = K$.

Now $\mathcal{P}$ is minimal, for if $\mathcal{P}' \subseteq \mathcal{P}$, then by Th. 4.1 there is a homomorphism $f: R_{\mathcal{P}} \to R_{\mathcal{P}'}$ which must be injective, because $R_{\mathcal{P}}$ is a field. Thus f is an isomorphism and $\mathcal{P}' = \mathcal{P}$, as claimed. ∎

We now have the following characterization of fully inverting maps to R-fields:

THEOREM 4.5.3. *Let R be any ring. Then there is an epic R-field K with a fully inverting map to K if and only if the following two conditions are satisfied:*

(i) *$1 \neq 0$ and the diagonal sum of full matrices over R is full,*

(ii) *the determinantal sum of non-full matrices over R, where defined, is non-full.*

Moreover, K is then the universal localization R_Φ, *where* Φ *is the set of all full matrices over R.*

Proof. Suppose that K as described exists. Then its singular kernel consists of all non-full matrices, so then the complement of Φ in $\mathfrak{M}(R)$ is the least matrix ideal $\mathcal{N}$; thus $\mathcal{N}$ is a prime matrix ideal and (i), (ii) hold. Conversely, assume (i), (ii); then the set $\mathcal{N}$ of all non-full matrices is a matrix ideal by (ii), necessarily the least, and it is prime, by (i). Thus we have an honest map to an epic R-field K, which is fully inverting, so $K = R_\Phi$, by Lemma 5.2. ∎

We next try to find a class of rings to which this result can be applied. Over any ring R we define the inner rank of a matrix A as follows. If A is $m \times n$, consider all factorizations

$$A = PQ, \text{ where } P \text{ is } m \times r \text{ and } Q \text{ is } r \times n. \qquad (2)$$

When r has the least possible value, (2) is called a *rank factorization* of A and r is called the *inner rank* or simply the *rank* of A, written rA. We note that an element of R, as 1×1 matrix, has rank 1 precisely when it is non-zero, while 0 has rank 0. It is clear that for an $m \times n$ matrix A, $rA \leq \min(m, n)$ and a full matrix, as defined in 1.4, is just a square matrix of rank equal to its number of rows. More specifically, an $m \times n$ matrix A is called *left full* if $rA = m$ and *right full* if $rA = n$; it is clear that if A is left full, then $m \leq n$, and if it is right full, then $m \geq n$. The following necessary condition for a matrix to be full is often useful.

PROPOSITION 4.5.4. *An $m \times n$ matrix over any ring with an $r \times s$ block of zeros cannot be left full unless $r + s \leq n$. In particular, an $n \times n$ matrix with an $r \times s$ block of zeros, where $r + s > n$, cannot be full.*

A matrix of the sort described in the last sentence is sometimes called *hollow*.

Proof. Let A be $m \times n$, with an $r \times s$ block of zeros, in the top-right-hand corner, say. We have the factorization

$$A = \begin{pmatrix} P & 0 \\ R & S \end{pmatrix} = \begin{pmatrix} P & 0 \\ 0 & I \end{pmatrix} \begin{pmatrix} I & 0 \\ R & S \end{pmatrix};$$

here P is $r \times (n - s)$ and S is $(m - r) \times s$. So A has been expressed as a product of an $m \times (m + n - r - s)$ by an $(m + n - r - s) \times n$ matrix, and if A is left full, we must have $m + n - r - s \geq m$, i.e. $r + s \leq n$. Now the second part is also clear. ∎

In n-firs the inner rank has the following important property:

PROPOSITION 4.5.5. *If R is an n-fir and $P \in {}^rR^n$, $Q \in {}^nR^s$ are such that*

$$PQ = 0, \tag{3}$$

then

$$rP + rQ \leqslant n. \tag{4}$$

Proof. By Cor. 1.6.2 we can trivialize the relation (3). Let $PU = (P' \quad 0)$, $U^{-1}Q = (0 \quad Q')^T$, where P' is $r \times n_1$, Q' is $n_2 \times s$ and $n_1 + n_2 = n$. Clearly we have $rP \leqslant n_1$, $rQ \leqslant n_2$, therefore (4) holds. ∎

A ring R is called a *Sylvester domain* if R is non-trivial and for any matrices A, B such that the number of columns of A equals the number of rows of B, equal to n, say, the following *nullity condition* holds:

$$AB = 0 \Rightarrow rA + rB \leqslant n. \tag{5}$$

By Prop. 5.5, every semifir is a Sylvester domain. By applying (5) to elements of R, we see that a Sylvester domain is indeed an integral domain. More generally we see that every left full matrix is left regular, and similarly on the right. The reason for the name is that these rings satisfy *Sylvester's law of nullity* for the inner rank:

COROLLARY 4.5.6. *Let R be a Sylvester domain and $A \in {}^rR^n$, $B \in {}^nR^s$. Then*

$$rA + rB \leqslant n + r(AB).$$

Proof. Write $r(AB) = r$ and take a rank factorization $AB = PQ$, so that Q has r rows. Then we have

$$(A \quad P)\begin{pmatrix} B \\ -Q \end{pmatrix} = 0,$$

hence by Prop. 5.5, $n + r \geqslant r(A \quad P) + r(B \quad -Q)^T \geqslant rA + rB$. ∎

We shall need some further consequences of the nullity condition:

LEMMA 4.5.7. *Let R be a Sylvester domain. Then*
  (i) *for any matrices A, B over R, $r(A \oplus B) = rA + rB$,*
  (ii) *if A, B, C are matrices over R with the same number of rows, and if $r(A \quad B) = r(A \quad C) = rA$, then $r(A \quad B \quad C) = rA$.*

Proof. (i) We take rank factorizations $A = PQ$, $B = P'Q'$; then

$$\begin{pmatrix} A & 0 \\ 0 & B \end{pmatrix} = \begin{pmatrix} P & 0 \\ 0 & P' \end{pmatrix} \begin{pmatrix} Q & 0 \\ 0 & Q' \end{pmatrix},$$

and hence $r(A \oplus B) \leqslant rA + rB$. We note that no hypothesis was necessary so far. In the other direction, let

$$\begin{pmatrix} A & 0 \\ 0 & B \end{pmatrix} = \begin{pmatrix} P \\ P' \end{pmatrix} (Q \quad Q')$$

be a rank factorization of $A \oplus B$, where the right-hand side is partitioned so that P and Q are square. Since $PQ' = 0$, we have

$$r(A \oplus B) \geqslant rP + rQ' \geqslant r(PQ) + r(P'Q') = rA + rB.$$

(ii) Let us take suitably partitioned rank factorizations $(A \quad B) = D(E, \quad E')$, $(A \quad C) = F(G \quad G')$. By hypothesis, $A = DE = FG$ are also rank factorizations of A, therefore

$$rA \leqslant rD, rG. \tag{6}$$

It follows that the number of columns of $(D \quad F)$ is $2rA$ and

$$(D \quad F) \begin{pmatrix} E \\ -G \end{pmatrix} = 0,$$

hence by the nullity condition,

$$2rA \geqslant r(D \quad F) + r(E \quad -G)^{\mathrm{T}} \geqslant rD + rG \geqslant r(DE) + r(FG) = 2rA.$$

Hence equality holds throughout, and bearing in mind (6), we find that $rA = rD = rG$. Further, we have $rA = r(D \quad F)$, so there is a rank factorization $(D \quad F) = H(J \quad K)$, where the number of columns of H is $rA = r(D \quad F)$. Now

$$(A \quad B \quad C) = (DE \quad DE' \quad FG') = (HJE \quad HJE' \quad HKG')$$

$$= H(JE \quad JE' \quad KG');$$

therefore $r(A \quad B \quad C) \leqslant rH \leqslant rA$ and (ii) follows. $\blacksquare$

We can now show that any Sylvester domain has a universal field of fractions; in fact we have a somewhat more precise result:

THEOREM 4.5.8. *In any Sylvester domain R the following equivalent conditions are satisfied:*
  (a) *the localization R_Φ at the set Φ of all full matrices is a field,*
  (b) *the set of all non-full matrices over R is a prime matrix ideal.*
Moreover, R_Φ is then the universal field of fractions of R.

Proof. Let $\mathscr{P}$ be the set of all non-full matrices over R. If $\mathscr{P}$ is a prime matrix ideal, then the conditions of Th. 5.3 are fulfilled, so R_Φ is then a field, and this shows that (b) $\Rightarrow$ (a). Conversely, if (a) holds, then the singular kernel of the map $R \to R_\Phi$ is just $\mathcal{P}$, so (b) holds. Next we show that (a) holds in any Sylvester domain, by verifying (i), (ii) of Th. 5.3. Condition (i) follows from Lemma 5.7 (i); to prove (ii), let $A = (a, C)$, $B = (b, C)$ be non-full, where $a, b \in {}^nR$, $C \in {}^nR^{n-1}$ and put $F = (a + b, C)$. If C is right full, then $\mathrm{r}C = n - 1 = \mathrm{r}A = \mathrm{r}B$, hence by Lemma 5.7 (ii), $\mathrm{r}F = \mathrm{r}C$ and so F is non-full. If C is not right full, then $\mathrm{r}C < n - 1$ and again $\mathrm{r}F < n$, so F is non-full in any case. Thus the conditions of Th. 5.2 are satisfied and (a) follows. When this holds, $\mathcal{P}$ is the least prime matrix ideal and so R_Φ is the universal field of fractions of R, again by Th. 5.3. ∎

We remark that the condition of being a Sylvester domain is necessary as well as sufficient for the conditions (a), (b) to hold (see FR, Th. 7.5.10, p. 417, or Ex. 5 below). As we have seen, any semifir is a Sylvester domain, hence we obtain

COROLLARY 4.5.9. *For any semifir (and in particular, any fir) R, the localization R_Φ at the set Φ of all full matrices is a universal field of fractions for R, with a fully inverting homomorphism $R \to R_\Phi$.* ∎

One useful consequence of this result is the following simple criterion for the extendibility of homomorphisms to the field of fractions:

THEOREM 4.5.10. *Let R, S be semifirs, or more generally, Sylvester domains, and $\mathcal{U}(R)$, $\mathcal{U}(S)$ their universal fields of fractions. Given a homomorphism $f: R \to S$, f extends to a homomorphism $f': \mathcal{U}(R) \to \mathcal{U}(S)$, necessarily unique, if and only if f is honest. In particular, every isomorphism between R and S extends to a unique isomorphism between $\mathcal{U}(R)$ and $\mathcal{U}(S)$.*

Proof. Denote by Φ, Ψ the set of all full matrices over R, S respectively. If f is honest, then $\Phi f \subseteq \Psi$, and so the mapping $R \to S \to S_\Psi$ is Φ-inverting. Hence there is a unique homomorphism $f': R_\Phi \to S_\Psi$ such that the diagram shown commutes, which means that f can be extended

in just one way. Conversely, if an extension exists, then any full matrix A over R is inverted over R_Φ and so is mapped to an invertible matrix over S_Ψ. But this is the image of Af, which must therefore be full; hence f is honest, as claimed. ∎

The rest follows since an isomorphism is always honest. ∎

For Sylvester domains Malcolmson's criterion can still be simplified. If R is a Sylvester domain, then any product of full matrices is full, and likewise for diagonal sums of full matrices, by Th. 5.8. Suppose now that a is a matrix block representing zero, $[a] = 0$. Then the product in (11) on p. 167 is non-full, but the left and right factors are full, hence the middle factor $a \oplus I$ is non-full, and hence so is a itself. Conversely, when a is non-full, then so is the product. Thus we obtain

THEOREM 4.5.11. *Let R be a Sylvester domain. Then any matrix block a over R represents zero if and only if a is not full.* ■

In the next section we shall see that like semifirs, Sylvester domains are projective-free. It can also be shown that a Sylvester domain has weak global dimension at most two but we shall not do so here (see FR, Cor. 5.5.5, p. 256). An example of a Sylvester domain of global dimension exactly 2 is the polynomial ring in two variables over a commutative field: $k[x, y]$ (see FR, Th. 5.5.12, p. 260). Thus we have found fields of fractions for a class of rings that are (i) projective-free and (ii) of weak global dimension at most 2. When we drop the first condition, some restriction has to be placed on the projective modules, for a field of fractions to exist. In the next section we shall examine the consequences of having a rank function defined on projectives, to generalize the rank of free modules.

Exercises

1. Show that an Ore domain is a Sylvester domain if and only if every full matrix is regular.

2. Show that $R = k[x]/(x^2)$ (where k is a commutative field), has a universal R-field, but no field of fractions. What are the conditions for a commutative ring R to have (i) a universal R-field, (ii) a universal field of fractions?

3. Show that the homomorphism of $k\langle z_0, z_1, \ldots \rangle$ into $k\langle x, y \rangle$ defined by $z_n \mapsto y^n x$ is injective but not honest. Show also that the homomorphism f defined by $z_0 f = y$, $z_{n+1} f = (z_n f)x - x(z_n f)$ is honest (see Cohn [90]).

4. Let A be a full matrix such that when the first column is omitted, the result is not left full. Find a factorization $A = B(1 \oplus C)$, where B and C are both full. Use this result to show that when the set of all full matrices is closed under diagonal sums and products, where defined, then the inner rank of a matrix is the maximum of the orders of its full submatrices.

5. Verify that the conditions of the last exercise are satisfied whenever R has an honest homomorphism to a field. Deduce that such a homomorphism preserves the inner rank. Hence show that any ring satisfying (a) or (b) of Th. 5.8 is a Sylvester domain.

$6°$. Let R be a semifir and S a subset of $R^\times$ which contains with any element c all elements similar to c and all factors of c. Is the ring $\langle R; c = 0$ for all $c \in S \rangle$ embeddable in a field?

$7°$. Let R be a fir and $\mathcal{P}$ a prime matrix ideal. Show that $\mathcal{Q} = \bigcap \mathcal{P}^n$ is again prime (where $\mathcal{P}^n$ is the matrix ideal generated by $A_1 \oplus \ldots \oplus A_n$, $A_i \in \mathcal{P}$). Is $\mathcal{Q}$ always the least prime matrix ideal of R? (Note that a proper ideal $\mathfrak{a}$ in any fir satisfies $\bigcap \mathfrak{a}^n = 0$, see FR, Cor. 5.11.4, p. 296, also 9.5 below.)

4.6 Projective rank functions and hereditary rings

In this section we shall examine conditions under which hereditary rings have fields of fractions; the results will not be used elsewhere in the book, so this section can be omitted on a first reading.

Given any ring R, let M be an R-module and take a presentation of M:

$$G \xrightarrow{\alpha} F \to M \to 0,$$

where F, G are free R-modules. If M is finitely presented, F and G may be taken free of finite rank and the map α is then represented by a matrix. If M is finitely generated projective, then F splits over im α and it follows that G may then also be taken to be finitely generated; thus any finitely generated projective module is actually finitely presented. We also recall that M is projective if and only if $\alpha: G \to F$ can be chosen so that $\alpha': F \to G$ exists satisfying $\alpha\alpha'\alpha = \alpha$ (see A.3, Prop. 6.6.1, p. 240). We shall mainly be interested in finitely generated projective modules, where this description takes the following simple form:

PROPOSITION 4.6.1. *Let R be any ring and P a projective right R-module with an n-element generating set. Then there is an idempotent $n \times n$ matrix E such that P is isomorphic to the cokernel of the endomorphism of ^nR defined by $I - E$. Conversely, the cokernel of the map defined by any idempotent matrix is finitely generated projective.*

Proof. We have $P = \operatorname{coker} \alpha$, where $\alpha: G \to F$ satisfies $\alpha\alpha'\alpha = \alpha$ for some $\alpha': F \to G$. Since P is generated by n elements, F can be taken as ^nR and $\alpha'\alpha$ is an idempotent endomorphism of ^nR, represented by an idempotent matrix; since im $\alpha =$ im $\alpha'\alpha$, it follows that $P = \operatorname{coker} \alpha'\alpha$.

Conversely, an idempotent $n \times n$ matrix corresponds to an idempotent endomorphism α of F, hence $F = \operatorname{im} \alpha \oplus \ker \alpha$, so $P \cong \ker \alpha$ is projective. ■

If P is associated with the idempotent matrix E, then P is the cokernel defined by $E' = I - E$. Now $^nR \cong \ker E' \oplus \operatorname{im} E'$, so $\operatorname{coker} E' = \ker E'$, i.e. P consists of all $u \in {}^nR$ such that $(I - E)u = 0$, in other words, $u = Eu$. Hence $P = E \cdot {}^nR$. We shall also want to describe the isomorphisms of projective modules in terms of the idempotent matrices representing them.

PROPOSITION 4.6.2. *Let R be any ring and P, Q finitely generated projective R-modules represented by idempotent matrices $E \in R_m$, $F \in R_n$ respectively. Then $P \cong Q$ if and only if there exist $A \in {}^mR^n$, $B \in {}^nR^m$ such that $AB = E$, $BA = F$.*

Proof. By definition, P is generated by the columns of E and Q is generated by the columns of F. Let $\theta: P \to Q$ be an ismorphism and suppose that θ maps Eu to Bu, while θ^{-1} maps Fv to Av. Applying θ to the columns of E and then θ^{-1}, we obtain $E = AB$; similarly, applying θ^{-1} to the columns of F and then θ, we find that $F = BA$. Conversely, if $E = AB$, $F = BA$, then $Eu \mapsto Bu$, $Fv \mapsto Av$ are mutually inverse homomorphisms between P and Q. ■

Two idempotent matrices $E \in R_m$, $F \in R_n$ are said to be *isomorphic* if there exist $A \in {}^mR^n$, $B \in {}^nR^m$ such that $AB = E$, $BA = F$. We note that on replacing A, B by EAF, FBE we may assume that in addition, $EA = A = AF$, $FB = B = BE$.

As an application of these ideas we show that Sylvester domains are projective-free:

THEOREM 4.6.3. *Every Sylvester domain is projective-free.*

Proof. Let R be a Sylvester domain; as subring of a field, R is weakly finite, in particular R has IBN, so to show that R is projective-free, it only remains to verify by Prop. 6.2, that every idempotent matrix is isomorphic to a unit matrix. Let E be an $n \times n$ idempotent matrix and put $rE = r$, $r(I - E) = s$. Since $E(I - E) = 0$, we have $r + s \leq n$. Taking rank factorizations $E = PQ$, $I - E = P'Q'$, we have

$$ (P \quad P') \binom{Q}{Q'} = PQ + P'Q' = I. $$

Hence $r + s = n$ and by weak finiteness,

$$\begin{pmatrix} Q \\ Q' \end{pmatrix} (P \quad P') = I,$$

thus $QP = I_r$ and this shows E to be isomorphic to I_r, hence the projective module defined by E is free. ∎

For any ring R consider the category $\mathcal{P}_R$ of all finitely generated projective right R-modules and their homomorphisms. The set $\mathcal{P}(R)$ of isomorphism classes can be defined as a commutative monoid by denoting the class of P by $[P]$ and writing

$$[P] + [Q] = [P \oplus Q]. \tag{1}$$

The universal group of $\mathcal{P}(R)$ is just the *Grothendieck group* of projectives, $\mathbf{K}_0(R)$. Of course we obtain the same monoid by using finitely generated projective *left* R-modules, by the duality $P \mapsto P^* = \mathrm{Hom}_R(P, R)$. The image of $\mathcal{P}(R)$ in $\mathbf{K}_0(R)$ is the monoid $\bar{\mathcal{P}}(R)$ of *stable isomorphism classes*, where $P, P' \in \mathcal{P}_R$ are called *stably isomorphic*, in symbols $P \sim P'$, if

$$P \oplus {}^nR \cong P' \oplus {}^nR$$

for some $n \geq 0$. For any two stably isomorphic modules are clearly identified in $\bar{\mathcal{P}}(R)$ and the monoid of such classes admits cancellation, and so has $\mathbf{K}_0(R)$ as group of fractions. The stable isomorphism class of P will be denoted by $\overline{[P]}$.

The correspondence $R \mapsto \mathcal{P}(R)$ is a functor from rings to monoids, for, given a homomorphism $f: R \to S$ and a projective R-module P, we have an S-module $P \otimes_R S$, which is again projective: if $P \oplus P' \cong {}^nR$, then $P \otimes S \oplus P' \otimes S \cong {}^nR \otimes S \cong {}^nS$. Thus f induces a map

$$\mathcal{P}(R) \to \mathcal{P}(S), \tag{2}$$

which is easily seen to be a monoid homomorphism. Moreover, stably isomorphic modules have stably isomorphic images, so that we also have a homomorphism from $\bar{\mathcal{P}}(R)$ to $\bar{\mathcal{P}}(S)$. A projective S-module is said to be *induced* from R if it occurs in the image of the map (2).

For any set $\mathcal{G}$ of objects in $\mathcal{P}_R$ an object P of $\mathcal{P}_R$ is called *subordinate* to $\mathcal{G}$ if it is a direct summand of a direct sum of members of $\mathcal{G}$; e.g. every projective is subordinate to $\{R\}$.

The monoid $\bar{\mathcal{P}}(R)$ can be used to define a preordering of $\mathbf{K}_0(R)$. Given $\alpha, \beta \in \mathbf{K}_0(R)$, we write $\alpha \leq \beta$ to mean: $\beta - \alpha \in \bar{\mathcal{P}}(R)$. This defines a preordering preserved by addition, because $\bar{\mathcal{P}}(R)$ is a monoid; the next result gives conditions for the preordering to be a partial ordering:

PROPOSITION 4.6.4. *For any ring R, the preordering defined on* $\mathbf{K}_0(R)$ *is a partial ordering, provided that R is weakly finite.*

Proof. The condition for a partial ordering is antisymmetry:

$$\overline{[P]} \leqslant \overline{[Q]}, \overline{[Q]} \leqslant \overline{[P]} \quad \Rightarrow \overline{[P]} = \overline{[Q]}. \tag{3}$$

The hypothesis in (3) states $\overline{[P \oplus S]} = \overline{[Q]}$, $\overline{[Q \oplus T]} = \overline{[P]}$, hence $\overline{[P \oplus S \oplus T]} = \overline{[P]}$; adding P', where $P \oplus P' = {}^nR$, we have $\overline{[S \oplus T]} = 0$, hence $S \oplus T \oplus {}^mR \cong {}^mR$; by weak finiteness, $S \oplus T = 0$, hence $S = T = 0$ and $P \cong Q$. ∎

We now turn to the question of generalizing the notion of rank. In the simplest case, that of a field, the dimension of a vector space is an important invariant, in fact the only one, since two spaces of the same dimension are isomorphic. For general modules there may be no numerical invariant, but for many rings there is a rank function on projective modules. Generally, for any ring R, by a *rank function* on projectives we shall understand a function ρ on the finitely generated projective right R-modules with non-negative real values, such that
(R.1) $P \cong P' \Rightarrow \rho P = \rho P'$,
(R.2) $\rho(P \oplus Q) = \rho P + \rho Q$,
(R.3) $\rho R = 1$.
If $\rho P \neq 0$ for $P \neq 0$, ρ is said to be *faithful*. We note that the relation $P \oplus S \cong Q \oplus S$ implies $\rho P = \rho Q$, by (R.1–2). Hence ρ is constant on stable isomorphism classes and so may be regarded as a function on $\overline{\mathscr{P}}(R)$, and it extends to a homomorphism from $\mathbf{K}_0(R)$ to R which is 1 on R. Conversely, every homomorphism from $\mathbf{K}_0(R)$ to R which is order-preserving and takes the value 1 on R defines a rank function on $\overline{\mathscr{P}}(R)$.

Our first aim will be to determine when a ring has a rank function, but some preparation is necessary. We begin by observing that every rank function on R also defines a rank function on certain homomorphic images of R. For the proof we shall need the notion of a trace ideal. For any right R-module M we define its *trace ideal* $\tau(M)$ as the set

$$\tau(M) = \left\{ \sum (f, x) \,|\, x \in M, f \in M^* = \mathrm{Hom}_R(M, R) \right\}. \tag{4}$$

Bearing in mind that M is a right and M^* a left R-module, we can easily verify that $\tau(M)$ is an ideal in R; e.g. for a non-zero free module F, $\tau(F) = R$. More generally, for a projective right ideal $\mathfrak{a}$, we can take f in (4) as the inclusion map, and this shows that $\tau(\mathfrak{a}) \supseteq \mathfrak{a}$. For a projective R-module P we can, by the dual basis lemma (see e.g. A.2, Prop. 4.5.5,

p. 149), write every $x \in P$ as $x = \sum u_i(\alpha_i, x)$, where the u_i form a generating set of P and $\alpha_i \in P^*$. Hence we have, for $x \in P$,

$$(f, x) = \sum (f, u_i(\alpha_i, x)) = \sum (f, u_i)(\alpha_i, x).$$

This shows that $\tau(P)^2 = \tau(P)$, thus the trace ideal of a projective module is idempotent.

More generally we can, for any set $\mathscr{I}$ of finitely generated projective right R-modules define $\tau(\mathscr{I})$ as the union of all the $\tau(P)$, where P ranges over all projectives subordinate to $\mathscr{I}$. Our next result describes the effect of dividing by a trace ideal.

PROPOSITION 4.6.5. *Given a ring R and a set $\mathscr{I}$ of finitely generated projective right R-modules, put $\mathfrak{t} = \tau(\mathscr{I})$. Then for any $Q \in \mathscr{P}_R$, $Q \otimes_R (R/\mathfrak{t}) = 0$ if and only if Q is subordinate to $\mathscr{I}$. Further, the monoid of induced projectives over $R/\mathfrak{t}$ is the quotient of $\mathscr{P}(R)$ by the relation $P \sim P'$ if and only if $P \oplus Q \cong P' \oplus Q'$, where Q, Q' are subordinate to $\mathscr{I}$.*

Proof. If $Q \otimes (R/\mathfrak{t}) = 0$, then every element of Q lies in the image of a map from a sum of members of $\mathscr{I}$ to Q. Since Q is finitely generated, there is a surjection from a sum of members of $\mathscr{I}$ to Q, hence Q is subordinate to $\mathscr{I}$. The converse is clear.

Now let $\alpha: P \to P'$ be a map inducing an isomorphism between $P \otimes (R/\mathfrak{t})$ and $P' \otimes (R/\mathfrak{t})$. Then there is a surjection $\alpha \oplus \beta: P \oplus Q \to P'$, where Q is subordinate to $\mathscr{I}$, and this map splits over $\ker(\alpha \oplus \beta)$:

$$P \oplus Q \cong P' \oplus \ker(\alpha \oplus \beta).$$

Moreover, $\ker(\alpha \oplus \beta)$ becomes zero over $R/\mathfrak{t}$ and so is subordinate to $\mathscr{I}$. Conversely, if $P \oplus Q \cong P' \oplus Q'$, with Q, Q' subordinate to $\mathscr{I}$, then clearly P, P' become isomorphic over $R/\mathfrak{t}$. ∎

We next prove a lemma on extending homomorphisms of preordered groups, analogous to the Hahn–Banach theorem; this will enable us to extend rank functions.

Consider the category whose objects are preordered groups G with an *order-unit*, (G, u), i.e. a distinguished element u such that G is the convex hull of the subgroup generated by u: for any $x \in G$ there exists $n \geq 0$ such that $-nu \leq x \leq nu$. The morphisms are homomorphisms preserving the preorder and the order-unit. For example, the additive group of real numbers with the usual order and the order-unit 1 is an object $(\mathbf{R}, 1)$ in this category. Any morphism $\lambda: (G, u) \to (\mathbf{R}, 1)$, i.e. a homomorphism λ from G to $\mathbf{R}$ such that $u\lambda = 1$, is called a *state*. Given (G, u), we write $G_+ = \{x \in G | x \geq 0\}$ for the positive cone of G and note that G is generated as a group, by G_+, for if $x \in G$ and $-nu \leq x \leq nu$, then $x + nu \geq 0$ and so $x = (x + nu) - nu$, where $x + nu, nu \in G_+$.

LEMMA 4.6.6. *Let G be a preordered abelian group with order-unit u and let (H, u) be a subobject of (G, u), i.e. H is a subgroup of G, containing u, with the preorder induced from G. Then every state on (H, u) extends to a state on (G, u).*

Proof. Let ϕ be a state on (H, u) and let $\mathcal{F}$ be the family of all subgroups containing H with states extending ϕ. $\mathcal{F}$ has a natural partial order: $(K, \lambda) \leq (K', \lambda')$ if $K \subseteq K'$ and λ' extends λ. It is clear that $\mathcal{F}$ is inductive, and so by Zorn's lemma it has a maximal element (K, λ). We shall complete the proof by showing that $K = G$.

If $G_+ \subseteq K$, then $G \subseteq K$, so when $K \subset G$, we can find $t \in G_+ \backslash K$. We define two real numbers α, β as follows:

$$\alpha = \sup \{\lambda(x)/n | x \in K \text{ and } nt \geq x \text{ for some } n \geq 0\},$$

$$\beta = \inf \{\lambda(x)/n | x \in K \text{ and } nt \leq x \text{ for some } n \geq 0\}.$$

We claim that

$$0 \leq \alpha \leq \beta < \infty. \tag{5}$$

Since $t \geq 0 \in K$, we have $0 = \lambda(0)/1 \leq \alpha$. Next $t \leq ru$ for some $r > 0$, hence $\beta \leq \lambda(ru)/1 = r < \infty$. Now take $p, q \in K$ and positive integers m, n such that $p \leq mt$ and $nt \leq q$. Then $np \leq mnt \leq mq$ and so $n\lambda(p) \leq m\lambda(q)$, hence $\lambda(p)/m \leq \lambda(q)/n$. Now $\sup(\lambda(p)/m) = \alpha$, $\inf(\lambda(q)/n) = \beta$, hence $\alpha \leq \beta$ and (5) follows.

We now take any $\gamma \in \mathbf{R}$ such that $\alpha \leq \gamma \leq \beta$ and show that

If $x \geq nt$ for some $x \in K$, $n \in \mathbf{Z}$, then $\lambda(x) \geq n\gamma$. $\tag{6}$

If $n = 0$, then $x \geq 0$ and so $\lambda(x) \geq 0$. If $n > 0$, then $x \geq nt$, so $\lambda(x)/n \geq \beta \geq \gamma$ and therefore $\lambda(x) \geq n\gamma$. If $n < 0$, then $-x \leq (-n)t$, hence $\lambda(-x)/(-n) \leq \alpha \leq \gamma$ and again $\lambda(x) \geq n\gamma$; so (6) holds for all $n \in \mathbf{Z}$.

Suppose now that $x + nt = 0$. Then by (6), $\lambda(x) + n\gamma \geq 0$, but also $(-x) + (-n)t = 0$, hence $\lambda(-x) + (-n)\gamma \geq 0$, and so $\lambda(x) + n\gamma = 0$. This shows that the map $\mu: K + \mathbf{Z}t \to \mathbf{R}$ given by

$$\mu(x + nt) = \lambda(x) + n\gamma, \text{ for } x \in K, n \in \mathbf{Z},$$

is well-defined. Moreover, (6) shows that $(K + \mathbf{Z}t) \cap G_+$ is mapped to $\mathbf{R}_+$, so μ preserves the preorder; further, $\mu(u) = \lambda(u) = 1$, and this shows that μ is a state on $(K + \mathbf{Z}t, u)$ extending λ. This contradicts the maximality of (K, λ), so $K = G$ and λ is the required extension of ϕ. ∎

For example, let ρ be a rank function on R and let $\mathscr{I}$ be the set of all projectives on which ρ is zero. Then ρ is defined on the monoid of induced projectives over $R/\tau(\mathscr{I})$, by Prop. 6.5, and by Lemma 6.6 it can be extended to a rank function ρ' on $R/\tau(\mathscr{I})$. If every projective $(R/\tau(\mathscr{I}))$-module is induced from one of R, then it follows that ρ' is faithful. E.g. this is so when R is v. Neumann regular (see Goodearl [79], 16.7) or when R is weakly semihereditary (see Prop. 6.9 below), but it does not hold in general. However, for any ring R with a rank function ρ we can define t_1 as the trace ideal of all the projectives on which ρ vanishes, and on R/t_1 there is an induced rank function ρ_1; moreover, since $\rho R = 1$, t_1 is a proper ideal. If ρ_1 is not faithful, we can take the trace ideal of all the projectives on which ρ_1 vanishes; it corresponds to an ideal $t_2 \supseteq t_1$ in R. Continuing in this way, we obtain an ascending sequence (t_n) of proper ideals, and rank functions ρ_n on R/t_n, which all correspond to each other. Since all the t_n are proper, so is their union t_ω and there is a rank function ρ_ω defined on R/t_ω. If ρ_ω is not faithful, we can continue this process transfinitely. For some ordinal α of cardinal at most $|R|$ the chain becomes stationary. If t is the corresponding ideal, then R/t is a non-zero ring with a faithful rank function. Thus we have proved

PROPOSITION 4.6.7. *Let R be any ring with a rank function ρ. Then there is a proper ideal t of R such that R/t has a faithful rank function which on projective modules induced from R agrees with ρ.* ∎

We can now achieve our aim of describing the rings which have a rank function; the necessary and sufficient condition turns out to be UGN (see 1.4).

THEOREM 4.6.8. *For any ring R the following conditions are equivalent:*

(a) *R has a rank function on projectives,*
(b) *R has a non-zero homomorphic image which is weakly finite,*
(c) *R has unbounded generating number.*

Proof. (a) ⇒ (b). If R has a rank function ρ, then by Prop. 6.7, a non-zero homomorphic image R' of R has a faithful rank function ρ'. We claim that R' is weakly finite; for if ${}^nR' \cong {}^nR' \oplus K$, then $\rho'(K) + n = n$, hence $\rho'(K) = 0$ and so $K = 0$, because ρ' is faithful, and this shows R'

to be weakly finite. To show (b) ⇒ (c), let R satisfy (b) and suppose that $\bar{R}$ is a non-zero weakly finite image of R. Then $\bar{R}$ has UGN, hence so does R, i.e. (c) holds. Finally, assume that R satisfies (c); this means that $^n R \cong {}^m R \oplus K$ implies $n \geq m$. In terms of the order-unit u of $\mathbf{K}_0(R)$ this states that

$$mu \leq nu \Rightarrow m \leq n. \tag{7}$$

It follows that the subgroup $\langle u \rangle$ generated by u is infinite cyclic, so we have a homomorphism $f: \langle u \rangle \to \mathbf{R}$ defined by $f(nu) = n$. To show that f is a state we must verify that $nu \geq 0 \Rightarrow f(nu) \geq 0$; but this holds by (7) with $m = 0$. Now f can be extended to a state on $\mathbf{K}_0(R)$ and hence a rank function on R by Lemma 6.6, and this shows that R satisfies (a). ∎

This result shows that most rings normally encountered have a rank function. For example, every projective-free ring clearly has a unique rank function. We shall find that there are other rings with a unique rank function, but the rank functions themselves will have better properties for rings that are hereditary. A weaker condition that is left–right symmetric is often useful. A ring R is said to be *weakly semihereditary* if for any two maps between finitely generated projective R-modules

$$\alpha: P \to P', \quad \beta: P' \to P'', \tag{8}$$

such that $\alpha\beta = 0$, we can write $P' = P_1' \oplus P_2'$ such that $\mathrm{im}\,\alpha \subseteq P_1' \subseteq \ker \beta$. In terms of matrices this condition can be restated as follows: For any $A \in {}^r R^n$, $B \in {}^n R^s$ such that $AB = 0$, there exists an idempotent $n \times n$ matrix E such that $AE = A$, $EB = 0$. By applying the duality * (or replacing E by $I - E$ in the matrix condition) we see that this condition is left–right symmetric. In a right semihereditary ring R, if α, β are as in (8), then $\mathrm{im}\,\beta$ is a finitely generated submodule of P'', hence projective and so P' splits over $\ker \beta$: $P' = \mathrm{im}\,\beta \oplus \ker \beta$. Since $\alpha\beta = 0$, we have $\mathrm{im}\,\alpha \subseteq \ker \beta$, and this shows R to be weakly semihereditary. Thus every right semihereditary ring is weakly semihereditary and by symmetry the same holds for left semihereditary rings.

For a weakly semihereditary ring it has been shown (by G. M. Bergman [72], see e.g. FR, Th. 0.3.7, p. 14), that every projective module is a direct sum of finitely generated modules. This generalization of Kaplansky's theorem uses the latter's decomposition theorem of a projective into countably generated modules.

In a weakly semihereditary ring we can pass from a rank function to a faithful rank function on a homomorphic image in a single step. This follows from the next result, which shows every projective module of the image to be induced.

PROPOSITION 4.6.9. *Let R be a weakly semihereditary ring and* t *the trace ideal of a set* $\mathcal{I}$ *of finitely generated projective R-modules. Then R/t is again weakly semihereditary and every finitely generated projective (R/t)-module is induced from R.*

Proof. We shall use a bar to indicate the map $R \to R/t$, thus $\bar{R} = R/t$ and $\bar{P} = P \otimes \bar{R}$ for $P \in \mathcal{P}_R$. Let $\bar{\alpha}: \bar{P} \to \bar{P}'$, $\bar{\beta}: \bar{P}' \to \bar{P}''$ be maps between finitely generated projective $\bar{R}$-modules such that $\bar{\alpha}\bar{\beta} = 0$. Then $\overline{\alpha\beta} = 0$, hence $\alpha\beta$ can be factored by a map to a projective subordinate to $\mathcal{I}$, say $\alpha\beta = \gamma\delta$, where $\gamma: P \to Q$, $\delta: Q \to P''$. Over R we have

$$(\alpha \quad \gamma)\begin{pmatrix} \beta \\ -\delta \end{pmatrix} = 0, \quad \bar{\gamma} = 0 = \bar{\delta}.$$

Since R is weakly semihereditary, we have $P' \oplus Q \cong P_1 \oplus P_2$, where $\operatorname{im}(\alpha \quad \gamma) \subseteq P_1 \subseteq \ker(\beta \quad -\delta)^{\mathrm{T}}$. Hence $\bar{P}' \cong \bar{P}_1 \oplus \bar{P}_2$ and $\operatorname{im} \bar{\alpha} \subseteq \bar{P}_1 \subseteq \ker \bar{\beta}$, and this shows that the weakly semihereditary condition holds between induced projectives.

Now any finitely generated projective $\bar{R}$-module may be defined by an idempotent matrix over $\bar{R}$, and this corresponds to a map $e: {}^nR \to {}^nR$ such that $\bar{e}^2 = \bar{e}$, i.e. $\bar{e}(1 - \bar{e}) = 0$. By the previous argument, applied to the maps $\bar{e}$, $1 - \bar{e}$, we have ${}^n\bar{R} \cong \bar{P}_1 \oplus \bar{P}_2$, where $\operatorname{im} \bar{e} \subseteq \bar{P}_1 \subseteq \ker(1 - \bar{e})$. But $\operatorname{im} \bar{e} = \ker(1 - \bar{e})$, hence $\operatorname{im} \bar{e} = \bar{P}_1$ and this shows that all finitely generated projective $\bar{R}$-modules are induced. By the first part $\bar{R}$ is weakly semihereditary, as we had to show. ∎

COROLLARY 4.6.10. *Let R be a weakly semihereditary ring with a rank function ρ and denote by t the trace ideal of all the projectives on which ρ vanishes. Then there is a unique rank function $\bar{\rho}$ induced on R/t by ρ and $\bar{\rho}$ is faithful.*

Proof. It is clear that ρ defines a unique rank function $\bar{\rho}$ on the induced projective (R/t)-modules, but by Prop. 6.9 this includes all finitely generated projectives, so $\bar{\rho}$ is uniquely defined. Moreover, if $\bar{\rho}\bar{P} = 0$, then $\rho P = 0$, hence $\bar{P} = 0$ by the definition of t. It follows that $\bar{\rho}$ is faithful. ∎

For each rank function ρ on projectives there is an inner rank on matrices, defined as before: Given a map $\alpha: P \to Q$ between projectives, the *inner ρ-rank* of α, $\rho(\alpha)$, is the infimum of $\rho(P')$ for all projectives P' such that α can be taken via P':

$$\rho(\alpha) = \inf \{\rho(P') | \alpha = \beta\gamma \text{ for } \beta: P \to P', \gamma: P' \to Q\}.$$

If 1_P is factored by maps via P', then P is a direct summand of P', hence

$\rho(1_P) = \rho(P)$. Further it is clear that $\rho(\alpha) \leqslant \min\{\rho P, \rho Q\}$. The map $\alpha: P \to Q$ is called *left full* if $\rho(\alpha) = \rho P$, *right full* if $\rho(\alpha) = \rho Q$ and *full* if it is left and right full. If more than one rank function is involved, we shall speak of ρ-full maps etc., to avoid confusion. In particular, for an $m \times n$ matrix A with $m \leqslant n$ say, we have $\rho A \leqslant m$, with equality if and only if A is left ρ-full.

For an idempotent matrix E the inner ρ-rank clearly equals the ρ-rank of the projective module defined by E, hence for any matrix A the inner ρ-rank may also be defined as

$$\rho A = \inf\{\rho E \,|\, E \text{ idempotent and } A = BEC\}. \tag{9}$$

We have seen that for the existence of universal fields of fractions the nullity condition played an important role. Such a condition can be considered more generally; if the nullity condition holds for ρ,

Given $\alpha: P \to P'$, $\beta: P' \to P''$, if $\alpha\beta = 0$, then $\rho(\alpha) + \rho(\beta) \leqslant \rho P'$,

then ρ is said to be a *Sylvester rank function*. As before, the law of nullity can be recovered from this special case: if $\alpha: P \to P'$, $\beta: P' \to P''$, then for any $\gamma: P \to Q$, $\delta: Q \to P''$ we have

$$\rho(\alpha \quad \gamma) + \rho(\beta \quad -\delta)^{\mathrm{T}} \leqslant \rho P' + \rho Q,$$

hence $\rho(\alpha) + \rho(\beta) \leqslant \rho P' + \rho Q$, and taking the infimum as Q varies, we obtain

$$\rho(\alpha) + \rho(\beta) \leqslant \rho P' + \rho(\alpha\beta).$$

It is of interest that this condition holds in all weakly semihereditary rings:

PROPOSITION 4.6.11. *In a weakly semihereditary ring every rank function is Sylvester.*

Proof. Let $\alpha: P \to P'$, $\beta: P' \to P''$ satisfy $\alpha\beta = 0$. Then there exists a decomposition $P' \cong P_1 \oplus P_2$ such that $\operatorname{im} \alpha \subseteq P_1 \subseteq \ker \beta$. Hence $\rho(\alpha) \leqslant \rho P_1$, $\rho(\beta) \leqslant \rho P_2$ and so $\rho(\alpha) + \rho(\beta) \leqslant \rho P_1 + \rho P_2 = \rho P'$. ∎

Homomorphisms to a field form an important source of rank functions. Thus let R be any ring with an R-field K, and for $P \in \mathscr{P}_R$ put $\bar{P} = P \otimes K$. Every projective K-module T is free of uniquely determined rank $\mathrm{r}T$ and we can use it to define a rank function on $\mathscr{P}_R$ by the rule

$$\rho P = \mathrm{r}\bar{P}.$$

Clearly $P \cong P'$ implies $\rho P = \rho P'$ and $\rho R = 1$, and since $P \cong P_1 \oplus P_2$ implies $\bar{P} \cong \bar{P}_1 \oplus \bar{P}_2$, we see that $\rho(P_1 \oplus P_2) = \rho P_1 + \rho P_2$. Thus ρ is indeed a rank function; moreover, it is $\mathbf{Z}$-valued. If K is actually a field of fractions, then the sequence $0 \to R \to K$ is exact and tensoring with P we obtain the exact sequence $0 \to P \to \bar{P}$; hence ρ is then faithful. Conversely, if ρ is faithful, then $\rho E > 0$ for every non-zero idempotent matrix E, and so for $a \in R$, $a \neq 0$, we have

$$\rho a = \inf \{\rho E | a = bEc,\ E \text{ idempotent}\} \neq 0,$$

because ρ is discrete-valued; so R is embedded in K. This proves

THEOREM 4.6.12. *Let R be any ring with an epic R-field K. Then the function on projectives defined by pullback from the dimension over K is a $\mathbf{Z}$-valued rank function on projectives, which is faithful if and only if K is a field of fractions of R.* ∎

For the proof of the next result we note that the properties of the inner rank established in Lemma 5.7 for Sylvester domains hold for general $\mathbf{Z}$-valued Sylvester rank functions, with the same proof:

$$\text{For any matrices } A, B, \rho(A \oplus B) = \rho A + \rho B. \tag{10}$$

If A, B, C are any matrices with the same number of rows and $\rho(A \quad B) = \rho(A \quad C) = \rho A$, then

$$\rho(A \quad B \quad C) = \rho A. \tag{11}$$

PROPOSITION 4.6.13. *Let R be a ring with a $\mathbf{Z}$-valued Sylvester rank function ρ. Then the set $\mathcal{P}$ of all matrices that are not ρ-full is a prime matrix ideal, $R_{\mathcal{P}}$ is a field and the natural homomorphism λ from R to $R_{\mathcal{P}}$ is rank-preserving in the sense that $\mathrm{r}(A\lambda) = \rho A$. The kernel of λ is the ideal generated by the entries of idempotent matrices E satisfying $\rho E = 0$.*

Proof. We have to verify (MI.1–6) for $\mathcal{P}$. (MI.1) is clear and (MI.3) follows by (11), (MI.2) and (MI.6) follow by (10) and (MI.4–5) follow from (10) and the fact that $\rho I_n = n$. Thus $\mathcal{P}$ is a prime matrix ideal. Let $K = R/\mathcal{P}$ be the corresponding epic R-field; we claim that the natural map $R \to K$ preserves the rank ρ. Given $A \in {}^m R^n$, let r be the greatest integer for which A has a ρ-full $r \times r$ submatrix. Then every $(r + 1) \times (r + 1)$ submatrix of A has ρ rank at most r; by induction, using (11), we can add columns and find that every $(r + 1) \times n$ submatrix has ρ-rank at most r. By symmetry we can add rows and so find that $\rho A = r$. Thus if $\rho A = r$, then there is a ρ-full $r \times r$ submatrix, but no

$(r + 1) \times (r + 1)$ submatrix can be ρ-full. Hence over K some $r \times r$ submatrix of A is non-singular, but every $(r + 1) \times (r + 1)$ submatrix is singular; this shows the natural map $R \to K$ to be rank-preserving. The kernel consists of all 1×1 matrices of ρ-rank 0, and by (9) this is the ideal generated by the entries of idempotent matrices of zero ρ-rank.

Now $R_{\mathscr{P}}$ is a local ring, hence projective-free, so the residue-class map $R_{\mathscr{P}} \to K$ preserves the unique rank function. If $A \in \mathfrak{M}(R)$ is full over $R_{\mathscr{P}}$, it is full over K and so is inverted; it follows that the map $R \to R_{\mathscr{P}}$ also preserves the ρ-rank, so any member of $\mathscr{P}$, being non-ρ-full, maps to a non-ρ-full matrix over $R_{\mathscr{P}}$, so by Lemma 5.2 (ii), $R_{\mathscr{P}}$ is a field and $\mathscr{P}$ is a minimal prime matrix ideal. ∎

If we apply this result to weakly semihereditary rings, we obtain a classification of certain epic R-fields defined by rank functions:

THEOREM 4.6.14. *Let R be a weakly semihereditary ring. Then there are natural bijections between the following sets:*
 (a) **Z**-*valued rank functions on R,*
 (b) *minimal prime matrix ideals over R,*
 (c) *universal localizations that are fields.*
Further, for any rank function ρ, the corresponding epic R-field will be a field of fractions if and only if ρ is faithful.

Proof. By Prop. 6.11 any rank function on R is Sylvester, so for a **Z**-valued rank function ρ we can apply Prop. 6.13 to deduce that the set of all non-ρ-full matrices is a minimal prime matrix ideal. By following the proof of Prop. 6.13 we see that any minimal prime matrix ideal is the singular kernel of an epic R-field which is of the form $R_{\mathscr{P}}$. Finally, any epic R-field K gives rise to a **Z**-valued rank function; if the singular kernel is $\mathscr{P}$, then the minimal prime matrix ideal corresponding to this rank function is just the minimal prime matrix ideal contained in $\mathscr{P}$. The last part is clear. ∎

This proof shows incidentally that over a weakly semihereditary ring every prime matrix ideal contains a unique minimal one.

Exercises

1. Show that every finitely related projective module (over any ring) can be written as a direct sum of a finitely presented projective module and a free module.

2. Show that the trace ideal of a non-zero projective module is a non-zero. Let M be a finitely presented module, presented as cokernel of a map defined by a regular matrix. Verify that M is bound, and hence has zero trace ideal.

3. In the proof of Lemma 6.6, show that $\alpha = \beta$.

4°. Show that a (left or) right semihereditary local ring is a semifir. Is the same true for a weakly semihereditary local ring?

5. Show that a ring R has free module type $(1, 2)$ (see 1.4) if and only if R contains an idempotent e such that e is isomorphic to 1 and to $1 - e$. Deduce that the localization of R at the set of all right invertible elements is zero.

6°. The proof of Th. 6.8 shows that a ring with a faithful rank function is weakly finite. Does every weakly finite ring have a faithful rank function?

4.7 Normal forms for matrix blocks over firs

In the field of fractions K of a commutative integral domain R the elements have the simple form ab^{-1}, but there is generally no convenient normal form; the best we can say is that

$$ab^{-1} = a'b'^{-1} \tag{1}$$

if and only if $ab' = ba'$. When R is a Bezout domain we can take a, b to be coprime and then (1) holds if and only if $(a', b') = u(a, b)$ for a unit u in R.

In the case of a general ring R with an epic R-field K, we can compare the blocks (or systems of equations) determining a given element of K. The result is rather more complicated than in the commutative case (see FR, 7.6), but again there are simplifications when R is a fir and K the universal field of fractions, which will be presented below.

Thus let R be a fir and U its universal field of fractions. We shall now write a typical element of U as $f = c - uA^{-1}v$, so the corresponding block has the form

$$\alpha = \begin{pmatrix} u & c \\ A & v \end{pmatrix} \tag{2}$$

where A is a full matrix over R, $n \times n$ say and $u \in R^n$, $v \in {}^nR$, $c \in R$; n is called the *order* of the block. Since the matrices inverted in U are all the full matrices, the block (2) is admissible (for f) whenever A is full. The least order of any block representing f is sometimes called the *depth* of f.

We note the following changes which can be made to this block without affecting f:

(i) *We can replace (A, v) by $P(A, v) = (PA, Pv)$, where P is any full matrix. For we have $c - u(PA)^{-1}Pv = c - uA^{-1}v$. Similarly we can replace $\begin{pmatrix} u \\ A \end{pmatrix}$ by $\begin{pmatrix} uQ \\ AQ \end{pmatrix}$, where Q is a full matrix.*

(ii) *If we replace A by $A \oplus I$, then for any $p, q \in R$,*

$$\begin{pmatrix} u & p & c + pq \\ A & 0 & v \\ 0 & 1 & q \end{pmatrix} \quad and \quad \begin{pmatrix} u & c \\ A & v \end{pmatrix} \quad represent\ the\ same\ element.$$

In fact the matrix on the left can be reduced to the form

$$\begin{pmatrix} u & 0 & c \\ A & 0 & v \\ 0 & 1 & 0 \end{pmatrix}$$

by means of operations of the form (i), and the latter is equivalent to the form (2) by the operation (E.2) introduced in 4.3.

Likewise the operations inverse to (i), (ii), where possible, can be applied without affecting f. This shows that in (2) A may be replaced by any matrix stably associated to A. But in (i), P, Q need not be invertible over R, as long as they are full. This suggests the following definition. We recall that a matrix A over any ring is called *left prime* if any square left factor of A necessarily has a right inverse; *right prime* matrices are defined similarly. Now a block (2) and the corresponding representation of f is said to be *reduced* if $(A \quad v)$ is left prime and $\begin{pmatrix} u \\ A \end{pmatrix}$ is right prime. By the chain condition in firs we can by applying the inverse of (i) reduce any block (2) to one which is reduced. For such a block we have the following uniqueness theorem.

THEOREM 4.7.1. *Let R be a semifir and U its universal field of fractions. Given $f \in U$, if*

$$\begin{pmatrix} u & c \\ A & v \end{pmatrix} \quad and \quad \begin{pmatrix} u' & c' \\ A' & v' \end{pmatrix} \tag{3}$$

are two reduced admissible blocks representing $f \in U$, then A and A' are stably associated. Further, when R is a fir, then any element of U can be represented by a reduced admissible block.

Proof. The last part is clear, since any block can be replaced by a reduced block, by the chain condition in R (see FR, Th. 1.2.3, p. 72 and Lemma 7.6.5, p. 425). Suppose now that the two blocks in (3) representing f are both reduced, of orders m and n say. Then the block

$$\begin{pmatrix} u & u' & e \\ A & 0 & v \\ 0 & A' & -v' \end{pmatrix}$$

where $e = c - c'$, represents zero, as we saw in 4.3. Since R is a semifir, it follows by Th. 5.11 that we have a factorization

$$\begin{pmatrix} u & u' & e \\ A & 0 & v \\ 0 & A' & -v' \end{pmatrix} = m \begin{array}{c} 1 \\ \begin{pmatrix} d \\ E \\ F \end{pmatrix} \end{array} (G \quad H \quad k) \begin{array}{c} m + n, \\ \end{array} \qquad (4)$$
$$ m + n$$

where the appended letters indicate the block sizes. By (4), $EH = 0$, hence by suitable internal modification we find

$$E = (E' \quad 0), \qquad H = \begin{pmatrix} 0 \\ H'' \end{pmatrix},$$

where E' is $m \times s$ and H'' is $(m + n - s) \times n$. Further, $A = EG$, and since A is full, we have $s \geqslant m$. Similarly, $A' = FH$, and so $m + n - s \geqslant n$, therefore $s = m$. We now partition (and rename) (4) as

$$\begin{pmatrix} u & u' & e \\ A & 0 & v \\ 0 & A' & -v' \end{pmatrix} = m \begin{array}{c} 1 \\ \begin{pmatrix} p & p' \\ P_1 & 0 \\ P_2 & P_2' \end{pmatrix} \end{array} \begin{array}{c} m & n & 1 \\ \begin{pmatrix} Q_1 & 0 & q \\ Q_1' & Q_2' & q' \end{pmatrix} \end{array} \begin{array}{c} m \\ n \end{array}. \qquad (5)$$

Thus $(A \quad v) = P_1(Q_1 \quad q)$; since $(A \quad v)$ is left prime, P_1 is invertible and by another internal modification may be taken to be I. Similarly $(A' \quad u')^T$ is right prime and so we can take $Q_2' = I$. Now (5) becomes

$$\begin{pmatrix} u & u' & e \\ A & 0 & v \\ 0 & A' & -v' \end{pmatrix} = \begin{pmatrix} p & u' \\ I & 0 \\ P_2 & A' \end{pmatrix} \begin{pmatrix} A & 0 & v \\ Q_1' & I & q' \end{pmatrix}. \qquad (6)$$

It follows that

$$P_2 A = -A' Q_1'. \qquad (7)$$

Since $u = pA + u'Q_1'$, any common right factor of A and Q_1' is also a common right factor of A and u and so must be invertible. Similarly on the left, so (7) is coprime, hence comaximal and it follows by Th. 1.5.1 that A and A' are stably associated. ∎

When just one of the blocks is reduced, we can still obtain some information:

COROLLARY 4.7.2. *Let R be a semifir with universal field of fractions*

*U and consider two admissible blocks (2) for the same element of U. If the
first block is reduced, then A is stably associated to a factor of A'.*

Proof. We again have a factorization (5), where now P_1 can be taken to
be I; thus we have

$$\begin{pmatrix} u & u' & e \\ A & 0 & v \\ 0 & A' & -v' \end{pmatrix} = \begin{pmatrix} p & p' \\ I & 0 \\ P_2 & P_2' \end{pmatrix} \begin{pmatrix} A & 0 & v \\ Q_1' & Q_2' & q' \end{pmatrix},$$

where the dimensions are as before, because both A and A' are full.
Hence we have

$$P_2 A = -P_2' Q_1', \quad A' = P_2' Q_2'. \tag{8}$$

Here the first relation is right prime, because $u = pA + p'Q_1'$. It follows
that A is stably associated to a right factor of P_2' and hence to a factor of
A', by the second equation (8). ∎

If a ring R has a universal field of fractions U, its singular kernel is the
unique least prime matrix ideal of R and so is contained in the singular
kernel of any other R-field. This is so in particular when R is a fir; in that
case we can always find a universal denominator.

THEOREM 4.7.3. (Universal denominators). *Let R be a semifir and
U its universal field of fractions. If an element f of U can be defined by a
reduced block α and there is a block for f which is admissible over an epic
R-field K, then the block α is also K-admissible. In particular this holds
for all elements of U when R is a fir.*

Proof. Let $f \in U$ and let α be a reduced admissible block for f, and
suppose that α' is any K-admissible block for f. Then α' is also
U-admissible, by Cor. 4.2, and if the denominators in α, α' are A, A'
respectively, then A is stably associated to a factor of A', by Cor. 7.2.
Since $A' \notin \text{Ker } K$, we have $A \notin \text{Ker } K$, so α is K-admissible, as claimed.
By Th. 7.1, a reduced block for f always exists, when R is a fir. ∎

The denominator of a reduced admissible block for f is called a
universal denominator for f.

Exercises

1. Let R be a semifir and U its universal field of fractions. Show that for $p \in U$,
the depth $d(p)$ satisfies the relation $d(p + q) \le d(p) + d(q)$, with equality

whenever p, q have denominators (in some block representation) that are totally coprime, i.e. have no similarity class of factors in common.

2. Let R be a ring with an epic R-field K, and let Σ be the complement of the singular kernel. Show that the solution u_n of a system $Au = 0$ of order n (with $u_0 = 1$) is unaffected if A is multiplied on the left by a matrix in Σ, or on the right by a matrix of the form $1 \oplus Q \oplus 1$, where $Q \in \Sigma$, or by adding a right multiple of columns 1 to $n - 1$ to the first or last columns.

3. Let R be a semifir and U its universal field of fractions. A system $Au = 0$ is said to be *reduced* if A is left prime and its core is right prime. Show that two reduced systems determine the same element of U if and only if we can pass from one to the other by the operations listed in Ex. 2 and their inverses.

4. Let R be a semifir which is a local ring but not principal, with universal field of fractions U. By considering elements of the form $\sum a_i^{-1}$ show that there are elements of any prescribed depth in U.

Notes and comments

Until about 1970 the only methods of embedding rings in skew fields, apart from Ore's method, were topological (see 2.4, 2.6). In Cohn [71] it was proved that any fir could be embedded in a field by inverting all full matrices, and this was shown to be a universal field of fractions in Cohn [72]. Matrix ideals were developed in the 1971 Tulane Lecture Notes (Cohn [72']), and the key result, that every prime matrix ideal over a ring R occurs as singular kernel of an epic R-field, was proved later that year (Cohn [71"]), a connected account being given in the first edition of FR. This proof constructed the field as a group and then used Lemma 1.1.1. A more direct method was found by Malcolmson [78] and independently by Gerasimov [79]; this (in a slightly simplified form) is also the method used here in 4.3.

The field spectrum of a general ring was defined as a topological space by Cohn [72'], in analogy to the commutative case, where it reduces to the familiar prime spectrum of a ring. The prime spectrum of a commutative ring was characterized as a topological space by a set of axioms by Hochster [69]; in Cohn [79] it was shown that Spec (R) for any ring R satisfies Hochster's axioms and that an affine scheme on R can be constructed. The theorem on universal denominators (Th. 7.3) can then be used to show that for a fir every global section of this scheme is rational (see FR, p. 488); whether every rational section is integral is not known, even for the case of free algebras. This question is related to the existence of fully algebraically closed fields (see 8.1).

The route to the universal field of fractions followed in 4.5 aims to bring out clearly at which points the various hypotheses are needed. It leads naturally to Sylvester domains, which were introduced by Dicks and Sontag [78]. The law of nullity for matrices over a commutative field was established by Sylvester [1884]. The properties of the rank function in 4.6 are taken from Schofield [85], while the construction of rank functions (Lemma 6.6) is taken from Cohn [90']. The equivalence (b) ⇔ (c) in Th. 6.8 was first proved generally by Malcolmson [80]; for regular rings it goes back to Goodearl [79], Th. 18.3. A direct proof is given in FR, Prop. 0.2.2.

A study of numerators and denominators is made in Cohn [82], where the theorem on universal denominators (Th. 7.3) is also proved, using systems of matrix equations. It has been replaced here by the slightly simpler proof in terms of matrix blocks, using the normal form theorem (Th. 7.1), which was obtained in discussions with M. L. Roberts.

5

Coproducts of fields

One of the main results of Ch. 4 stated that every semifir has a universal field of fractions. This is now applied to show that every family of fields all having a common subfield can be embedded in a universal fashion in a field, their *field coproduct*. We begin in 5.1 by explaining the coproduct construction for groups (where it is relatively simple) and for rings, and derive some of the simpler consequences when the common subring is a field. When the factors themselves are fields, an elaboration of these results will show the ring coproduct of fields to be a fir (by an analogue of the weak algorithm, see Cohn [60, 61]), but we shall not follow this route, since it will appear as a consequence of more general later results.

The study of coproducts requires a good deal of notation; some of this is introduced in 5.2 and is used there to define the module induced by a family of modules over the factor rings and compute its homological dimension. In 5.3 we prove the important coproduct theorems of Bergman [74]: If R is the ring coproduct of a family (R_λ) of rings, taken over a field K, then (i) the global dimension of R is the supremum of the global dimensions of the factors (or possibly 1 if all the factors have global dimension 0) (Th. 3.5), (ii) the monoid of projectives $\mathcal{P}(R)$ is the coproduct of the $\mathcal{P}(R_\lambda)$ over $\mathcal{P}(K)$ (Th. 3.8). As an immediate consequence we have the theorem that a coproduct of firs over a field is a fir (Th. 3.9), which tells us in particular that a coproduct of fields is a fir.

The coproduct may be regarded as a special kind of tree product. We shall not take up this topic in general, but confine our attention to one important case, the HNN-extension. This construction, first carried out for groups by Higman, Neumann and Neumann [49], is described for fields in 5.5 and for rings in 5.6. It allows us to embed any ring in a simple ring and any field in a homogeneous field.

In 5.7 we return to the Bergman coproduct theorems and show how they can be used to determine the effect of adjoining elements or of localizing on the global dimension and the monoid of projectives. This leads to a simple method of constructing rings to given specifications, for example it provides a source of integral domains not embeddable in fields. Next we turn to derivations and in 5.8 we define the universal derivation bimodule of a ring and examine its relation to the corresponding bimodule formed from a homomorphic image. As an application we show that the number of generators of a free field is an invariant.

The results of 5.3 and 5.7 are used in 5.9 to construct field extensions of different left and right degrees, taking first the case where one degree is finite and the other infinite and then the (rather harder) case where both are finite.

The final section 5.10 examines the special case in which the coproduct of fields is a principal ideal domain. This is so precisely when there are just two factors, both of degree 2. It may be viewed as an analogue of the situation in groups, where a free product of two cyclic groups of order 2 is the infinite dihedral group.

5.1 The coproduct construction for groups and rings

Let $\mathscr{A}$ be any category; we recall the definition of a coproduct. Given a family (A_λ) of objects in $\mathscr{A}$, suppose that an object S exists with maps $u_\lambda\colon A_\lambda \to S$ such that for any family of maps $\varphi_\lambda\colon A_\lambda \to X$ to the same object X there is a unique map $f\colon S \to X$ satisfying $\varphi_\lambda = u_\lambda f$. Then S with the maps u_λ is called the *coproduct* of the family (A_λ) and is denoted by $\coprod A_\lambda$, or by $A_1 \amalg \ldots \amalg A_n$ in the case of a finite family.

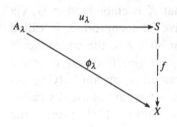

Thus we have a bijection

$$\mathscr{A}\left(\coprod A_\lambda, X\right) \cong \prod \mathscr{A}(A_\lambda, X).$$

EXAMPLES. The coproduct of sets is the disjoint union; for abelian groups we obtain the direct sum (also called the restricted direct product), for general groups we have the free product, but this case will be taken up in more detail in a moment.

Often we need an elaboration of this idea. Let K be a fixed object in $\mathscr{A}$ and consider the *comma category* $(K, \mathscr{A})$: its objects are arrows $K \to A$ ($A \in \mathrm{Ob}\,\mathscr{A}$) and its morphisms are commutative triangles:

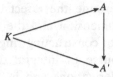

This category has the initial object $K \xrightarrow{1} K$; it reduces to $\mathcal{A}$ when K is an initial object of $\mathcal{A}$. Now the coproduct in $(K, \mathcal{A})$ is called the *coproduct over K*. E.g. for two objects in $(K, \mathcal{A})$ the coproduct is just the pushout:

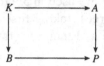

Let us consider coproducts over a fixed group K in the category of groups. This means that we have a family of groups (G_λ) and homomorphisms $i_\lambda: K \to G_\lambda$ and the coproduct C with maps $u_\lambda: G_\lambda \to C$ is a sort of 'general pushout'. Its existence is clear: we take presentations for all the groups G_λ (on disjoint generating sets) and then obtain a presentation for C by taking the union of the presentations of the G_λ together with all relations of the form $xi_\lambda = xi_\mu$ ($x \in K$), which identify the images of K. Clearly any element of K mapped to 1 by any i_λ must be mapped to 1 by every u_λ, so by modifying K and the G_λ we may as well assume that each i_λ is injective; this means that K is embedded in G_λ via i_λ. If in this situation all the u_λ are injective, the coproduct is called *faithful*. If moreover, $G_\lambda u_\lambda \cap G_\mu u_\mu = Ki_\lambda$ for all $\lambda \neq \mu$, the coproduct is called *separating*. These definitions apply quite generally for concrete categories (i.e. categories where the objects have an underlying set structure). A faithful and separating coproduct of groups is usually called a *free product*, and a basic result, due to Schreier [27], asserts the existence of free products of groups. Although this result will not be directly needed here, we shall outline the proof, since it is very similar to the corresponding proof for the coproduct of fields, which we shall soon meet.

THEOREM 5.1.1 (Schreier). *The coproduct of groups (over a fixed group) is faithful and separating.*

Proof (outline). Let $(G_\lambda; \lambda \in \Lambda)$ be the family of groups, where G_λ has a subgroup K_λ which is isomorphic to the base group K. We can write down a normal form for the elements of the coproduct C as follows. Choose a left transversal for K_λ in G_λ of the form $S_\lambda \cup \{1\}$, thus $G_\lambda = K_\lambda \cup S_\lambda K_\lambda$. Then every element of C can be written in just one way as

$$s_1 s_2 \ldots s_n c, \quad \text{where } n \geqslant 0, s_i \in S_{\lambda_i}, \lambda_{i-1} \neq \lambda_i, c \in K, \tag{1}$$

where the different K_λ are identified with K for convenience. It is clear how to write any element of C in this form. The uniqueness can then be proved by defining a multiplication of the expressions (1), which consists in a set of rules reducing the formal product of two such expressions to normal form and verifying that one obtains a group in this way. A quicker way (v.d. Waerden [48]) is to define for each $\lambda \in \Lambda$ a group action of G_λ on the set of expressions (1):

$$s_1 \ldots s_n c \cdot g = \begin{cases} s_1 \ldots s_n s_p c' & \text{if } \lambda_n \neq \lambda \text{ and } cg = s_p c' \text{ in } G_\lambda, \\ s_1 \ldots s_{n-1} s_p c' & \text{if } \lambda_n = \lambda \text{ and } s_n cg = s_p c' \text{ in } G_\lambda, \end{cases}$$

with the understanding that s_p is to be omitted if $cg \in K_\lambda$, or in the second case, if $s_n cg \in K_\lambda$. Now it can be verified that these group actions agree on K_λ and hence combine to give a C-action on the set of elements (1). For $n = 0$ and $c = 1$, (1) reduces to 1 and by considering the effect of the group action on 1 we see that the coproduct is faithful and separating. ∎

Our aim in this chapter is to study the coproduct of rings. Let K be a fixed ring and consider K-rings; we shall usually want our K-ring R to be *faithful*, i.e. the natural map $K \to R$ should be injective. For example, if K is a field, any non-zero K-ring is faithful; this is the case we shall mainly be concerned with, but at first we place no restriction on K. As in the case of groups, we can establish the existence of the coproduct of a family of K-rings (R_λ) by taking a presentation for each K-ring R_λ in our family and writing all these presentations together, as well as the relations identifying the images of K; but the coproduct need not be faithful or separating. Before finding conditions ensuring this we look at some examples; we shall use * to denote the coproduct of rings.

1. Let k be a commutative field, $K = k[x]$, where x is a central indeterminate, $R = k(x)$ the rational function field, $S = k[x, y; xy = 0]$. In $R \underset{K}{*} S$ we have $y = 1 \cdot y = x^{-1} \cdot xy = 0$, so S is not faithfully represented in the coproduct.

2. The inclusion $\mathbf{Z} \subseteq \mathbf{Q}$ is an epimorphism of rings, and by Prop. 4.1.1, we have $\mathbf{Q} \underset{\mathbf{Z}}{*} \mathbf{Q} = \mathbf{Q}$. Hence the coproduct is faithful, but not separating. More generally this applies for any ring epimorphism.

When the coproduct is faithful and separating, we shall sometimes call it the *free product* (as for groups). Thus the question is: When does the free product of rings exist? We begin by giving a necessary condition; for simplicity we limit ourselves to two factors:

Let R_1, R_2 be faithful K-rings. If their free product is to exist, we must have

$$c_1 a, a c_2 \in K \Rightarrow c_1 a \cdot c_2 = c_1 \cdot a c_2 \text{ in } K, \text{ for all } c_i \in R_i, a \in K.$$

Clearly this also holds with R_1 and R_2 interchanged, and more generally, for matrix equations. If we take *all* implications of a suitable form we can obtain necessary and sufficient conditions for the free product to exist, but they will not be in a very explicit form. Syntactic criteria of a rather different form have been obtained by P. D. Bacsich and D. R. Hughes [74]; see also W. A. Hodges [93], 8.6.

We now come to a simple sufficient condition for the existence of free products of rings, which will be enough for all the applications we have in mind. In essence it states that the free product of a family (R_λ) of faithful K-rings exists provided that each quotient module R_λ/K is free as right K-module.

We shall first consider the case of two factors, where the coproduct takes a particularly simple form. Thus we have $R = R_1 \underset{K}{*} R_2$; we shall use the following convention. For any positive integer v, R_v is to mean R_1 if v is odd and R_2 if v is even. Further we abbreviate the quotient module R_v/K as $\bar{R}_v$. By assumption, $\bar{R}_v$ is free as right K-module and we shall take a right K-basis of $\bar{R}_v$ and lift it to R_v; in this way we get a right K-basis of R_v including 1, which we shall write as $T_v \cup \{1\}$ ($v = 1, 2$).

Next we define a filtration on R. By definition every element of R may be written as a finite sum of products

$$a_1 a_2 \ldots a_n, \tag{2}$$

where a_v is the image of an element of R_v under the natural homomorphism $R_v \to R$. We put $A_0 = K$ and let A_n be the set of all sums of such products (2) with n factors, for $n = 1, 2, \ldots$. It is clear that A_n is a K-bimodule and provided that $R_v \neq K$ for $v = 1, 2$, we have

$$A_0 \subset A_1 \subset A_2 \subset \ldots, \quad A_i A_j \subseteq A_{i+j}, \quad \bigcup A_n = R. \tag{3}$$

Thus (A_n) is a filtration of R. Since $\bar{R}_n$ is free as right K-module, we have $R_n = K \oplus T_n K$ and A_n is a free right K-module with basis $t_1 \ldots t_v$, where $t_1 \in T_1 \cup \{1\}$, $t_i \in T_i$ for $i > 1$ and $v = 1, 2, \ldots, n$. The fact that t_i and t_{i+1} do not lie in the same factor R_v ensures that there is no cancellation between them. Thus A_n/A_{n-1} has a basis consisting of all products $t_1 \ldots t_n$, where $t_1 \in T_1 \cup \{1\}$, $t_i \in T_i$, $i = 2, \ldots, n$, and it follows that

$$A_n/A_{n-1} \cong R_1 \otimes \bar{R}_2 \otimes \bar{R}_3 \otimes \ldots \otimes \bar{R}_n. \tag{4}$$

This shows in particular that R_1, R_2 are embedded in R, and identifying them with their images, we see that $R_1 \cap R_2 = K$. Thus we have

THEOREM 5.1.2. *Let K be any ring and R_1, R_2 faithful K-rings such* *that R_ν/K is free as right K-module ($\nu = 1, 2$). Then the coproduct $R_1 \underset{K}{*} R_2$ is faithful and separating.* ■

It will be convenient to have a second filtration (H_n) on R. We put $H_0 = K$, $H_{-1} = 0$ and for $n = 1, 2, \ldots$ define H_n as the set of all sums of products (2) with n factors, where $a_i \in R_1 \cup R_2$. The H_n are again K-bimodules and the conditions for a filtration are easily checked:

$$0 = H_{-1} \subset H_0 \subset H_1 \subset \ldots, \qquad H_i H_j \subseteq H_{i+j}, \qquad \bigcup H_n = R.$$

With each $a \in R^\times$ we associate a non-negative integer $h(a)$, its *height*, by the rule

$$h(a) = n \text{ if and only if } a \in H_n \backslash H_{n-1}.$$

We also assign a height to 0 by putting $h(0) = -\infty$. It is clear that we have the inclusions

$$H_0 \subset A_1 \subset H_1 \subset A_2 \subset \ldots \subset H_{n-1} \subset A_n \subset H_n \subset \ldots,$$

and from the relations between H_n and A_n we obtain the natural isomorphisms

$$A_n/H_{n-1} \cong \bar{R}_1 \otimes \bar{R}_2 \otimes \ldots \otimes \bar{R}_n, \tag{5}$$

$$H_n/A_n \cong \bar{R}_2 \otimes \bar{R}_3 \otimes \ldots \otimes \bar{R}_{n+1}, \tag{6}$$

and

$$H_n/H_{n-1} \cong A_n/H_{n-1} \oplus H_n/A_n. \tag{7}$$

Suppose that n is even; then the last factor on the right of (5) is $\bar{R}_2$ and we shall denote the submodule of H_n corresponding to the first term on the right of (7) by H_n^{12} and the submodule corresponding to the second term by H_n^{21}; when $n = 0$, $H_0^{12} = H_0^{21} = H_0$ by convention. When n is odd, we can form H_n^{11} and H_n^{22} in the same way; thus $H_n^{\mu\nu}$ is the sum of products of at most n factors, the first one in R_μ and the last in R_ν. It is clear that if $H_n^{\mu\nu}$ is defined, then $\mu + \nu \equiv n + 1 \pmod{2}$, so for given n, each of μ, ν determines the other and we shall often write simply $H_n^{\mu\cdot}$ or $H_n^{\cdot\nu}$ for $H_n^{\mu\nu}$. The elements of $H_n^{\mu\nu}$ which are of height n (i.e. those corresponding to a non-zero element on the right of (7)) are called *left μ-pure* or *right ν-pure* or more precisely, *pure of type* (μ, ν), where we may also write $(\mu, \cdot)$ or $(\cdot, \nu)$ instead of (μ, ν). As an example, if c is any element of R of positive height n, then we have a decomposition

$$c = c' + c'', \quad c' \in H_n^{1\cdot}, \quad c'' \in H_n^{2\cdot}. \tag{8}$$

Here c', c'' are not unique, but either is determined up to an element of H_{n-1}, by (7).

By a *monomial unit* in R we understand a unit which can be expressed as a monomial (2). Soon we shall see that in a coproduct of integral domains over a field every unit is a monomial unit. An element c of R is said to be *right reduced* if $h(cu) \geqslant h(c)$ for every monomial unit u; *left reduced* is defined similarly.

Our next result provides information on the height of a product:

THEOREM 5.1.3. *Let K be a field and $R = R_1 \underset{K}{*} R_2$ the coproduct of two integral domains R_1, R_2 over K. Then for any two elements a, b of positive height in R we have*

$$h(ab) \leqslant h(a) + h(b) \text{ with equality unless } a \text{ is right } v\text{-pure and}$$

$$b \text{ is left } v\text{-pure } (v = 1, 2). \tag{9}$$

Further, if a is right v-pure and b is left v-pure $(v = 1, 2)$, then

$$h(ab) \leqslant h(a) + h(b) - 1, \tag{10}$$

with equality if either a is right reduced or b is left reduced.

Proof. Write $h(a) = r$, $h(b) = s$; it is clear that $h(ab) \leqslant r + s$. If this inequality is strict, then

$$ab \equiv 0 \ (\mathrm{mod}\ H_{r+s-1}).$$

Let us write $a = a' + a''$, where $a' \in H_r^{\cdot 1}$, $a'' \in H_r^{\cdot 2}$ and $b = b' + b''$, where $b' \in H_s^{1 \cdot}$, $b'' \in H_s^{2 \cdot}$. Then

$$ab \equiv a'b'' + a''b' \equiv 0 \ (\mathrm{mod}\ H_{r+s-1}).$$

Since the two summands are (zero or) pure of different types, we may equate them to zero separately:

$$a'b'' \equiv a''b' \equiv 0 \ (\mathrm{mod}\ H_{r+s-1}).$$

Suppose that $a' \notin H_{r-1}$ and write $a' = \sum a_i \lambda_i + a^*$, where $\lambda_i \in K$, $a^* \in H_{r-1}$ and the a_i are right linearly independent $(\mathrm{mod}\ H_{r-1})$ over K. We then have $\sum a_i \lambda_i b'' \equiv 0 \ (\mathrm{mod}\ H_{r+s-1})$, hence $\lambda_i b'' \in H_{s-1}$. Since $a' \notin H_{r-1}$, not all the λ_i vanish, say $\lambda_1 \neq 0$. Then $b'' = \lambda_1^{-1} \cdot \lambda_1 b'' \in H_{s-1}$ and so either $a' \in H_{r-1}$ or $b'' \in H_{s-1}$; similarly, either $a'' \in H_{r-1}$ or $b' \in H_{s-1}$. If $a' \in H_{r-1}$, then $a'' \notin H_{r-1}$ (because $h(a) = r$), and so $b' \in H_{s-1}$; this means that a is right 2-pure and b is left 2-pure. Similarly, if $b'' \in H_{s-1}$, then b is left 1-pure and a is right 1-pure, so (9) is established.

Suppose now that a is right 1-pure and b is left 1-pure; then (10) clearly holds. Assume further that b is left reduced. We can write a as a finite sum

$$a = \sum a_i'' x_i', \quad \text{where } a_i'' \in H_{r-1}^{\cdot 2}, \, x_i' \in R_1.$$

For a is congruent (mod H_{r-1}) to such an expression, but any element of H_{r-1} may also be written in this form. Now if $h(ab) < r + s - 1$, then

$$ab \equiv \sum a_i'' x_i' b \equiv 0 \, (\text{mod } H_{r+s-2}).$$

Let us write $a_i'' = \sum a_h \mu_{hi} + a_i^*$, where $\mu_{hi} \in K$, $a_i^* \in H_{r-2}$ and the a_h are right linearly independent (mod H_{r-2}) over K. This implies that each a_h is right 2-pure and

$$\sum \mu_{hi} x_i' b \equiv 0 \, (\text{mod } H_{s-1}) \text{ for all } h. \tag{11}$$

We have

$$a = \sum a_i'' x_i' = \sum a_h \mu_{hi} x_i' + \sum a_i^* x_i'.$$

Let us put $x_h = \sum \mu_{hi} x_i'$, $a^* = \sum a_i^* x_i'$; then $x_h \in R_1$, $a^* \in H_{r-1}^{\cdot 1}$ and

$$a = \sum a_h x_h + a^*. \tag{12}$$

The congruences (11) now take the form

$$x_h b \equiv 0 \, (\text{mod } H_{s-1}) \text{ for all } h. \tag{13}$$

By symmetry b may be expressed as

$$b = \sum y_k b_k + b^*,$$

where $y_k \in R_1$, b_k is left 2-pure of height $s - 1$ and $b^* \in H_{s-1}^{1\cdot}$. Further, the argument that led to (13) now shows that

$$x_h y_k \equiv 0 \, (\text{mod } H_0) \text{ for all } h, k. \tag{14}$$

Since $h(a) = r$, (12) shows that the x_h are not all zero, say $x_1 \neq 0$. Similarly we may suppose that $y_1 \neq 0$; then $u = x_1 y_1 \in K$ by (14), hence $x_1 y_1 u^{-1} = 1$ and x_1 is a unit in R_1; similarly y_1 is a unit, so by (13), b is not left reduced, which contradicts the hypothesis. The same argument gives a contradiction if a is right reduced, and the conclusion follows. ∎

COROLLARY 5.1.4. *Let K, R_1, R_2 and $R = R_1 \underset{K}{*} R_2$ be as in Th. 1.3. Given a right reduced element a in R, we have*

$$h(ab) \geq h(a) \text{ for all } b \in R^\times. \tag{15}$$

Moreover, R is an integral domain and every unit in R is a monomial unit.

Proof. If $b \in K^\times$, then $h(ab) = h(a)$ and (15) holds. Otherwise $h(b) > 0$ and (15) is a consequence of (10). In particular, since every non-zero element is right associated to a right reduced element, R cannot have zero-divisors. Finally, if a is a unit of positive height and right reduced, then $h(a) > h(a \cdot a^{-1})$, which contradicts (15), hence such an a has zero height. This shows that every unit in R is a monomial unit. ∎

Exercises

1. Show that any category with coproducts has an initial object.

2. Let A, B be monoids containing submonoids isomorphic to a group C. Show that the coproduct of A and B over C is faithful and separating (for an appropriate definition of the latter).

3. Show that in a free product of groups any element of finite order has a conjugate lying in one of the factors.

4. Let k be a commutative field and $R_1 = k\langle x_1, x_2, y_1, y_2 \rangle$ be a free algebra. Show that the subalgebra S_1 generated by the elements $t_{ij} = x_i y_j$ $(i, j = 1, 2)$ is free on these four generators. Let R_2 be the k-algebra generated by six elements a, b, t_{ij} $(i, j = 1, 2)$ with the defining relations $t_{11} - at_{21} - t_{12}b + at_{22}b = 0$. Show that the subalgebra S_2 generated by the t_{ij} is again free on these generators. Verify that R_2 is an integral domain (like R_1) and that their coproduct $R_1 * R_2$ over the subalgebra generated by the t_{ij} is faithful and separating, but has zero-divisors.

5. Let A be the skew group ring of the infinite cyclic group with generator t over **C** with commutation rule $it = t^{-1}i$. Express A as coproduct of two copies of **C** over **R**.

6. Let K be a field, R_λ a family of integral domains which are K-rings and $R = \underset{K}{*} R_\lambda$ their coproduct. Show that any (left or) right algebraic element over K is conjugate to an element in one of the factors. Give an example of an element conjugate to an algebraic element in R_λ which is not itself algebraic.

7. (After B. H. Neumann [54]) Let k be a commutative field and put $K_1 = k\langle x, y; xyx = y \rangle$, $K_2 = k\langle y, z; yzy = z \rangle$, $K_3 = k\langle z, x; zxz = x \rangle$. Show that the ring coproduct R of K_1, K_2, K_3 amalgamating $k(x)$ in K_1 and K_3, $k(y)$ in K_1 and K_2 and $k(z)$ in K_2 and K_3 is faithful and separating, but is not an integral domain. (Hint. Form $k(\xi, \eta, \zeta)$ as a field in three central indeterminates and successively adjoin x with $I(x)$: $\zeta \mapsto \zeta^{-1}$, $x^2 = \xi$, then y with $I(y)$: $x \mapsto x^{-1}$, $y^2 = \eta$, then z with $I(z)$: $x \mapsto x\zeta^{-1}$, $y \mapsto y^{-1}$, $z^2 = \zeta$; now verify that $(xyz)^2 = 1$.)

5.2 Modules over coproducts

We now begin a more detailed study of coproducts of arbitrary families and first introduce a systematic notation. We write R_0 instead of K and take a family (R_λ) of R_0-rings, where λ runs over a set Λ. We shall write $\Lambda_0 = \Lambda \cup \{0\}$ (assuming of course that $0 \notin \Lambda$), and write λ, λ' etc. for a general element of Λ and μ, μ' etc. for a general element of Λ_0. We abbreviate $*_{R_0}$, $\otimes_{R_\mu}$ as $*_0$, $\otimes_\mu$ and call R_0 the *base ring* and R_λ the *factor rings* of the coproduct $R = *_0 R_\lambda$. Further, we shall often write $M_\mu^\otimes$ for $M_\mu \otimes_\mu R$, where M_μ is an R_μ-module.

By an *induced module* over the coproduct R we understand an R-module of the form $M = \bigoplus_\mu M_\mu^\otimes$, where M_μ is an R_μ-module. An induced R-module of the form $M_0^\otimes$ or an R_λ-module of the form $M_0 \otimes_0 R_\lambda$ is called a *basic module*. Thus for example, every free R_λ-module is basic, $R_\lambda^n = R_0^n \otimes_0 R_\lambda$ and every free R-module is induced, as well as basic: $R^n = R_\mu^{\otimes n}$.

It is easily seen that an induced module $M = \bigoplus M_\mu^\otimes$ has the following universal property:

Given an R-module N and a family of R_μ-homomorphisms f_μ: $M_\mu \to N$ ($\mu \in \Lambda_0$), there is a unique R-homomorphism f: $M \to N$ such that the triangle shown commutes for all $\mu \in \Lambda_0$.

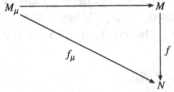

For the map f_μ: $M_\mu \to N$ extends to a map from $M_\mu^\otimes$ to N by the universal property of the tensor product and these maps combine to give a map of the direct sum $M = \bigoplus M_\mu^\otimes$ into N.

We remark that for this property to hold, R may be any ring with homomorphisms $R_\lambda \to R$ which all induce the same map on R_0, in other words, R may be any homomorphic image of the coproduct. However, when R is the coproduct (and the R_λ satisfy the conditions of Th. 1.2), we shall prove that the maps $M_\mu \to M$ are injective; this will also provide another proof of Th. 1.2.

Thus we shall assume that each R_λ is a faithful R_0-ring, and identifying R_0 with its image in R_λ, we assume that R_λ/R_0 is free as right R_0-module. We take again an R_0-basis of R_λ/R_0 and lift it to R_λ to get an R_0-basis of R_λ of the form $T_\lambda \cup \{1\}$. The union of the T_λ will be written T. Next let M_μ be a right R_μ-module, free as R_0-module, and write $M = \bigoplus M_\mu^\otimes$. For each M_μ we have an R_0-basis S_μ and we write S for the union of the S_μ. We observe that when R_0 is a field, all assumptions are satisfied, provided that the factors R_λ are all non-zero. The members of T_λ or S_λ are said to be *associated* with the index λ; elements of S_0 are not associated with any index.

We claim that the induced module $M = \bigoplus M_\mu^\otimes$ is a free R_0-module with a basis consisting of all products of the form

$$st_1 \ldots t_n, \quad s \in S, \quad t_i \in T, \quad n \geq 0, \tag{1}$$

where no two successive factors are associated with the same index. These products (1) are called *monomials* and the set of all such monomials is denoted by U.

Let F be the free R_0-module on U as basis. To prove our claim it will be enough to verify that F has the universal property of induced modules, but we first have to define the R-module structure on F. For this it is enough, by the definition of R as a coproduct, to define for each λ an R_λ-module structure on F which extends the given R_0-module structure. A monomial $u \in U$ is said to be *associated* with λ if its last factor (an element of S or T) is associated with λ. The set of all elements of U *not* associated with λ is denoted by U^λ; these are the monomials which when multiplied by an element of T_λ on the right, give again a monomial. For any fixed $\mu \in \Lambda_0$ we may write F as

$$F = S_\mu R_0 \oplus (U \backslash S_\mu) R_0; \tag{2}$$

here the first term on the right may be identified with M_μ and so may be given an R_μ-module structure. For $\mu = 0$ the second term on the right of (2) is a free R_0-module. We claim that for $\lambda \neq 0$, $(U \backslash S_\lambda) R_0$ has a natural structure as free R_λ-module. To see this we observe that $U \backslash S_\lambda$ is the disjoint union of U^λ and $U^\lambda T_\lambda$, hence

$$(U \backslash S_\lambda) R_0 = U^\lambda R_0 \oplus U^\lambda T_\lambda R_0,$$

and since $T_\lambda \cup \{1\}$ is an R_0-basis of R_λ, we find that $(U \backslash S_\lambda) R_0$ becomes a free R_λ-module with basis U^λ in this way. Thus F has an R_λ-module structure for each λ, extending the given R_0-module structure, and hence F is an R-module.

It remains to verify the universal property of the induced module $\bigoplus M_\mu^\otimes$ for F. Let $f_\mu: M_\mu \to N$ ($\mu \in \Lambda_0$) be any family of R_0-linear maps of M_μ into N and define $f: F \to N$ as an R_0-linear map on the basis U by the rule

$$(st_1 \ldots t_n)f = (sf_\mu)t_1 \ldots t_n, \quad \text{if } s \in S_\mu.$$

It is clear that f is R_μ-linear for all μ and hence R-linear. Thus we have proved

LEMMA 5.2.1. *Let $R = \underset{0}{*} R_\lambda$ be the coproduct of a family (R_λ) of faithful R_0-rings, where R_λ / R_0 is free as right R_0-module with the image of T_λ as basis, and let $M = \bigoplus M_\mu^\otimes = \bigoplus M_\mu \otimes_\mu R$ be an induced R-module, where each M_μ is free as right R_0-module with basis S_μ. Then M is a free right R_0-module on the set U of monomials as basis, where U is defined as above in terms of the bases $T = \bigcup T_\lambda$ and $S = \bigcup S_\mu$. Moreover, for each $\lambda \in \Lambda$, M_λ is embedded in M by the canonical map and there is a direct* sum decomposition $M = M_\lambda \oplus M'_\lambda$, where M'_λ is a free R_λ-module on U^λ as basis. ∎

For basic R-modules we shall need a slight variant of this result.

LEMMA 5.2.2. *Let R_λ, R, T be as in Lemma 2.1 and let M_0 be a free right R_0-module with basis Q_0. Suppose further that for each $\lambda \in \Lambda$, another R_0-basis $Q_{0\lambda}$ of M_0 is given. Then $M_0^\otimes = M_0 \otimes R$ is a free R_0-module with a basis consisting of all products*

$$qt_1 \ldots t_n, \; t_i \in T, \; n \geqslant 0,$$

where no two successive t_i are associated with the same index and if t_1 is associated with λ, then $q \in Q_{0\lambda}$, while for $n = 0$, $q \in Q_0$.

This follows in the same way as Lemma 2.1; we just have to use the basis $Q_{0\lambda}$ in defining the R_λ-module structure on $M_0^\otimes$. ∎

If we apply Lemma 2.1 with $M_\lambda = R_\lambda$ we see that the coproduct is faithful. To verify that it is separating, we take $M_0 = R_0$, $M_\lambda = 0$ ($\lambda \in \Lambda$); then we find that $M = R$ has an R_0-basis consisting of all monomials $t_1 t_2 \ldots t_n$, where $t_i \in T$ and no two neighbouring factors are associated with the same λ. Thus if $1, 2 \in \Lambda$, $1 \neq 2$, then the R_0-module $R_1 \oplus R_2$ has as basis $T_1 \cup T_2 \cup \{1\}$, hence $R_1 \cap R_2 = R_0$ and this shows the coproduct to be separating. Thus we have proved

THEOREM 5.2.3. *Let R_0 be any ring, $\{R_\lambda | \lambda \in \Lambda\}$ a family of faithful R_0-rings and $R = \underset{0}{*} R_\lambda$ their coproduct. If the quotient modules R_λ/R_0 are free as right R_0-modules, then the coproduct is faithful and separating and for each $\mu \in \Lambda_0 = \Lambda \cup \{0\}$, the quotient R/R_μ (hence also R itself) is free as right R_μ-module.* ∎

When R_0 is a field, the case of main importance to us, all the conditions are clearly satisfied, because every module over a field is free, so we obtain

COROLLARY 5.2.4. *The free product of any family (R_λ) of non-zero rings over a field R_0 exists and is left and right free over each R_λ.* ∎

Th. 2.3 provides a means of finding the homological dimension hd_R of induced R-modules, for a coproduct R:

PROPOSITION 5.2.5. *Let $R = \underset{0}{*} R_\lambda$ be a coproduct of non-zero R_0-rings*

R_λ, where R_0 is a field, and let $M = \oplus M_\mu^\otimes = \oplus M_\mu \otimes_\mu R$ be an induced R-module. Then

$$\mathrm{hd}_R M = \sup_\mu \{\mathrm{hd}_{R_\mu} M_\mu\}.$$

Proof. Clearly it suffices to show that $\mathrm{hd}_R(M_\mu^\otimes) = \mathrm{hd}_{R_\mu} M_\mu$. Now R is left free, hence left flat, over R_μ, therefore $- \otimes_\mu R$ converts a projective R_μ-resolution of M_μ to a projective R-resolution of $M_\mu^\otimes$ and so $\mathrm{hd}_R(M_\mu^\otimes) \leq \mathrm{hd}_{R_\mu} M_\mu$. To show that equality holds, we observe that since R is right R_μ-projective, any R-projective resolution of $M_\mu^\otimes$ is also R_μ-projective, hence $\mathrm{hd}_R(M_\mu^\otimes) \geq \mathrm{hd}_{R_\mu} (M_\mu^\otimes)$, but the latter equals $\mathrm{hd}_{R_\mu} M_\mu$, because M_μ and $M_\mu^\otimes$ differ only by an R_μ-free summand, by Lemma 2.1. ∎

Exercises

1. Verify that the coproduct $R = \underset{0}{*} R_\lambda$ over a field R_0 is flat as (left or) right R_λ-module; deduce the analogue of Prop. 2.5 for the weak dimension.

2. Let $L = E \underset{K}{\circ} F$ be any field coproduct. If an element $\lambda \in K$ has a square root in E and in F but not in K, say $\lambda = a^2 = b^2$ ($a \in E$, $b \in F$), show that $b = c^{-1}ac$, where $c = a + b$. Suppose that K contains a primitive fourth root of 1; show that any root of $x^4 - \lambda = 0$ in L is conjugate to an element in either E or F.

3. Let k be a commutative field with a primitive cube root of 1 and put $E = k(x)$, $F = k(y)$, $K = k(t)$ for indeterminates x, y, t, and put $L = E \underset{K}{\circ} F$, where K is embedded in E by mapping $t \mapsto x^2$ and in F by mapping $t \mapsto y^3$. Show that L contains an element of degree 6 over K. (Hint. Use Ex. 7 of 2.2. Explicitly put $z = xy^{-1}$, $q = z^2 - y$, $v = q^{-1}zq - z$ and verify that $v^6 = t$. Note that this result is in contrast with the situation in groups, where a theorem of Kurosh [34] states that any subgroup of a free product of groups is a free product of a free group and of groups that are conjugate to subgroups of the factors.)

5.3 Submodules of induced modules over a coproduct

From now on we shall assume that the base ring R_0 is a field. In that case the hypotheses of Th. 2.3 all hold, and for any family (R_λ) of non-zero R_0-rings the coproduct $R = \underset{0}{*} R_\lambda$ is faithful and separating. We shall continue to use the notation of 5.2; thus $T = \bigcup T_\lambda$, where $T_\lambda \cup \{1\}$ is an R_0-basis of R_λ, $R_\lambda = R_0 \oplus T_\lambda R_0$, and if $M = \oplus M_\mu^\otimes$ is an induced module, we have an R_0-basis S_μ of M_μ and $S = \bigcup S_\mu$. Further, M itself has the R_0-basis U of monomials, thus $M = M_\lambda \oplus U^\lambda R_\lambda$ for any $\lambda \in \Lambda$, hence for any $u \in U^\lambda$ we have a projection map $p_\lambda(u): M \to R_\lambda$, which maps $x \in M$ to the coordinate in R_λ of the monomial u in x. Similarly,

$M = UR_0$ and we have a projection map $p_0(u): M \to R_0$. We shall write U^0 for U, so that there is a projection map p_μ for each element in U^μ.

For any subset X of M we define its μ-*support* as

$$\mathcal{D}_\mu(X) = \{u \in U^\mu | xp_\mu(u) \neq 0 \text{ for some } x \in X\}.$$

In particular this defines the μ-support of an element of M. The 0-support is also called the *support*: $\mathcal{D}_0(X) = \mathcal{D}(X)$. We note that any finitely generated R_μ-submodule of M has finite μ-support. From the definition it is also clear that the λ-support of an element x of M is empty precisely when $x \in M_\lambda$, while its 0-support is empty precisely when $x = 0$.

The *degree* of a monomial $st_1 \ldots t_n$ is defined to be n. We choose a well-ordering of Λ and of S, T and then well-order U by degree, with monomials of the same degree ordered lexicographically, reading from left to right. Given $x \in M$, if $x \neq 0$, we can write it as an R_0-linear combination of the monomials in its support. The greatest such term, in the well-ordering of U, is called its *leading term*, its degree the *degree* of x, written $\deg x$. If $\deg x = n$ and all terms of degree n are associated with λ, x is called λ-*pure*; thus if x is not λ-pure, there are terms of degree n whose monomial is in U^λ. The greatest such term is called the *leading co-λ term*. If x is λ-pure for some λ, it is called *pure*; otherwise it is called *impure* or also 0-*pure* and its leading term is then called the *leading co-0 term*. We note that a λ-pure element of degree n has a λ-support consisting of monomials of degree $< n$, while a 0-pure element has a 0-support consisting of monomials of degree $\leq n$.

We shall need to know the effect of acting on M with the basis T; it will be enough to look at the case when there is no interaction:

LEMMA 5.3.1. *Given an induced R-module $M = \bigoplus M_\mu^\otimes$ over a coproduct $R = \ast_0 R_\lambda$, let $y \in M$ be not λ_1-pure, with leading co-λ_1 term u which occurs in y with coefficient 1, thus $yp_{\lambda_1}(u) = 1$. If $t_i \in T_{\lambda_i}$ for $i = 1$, $2, \ldots$, n and successive λ_i are distinct, then $yt_1 \ldots t_n$ is λ_n-pure and has leading term $ut_1 \ldots t_n$, again with coefficient 1.*

Proof. By induction it is enough to prove the result for $n = 1$. We shall write t for t_1 and λ for λ_1; thus we have to show that yt is λ-pure with leading term ut, with coefficient 1. If $y = u$, this is clear; if $y \neq u$, then y contains a term vc, where v is a monomial $\neq u$ and $c \in R_0$, thus $vct \in vR_0$. If v (unlike u) is associated with λ, then vct has degree $< \deg u + 1 = \deg ut$, so ut comes after vct. Otherwise $v \in U^\lambda$ and then $v < u$ because u is the leading co-λ term of y. The product vct will be an R_0-linear combination of terms vt', $t' \in T_\lambda \cup \{1\}$; by the ordering of U these terms are $< ut$, so ut, which occurs with coefficient 1, is the leading term of yt. Further, the terms ut' with $t' \neq 1$ are like ut associated with λ, therefore yt is λ-pure. ∎

In studying the submodules of an induced module we shall be particularly interested in those that are themselves induced. The next result gives a sufficient condition:

PROPOSITION 5.3.2. *Given an induced R-module* $M = \bigoplus M_\mu^\otimes$ *over a coproduct* $R = \underset{0}{*} R_\lambda$, *let* L_μ *be an* R_μ-*submodule of M with the following properties. For each* $\mu \in \Lambda_0$,

(A_μ) *All members of* L_μ *are* μ-*pure,*

and for any $\mu, \mu' \in \Lambda_0$,

$(B_{\mu\mu'})$ *The* μ'-*support of* $L_{\mu'}$ *contains no monomial u which is also leading co-μ' term of a (non-μ'-pure) element xa, where* $x \in L_\mu$, $a \in R$, *and if* $\mu' = \mu$, *then also* $\deg xa > \deg x$.

Then the natural map $\bigoplus L_\mu^\otimes \to \sum L_\mu R \subseteq M$ *is an isomorphism.*

A family of modules (L_μ) in M satisfying A_μ and $B_{\mu\mu'}$ is said to be *well-positioned*.

Proof. Given $\mu \in \Lambda_0$, for each monomial u that occurs as leading term of some element of L_μ we choose an element $q \in L_\mu$ having u with coefficient 1 as leading term and write Q_μ for the set of all such q. By the well-ordering of U and by A_μ it follows that Q_μ is an R_0-basis of L_μ.

The leading term of $q \in Q_\mu$ will also be called its 'key term'. When $u \neq 0$, this is also the leading co-μ' term of q for all $\mu' \neq \mu$. An element of L_0, being 0-pure, has likewise a leading co-λ term for all $\lambda \in \Lambda$, but for some λ this will not equal its leading term (if the latter is λ-pure), and we shall need to modify Q_0 in this case. For any λ and any monomial u which is leading co-λ term of a member of L_0 we choose $q \in L_0$ having u as leading co-λ term, with coefficient 1, and denote the set of all such q by $Q_{0\lambda}$. Each such $Q_{0\lambda}$ is again an R_0-basis of L_0. By the 'key term' of a member of $Q_{0\lambda}$ we shall mean its leading co-λ term. Now let V be the set of all products

$$qt_1 \ldots t_n, \quad \text{where } n \geqslant 0,\ t_i \in T_{\lambda_i} \text{ and } \lambda_i \neq \lambda_{i+1} \text{ and either}$$

(i) $n = 0$ and $q \in Q_\mu$ for some μ, or
(ii) $n \geqslant 1$, $q \in Q_{0\lambda}$ and $\lambda = \lambda_1$ or   (1)
(iii) $n \geqslant 1$, $q \in Q_\lambda$ and $\lambda \neq \lambda_1$.

If $qt_1 \ldots t_n$ is as in (1) and the key term of q is u, then the leading term of $qt_1 \ldots t_n$ is $ut_1 \ldots t_n$, by Lemma 3.1. We claim that the members of V have distinct leading terms and so are R_0-linearly independent. If two distinct elements of V have the same leading term, then we have an equation of the form

$$ut_1 \ldots t_m = u' t'_1 \ldots t'_n \text{ in } U.$$

Here we may take $m \geqslant n$, say $m = n + r$. Then this equation reduces to

$$ut_1 \ldots t_r = u'.$$

Let $q \in L_\mu$, $q' \in L_\mu$, correspond to u, u' respectively and consider the following cases:

Case 1. $r > 0$. If $\mu' \neq 0$, then $t_r \in T_{\mu'}$, so the μ'-support of $q' \in T_{\mu'}$ contains $ut_1 \ldots t_{r-1}$ which is also the leading co-μ' term of the non-μ'-pure element $qt_1 \ldots t_{r-1}$ and this contradicts $B_{\mu\mu'}$. If $\mu' = 0$, then since the support of $q' \in L_0$ contains $ut_1 \ldots t_r$ which is also the leading term of the pure element $qt_1 \ldots t_r$, we again have a contradiction to $B_{\mu\mu'}$.

Case 2. $r = 0$. Then $u = u'$ is associated with λ, say, and $q, q' \in Q_\lambda \cup Q_{0\mu}$, where μ is the index associated with $t_1 = t'_1$ if $m > 0$, while for $m = 0$, $Q_{0\mu} = Q_0$. By the construction of the Qs, if $q \neq q'$, they cannot belong to the same set, say $q \in Q_\lambda$, $q' \in Q_{0\mu}$. Then the support of $q' \in L_0$ contains a monomial u which is also the leading term of a pure element $q \cdot 1$, where $q \in L_\lambda$, and this contradicts $B_{\lambda 0}$.

This shows that the elements of V have distinct leading terms and so V is an R_0-basis of $\sum L_\mu R$, as we wished to show. ∎

With the help of this result we can show that every submodule of an induced module is itself induced:

THEOREM 5.3.3. *Let* $R = \underset{0}{*} R_\lambda$ *be the coproduct of a family* (R_λ) *of non-zero rings over a field* R_0. *Then every submodule of an induced module has an induced module structure. More precisely, if* $M = \bigoplus M_\mu^\otimes = \bigoplus M_\mu \otimes_\mu R$, *where* M_μ *is an* R_μ-*module, then for any* R-*submodule* L *of* M *there is an* R_μ-*submodule* L_μ *of* L *(for all* μ) *such that the natural map*

$$\bigoplus L_\mu^\otimes \to L \tag{2}$$

is an isomorphism.

Proof. Let M be as stated and let L be any R-submodule of M. For each $\lambda \in \Lambda$ let V^λ be the set of monomials occurring as leading co-λ terms of non-λ-pure elements of L, and denote by L_λ the set of all elements whose λ-support does not contain any element of V^λ. Clearly L_λ is an R_λ-submodule of L whose members are all λ-pure. Let V^0 be the set of monomials occurring as leading terms of pure elements of L and let L_0 be the set of all elements of L whose support contains no element of V^0. Again it is clear that L_0 is an R_0-module, whose elements are 0-pure (i.e.

impure). Thus A_μ holds, and $B_{\mu\mu'}$ also follows from the definition of the L_μ, hence this family is well-positioned, and by Prop. 3.2 the map (2) is injective. We put $L' = \sum L_\mu R$ and complete the proof by showing that $L' = L$. If $L' \subset L$, choose $y \in L \backslash L'$ so as to minimize the leading term in $\bigcup \mathscr{D}_\lambda(y)$. If, for some λ, y is λ-pure, then since $y \notin L_\lambda$, the λ-support of y contains a monomial $u \in V^\lambda$, where u is the leading co-λ term of some $x \in L$, so x has a term $u\alpha$, $\alpha \in R_0$. Let u be chosen to be the greatest; the monomials in $D_\lambda(y)$ have degree less than $\deg y$, hence $\deg x = \deg u < \deg y$ and so $x \in L'$. Further, there exists $c \in R_\lambda$ such that $u \notin D_\lambda(y - xc)$, and $y - xc$ is either λ-pure with all monomials in its λ-support $< u$, or of lower degree than y. Hence $y - xc \in L'$, by the choice of y, and so $y = (y - xc) + xc \in L'$. If y is impure, then since $y \notin L_0$, some monomial in $D(y)$ occurs in the leading term of some pure element x of L. Let u be the greatest such monomial; then $x \in L'$ and for some $c \in L_0$, $y - xc$ has a leading term $< u$, so again $y - xc \in L'$ and hence $y \in L'$. Therefore no such y can exist and $L' = L$, so (2) is an isomorphism. ∎

Sometimes the following more precise statement is useful:

COROLLARY 5.3.4. *Let R and M be as in Th. 3.3 and suppose that (L_μ) is a well-positioned system of R_μ-submodules of the induced R-module M, and $L = \sum L_\mu R$. Then $L \cap M_\mu \subseteq L_\mu$ for all μ; further, if $\sum L_\mu R = M$, then $L_\mu = M_\mu$.*

Proof. Let $x \in L$ be of degree 1 as member of M; writing $x = \sum v x_v$ ($x_v \in R_0$), we see that each v occurring in x must also be of degree 1, i.e. lie in $\sum M_\mu$ and so be of length 1 as member of V, i.e. lie in some L_μ. If further, $x \in M_{\mu'}$ for some μ', then x cannot involve terms from $L_{\mu''}$ for $\mu'' \neq \mu'$, because the leading co-μ' term of the $L_{\mu''}$-part of x cannot be cancelled by any other terms; therefore $x \in L_{\mu'}$. This shows that $L \cap M_\mu \subseteq L_\mu$. Suppose now that $\sum L_\mu R = M$; then our conclusion states that $M_\mu \subseteq L_\mu$ and if this inclusion were proper, we would have a proper inclusion of induced modules $M \subset L$, which is impossible, therefore $L_\mu = M_\mu$ as claimed. ∎

Th. 3.3 leads to a formula for the global dimension of a coproduct:

THEOREM 5.3.5. *Let (R_λ) be a family of non-zero R_0-rings, where R_0 is a field, and let $R = \underset{0}{*} R_\lambda$. Then the global dimension of R is given by*

$$\text{r.gl.dim.} R = \sup_\lambda \{\text{r.gl.dim.} R_\lambda\}, \tag{3}$$

whenever the right-hand side is positive. When the right-hand side is 0, *then* r.gl.dim. $R = 1$ *unless there is only one factor* $\neq R_0$.

Proof. By Prop. 2.5, r.gl.dim. $R \geq \sup_\lambda \{$r.gl.dim.$R_\lambda\}$. To prove that equality holds, it will be enough to show that for any submodule M of a free R-module F,

$$\mathrm{hd}_R(M) \leq \sup_\lambda (\mathrm{r.gl.dim.}\,R_\lambda) - 1, \tag{4}$$

or 0, if the right-hand side is negative. Now F is induced, hence by Th. 3.3, so is M, say $M = \bigoplus M_\mu^\otimes$, where M_μ is an R_μ-submodule of M. But F is free as R_μ-module, hence

$$\mathrm{hd}_{R_\mu}(M_\mu) \leq \max \{(\mathrm{r.gl.dim.}\,R_\mu) - 1, 0\},$$

and now (4) follows by Prop. 2.5. This proves (3) when the right-hand side is positive; now the rest is clear. ∎

COROLLARY 5.3.6. *With the notation of Th. 3.5, every projective R-module is induced by projective R_μ-modules.*

For if P is a projective R-module, then P is a submodule of a free R-module, so by Th. 3.3, $P \cong \bigoplus P_\mu^\otimes$ and each P_μ is projective as R_μ-module, by Prop. 2.5, because P is R_μ-projective. ∎

The main theorems, 2.3 and 3.3, can also be proved when R_0 is a matrix ring over a field or more generally for any semisimple ring (see Bergman [74]) with only minor modifications, but we shall not need these cases (see Ex. 1).

Given a homomorphism $f: M \to N$ of induced R-modules, if $M = \bigoplus M_\mu^\otimes$ is finitely generated and the system of R_μ-submodules $M_\mu f \subseteq N$ is not well-positioned, then we can modify f so as to obtain a well-positioned system, by means of certain types of automorphisms of induced modules, which we now define.

Consider a homomorphism

$$f: \bigoplus M_\mu^\otimes \to \bigoplus N_\mu^\otimes. \tag{5}$$

If f arises from a family of R_μ-linear maps $f_\mu: M_\mu \to N_\mu$, we shall call it *induced*. Under the injectivity conditions of Th. 3.3 this is the case precisely when f maps M_μ into N_μ for each μ.

Among the isomorphisms between induced modules there is a type arising from the fact that a free R-module can be written as an induced module in more than one way. For any $\mu, \mu' \in \Lambda_0$ we have $R_\mu^\otimes \cong R \cong R_{\mu'}^\otimes$, hence

$$(M_\mu \oplus R_\mu)^{\otimes} \oplus M_{\mu'}^{\otimes} \cong M_\mu^{\otimes} \oplus (M_{\mu'} \oplus R_{\mu'})^{\otimes}. \qquad (6)$$

An isomorphism (5) arising by a transfer of terms as in (6) is called a *free transfer isomorphism*.

Secondly there is the transvection automorphism, familiar from the study of linear groups. Let $e: M_{\mu'} \to R_{\mu'}$ be an $R_{\mu'}$-linear functional, extended to M so as to annihilate M_μ for $\mu \neq \mu'$. Next, for any μ'' and $x \in M_{\mu''}$ there is a map $\alpha(x): R \to M$ defined by $1 \mapsto x \in M_{\mu''}^{\otimes} \subseteq M$. Clearly $\alpha(x)e: R \to R$ vanishes if $\mu' \neq \mu''$ and this will hold even if $\mu' = \mu''$ if we add the condition $x \in \ker e$. Now for any $a \in R$ let $\lambda(a): R \to R$ be left multiplication by a; then the map $e\lambda(a)\alpha(x): u \mapsto xa(ue)$ is nilpotent and so $\theta = 1_M - e\lambda(a)\alpha(x)$ is an automorphism of M. Such an automorphism will be called a *transvection*, *μ-based* in case $\mu' = \mu'' = \mu$.

With the help of these isomorphisms we can transform any homomorphism of induced modules so as to obtain a well-positioned family in the image module.

PROPOSITION 5.3.7. *Let $R = \underset{0}{*} R_\lambda$ be the coproduct of a family (R_λ) of non-zero R_0-rings, where R_0 is a field, and let $f: M \to N$ be a homomorphism of induced R-modules, where M is finitely generated. Then there exists an induced R-module M' and an isomorphism $\alpha: M' \to M$ which is a finite product of free transfer isomorphisms and transvections such that the system $(M'_\mu \alpha f)$ of submodules of N is well-positioned, and αf is then an induced homomorphism.*

Proof. Let $M = \oplus M_\mu^{\otimes}$ and suppose that the family $(M_\mu f)$ is not well-positioned in N. To remedy this defect, we shall modify f and M; we assign to every map as in (5) an index in a certain well-ordered set and show that each adjustment of f lowers this index. By induction, a finite sequence of these adjustments will reduce f to a map $f': M' \to N$ such that $(M'_\mu f')$ is well-positioned; since f' is determined by its effect on the M'_μ, f' is then induced.

To define the index, consider $\Lambda_0 \times U$, where U is well-ordered as before and Λ_0 has a well-ordering with 0 as first element. We well-order $\Lambda_0 \times U$ first by the degree of the second factor and for a given degree lexicographically from left to right. Next let H be the set of almost everywhere zero functions from $\Lambda_0 \times U$ to $\mathbf{N}$, well-ordered lexicographically reading from highest to lowest in $\Lambda_0 \times U$.

For any map f as in (5) we define its index $h_f \in H$ by the rule

$$(\mu, u)h_f = \begin{cases} 1 \text{ if } u \in \mathcal{D}_\mu(M_\mu f)(\text{and hence } u \in U^\mu), \\ 0 \text{ otherwise.} \end{cases}$$

We now show how to diminish h_f if $(M_\mu f)$ is not well-positioned. Suppose first that (A_μ) fails for some μ, so $M_\mu f$ contains an element xf of degree n say, which is not μ-pure. Let $u \in U^\mu$ be the μ-leading term of xf and consider the restriction $p'_\mu(u) = p_\mu(u)|L_\mu$, where $p_\mu(u)$ is the projection defined at the beginning of this section. It follows that $xfp'_\mu(u)$ is non-zero and lies in R_0; this is clear when $\mu = 0$, while for $\mu \neq 0$ a coefficient from $R_\mu \backslash R_0$ applied to u would give a term of degree $n + 1$. Since R_0 is a field, this coefficient is invertible and so the map $fp_\mu(u): M_\mu \to R_\mu$ splits over its kernel M'_μ and we get a decomposition $M_\mu = M'_\mu \oplus xR_\mu$. Now x must be μ'-pure for some $\mu' \neq \mu$; we shall transfer the free summand xR_μ from M_μ to $M_{\mu'}$. In detail, we define a free transfer isomorphism $\iota: M' \to M$, where M'_μ is as above and $M'_{\mu'} = M_{\mu'} \oplus xR_{\mu'}$, while $M'_{\mu''} = M_{\mu''}$ for $\mu'' \neq \mu, \mu'$. We now put $f' = \iota f: M' \to N$ and note that the μ-support of $M'_\mu f'$ no longer contains u. So the first place where $h_{f'}$ differs from h_f is either (μ, u) or (μ', u'), where $u \in u'R_{\mu'}$. If $\mu' \neq 0$, we have $(\mu, u) > (\mu', u')$ by looking at degrees and for $\mu' = 0$ this inequality still holds by the ordering of the first component. Hence $(\mu, u)h_f = 1 > 0 = (\mu, u)h_{f'}$, so f' has lower index than f.

If $B_{\mu\mu'}$ fails, let the μ'-support of $M_{\mu'}f$ contain a monomial u which is also the leading co-μ' term with coefficient 1 of a (non-μ'-pure) element $xfa = xaf$, where $x \in M_\mu$, $a \in R$ and if $\mu' = \mu$, then $\deg xa > \deg x$. For each $y \in M_{\mu'}$, there is a unique element $y\varphi = y - xa(yfp_{\mu'}(u))$ such that $u \notin \mathcal{D}_{\mu'}(y\varphi)$. Indeed, the map $fp_{\mu'}(u): M_{\mu'} \to R_{\mu'}$ is $R_{\mu'}$-linear. We extend the map $\varphi: M_{\mu'} \to M$ to M by defining it as the identity on $M_{\mu''}$ for $\mu'' \neq \mu'$ and claim that it is a transvection. This will follow if we show that $xfp_{\mu'}(u) = 0$; for $\mu' \neq \mu$ this is clear, while for $\mu' = \mu$ we have $\deg xa > \deg x$, so $u \notin \mathcal{D}_\mu(xf)$. Now the first place where $h_{\varphi f}$ differs from h_f is (μ', u) and it is clear that $h_{\varphi f} < h_f$, so we have again lowered the index. By induction we find that after a finite number of such adjustments of f the family $(M_\mu f)$ is well-positioned. ∎

This result allows us to give a precise description of the finitely generated projective modules over a coproduct. We recall the functor $\mathcal{P}$ from rings to monoids which was introduced in 4.6. Our aim will be to show that $\mathcal{P}$ preserves coproducts over a field as base ring. We remark that over any coproduct $R = \underset{0}{*} R_\lambda$ of rings we have, for any R-module M induced by a family (M_μ) of projectives, $\sum[M_\mu] = \sum[M'_\mu]$ if (M'_μ) is obtained from (M_μ) by a free transfer isomorphism or a transvection.

THEOREM 5.3.8. *Let $R = \underset{0}{*} R_\lambda$ be the coproduct of a family (R_λ) of non-zero R_0-rings over a field R_0. Then the induced map*

$$\coprod_{\mathcal{P}(R_0)} \mathcal{P}(R_\lambda) \to \mathcal{P}(R) \tag{7}$$

is an isomorphism.

Proof. We have homomorphisms $\mathcal{P}(R_\lambda) \to \mathcal{P}(R)$ which agree on the distinguished element $[R_\lambda]$ and hence give a homomorphism (7). We first show that (7) is injective.

Suppose that two elements $\sum[L_\mu]$, $\sum[M_\mu]$ of the left-hand side of (7) have the same image, so that there is an isomorphism

$$f: \oplus L_\mu^\otimes \to \oplus M_\mu^\otimes. \tag{8}$$

By Prop. 3.7 there is an induced R-module L' with an isomorphism α: $L' \to L$ such that the system $(L'_\mu \alpha f)$ is well-positioned. Now by Cor. 3.4, $L'_\mu \alpha f = M_\mu$ and so $\sum[L_\mu] = \sum[L'_\mu] = \sum[M_\mu]$ by the above remark, and this shows (7) to be injective.

To establish surjectivity, let P be a finitely generated projective R-module, say $P \oplus Q \cong {}^n R$. By Cor. 3.6, $P \cong \oplus P_\mu^\otimes$, $Q \cong \oplus Q_\mu^\otimes$, and it remains to show that the P_μ are finitely generated. There is an isomorphism $f: (\oplus R_\mu^{n_\mu})^\otimes \to \oplus(P_\mu \oplus Q_\mu)^\otimes$ such that $\sum n_\mu = n$. By Prop. 3.7 we can apply free transfer isomorphisms and transvections to the family $(R_\mu^{n_\mu})$ so as to obtain a family (L_μ) whose image is well-positioned. Clearly L_μ is again finitely generated and its image is $P_\mu \oplus Q_\mu$ by Cor. 3.4. Hence P_μ is finitely generated; thus (7) is surjective and hence an isomorphism. ∎

This result enables us to derive several consequences without difficulty:

THEOREM 5.3.9. *The coproduct of a family of firs over a field is a fir. In particular, the coproduct of fields (over a field) is a fir.*

Proof. By Th. 3.5, the right ideals of the coproduct are projective, and by Th. 3.8, all projectives are free of unique rank, hence all right ideals are free of unique rank. Similarly for left ideals, so the coproduct is a fir. ∎

A similar result holds for n-firs:

PROPOSITION 5.3.10. *Let R_0 be a field and n a positive integer. Then the coproduct of any family of n-firs over R_0 is an n-fir. Moreover, the coproduct of any family of semifirs over R_0 is a semifir.*

Proof. Let $R = \underset{0}{*} R_\lambda$, where each R_λ is an n-fir. For any map ${}^n R \to R$ the image can by Th. 3.3 be written as $\oplus M_\mu^\otimes$, so there is an induced surjection $\alpha: \oplus(R_\mu^{n_\mu})^\otimes \to \oplus M_\mu^\otimes$, where $\sum n_\mu = n$. By free transfer isomorphisms and transvections we obtain a family $(R_\mu^{m_\mu})$ with $\sum m_\mu = n$, such that M_μ is the image of $R_\mu^{m_\mu}$; hence M_μ can be generated by m_μ elements, and as submodule of the projective R_μ-module R, M_μ is free of rank at most m_μ, so $\oplus M_\mu^\otimes$ is free of rank at most n. By repeating this argument with an isomorphism, we see that the rank must be unique. Hence R is an n-fir. If each factor is a semifir, it is an n-fir for all n, so R is then an n-fir for all n, and so R is then a semifir. ∎

For $n = 1$ this tells us that the coproduct of integral domains is an integral domain. This was already proved in Cor. 1.4; the rest of that corollary can also be proved in this way:

COROLLARY 5.3.11. *Let R_0 be a field and (R_λ) a family of integral domains which are R_0-rings. Then the coproduct $R = \underset{0}{*} R_\lambda$ is an integral domain and any unit in R is a monomial unit.*

Proof. We have seen in Prop. 3.10 that R is an integral domain. Now each unit $u \in R$ defines an automorphism $x \mapsto ux$ of $R = R_0^\otimes$. Here R is free on one generator; any free transfer isomorphism just amounts to renaming the generator, while a surjection is a unit in some R_μ. The only transvection is the identity map, since it must be μ-based for some μ, but R_μ is an integral domain. Hence the result follows by Prop. 3.7. ∎

In 4.5 we saw that any semifir (and in particular, any fir) has a universal field of fractions in which all full matrices are inverted, so it follows that the coproduct of fields E_λ over a field K has a universal field of fractions. This field will be called the *field coproduct* or simply *coproduct* of the E_λ, written $\underset{K}{\circ}E_\lambda$ or in the case of two factors E, F, $E \underset{K}{\circ} F$.

Exercises

1. Let k be a commutative field and $A = k\langle a, b, c, d, d'\rangle$ a free algebra. Show that there is a faithful A-ring R in which the equations $xa = c$, $xb = d$ have a solution $x = u$ and a faithful A-ring S in which the equations $ax = b$, $cx = d'$ have a solution $x = v$. Show that the homomorphism $A \underset{A}{*} A = A \to R \underset{A}{*} S$ induced by the inclusions $A \to R$, $A \to S$ is not injective.

2. (Schofield) Let E, F be fields containing a common subfield K and suppose that E is not finitely generated over K. Show that the centre of the field coproduct $E \underset{K}{\circ} F$ is the centre of K. (Hint. Given c in the centre, verify that

$c \in E' \underset{K}{\circ} F$ for a proper subfield E' of E; now write $E \underset{K}{\circ} F = E \underset{E'}{\circ} (E' \underset{K}{\circ} F)$ and show successively that c lies in E', K.)

3. (J. W. Kerr [82]) Let D_0 be a field with centre k and define for $n \geqslant 1$, D_n as the field coproduct over k of D_{n-1} and $k\langle x_n, y_n, z_n \rangle$. Take any subalgebra A_0 of D_0 and define A_n as the subring of D_n generated over A_{n-1} by x_n, y_n, $a^{-1} z_n$ for all $a \in A_{n-1}^{\times}$. Show that the union $A = \bigcup A_n$ is a right Ore domain, but the power series ring $A[[t]]$ is not right Ore. (Hint. Examine $\sum x_n t^n$ and $\sum y_n t^n$.)

5.4 The tensor ring on a bimodule

For any commutative field k the *free k-algebra* on a set X, $k\langle X \rangle$, may be defined by the following universal mapping property: $k\langle X \rangle$ is generated by X as k-algebra and any mapping of X into a k-algebra A can be extended to a unique k-algebra homomorphism of $k\langle X \rangle$ into A. The elements of $k\langle X \rangle$ can be uniquely written as

$$\sum a_{i_1 \ldots i_r} x_{i_1} \ldots x_{i_r}, \quad \text{where } x_i \in X, a_{i_1 \ldots i_r} \in k.$$

As is easily seen, $k\langle X \rangle$ may also be represented as a coproduct:

$$k\langle X \rangle = \underset{k}{*} \, k[x],$$

where x runs over X. Since each $k[x]$ is a principal ideal domain (and hence a fir), it follows from Th. 3.9 that $k\langle X \rangle$ is a fir. More generally, we can show in the same way that for any field D the tensor D-ring on any set is a fir:

THEOREM 5.4.1. *Let D be a field and k a central subfield. Then for any set X the tensor D-ring on X over k, $D_k\langle X \rangle$, is a fir.*

Proof. We have just seen that $k\langle X \rangle$ as a coproduct of PIDs is a fir, so the result follows because $D_k\langle X \rangle = D \underset{k}{*} k\langle X \rangle$. ∎

Since every fir has a universal field of fractions (Cor. 4.5.9), we have universal fields of fractions for $k\langle X \rangle$ and $D_k\langle X \rangle$, which will be denoted by $k \langle\!\langle X \rangle\!\rangle$ and $D_k \langle\!\langle X \rangle\!\rangle$ and called *free fields*. To elucidate the relation between them we need a lemma:

LEMMA 5.4.2. *Let R, S be semifirs over a field K. Then*
  (i) *the inclusion $R \to R \underset{K}{*} S$ is honest, and*
  (ii) *if $\mathcal{U}(R)$ denotes the universal field of fractions of R, then the natural map $R \underset{K}{*} S \to \mathcal{U}(R) \underset{K}{*} S$ is honest.*

Proof. (i) Consider the natural homomorphisms

$$R \to R * S \to \mathcal{U}(R) * S.$$

Any full matrix A over R is invertible over $\mathcal{U}(R)$, hence also over $\mathcal{U}(R) * S$ and so A is full over $R * S$, as we had to show.

(ii) By (i) any full matrix over R is full over $R * S$, hence we have a homomorphism $\mathcal{U}(R) \to \mathcal{U}(R * S)$ (Th. 4.5.10), and it follows that we have a homomorphism $\mathcal{U}(R) * S \to \mathcal{U}(R * S)$. Thus we have mappings

$$R * S \to \mathcal{U}(R) * S \to \mathcal{U}(R * S).$$

Now any full matrix over $R * S$ is invertible over $\mathcal{U}(R * S)$ and hence is full over $\mathcal{U}(R) * S$. ∎

We can now describe the relation between different free fields:

PROPOSITION 5.4.3. *Let $D \subseteq E$ be any fields and let k be a central subfield of E that is also contained in D. Then*

$$E_k \langle\!\langle X \rangle\!\rangle \cong E \underset{D}{\circ} D_k \langle\!\langle X \rangle\!\rangle. \tag{1}$$

Hence there is a natural embedding

$$D_k \langle\!\langle X \rangle\!\rangle \to E_k \langle\!\langle X \rangle\!\rangle. \tag{2}$$

In particular, taking $D = k$, we find that $k \langle\!\langle X \rangle\!\rangle$ is embedded in $E_k \langle\!\langle X \rangle\!\rangle$ and

$$E_k \langle\!\langle X \rangle\!\rangle \cong E \underset{k}{\circ} k \langle\!\langle X \rangle\!\rangle. \tag{3}$$

Proof. We have the natural isomorphisms

$$E \underset{D}{*} D_k \langle X \rangle \cong E \underset{D}{*} D \underset{k}{*} k \langle X \rangle \cong E \underset{k}{*} k \langle X \rangle \cong E_k \langle X \rangle.$$

Now the natural homomorphism $E_k \langle X \rangle \to E \underset{D}{*} D_k \langle\!\langle X \rangle\!\rangle$ is honest, by Lemma 4.2 (ii), hence we have an embedding

$$E_k \langle\!\langle X \rangle\!\rangle \to E \underset{D}{\circ} D_k \langle\!\langle X \rangle\!\rangle. \tag{4}$$

But the right-hand side is a field generated by E and X, hence it is generated as a field by the image of $E_k \langle X \rangle$, so (4) is surjective and we have the natural isomorphism (1). Now (2) and (3) are immediate consequences. ∎

Under suitable conditions this result continues to hold when there is a change of ground field (see 6.4 below). For the moment we shall merely deal with the case of a simple transcendental extension.

PROPOSITION 5.4.4. *Let E be a field with a central subfield k and form $E(t)$ with a central indeterminate t. Then there is a natural embedding*

$$E_k \langle\!\langle X \rangle\!\rangle \to E(t)_{k(t)} \langle\!\langle X \rangle\!\rangle. \tag{5}$$

Proof. The assertion will follow if we can show that the natural homomorphism

$$E_k \langle X \rangle \to E(t)_{k(t)} \langle X \rangle \tag{6}$$

is honest. Let us write $F = E_k \langle\!\langle X \rangle\!\rangle$; there is a natural inclusion $E \to F$, hence a mapping $E(t) \to F(t)$ and so an $E(t)$-ring homomorphism

$$E(t)_{k(t)} \langle X \rangle \to F(t). \tag{7}$$

Now let A be a full matrix over $E_k \langle X \rangle$. Then A is invertible over F, hence also over $F(t)$ and by (7) it is full over $E(t)_{k(t)} \langle X \rangle$. This shows (6) to be honest and now the embedding (5) follows. ∎

The fact that the free algebra is a fir also follows from the existence of the weak algorithm, as explained in Ch. 2 of FR. The weak algorithm can actually be applied to a wider class of rings, to which we now turn. We begin with a general definition.

Let R be a ring and M an R-bimodule. We put

$$M^n = M \otimes M \otimes \ldots \otimes M, \quad n \text{ factors, taken over } R;$$

thus $M^1 = M$ and by convention, $M^0 = R$. It is clear that $M^r \otimes M^s \cong M^{r+s}$; hence we have a multiplication on the direct sum

$$\mathbf{T}(M) = \oplus M^n, \tag{8}$$

which is associative and so turns it into an R-ring. This ring is called the *tensor R-ring* on M; usually we shall single out a central subfield k of R and assume that the left and right actions of k on M agree. Then $\mathbf{T}(M)$ will be a k-algebra, which will be denoted by $R_k \langle M \rangle$; it will usually be clear from the context whether the tensor ring on a set or a bimodule is intended. From (8) we see that $R_k \langle M \rangle$ is a graded ring, with M^n as component of degree n, and there is a natural bimodule homomorphism $M \to R_k \langle M \rangle$, mapping M to M^1, with the usual universal property for maps of M to R-rings. When $R = D$ is a field, the tensor ring $D_k \langle M \rangle$ can be shown to possess a weak algorithm relative to the grading; consequently it is a fir, for any D-bimodule M. We shall not give the proof here, which uses the weak algorithm, as the result will not be needed (see FR, 2.4), but confine ourselves to showing that the tensor ring $D_k \langle M \rangle$ is a semifir. Somewhat more generally we shall establish the result when k is replaced by a (possibly skew) subfield. Thus we shall prove

THEOREM 5.4.5. *Let D be a field with a subfield F and let M be a D-bimodule with an F-centralizing generating set. Then the tensor ring $D_F\langle M \rangle$ is a semifir.*

Proof. Put $R = D_F\langle M \rangle$, let $\{u_\nu\}$ be a right F-basis of D and $\{v_\lambda\}$ an F-centralizing generating set of M, which we may take to be left and right linearly independent over D, without loss of generality, by omitting superfluous terms. Any element of M has the form $m = \sum u_\nu v_\lambda a_{\nu\lambda}$, with uniquely determined coefficients $a_{\nu\lambda}$ in D. Since $R = \bigoplus M^n$, it follows that every element of R can be uniquely written as

$$f = c + \sum u_\nu v_\lambda f_{\nu\lambda}, \quad \text{where } c \in D, f_{\nu\lambda} \in R.$$

Now assume that we have a relation in R:

$$\sum_1^n a_i b_i = 0.$$

To show that this relation can be trivialized it is enough to do this in a given degree; thus we may assume that the a_i, b_i are homogeneous and that $\deg a_i + \deg b_i = r > 0$. We shall use double induction, on n and r. If each a_i has positive degree, we can write $a_i = \sum u_\nu v_\lambda a_{\nu\lambda i}$; equating cofactors of $u_\nu v_\lambda$, we find

$$\sum a_{\nu\lambda i} b_i = 0,$$

and now the result follows by induction on r. There remains the case where some a_i, say a_1, has degree 0. Thus $a_1 \in D$ and either $a_1 = 0$ and we can use induction on n, or $a_1 \neq 0$, in which case we can replace a_2 by $a_2 - a_1 \cdot a_1^{-1} a_2 = 0$ and b_1 by $b_1' = b_1 + a_1^{-1} a_2 \cdot b_2$; thus we obtain

$$\sum a_i b_i = a_1 b_1' + a_3 b_3 + \ldots + a_n b_n = 0,$$

so we have again diminished n and can apply induction to complete the proof. ∎

In particular, this result shows the tensor ring on a set X, $D_F\langle X \rangle$, to be a semifir, since it can be expressed as tensor ring on the D-bimodule generated by the F-centralizing set X. The same proof shows $D_F\langle\langle X \rangle\rangle$ to be a semifir.

Sometimes a slight variant of the above construction is needed, leading to a filtered ring. By a *pointed R-bimodule* we shall understand an R-bimodule M with a subbimodule isomorphic to R, which is complemented as right R-module, thus $M = R \oplus N$. A map $f: M \to M'$ between pointed bimodules is then a homomorphism which forms a commutative triangle with the canonical maps from R. Given any pointed R-bimodule $M = R \oplus N$, we put $\bar{M} = M/R$; thus $\bar{M}$ is an R-bimodule which is isomorphic to N as right R-module. We now define a sequence of R-bimodules in terms of M recursively as follows: $M^0 = R$, $M^1 = M$ and if M^{n-1} has been defined, we put

$$M^n = R \oplus (N \otimes M^{n-1}). \tag{9}$$

The right R-module structure is clear, since R and M^{n-1} are bimodules; the left R-module structure is defined as follows: If $u \in N$, $v \in M^{n-1}$ and $a \in R$, then $au = a_1 + u_1$, say, where $a_1 \in R$, $u_1 \in N$. Then we put

$$a(u \otimes v) = a_1 v + u_1 \otimes v.$$

It is easily verified that with this definition M^n is a pointed R-bimodule with M^{n-1} as subbimodule. In this way we obtain a direct system of R-bimodules

$$R = M^0 \to M^1 \to M^2 \to \ldots,$$

whose direct limit is a ring in a natural way, since we have a map $M^r \otimes M^s \to M^{r+s}$. This ring is denoted by $R_k \langle M; R \rangle$ and is called the *filtered tensor ring* on the pointed R-bimodule M. Like all universal constructions this tensor ring has a universal property:

THEOREM 5.4.6. *Let R be a ring and a k-algebra. Then (i) for any R-bimodule M, the natural homomorphism $M \to R_k \langle M \rangle$ is universal for bimodule homomorphisms from M to R-rings, (ii) for any pointed R-bimodule M the natural homomorphism $M \to R_k \langle M; R \rangle$ is universal for pointed bimodule maps from M to R-rings.*

The proof is a straightforward verification, which may be left to the reader. ∎

Our aim will be to find conditions for the natural homomorphism to be injective. The main condition is that M should be (left and right) flat. We shall not treat this problem in its most general form, but confine our attention to the case used later, where R is an integral domain. To allow induction arguments to be used, we must then show that the tensor ring is again an integral domain. We begin by examining the tensor product of flat modules. Let us recall that a right R-module M is called *flat* if $M \otimes -$ preserves exactness, i.e.

$$0 \to U' \to U \text{ exact } \Rightarrow 0 \to M \otimes U' \to M \otimes U \text{ exact.} \tag{10}$$

It follows immediately that if M is a flat right R-module and N an R-bimodule which is right flat, then $M \otimes N$ is again right flat. We also recall the following standard criterion for flatness (see A.3, 6.6): M_R is flat if and only if, for any relation $\sum u_i a_i = 0$ ($u_i \in M$, $a_i \in R$), there exist $v_j \in M$, $b_{ji} \in R$ such that $u_i = \sum v_j b_{ji}$, $\sum b_{ji} a_i = 0$. Further we recall that a right R-module M is called *torsion-free* if $ur = 0$ for $u \in M$, $r \in R$ implies $u = 0$ or $r = 0$.

LEMMA 5.4.7. *Let R be an integral domain. Then any flat R-module is torsion-free. Further, if U_R is flat and $_RV$ is left torsion-free, then in $U \otimes V$, $u \otimes v = 0$ implies $u = 0$ or $v = 0$.*

Proof. Let U_R be flat and suppose that $ur = 0$. By flatness there exist $u_1, \dots, u_n \in U$, $x_1, \dots, x_n \in R$ such that $u = \sum u_i x_i$, $x_i r = 0$. If $r \neq 0$, then $x_i = 0$ for all i and so $u = 0$; thus U is torsion-free. Suppose now that U_R is flat and $_RV$ torsion-free and that $u \otimes v = 0$ in $U \otimes V$. If $v \neq 0$, then Rv is a free submodule of V and $U \otimes Rv$ is embedded in $U \otimes V$, hence $u \otimes v = 0$ in $U \otimes Rv$ and so $u = 0$. ∎

When R is a PID, then conversely, every torsion-free R-module is flat, but for general integral domains this need not hold.

With the help of this lemma we can show that the tensor ring on a flat bimodule over an integral domain is again an integral domain.

THEOREM 5.4.8. *Let R be a k-algebra which is an integral domain, and let M be an R-bimodule which is left and right flat. Then the tensor R-ring $R_k \langle M \rangle$ is again an integral domain.*

Proof. It is clear that each homogeneous component M^n of $R_k \langle M \rangle$ is left and right flat and so torsion-free. Let $a, b \in R_k \langle M \rangle$, suppose that $a, b \neq 0$ and write each as a sum of homogeneous components: $a = a_r + a_{r-1} + \dots$, $b = b_s + b_{s-1} + \dots$, where a_i, b_i lie in M^i and $a_r, b_s \neq 0$. By Lemma 4.7, $a_r b_s \neq 0$ and this is the component of degree $r + s$ of ab, hence $ab \neq 0$. ∎

For filtered rings on a pointed bimodule there is a corresponding result. We recall that with a filtered ring (R_n), where $R_n \subseteq R_{n+1}$, there is associated a graded ring $\bigoplus (R_n / R_{n-1})$, and the filtered ring is an integral domain, provided that the associated graded ring is one (Prop. 2.6.1).

COROLLARY 5.4.9. *Let R be a k-algebra which is an integral domain and let M be a pointed R-bimodule such that $\bar{M} = M/R$ is left and right flat. Then the filtered tensor ring $R_k \langle M; K \rangle$ is again an integral domain.*

Proof. Let us write $\bar{M}^n = \bar{M} \otimes \ldots \otimes \bar{M}$ (n factors); then

$$M^n/M^{n-1} \cong \bar{M}^n. \tag{11}$$

To prove the result we shall use induction on n. For $n = 1$ it holds by definition, so take $n > 1$. By the induction hypothesis we have the exact sequence

$$0 \to M^{n-2} \to M^{n-1} \to \bar{M}^{n-1} \to 0.$$

If we tensor on the left with $\bar{M} = N$, then since N is right flat, we have the exact sequence

$$0 \to N \otimes M^{n-2} \to N \otimes M^{n-1} \to \bar{M}^n \to 0.$$

By the definition (9), $\overline{M^n} = N \otimes M^{n-1}$ and so

$$M^n/M^{n-1} \cong \overline{M^n}/\overline{M^{n-1}} \cong \bar{M}^n,$$

which proves (11).

Thus we have a filtered ring whose associated graded ring is the tensor ring $R_k\langle \bar{M} \rangle$; the latter is an integral domain, by Th. 4.8, hence so is the filtered ring, as we wished to show. ∎

Let us return to the tensor ring on a bimodule. An important example is the D-bimodule $M = D \otimes_k D$ for the field D with central subfield k. The tensor D-ring $D_k\langle M \rangle$ has the following universal property: Given any D-ring R and any element $c \in R$, there is a unique D-ring homomorphism from $D_k\langle M \rangle$ to R such that

$$1 \otimes 1 \mapsto c. \tag{12}$$

For the map (12) can be extended to a bimodule homomorphism $M \to R$ and hence to a D-ring homomorphism from $D_k\langle M \rangle$ to R. This shows that we have an isomorphism

$$D_k\langle D \otimes_k D \rangle \cong D_k\langle x \rangle, \quad 1 \otimes 1 \mapsto x.$$

If we examine the proof, we see that it is not necessary to assume k to be central, or even commutative. Thus if K denotes an arbitrary subfield of D, we have $D_K\langle D \otimes_K D \rangle \cong D_K\langle x \rangle$. More generally, for any set X we have $D_K\langle X \rangle \cong D_K\langle {}^X(D \otimes_K D) \rangle$, where ${}^X M$ denotes the direct sum of copies of M indexed by X. By applying Th. 4.1 (which clearly holds for general subfields of D) we derive the following consequence:

PROPOSITION 5.4.10. *Let D be a field and K a subfield. Then for any set X, the tensor ring $D_K\langle {}^X(D \otimes_K D) \rangle$ is a fir.* ∎

Exercises

1. Let H be a semifir with universal field of fractions D and let K be a subfield of H. Show that the mapping $H_K\langle X \rangle \to D_K\langle X \rangle$ induced by the inclusion $H \subseteq D$ is honest.

2. Show that over a commutative field every idempotent 2×2 matrix $\neq 0, I$ has trace 1. Over the free field $k(x, y)$ construct an idempotent 2×2 matrix $\neq 0, I$ with trace $1 + x - x'$, where x' is a conjugate of x.

3. Let k be any commutative field and $D = k(x) \underset{k}{\circ} k(y)$ a field coproduct. Show that D has an involution $*$ such that $x^* = x$, $y^* = y^{-1}(x - x^2)$. Construct a self-adjoint idempotent 2×2 matrix $\neq 0, I$ over D with trace $1 + x - x'$, where x' is a conjugate of x.

4. Give an example of a module over the polynomial ring $k[x, y]$ which is torsion-free but not flat.

5.5 HNN-extension of fields

Although Schreier had discussed free products of groups in 1927, it was not until 20 years later that significant applications were made, notably in the classic paper by Higman, Neumann and Neumann [49]. Their main result was the following

HNN THEOREM. *Let G be any group with two subgroups A, B which are isomorphic, say $f: A \to B$ is an isomorphism. Then G can be embedded in a group H containing also an element t such that*

$$t^{-1}at = af \text{ for all } a \in A. \quad \blacksquare$$

The group H is usually denoted by $\langle G, t; \ t^{-1}at = af, a \in A \rangle$ and is called an *HNN-extension*. We shall omit the proof (but see Ex. 1).

We observe that the above theorem would be trivial if f were an automorphism of the whole of G: then H would be the split extension of G by an infinite cycle inducing f. But for proper subgroups A, B the result is non-trivial and (at first) surprising. It has many interesting and important consequences for groups and it is natural to try and prove an analogue for fields. What is needed is a coproduct in the category of fields. However, we shall not adopt a categorical point of view: as we saw, the morphisms in the category of fields are all monomorphisms and this is somewhat restrictive. Over a fixed ring, it is true, we have defined specializations, but it would be more cumbersome to define them without a ground ring, and not really helpful.

In this section we shall prove an analogue of the HNN theorem using the field coproduct introduced in 5.4. But we shall also need some auxiliary results on subfields of coproducts. It will be convenient to regard all our fields as algebras over a given commutative field k; this just amounts to requiring k to be contained in the centre of each field occurring.

THEOREM 5.5.1. *Let K be a field and A, B subfields of K, isomorphic under a mapping $f: A \to B$, where K, A, B are k-algebras and f is k-linear. Then K can be embedded in a field L, again a k-algebra, in which A and B are conjugate by an inner automorphism inducing f, thus L contains $t \neq 0$ such that*

$$t^{-1}at = af \text{ for all } a \in A. \tag{1}$$

The field L is again called an *HNN-extension* and is denoted by $K_k \langle t, t^{-1}; t^{-1}at = af, a \in A \rangle$.

Proof. We shall give two proofs, one using the weak algorithm and so depending on results of FR, and a second one using only results proved here.

Define K as right A-module by the usual multiplication and as left A-module by

$$a \cdot u = (af)u, \quad a \in A, u \in K. \tag{2}$$

We now form the K-bimodule $K \otimes_A K$ with the usual multiplication by elements of K. If we abbreviate $1 \otimes 1$ as t, this consists of all sums $\sum u_i t v_i$ ($u_i, v_i \in K$) with the defining relations

$$at = t \cdot af, \quad a \in A. \tag{3}$$

Now the tensor ring on this bimodule $K \otimes_A K$ satisfies the weak algorithm (see FR, Th. 2.6.2). Hence it is a fir and so it has a universal field of fractions L. Thus K has been embedded in a field L in which (3) holds. We remark that Prop. 4.10 cannot be used as it stands since $K \otimes K$ does not have the standard bimodule structure. However we can prove the result without the weak algorithm as follows.

We take a family (K_i) of copies of K indexed by $\mathbf{Z}$ and form their field coproduct amalgamating B in K_i with A in K_{i+1} along the isomorphism f. This can be done stepwise and taking the direct limit, we obtain a field F, say. In F we have the shift automorphism α which consists in replacing any element of K_i by the corresponding element of K_{i+1}. We now form $L = F(t; \alpha)$ and embed K in L by identifying it with K_0. Then for any $a \in A \subseteq K_0$ we have $at = t \cdot a^\alpha = t \cdot af$, so (2) again holds. ∎

Any field D which is a k-algebra is said to be *n-homogeneous* over k if for any elements $a_1, \ldots, a_n, b_1, \ldots, b_n \in D$ such that the map $a_i \mapsto b_i$ defines an isomorphism between the k-fields they generate, there exists $t \in D^\times$ such that $t^{-1}a_i t = b_i$ $(i = 1, \ldots, n)$; D is called *homogeneous* if it is n-homogeneous for all n. The construction of homogeneous fields is an easy consequence of Th. 5.1:

COROLLARY 5.5.2. *Every field D (over a central subfield k) can be embedded in a field (again over k) of the same cardinal or at least countable, which is homogeneous.*

Proof. Given finitely many as and bs such that $a_i \mapsto b_i$ defines an isomorphism, we can by Th. 5.1 extend D to include an element $t \neq 0$ such that $t^{-1}a_i t = b_i$, and the least such extension has the same cardinal as D or is countable. If we do this for all pairs of finite sets in D which define isomorphisms, we get a field D_1, still of the same cardinal as D or countable, such that any two finitely generated isomorphic subfields of D are conjugate in D_1. We now repeat this process, obtaining D_2, and if we continue thus, we get a tower of fields

$$D \subset D_1 \subset D_2 \subset \ldots.$$

Their union is a field L with the required properties, for if $a_1, \ldots, a_n$, $b_1, \ldots, b_n \in L$ and $a_i \mapsto b_i$ defines an isomorphism, we can find D_r to contain all the as and bs, hence they become conjugate in D_{r+1} and so are conjugate in L. ∎

A homogeneous field has the property that any two elements with the same (or no) minimal equation over k are conjugate. Hence we obtain

COROLLARY 5.5.3. *Every field K (over a central subfield k) can be embedded in a field L over k in which any two elements with the same minimal equation over k or both transcendental over k are conjugate.* ∎

Later we shall need an analogue of this result for matrices. We first establish a matrix version of Th. 5.1.

LEMMA 5.5.4. *Given a field K (over k) and $n \geq 1$, let E be a subfield of $\mathfrak{M}_n(K)$. If F_1, F_2 are subfields of E which are isomorphic under a map φ: $F_1 \to F_2$, then there is an extension L of K and a matrix $T \in \mathbf{GL}_n(L)$ such that*

$$x\varphi = T^{-1}xT \text{ for all } x \in F_1.$$

Proof. By Th. 5.1, E has an extension field E' with an element T inducing φ. Consider the coproduct $R = \mathfrak{M}_n(K) \underset{E}{*} E'$; this is of the form $\mathfrak{M}_n(G)$, where G is a ring containing K. By Th. 3.5, R and hence G is hereditary; by Th. 3.8 G is projective-free and so it is a fir. If L is its universal field of fractions, then L contains K and $\mathfrak{M}_n(L)$ contains the matrix T inducing φ. ∎

The scope of this lemma can still be extended as follows: Let F_1, F_2 be subfields of $\mathfrak{M}_n(K)$ as before, and suppose that there is a subfield F isomorphic to F_2 by an isomorphism $g: F_2 \to F$, such that F and F_i are contained in a common subfield E_i of $\mathfrak{M}_n(K)$, for $i = 1, 2$. By the lemma we can enlarge K to a field L containing an invertible matrix T_1 which induces the isomorphism $fg: F_1 \to F$ and then enlarge L to a field M containing an invertible matrix T_2 inducing the isomorphism $g: F_2 \to F$. Now $T_1 T_2^{-1}$ induces the isomorphism $f: F_1 \to F_2$.

We can now prove an analogue of Th. 5.1 for matrices, at least for the transcendental case. A square matrix A over a field K is said to be *totally transcendental* over the central subfield k if for every non-zero polynomial $f \in k[t]$, the matrix $f(A)$ is non-singular. Clearly if A is totally transcendental over k, then the field generated by A over k is a simple transcendental extension of k. The field K is said to be *matrix-homogeneous* over k if any two totally transcendental matrices of the same size over k are conjugate.

THEOREM 5.5.5. *Every field K with a central subfield k has an extension field L which is matrix-homogeneous over k.*

Proof. Let A, B be two $n \times n$ matrices over K, both totally transcendental over k. Then $k(A)$ is a purely transcendental extension of k, thus if u is a central indeterminate over K, we have $k(A) \cong k(u)$ and likewise $k(B) \cong k(u)$. We shall take $F_1 = k(A)$, $F_2 = k(B)$, $F = k(u)$. Consider the field $K((u))$ of formal Laurent series in u over K. We have

$$\mathfrak{M}_n(K((u))) \cong \mathfrak{M}_n(K)((u)), \qquad (4)$$

and F, F_1 are contained in the subfield $k(A)((u))$ of (4), while F, F_2 are contained in $k(B)((u))$. We can therefore apply Lemma 5.4 and the remark following it and obtain an extension field L of $K((u))$ with an invertible $n \times n$ matrix inducing the k-isomorphism $k(A) \to k(B)$ defined by $A \mapsto B$.

We can repeat this process for other pairs of totally transcendental matrices until we obtain a field $K_1 \supset K$ in which any two totally transcendental matrices over K of the same order are conjugate. If we repeat the construction for K_1 we obtain a tower of fields over k,

$$K \subset K_1 \subset K_2 \subset \ldots ,$$

whose union is a field L with the property that any two totally transcendental matrices of the same order are conjugate, so L is the required matrix-homogeneous field. ∎

This means, for example, that over L any totally transcendental matrix $n \times n$ A can be transformed to *scalar* (not merely diagonal) form. We need only choose a transcendental element α over k in L; then αI_n is totally transcendental, therefore $T^{-1}AT = \alpha I$ for some $T \in \mathbf{GL}_n(L)$. We shall return to this topic in Ch. 8.

Our next objective is to show that every countable field can be embedded in a two-generator field. This corresponds to a theorem of B. H. Neumann [54] for groups. We shall need some lemmas on field coproducts; first we examine a situation in which a subfield of a given field is a field coproduct. If D is a field with subfield E, we shall write $D_E\langle X \rangle$ for the tensor D-ring on X with E centralizing X. By Prop. 4.10 this is a fir; its universal field of fractions will be denoted by $D_E \langle\!\langle X \rangle\!\rangle$ and will also be called a *free field*.

LEMMA 5.5.6. *Let D be a field with subfield E and let x be an indeterminate. Then the free field $D_E\langle\!\langle x \rangle\!\rangle$ can be written in the form*

$$D_E\langle\!\langle x \rangle\!\rangle \cong F(x; \alpha), \tag{5}$$

where F is the field coproduct of countably many copies D_i of D $(i \in \mathbf{Z})$ over E, and α is the shift automorphism $D_i \to D_{i+1}$.

Proof. Let R be the ring coproduct of the countable family D_i over E; it is a fir, with universal field of fractions F, say. The subfields $x^{-i}Dx^i$ of $D_E\langle\!\langle x \rangle\!\rangle$ generate a subfield G which is an epimorphic image of R, hence G is an R-specialization of F. Now x induces an automorphism of G by conjugation and we can extend the specialization from F to G by forming the skew polynomial ring $F[x; \alpha]$ with the shift automorphism α. We thus have a specialization from $F(x; \alpha)$ to $D_E\langle\!\langle x \rangle\!\rangle$ which by the universal property of $D_E\langle\!\langle x \rangle\!\rangle$ must be an isomorphism. ∎

We shall also need a result on free sets in field coproducts. Given a field over k as central subfield, by a *free set* over k we understand a subset Y such that the subfield generated by Y is free, i.e. isomorphic to the free field $k\langle\!\langle Y \rangle\!\rangle$.

LEMMA 5.5.7. *Let E be a field, generated over the central subfield k by a family $\{e_\lambda\}$ of elements, and let U be a field containing a free family $\{u_\lambda\}$ over k. Then the elements $u_\lambda + e_\lambda$ form a free set in the field coproduct $U \underset{k}{\circ} E$.*

Proof. We may assume that $U = k\langle u_\lambda \rangle$; for by hypothesis U contains $k\langle u_\lambda \rangle = F$, say, and we have $E \underset{k}{\circ} U = (E \underset{k}{\circ} F) \underset{F}{\circ} U$. So if the family $\{u_\lambda\}$ is free in $E \underset{k}{\circ} F$, it is free in $E \underset{k}{\circ} U$. Now the field coproduct $D = E \underset{k}{\circ} U$ has the following universal property: Given any E-field F and any family $\{f_\lambda\}$ of elements of F (indexed by the same set), there is a unique specialization from D to F over E, with domain generated by E and the u_λ, which maps u_λ to f_λ. In particular, there are specializations from D to itself which map u_λ to $u_\lambda + e_\lambda$, or to $u_\lambda - e_\lambda$ respectively. On composing these mappings we obtain the identity mapping, $u_\lambda \mapsto u_\lambda + e_\lambda \mapsto u_\lambda - e_\lambda + e_\lambda = u_\lambda$, and similarly in the opposite order. Hence they are inverse to each other and so are automorphisms. It follows that the $u_\lambda + e_\lambda$ like the u_λ form a free set. ∎

We can now achieve our objective, the embedding theorem mentioned earlier; the proof runs parallel to the group case.

THEOREM 5.5.8. *Let E be a field, countably generated over a central subfield k. Then E can be embedded in a two-generator field over k.*

In essence the proof runs as follows: Suppose that E is generated by $e_0 = 0, e_1, e_2, \ldots$; we construct an extension field L generated by elements x, y, z over E satisfying

$$y^{-i}xy^i = z^{-i}xz^i + e_i \quad (i = 0, 1, \ldots).$$

Then L is in fact generated by x, y, z alone. If we now adjoin t such that $y = txt^{-1}$, $z = t^{-1}xt$, the resulting field is generated by x and t.

To give a formal proof, let F be the free field on x, y over k; it has a subfield U generated by $u_i = y^{-i}xy^i$ $(i = 0, 1, \ldots)$ freely, by Lemma 5.6, and similarly, let G be the free field on x, z over k, with subfield V freely generated by $v_i = z^{-i}xz^i$ $(i = 0, 1, \ldots)$. Now form $K = E \underset{k}{\circ} F$ and take the subfield W generated by $w_i = u_i + e_i$ $(i = 0, 1, \ldots)$, freely by Lemma 5.7. We note that $w_0 = u_0 + e_0 = u_0 = x$, so K is generated over k by x, y and the w_i $(i \geqslant 1)$.

Let L be the field coproduct of K and G, amalgamating W and V along the isomorphism $w_i \leftrightarrow v_i$. We note that $w_0 = x = v_0$ and that L is generated by x, y, z and the w_i or also by x, y, z and the v_i, or simply by x, y, z. Now L contains the isomorphic subfields generated by x, y and by z, x respectively, hence we can adjoin t to L to satisfy $t^{-1}xt = z$, $t^{-1}yt = x$, by Th. 5.1. It follows that we have an extension of L generated by x, t over k and it contains E. ∎

The following special case is already a consequence of Lemma 5.6, obtained by taking $K = k(y)$:

COROLLARY 5.5.9. *Any free field of countable rank, $D_k \{x_1, x_2, \ldots \}$, can be embedded in $D_k \{x, y\}$ by mapping x_r to $y^{-r}xy^r$ ($r = 1, 2, \ldots$).* ∎

From Th. 5.8 we also obtain the usual general form:

COROLLARY 5.5.10. *Every field over k can be embedded in a field L such that every countably generated subfield of L is contained in a two-generator subfield of L.*

Proof. Let E be the given field and E_λ a typical countably generated subfield (always over k). Then there is a two-generator field L_λ containing E_λ, by the theorem. Let M_λ be the field coproduct of E and L_λ over E_λ; if we do this for each countably generated subfield of E we get a family $\{M_\lambda\}$ of fields, all containing E. We form their field coproduct E' over E; in E' every countably generated subfield of E is contained in a two-generator subfield of E', namely E_λ is contained in L_λ. Now we repeat the process which led from E to E':

$$E \subset E' \subset E'' \subset \ldots \subset E^\omega \subset E^{\omega+1} \subset \ldots \subset E^\nu,$$

where $E^\alpha = \bigcup \{E^\beta | \beta < \alpha\}$ at a limit ordinal α, and where ν is the first uncountable ordinal. Then E^ν is a field in which every countably generated subfield is contained in some E^α ($\alpha < \nu$) and hence in some two-generator subfield of $E^{\alpha+1} \subset E^\nu$. ∎

At this point it is natural to ask whether there is a countable field, or one countably generated over k, containing a copy of every countable field of a given characteristic. As in the case of groups, the answer is 'no'. This is shown by the following argument (for which I am indebted to A. J. Macintyre).

For any field K, denote by $\mathscr{S}(K)$ the set of isomorphism types of finitely generated subgroups of $K^\times$. Clearly if K is countable, then so is $\mathscr{S}(K)$. Now D. B. Smith [70] has shown that there are $\mathfrak{c} = 2^{\aleph_0}$ isomorphism types of finitely generated orderable groups, i.e. there are $\mathfrak{c}$ groups which can be ordered and are distinct as groups. Further, every ordered group can be embedded in a field of prescribed characteristic, by the methods of 2.4, hence every countable ordered group can be embedded in a countable field. It follows that there are $\mathfrak{c}$ distinct sets $\mathcal{S}(K)$ as K runs over all countable fields of any given characteristic. Therefore these fields cannot all be embedded in a two-generator field, so no countable field can contain a copy of every countable field.

The methods of this section can also be used to determine the precise centre of a free field.

THEOREM 5.5.11. *Let D be a field with a central subfield k. Then for any set X the centre of the free field $D_k \langle\!\langle X \rangle\!\rangle$ is k, unless either (i) $D = k$ and $|X| = 1$ or (ii) the centre of D is larger than k and $X = \varnothing$.*

Proof. Suppose first that $D \supset k$ and $|X| = 1$, say $X = \{x\}$. By Lemma 5.6 the field $E = D_k \langle\!\langle x \rangle\!\rangle$ can be written in the form $E = F(x; \alpha)$, where F is the field coproduct of copies of D over k and α, the shift automorphism has infinite order. Thus F is a k-algebra and k is the fixed field of α, hence by Th. 2.2.10, E has the precise centre k. This proves the result when $D \supset k$ and $|X| = 1$. If $|X| > 1$, write $X = X' \cup \{x\}$, where $X' \neq \varnothing$; now $D_k \langle\!\langle X \rangle\!\rangle$ is the universal field of fractions of $D_k \langle X \rangle = D_k \langle X' \rangle \underset{k}{*} k[x]$, and this is $D_k \langle\!\langle X' \rangle\!\rangle_k \langle\!\langle x \rangle\!\rangle$; this reduces the problem to the case already treated. There still remains the case $X = \varnothing$, but then the exceptions reduce trivially to the case (ii). ∎

Exercises

1. Prove the HNN theorem for groups along the lines of the proof of Th. 5.1.

2. Let K be a field and E a subfield with an isomorphism $f: K \to E$. Using the skew function field, find an extension of K with an element inducing f.

3. Show that there is an infinite field in which any two elements $\neq 0, 1$ are conjugate. Show further that any two countable fields with this property are isomorphic. (Hint. Embed $\mathbf{F}_2(x)$ in a homogeneous field without adding algebraic elements.) Deduce that in such a field every element is a multiplicative commutator.

4. For any field K show that the group $\mathbf{PGL}_2(K)$ of fractional linear transformations $x \mapsto (ax + b)(cx + d)^{-1}$, $\begin{pmatrix} a & b \\ c & d \end{pmatrix}$ non-singular, is triply transitive on the projective line $K_\infty = K \cup \{\infty\}$. By using for K the field constructed in Ex. 3 obtain a field L for which $\mathbf{PGL}_2(L)$ is four-fold transitive on L_∞ (P. J. Cameron).

5°. Let D be a field with a central subfield k and in $P = D_k \langle\!\langle x \rangle\!\rangle$ consider the subfields $K_\alpha = (1 + \alpha x)^{-1} K (1 + \alpha x)$, for $\alpha \in k$. Is the subfield of P generated by the K_α their field coproduct? (G. M. Bergman) If true, this would provide an analogue to Cor. 5.10 for subfields generated over k by at most $|k|$ elements.

5.6 HNN-extensions of rings

In this section we shall prove analogues of the HNN theorem for rings. Instead of the embedding techniques of Ch. 4 we here need the results from 5.4 on filtered tensor rings. As before, all our rings will be algebras over a given commutative field k.

THEOREM 5.6.1. *Let R be an integral domain and A, B subrings of R, isomorphic under a mapping $f: A \to B$, where R, A, B are k-algebras and f is k-linear. Then R can be embedded in a ring S, again a k-algebra and an integral domain, containing an element $t \neq 0$ such that*

$$at = t \cdot af \quad \text{for all } a \in A, \tag{1}$$

provided that R is flat as right A-module and as left B-module.

Proof. Define R as right A-module by the usual multiplication and as left A-module by the rule

$$a \cdot u = (af)u \quad \text{for all } a \in A, u \in R. \tag{2}$$

We now form the R-bimodule $M = R \otimes_A R$ with the usual multiplication by elements of R as module action. If we abbreviate $1 \otimes 1$ as t, this bimodule consists of all sums $\sum u_i t v_i$ ($u_i, v_i \in R$) with the defining relations (1). For any left R-module U we have

$$M \otimes_R U = R \otimes_A R \otimes_R U = R \otimes_A U;$$

since R is right A-flat, it follows that M is right R-flat. Similarly we find that M is left R-flat, because R is left B-flat. By Th. 4.8, the tensor ring $R_k \langle M \rangle$ is an integral domain and the component of degree 0 is isomorphic to R. Thus R is embedded in an integral domain with an element $t \neq 0$ satisfying (1). ∎

To give an illustration, let R be a k-algebra which is an integral domain and let $a, b \in R$ be such that a, b are not algebraic over k. Then $k[a]$ and $k[b]$ are isomorphic subrings of R, with an isomorphism mapping a to b, and R is torsion-free as (left or right) $k[a]$- or $k[b]$-module, because R is an integral domain. But over a PID, torsion-free modules are flat, so all the hypotheses of Th. 6.1 are satisfied and we obtain

COROLLARY 5.6.2. *Let R be a k-algebra which is an integral domain and let a, b be elements of R not algebraic over k. Then R can be embedded in a k-algebra S containing an element $t \neq 0$ such that*

$$at = tb,$$

and S is again an integral domain. ■

Using filtered tensor rings, we can prove the following slight generalization, which will be used later.

THEOREM 5.6.3. *Let R be a k-algebra which is an integral domain with $k^{\times}$ as precise group of units. Given a, $b \in R \backslash k$ and any $c \in R$, there exists a k-algebra S containing R as well as an element $t \neq 0$ satisfying the equation*

$$at - tb = c. \qquad (3)$$

Moreover, S is again an integral domain with $k^{\times}$ as precise group of units.

Proof. Consider the direct sum $R \oplus R = (R, R)$ as right $k[a]$-module by right multiplication in R, and as left $k[a]$-module by the rule

$$a(x_1, x_2) = (ax_1 + cx_2, bx_2).$$

On the submodule $(R, 0)$ this is just the structure induced by left multiplication in R and we shall regard $(R, 0)$ as R-bimodule in the obvious way. We now form the tensor product of R with (R, R):

$$M = R \otimes_R R \oplus R \otimes_{k[a]} R. \qquad (4)$$

The first term on the right is just R; denoting the second term by N, we have $M = R \oplus N$ and if we write t for $1 \otimes 1$ in N, we see that N is generated by t as right R-module and $M = R \oplus N$ has a left R-module structure with $at = tb + c$. In this way M becomes a pointed R-bimodule. Further, $\bar{M} = M/R$ is generated by t as R-bimodule subject to $at = tb$; thus $\bar{M}$ is essentially the module used in the proof of Th. 6.1. To verify that it is flat we need to show that it is torsion-free as $k[a]$-module. We first show that a is not algebraic over k. Suppose that a satisfies an equation over k; since $a \neq 0$, we may divide by a if necessary, so as to obtain an equation with non-zero constant term:

$$\lambda_0 a^n + \lambda_1 a^{n-1} + \ldots + \lambda_{n-1} a + 1 = 0.$$

Hence $a(\lambda_0 a^{n-1} + \ldots + \lambda_{n-1}) = -1$; this shows a to be a unit, so it must lie in k, against the hypothesis.

Thus $k[a]$ is a PID and since R is an integral domain, it is torsion-free as right $k[a]$-module. Now over a PID any finitely generated torsion-free module is free (see e.g. A.1, 10.6), hence any row vector $(u_1, \ldots, u_n) \in R^n$ is equivalent over $\mathbf{GL}_n(k[a])$ to a vector $(u'_1, \ldots, u'_r, 0, \ldots, 0)$, where $u'_1, \ldots, u'_r$ are right linearly independent over $k[a]$. Hence any element of $\bar{M}$ can be written as $\sum u_i t v_i$, where we may assume the u_i to be right linearly independent over $k[a]$. It follows that $\sum u_i t v_i = 0$ if and only if $v_i = 0$ for all i. Thus if $\sum u_i t v_i p = 0$, where $p \in k[a]$ and the u_i are as before, then $v_i p = 0$ for all i, hence either $p = 0$ or $v_i = 0$ for all i, which shows $\bar{M}$ to be torsion-free and hence flat.

We now form the filtered tensor ring on M: $S = R_k\langle M; R \rangle$. This ring is an integral domain by Cor. 4.9, and it contains R as subring. Let $u \in S$ be a unit and write S_n for the filtration defined by M. If $u \in S_r \backslash S_{r-1}$ and $u^{-1} \in S_s \backslash S_{s-1}$, then $1 = uu^{-1} \in S_{s+r} \backslash S_{s+r-1}$, which is a contradiction, unless $r = s = 0$; but then $S_0 = R$ and this has $k^\times$ as group of units. Hence the group of units of S is precisely $k^\times$, as we had to show. ∎

By a repetition of this argument we obtain the following embedding theorem:

THEOREM 5.6.4. *Let R be a k-algebra which is an integral domain with $k^\times$ as precise group of units. Then R can be embedded in a k-algebra T which is an integral domain with $k^\times$ as precise group of units, such that the equation*

$$ax - xb = c \qquad (5)$$

has a solution for any a, b, $c \in T$ such that a, $b \notin k$.

Proof. By Th. 6.3, R can be embedded in a k-algebra with the same properties as R, in which a given equation (5) (with a, $b \notin k$) can be solved, and by induction we obtain a k-algebra T_1 containing R with the same properties as R, in which all equations (5) over R with a, $b \notin k$ can be solved. If we repeat the process, we obtain a chain

$$R \subset T_1 \subset T_2 \subset \ldots,$$

whose union is a k-algebra with the required properties. ∎

The algebra T constructed here is plainly a simple ring. But we can say a little more. Consider the following formula:

$$(P_n(a)) \ \exists \ x_1, \ldots, x_n, y_1, \ldots, y_n \sum_1^n x_i a y_i = 1. \qquad (6)$$

A simple ring may be defined as a non-trivial ring in which every non-zero element a satisfies $(P_n(a))$ for some n. If we can choose a fixed n for all $a \in R$ (this is the case if e.g. R is a full matrix ring over a field), then R will be called n-simple. It is easily seen that an ultrapower of a simple ring R is again simple precisely when R is n-simple, for some n. Now the equation (5) shows that the ring T of Th. 6.4 is 2-simple.

Exercises

1. Show that a 1-simple ring without idempotents other than 0 or 1 is a field.

2. Let R be a k-algebra in which every equation $ax - ya = c$ with $a \notin k$ has a solution. By writing this equation as $aR + Ra = R$ show that R is reduced (i.e. $a^2 = 0$ implies $a = 0$). Deduce that R is an integral domain.

3. Let R be a k-algebra which is an integral domain, with elements a, b that are not algebraic over k. Find an integral domain containing R as well as t, t^{-1} such that $at = tb$.

4. Show that a Lie ring can be embedded in a Lie division ring (defined as Lie ring in which $[a, x] = b$ has a solution for all $a \neq 0$), by embedding its universal associative envelope in a ring in which $ax - xa = b$ has a solution for all $a \neq 0$ (see Cohn [59']).

5.7 Adjoining generators and relations

Throughout this section all rings will be k-algebras, where k is a commutative field, and all mappings will be k-linear.

If R is any ring, we can adjoin an indeterminate x by forming the coproduct

$$R' = R \underset{k}{*} k[x]. \tag{1}$$

Since $k[x]$ is a PID, the new ring R' has the same global dimension as R, or global dimension 1 if R was semisimple, by Th. 3.5, while $\mathcal{P}(R') \cong \mathcal{P}(R)$, by Th. 3.8.

Secondly, let R be any ring and $f \in R$; then we can adjoin the relation $f = 0$ to R by forming its quotient by the ideal generated by f. This ring may be denoted by $R\langle f = 0 \rangle$. About this ring much less can be said; this is in the nature of things, since every ring can be obtained by imposing suitable relations on a free ring.

A third process consists in forming localizations, i.e. adjoining inverses. This again does not raise the homological dimension, though it may change the monoid of projectives, as we see by forming the field of fractions of a commutative integral domain which is not projective-free.

Our aim is to generalize these processes. We can think of x in (1) as providing a homomorphism between free right R-modules of rank 1 (by left multiplication). More generally, let P, Q be finitely generated projective right R-modules. To obtain the R-ring with a universal homomorphism from P to Q, let E, F be the idempotent matrices defining P, Q respectively. If E is $n \times n$ and F is $m \times m$, we may think of P as the submodule of ${}^nR$ spanned by the columns of E and similarly for Q and F. Any homomorphism $\alpha: P \to Q$ maps E to $A \in {}^mR^n$, and $A = E\alpha = (E\alpha)E = AE$, while $A = FA'$ for some $A' \in {}^mR^n$, hence $FA = A$. Thus we have

$$AE = A = FA. \tag{2}$$

Conversely, if A satisfies (2), then the map $E \mapsto A$ defines a homomorphism from P to Q. This makes it clear how to adjoin a universal homomorphism $\alpha: P \to Q$ to R: we adjoin mn indeterminates a_{ij} to R, whose matrix A is subject to the defining relations (2). Similarly, to adjoin a universal isomorphism $P \to Q$ we adjoin first the homomorphism α as before and then localize at α, by adjoining mn indeterminates b_{ji} to R, forming an $n \times m$ matrix B, subject to the defining relations

$$AB = F, \quad BA = E. \tag{3}$$

As we saw in 4.6, we may always assume the additional relations

$$EB = B = BF.$$

The rings so formed will be denoted by $R_k\langle \alpha: P \to Q \rangle$ and $R_k\langle \alpha, \alpha^{-1}: P \cong Q \rangle$ respectively. Consider first $T = R_k\langle \alpha: P \to Q \rangle$. Writing $P^\otimes = P \otimes_R T$, we can describe T as the R-ring with universal homomorphism $P^\otimes \to Q^\otimes$. If $\varphi: R \to R'$ is any ring homomorphism, then it is clear that the T-ring with universal homomorphism $P \otimes R' \to Q \otimes R'$ is $R' \otimes T$. To obtain a manageable form for T we shall replace R by a matrix ring $S = \mathfrak{M}_N(R)$, where N is a natural number chosen so large that $P \oplus Q$ is a direct summand of ${}^NR$. Then the matrices E, F are replaced by orthogonal idempotents e, f in S. We now have $P = eS$, $Q = fS$ and we need to find an S-ring with a universal map $eS \to fS$. Put $g = 1 - e - f$ and consider the map from $k \times k \times k$ to S given by

$$(a, b, c) \mapsto ea + fb + gc.$$

The projective module induced by the first factor k is eS, by the second it is fS, and the $(k \times k)$-ring with a universal map from $0 \times k$ to $k \times 0$ is the upper triangular matrix ring $\mathbf{T}_2(k) = \begin{pmatrix} k & k \\ 0 & k \end{pmatrix}$, where $k \times k$ embeds in the diagonal and e_{12} induces the universal map. It follows that the universal ring is T, defined by

$$\mathfrak{M}_N(T) \cong \mathfrak{M}_N(R) \underset{K}{*} (\mathbf{T}_2(k) \times k), \quad \text{where } K = k \times k \times k. \quad (4)$$

Our main interest lies in relating the global dimension and monoid of projectives of T to those of R. The answer is provided by

THEOREM 5.7.1. *Let R be a k-algebra, P, Q finitely generated projective right R-modules and let $T = R_k\langle \alpha: P \to Q \rangle$ be the R-ring with universal homomorphism from $P^\otimes$ to $Q^\otimes$. Then*

$$\text{r.gl.dim.} T = \max \{\text{r.gl.dim.} R, 1\}; \quad (5)$$

similarly for the left global dimension, and the ring homomorphism $R \to T$ induces an isomorphism

$$\mathcal{P}(R) \cong \mathcal{P}(T), \quad (6)$$

so all finitely generated projective T-modules are induced from R.

Proof. We have seen that T is given by (4) above. The first assertion follows from Th. 3.5, because gl.dim.$(\mathbf{T}_2(k) \times k) = 1$, and the second follows by Th. 3.8, because $\mathcal{P}(\mathbf{T}_2(k) \times k) \cong \mathcal{P}(k \times k \times k)$. The same result holds for the left global dimension, because we clearly have $R_k\langle \alpha: P \to Q \rangle \cong R_k\langle \beta: Q^* \to P^* \rangle$, where β is the dual of α. ∎

In a similar way we can deal with the adjunction of an isomorphism between projectives P and Q, yielding $R_k\langle \alpha, \alpha^{-1}: P \cong Q \rangle$. To obtain this ring we note that the $(k \times k)$-ring with a universal isomorphism between $0 \times k$ and $k \times 0$ is $\mathfrak{M}_2(k)$, where $k \times k$ embeds in the ring of diagonal matrices. Here e_{12} and e_{21} are the universal map and its inverse. Thus we use the same construction as before, but with $\mathfrak{M}_2(k)$ in place of $\mathbf{T}_2(k)$.

THEOREM 5.7.2. *Let R be a k-algebra and P, Q finitely generated projective right R-modules and let $T = R_k\langle \alpha, \alpha^{-1}: P \cong Q \rangle$ be the R-ring with a universal isomorphism $P^\otimes \cong Q^\otimes$. Then*

$$\text{r.gl.dim.} T = \text{r.gl.dim.} R, \quad (7)$$

unless the right-hand side is 0, when the left-hand side is either 0 or 1, and similarly for the left global dimension. Further, the ring homomorphism $R \to T$ induces a surjective homomorphism

$$\mathcal{P}(R) \to \mathcal{P}(T),$$

with kernel generated by the relation $[P] = [Q]$.

Proof. In the earlier notation we have

$$\mathfrak{M}_N(T) \cong \mathfrak{M}_N(R) \underset{K}{*} (\mathfrak{M}_2(k) \times k), \quad \text{where } K = k \times k \times k. \tag{8}$$

Now (7) follows as before, when r.gl.dim.$R \geqslant 1$. When r.gl.dim.$R = 0$, the left-hand side is at most 1; it can be either 1 or 0, as is shown by the cases (i) $R = k[t, t^{-1}]$, (ii) $R = k \times \mathfrak{M}_2(k)$, where we adjoin a universal isomorphism between $k \times 0$ and $0 \times \mathfrak{M}_2(k)e_{11}$ to obtain the ring $\mathfrak{M}_3(k)$.

To prove the second part, we note that $\mathcal{P}(\mathfrak{M}_2(k) \times k)$ is the quotient of $\mathcal{P}(k \times k \times k)$ by the relation $[k \times 0 \times 0] = [0 \times k \times 0]$, and the co-product on the right of (8) has as monoid of projectives the commutative coproduct of the corresponding monoids. Since $\mathcal{P}(\mathfrak{M}_N(T)) \cong \mathcal{P}(T)$ by Morita equivalence, the assertion follows. ∎

As a further application of 5.3 we consider the adjunction of an idempotent map on a finitely generated projective P. If we adjoin a universal idempotent map on the free module of rank 1 to the field k we obtain the ring $k \times k$, so the R-ring with universal idempotent $e \in \mathfrak{M}_n(R)$ satisfying $e(^nR) \cong P$, denoted by $R_k\langle e^2 = e: P \to P \rangle$, is the centralizer of the matrix units in $\mathfrak{M}_n(R) \underset{K}{*} (k \times k \times k)$, where $K = k \times k$ maps to $\mathfrak{M}_n(R)$ by $(a, b) \mapsto ea + (1 - e)b$ and to $k \times k \times k$ by $(a, b) \mapsto (a, a, b)$. Thus we have

$$\mathfrak{M}_n(T) \cong \mathfrak{M}_n(R) \underset{K}{*} (k \times k \times k), \quad \text{where } K = k \times k.$$

From this equation it is clear that the analogue of (7) holds unless the right-hand side is 0 or 1. To find $\mathcal{P}(T)$ we observe that we now have

$$[P^{\otimes}] = [Q'] + [Q''], \tag{9}$$

so $\mathcal{P}(T)$ is obtained by adding two generators $[Q']$, $[Q'']$ to $\mathcal{P}(R)$ with the defining relation (9). This yields

THEOREM 5.7.3. *Let R be a k-algebra, P a finitely generated projective right R-module and let $T = R_k\langle e^2 = e: P \to P \rangle$ be the R-ring with universal idempotent endomorphism e of P. Then*

$$\text{r.gl.dim.}T = \max\{\text{r.gl.dim.}R, 1\},$$

and similarly for the left global dimension, and $\mathcal{P}(T)$ is obtained from $\mathcal{P}(R)$ by adjoining two generators $[Q']$, $[Q'']$ with defining relation (9). ∎

A slightly different situation arises when we adjoin inverses of existing maps. This is essentially localization and we can show that a hereditary ring remains so on localization:

THEOREM 5.7.4. *Let R be a right hereditary ring and* Σ *a set of maps between finitely generated projective modules. Then the localization* R_Σ *is again right hereditary.*

Proof. We begin by proving

$$\text{Ext}_R^1(M, N) \cong \text{Ext}_{R_\Sigma}^1(M, N), \tag{10}$$

for any pair of R_Σ-modules M, N. The natural homomorphism $R \to R_\Sigma$ allows us to consider any R_Σ-module as an R-module; further it is clear that a left R_Σ-module M is characterized in $_R\mathcal{M}$ by the property that for any map $\alpha\colon P \to Q$ in Σ, the induced map $\alpha \otimes 1\colon P \otimes_R M \to Q \otimes_R M$ is bijective. It follows that if an R-module is an extension, in the category $_R\mathcal{M}$, of a pair of R_Σ-modules, then M is itself an R_Σ-module, and this is just expressed by (10).

Now R is right hereditary, hence $\text{Ext}_R^1(M, -)$ is right exact; by (10), so is $\text{Ext}_{R_\Sigma}^1(M, -)$ and this shows R_Σ to be right hereditary, as claimed. ∎

These methods also allow us to find out more about the matrix reduction functor introduced in 1.7. We shall need a lemma on maps with zero product:

LEMMA 5.7.5. *Let* $R = \underset{K}{*} R_\lambda$ *be a coproduct of K-rings, where K is a field. Given finitely generated projective R-modules P, P′, P″ and homomorphisms* $\alpha\colon P' \to P$, $\beta\colon P \to P''$ *such that* $\alpha\beta = 0$, *there exist finitely generated projective* R_λ-*modules* P_λ, P'_λ, P''_λ *almost all zero, a decomposition of P into induced projectives*

$$P = \bigoplus P_\lambda^\otimes$$

and maps $\alpha_\lambda\colon P'_\lambda \to P_\lambda$, $\beta_\lambda\colon P_\lambda \to P''_\lambda$ *such that* $\alpha_\lambda\beta_\lambda = 0$, *and there is a commutative diagram*

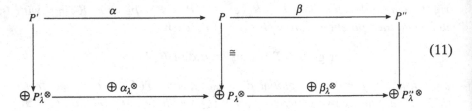

$$\tag{11}$$

Proof. By Th. 3.3, im β is an induced module, the map $P \to \text{im } \beta$ is a surjection of induced modules and by Prop. 3.7 we have a commutative diagram

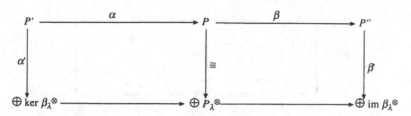

and it remains to show how to replace $\operatorname{im}\beta_\lambda$ and $\ker\beta_\lambda$ by finitely generated projective R_λ-modules. The image of α' is a finitely generated submodule of $\bigoplus\ker\beta_\lambda^{\otimes}$ and so lies in an R-submodule generated by a finite subset of $\bigcup\ker\beta_\lambda^{\otimes}$. Hence we can find finitely generated free R_λ-modules P'_λ for each λ, but almost all 0, with an induced map from $\bigoplus P'^{\otimes}_\lambda$ to $\bigoplus\ker\beta_\lambda^{\otimes}$ whose image contains $\operatorname{im}\alpha'$. Since P' is projective, we can lift α' from $\bigoplus\beta_\lambda^{\otimes}$ to $\bigoplus P'^{\otimes}_\lambda$ and so obtain a commutative diagram

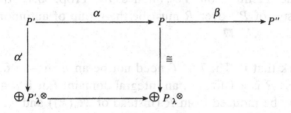

If we dualize this diagram, use the same argument to fill in the bottom right-hand corner and dualize once more, we obtain (11). ∎

We can now deal with the matrix reduction ring:

THEOREM 5.7.6. *Let R be any k-algebra and $n \geqslant 1$. Then the matrix reduction $T = \mathfrak{W}_n(R;k)$ satisfies*

$$\mathrm{r.gl.dim.}\,T = \mathrm{r.gl.dim.}\,R,$$

unless the right-hand side is 0. *Moreover, T is an $(n-1)$-fir and its group of units is $k^{\times}$.*

Proof. We have

$$\mathfrak{M}_n(T) = \mathfrak{M}_n(k) \underset{k}{*} R = S, \text{ say.}$$

The assertion on global dimensions follows from Th. 3.5, as before. Now let $m < n$ and take $a \in T^m$, $b \in {}^mT$ such that $ab = 0$. We may regard a, b as homomorphisms to and from $P^{\otimes}$, where P is induced from ${}^mk$ and $\otimes$ refers to the natural map $\mathfrak{M}_n(k) \to S$. By Lemma 7.5 we have a commutative diagram

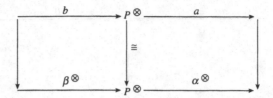

where α, β are defined over $\mathfrak{M}_n(k)$ and compose to 0 (reading from right to left, exceptionally). The middle term in the bottom row must be $P^\otimes$, because this is the only form an induced module can take; moreover, $\alpha\beta = 0$, so by a linear transformation this relation can be trivialized, and hence the same holds for $ab = 0$.

Finally, suppose that u is a unit in T; it defines an automorphism of $P^\otimes$, where $P \cong k$, but all free transfer automorphisms and transvections must be the identity on $P^\otimes$, hence by Prop. 3.7, the group of automorphisms of $P^\otimes$ over R must be the group of automorphisms of P over $\mathfrak{M}_n(k)$, i.e. $k^\times$. ∎

We remark that in Th. 7.6, T need not be an n-fir – in fact it will not be one unless R is a 1-fir, i.e. an integral domain. For if $a \in T^n$, $b \in {}^nT$, then $P^\otimes$ may be induced from R (instead of $\mathfrak{M}_n(k)$) and we can then no longer be sure that the relation $\alpha\beta = 0$ can be trivialized (see the examples below and Ex. 9).

However, when R is a fir, then so is T:

COROLLARY 5.7.7. *Let R be a k-algebra which is a fir, and let $n \geqslant 1$. Then the matrix reduction $T = \mathfrak{W}_n(R; k)$ is again a fir.*

For by Th. 7.6, T is hereditary and the monoid of projectives of $S = \mathfrak{M}_n(T) = \mathfrak{M}_n(k) \underset{k}{*} R$ is free on $(1/n)[S]$, hence $\mathcal{P}(T)$ is free on $[T]$, i.e. T is projective-free, and so it is a fir. ∎

The above results may be applied in various ways. For example, consider the ring obtained by adjoining a map with a one-sided inverse:

$$T = R_k\langle \alpha: P \to Q, \beta: Q \to P; \alpha\beta = 1_P \rangle.$$

This ring may be obtained by forming first $S = R_k\langle \varepsilon^2 = \varepsilon: P \to Q \rangle$ and then $T = S_k\langle \alpha, \alpha^{-1}: P \cong \operatorname{im} \varepsilon \rangle$. It follows that T has the same global dimension as R, except that gl.dim.T may be 1 when R is semisimple. Further, $\mathcal{P}(T)$ is obtained by adjoining one generator $[Q'']$ and one defining relation: $[P] + [Q''] = [Q]$.

As an illustration we shall consider the universal non-IBN and non-UGN rings. Let us denote by I the element of $\mathscr{P}(R)$ corresponding to $[R]$.

(i) IBN. For $n > m \geqslant 1$ the universal non-IBN ring of type (m, n) is

$$V_{m,n} = k\langle \alpha, \alpha^{-1}: k^m \cong k^n \rangle.$$

By Th. 7.2 this ring is hereditary and $\mathscr{P}(V_{m,n})$ is generated by I with defining relation $mI = nI$. Since $m < n$, the ring is an $(m - 1)$-fir, by an argument as in the proof of Th. 7.6; of course it is not an m-fir. When $m = n$, we can use the same results and find that $\mathscr{P}(V_{n,n})$ is free on I; thus $V_{n,n}$ is hereditary and projective-free, and so is a fir.

(ii) UGN. For $n > m \geqslant 1$ the universal non-UGN ring of type (m, n) is

$$U_{m,n} = k\langle \alpha: k^n \to k^m, \beta: k^m \to k^n; \alpha\beta = 1 \rangle,$$

(composing left to right). The above example shows $U_{m,n}$ to be hereditary with $\mathscr{P}(U_{m,n})$ generated by I and $[P]$ subject to the defining relation

$$mI = nI + [P].$$

Thus $U_{m,n}$ is an $(m - 1)$-fir, when $m < n$. When $n \leqslant m$, the same definition can be used, but the resulting ring then has UGN.

(iii) W. The universal weakly n-infinite ring is

$$W_n = k\langle \varepsilon^2 = \varepsilon: k^n \to k^n \rangle.$$

By Th. 7.3, W_n is hereditary and $\mathscr{P}(W_n)$ is generated by I and $[P]$ with the defining relation $nI = nI + [P]$.

Exercises

1. Let R be a right hereditary ring and Σ a set of maps between finitely generated projective R-modules. Show that the right localization $R_{(\Sigma)}$ (which consists in adjoining right inverses of all maps in Σ) is again right hereditary.

2. Show that any finitely generated abelian monoid is finitely related. (Hint. Consider the monoid algebra.)

3. Let A be a finitely generated abelian monoid with distinguished element $I \neq 0$ such that (i) I is a fundamental unit, i.e. for each $x \in A$ there exist $y \in A$ and $n \in \mathbf{N}$ such that $x + y = nI$, and (ii) A is *conical* (i.e. $x + y = 0$ implies $x = y = 0$). Show that there is a hereditary ring R (and k-algebra) such that $\mathscr{P}(R) \cong A$ as monoids with distinguished element I, such that for any k-algebra S and any homomorphism φ from A to $\mathscr{P}(S)$ (preserving fundamental units), there exists a k-algebra homomorphism $f: R \to S$ which induces φ (Bergman [74']).

4. Show that $R = \mathfrak{W}_2(\mathfrak{M}_3(k))$ is a hereditary ring with $\mathscr{P}(R)$ isomorphic to the

monoid generated by 1 and $1\frac{1}{2}$ in $\frac{1}{2}\mathbf{N}$. Deduce that R is an integral domain satisfying Klein's nilpotence condition, but that R has no R-field (Bergman [74']).

5. Show that (for $m < n$) $V_{m,n}$ and $U_{m,n}$ are not m-firs. (Hint. Verify that in an m-fir R, R^m has unique rank and cannot be generated by fewer than m elements.)

6. Give an example of a simple non-Ore integral domain. (Hint. Apply Th. 6.4 to $V_{m,n}$, where $n > m > 1$.)

7. (Schofield) Let K be a field, M a K-bimodule and N a subbimodule. Show that the natural inclusion $K\langle N \rangle \to K\langle M \rangle$ is honest.

8. (Dicks and Sontag [78]) Let $A = (a_{i\lambda})$ be an $m \times r$ matrix and $B = (b_{\lambda j})$ an $r \times n$ matrix and let R be the k-algebra on the $(m + n)r$ generators $a_{i\lambda}$, $b_{\lambda j}$ with defining relations $AB = 0$. Verify that R is an $(r - 1)$-fir. For m, n, $r \geqslant 1$ show that every full matrix is left regular if m, $n < r$, but R is a Sylvester domain if and only if $m + n \leqslant r$.

9. (Bergman [74']) Show that if R is an $(r - 1)$-fir but not an r-fir, then $\mathfrak{W}_n(R)$ is an $(nr - 1)$-fir but not an nr-fir.

5.8 Derivations

We now return to make a more detailed study of derivations. This will be useful in calculating invariants of free algebras and free fields. We shall now assume that all our rings are K-rings, where K is a k-algebra (and k a commutative field, as usual).

On tensor rings any derivation can be described by its effect on the bimodule, as our first result shows:

PROPOSITION 5.8.1. *Let* R, A, B *be* K-*rings*, U *an* R-*bimodule and* $T = \mathbf{T}(U)$ *the tensor ring on* U. *Further let* $\alpha: T \to A$, $\beta: T \to B$ *be homomorphisms. Then for any* (A, B)-*bimodule* M, *any* K-*linear homomorphism* $f: U \to M$ *extends to a unique* (α, β)-*derivation of* T *in* M.

Proof. We recall from 2.1 that an (α, β)-derivation δ of T in M may be described as a homomorphism from T to a triangular matrix ring:

$$T \to \begin{pmatrix} A & M \\ 0 & B \end{pmatrix}, \quad x \mapsto \begin{pmatrix} x^\alpha & x^\delta \\ 0 & x^\beta \end{pmatrix}. \tag{1}$$

For x in U, with xf in the $(1, 2)$-entry instead of x^δ, this defines a K-linear map from U to the triangular matrix ring in (1), which extends to a unique homomorphism from T, by universality. The images are again triangular matrices and the $(1, 2)$-entry has the form x^δ, where δ is an (α, β)-derivation extending f. ∎

For simplicity we shall confine ourselves in what follows to the special case where $A = B = R$ and $\alpha = \beta = 1$; the $(1, 1)$-derivations will just be derivations, with images in an R-bimodule. The general case is usually no harder but requires more notation.

For any R-bimodule M the derivations (over K) form a module $\mathrm{Der}_K (R, M)$ and it is clear that $\mathrm{Der}_K (R, -)$ is a functor, which is actually representable, i.e. there is an R-bimodule $\Omega_K(R)$, called the *universal derivation bimodule*, with a derivation δ of R in $\Omega_K (R)$ such that there is a natural isomorphism

$$\mathrm{Hom}_K (\Omega_K(R), M) \cong \mathrm{Der}_K (R, M),$$

given by the correspondence $\alpha \mapsto \delta\alpha$. It is not difficult to construct $\Omega_K(R)$ directly in terms of generators a^δ ($a \in R$) and defining relations expressing that δ is a derivation, but there is a more concrete realization, as the kernel of the multiplication map on R. We recall that the multiplication on R may be expressed as a K-bilinear map

$$m\colon R \otimes_K R \to R, \quad x \otimes y \mapsto xy.$$

In terms of it the universal derivation bimodule may be described as follows:

THEOREM 5.8.2. *For any K-ring R (where K is arbitrary) there is an exact sequence*

$$0 \to \Omega_K(R) \to R \otimes_K R \xrightarrow{m} R \to 0, \tag{2}$$

where m is the multiplication map and $\Omega_K(R)$ is generated as left (or right) R-module by the elements

$$x^\delta = x \otimes 1 - 1 \otimes x \quad (x \in R). \tag{3}$$

Proof. The map $m\colon x \otimes y \mapsto xy$ is clearly an R-bimodule homomorphism which is surjective and if its kernel is denoted by $\Omega_K(R)$, we obtain the exact sequence (2). The elements (3) all lie in this kernel and conversely, if $\sum x_i \otimes y_i \in \Omega_K(R)$, then $\sum x_i y_i = 0$ and so

$$\sum x_i \otimes y_i = \sum (x_i \otimes 1 - 1 \otimes x_i) y_i = \sum x_i (1 \otimes y_i - y_i \otimes 1),$$

which shows $\Omega_K(R)$ to be generated as left or right R-module by the elements (3).

It remains to identify $\Omega_K(R)$ as the universal derivation bimodule. The map $x \mapsto x^\delta$, where x^δ is given by (3), is easily seen to be a derivation of

R in $\Omega_K(R)$, and it is universal, for given any derivation d of R in an R-bimodule M, we can define a homomorphism f from $\Omega_K(R)$ to M as follows: If $\sum x_i \otimes y_i \in \Omega_K(R)$, then $\sum x_i y_i = 0$, hence $\sum x_i^d y_i + x_i y_i^d = 0$ and we may define f by the equations

$$\left(\sum x_i \otimes y_i\right) f = \sum x_i^d \cdot y_i = -\sum x_i \cdot y_i^d.$$

The first equation shows f to be right R-linear, the second shows it to be left R-linear; moreover, $(x\lambda)^d y = x^d \lambda y$ for $\lambda \in K$, so the expression $\sum x_i^d y_i$ is balanced and is uniquely determined by $\sum x_i \otimes y_i$. Now δf maps x to $(x \otimes 1 - 1 \otimes x) f = x^d$, hence $d = \delta f$ and f is unique, since it is given on the generating set $\{x \otimes 1 - 1 \otimes x\}$. ∎

By combining this result with Prop. 4.1.1 (c), we find

COROLLARY 5.8.3. *A ring homomorphism $K \to R$ is an epimorphism if and only if the only derivation on R over K is* 0.

For the criterion for an epimorphism reduces to $\Omega_K(R) = 0$, and this is clearly so if and only if the only derivation is zero. ∎

In the sequel we frequently have to deal with a family of ring homomorphisms $R_\lambda \to S$ and S-modules that are induced from R_λ-modules: $M \otimes_{R_\lambda} S$. It will be convenient to denote this induced module by $M^\otimes$, leaving the ring homomorphism to be inferred from the context (or indicated separately, in cases of doubt). Similarly for bimodules the induced module $S \otimes M \otimes S$ will be written $^\otimes M^\otimes$.

Our next object will be to relate the universal derivation bimodule to that of a residue-class ring. Here we shall need some basic properties of the Tor functor, assumed to be known to the reader (see e.g. A.3, Ch. 3). We begin with a lemma needed in the next proof.

LEMMA 5.8.4. *Let T be a ring, $\mathfrak{a}$ an ideal in T and $R = T/\mathfrak{a}$. Then*

$$\operatorname{Tor}_1^T(R, R) \cong \mathfrak{a} \otimes_T R \cong \mathfrak{a}/\mathfrak{a}^2. \qquad (4)$$

Proof. Consider the exact sequence

$$0 \to \mathfrak{a} \to T \to R \to 0. \qquad (5)$$

Applying $\otimes_T R$ and noting that $\operatorname{Tor}_1^T(T, R) = 0$, $R \otimes_T R \cong T \otimes_T R \cong R$, we obtain

$$0 \to \operatorname{Tor}_1^T(R, R) \to \mathfrak{a} \otimes_T R \to R \to R \to 0.$$

Here the last map is just the identity; this proves the first isomorphism in (4). To establish the second, we operate on the exact sequence (5) with $\mathfrak{a} \otimes_T$ and find an exact sequence

$$\mathfrak{a} \otimes \mathfrak{a} \to \mathfrak{a} \to \mathfrak{a} \otimes R \to 0.$$

Clearly the image of the first map is $\mathfrak{a}^2$, and hence (4) follows. ∎

THEOREM 5.8.5. *Let T be a K-ring, where K is a field, let $\mathfrak{a}$ be an ideal of T and write $R = T/\mathfrak{a}$. Then there is an exact sequence*

$$0 \to \mathfrak{a}/\mathfrak{a}^2 \xrightarrow{\alpha} {}^\otimes \Omega_K(T)^\otimes \xrightarrow{p} \Omega_K(R) \to 0,$$

where $\otimes$ is based on and p is induced by the natural homomorphism $T \to R$.

Proof. By Th. 8.2 we have the exact sequence

$$0 \to \Omega_K(T) \to T \otimes_K T \to T \to 0,$$

which is split exact as sequence of left T-modules. Tensoring on the left with $R = T/\mathfrak{a}$ over T, we obtain the exact sequence

$$0 \to {}^\otimes \Omega_K(T) \to R \otimes_K T \to R \to 0.$$

If we now operate with $\otimes_T R$ and bear in mind that $R \otimes_T R \cong R$, we obtain the exact sequence

$$\operatorname{Tor}_1^T(R \otimes_K T, R) \to \operatorname{Tor}_1^T(R, R) \to {}^\otimes \Omega_K(T)^\otimes \to R \otimes_K R \to R \to 0.$$

$$(6)$$

Now $R \otimes_K T$ is projective as right T-module, since for any T-module M,

$$\operatorname{Hom}_T(R \otimes_K T, M) \cong \operatorname{Hom}_K(R, \operatorname{Hom}_T(T, M)) \cong \operatorname{Hom}_K(R, M)$$

and this functor is exact in M because K is a field. It follows that $\operatorname{Tor}_1^T(R \otimes_K T, R) = 0$, so the first term in (6) may be replaced by 0. The term $\operatorname{Tor}_1^T(R, R)$ may be replaced by $\mathfrak{a}/\mathfrak{a}^2$, by Lemma 8.4, and the last map in (6) is the multiplication map m whose kernel we know to be $\Omega_K(R)$, so (6) may be replaced by the exact sequence

$$0 \to \mathfrak{a}/\mathfrak{a}^2 \to {}^\otimes \Omega_K(T)^\otimes \to \Omega_K(R) \to 0,$$

and this is the sequence we had to find. ∎

This result just expresses the fact that $\Omega_K(R)$ is obtained from $\Omega_K(T)$ by tensoring with T and dividing out by $\mathfrak{a}$. The kernel is $\mathfrak{a}/\mathfrak{a}^2$, because the derivation bimodule is a linearized form. Explicitly, if R is given by a presentation

$$R = K\langle X; \Phi \rangle,$$

then $\Omega_K(R)$ is the R-bimodule on generators x^δ ($x \in X$) and defining relations $\varphi^\delta = 0$ ($\varphi \in \Phi$), where φ^δ is the formal derivative of φ. If $\mathfrak{a}$ denotes the ideal of $k\langle X \rangle$ generated by Φ, then any element of $\mathfrak{a}^2$ has the form $\sum a_i b_i$, where $a_i, b_i \in \mathfrak{a}$; now $(\sum a_i b_i)^\delta = \sum a_i^\delta b_i + \sum a_i b_i^\delta$ and this vanishes as element of the bimodule $\Omega_K(R)$, in illustration of the fact that the kernel is $\mathfrak{a}/\mathfrak{a}^2$.

We also note the special case of an idempotent ideal:

COROLLARY 5.8.6. *Let T be a K-ring, where K is a field, let $\mathfrak{a}$ be an idempotent ideal of T and put $R = T/\mathfrak{a}$. Then $\Omega_K(R) \cong {}^\otimes\Omega_K(T)^\otimes$.*

This follows from Th. 8.5, since now $\mathfrak{a}/\mathfrak{a}^2 = 0$, by hypothesis. ∎

The results obtained so far allow us to calculate the universal derivation bimodule of a tensor ring:

PROPOSITION 5.8.7. *Let K be a field, U a K-bimodule and $T = K_k\langle U \rangle$ the tensor K-ring on U. Then the universal derivation bimodule for T is the T-bimodule induced from U:*

$$\Omega_K(T) \cong {}^\otimes U^\otimes,$$

where $\otimes$ is taken relative to $K \to T$ and the derivation extends the identity map on U.

Proof. The K-ring T is generated by U with no relations except those holding in U. Hence $\Omega_K(T)$ is generated by u^δ ($u \in U$) with linearity and $(au)^\delta = au^\delta$, $(ua)^\delta = u^\delta a$ for $u \in U$, $a \in K$. ∎

We can also express the universal derivation bimodule of a coproduct in terms of those of its factors:

THEOREM 5.8.8. *Let $\{R_\lambda\}$ be a family of K-rings, where K is a field. Then the universal derivation bimodule of the coproduct is the direct sum of those of its factors:*

$$\Omega_K(\underset{K}{*} R_\lambda) \cong \bigoplus({}^\otimes\Omega_K(R_\lambda)^\otimes),$$

where $\otimes$ refers to the homomorphism from R_λ to the coproduct.

Proof. A presentation of R_λ is obtained by taking presentations of all the R_λ and identifying K in all of them. Hence $\Omega_K(\underset{K}{*} R_\lambda)$ is generated as bimodule by the images induced by the bimodules $\Omega_K(R_\lambda)$ and there are no further relations. ∎

We shall also want to know the universal derivation bimodule for a localization:

THEOREM 5.8.9. *Let R be a K-ring, where K is a field, and let Σ be any collection of matrices over R. Then the universal derivation bimodule for the localization R_Σ is given by*

$$\Omega_K(R_\Sigma) \cong {}^{\otimes}\Omega_K(R)^{\otimes},$$

where $\otimes$ refers to $R \to R_\Sigma$.

Proof. Let us write $N = {}^{\otimes}\Omega_K(R)^{\otimes}$; we shall prove the theorem by extending the universal derivation $\delta: R \to \Omega_K(R)$ to a universal derivation of R_Σ in N. Since each derivation defines a homomorphism to a triangular matrix ring, we have a diagram

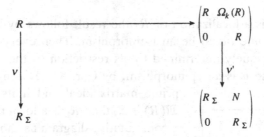

where v, v' are the canonical maps and the horizontal map is induced by δ. Now every matrix in Σ maps to an invertible matrix over the triangular matrix ring in the bottom right-hand corner, because it is invertible mod the nilpotent ideal $\begin{pmatrix} 0 & N \\ 0 & 0 \end{pmatrix}$. Hence there is a unique map completing the diagram to a commutative square; clearly this map defines a derivation $\delta_\Sigma: R_\Sigma \to N$ and it remains to show that this is a universal derivation.

Given a derivation of R_Σ over K to an R_Σ-bimodule M, $d: R_\Sigma \to M$, this defines a derivation of R in M which factors through $\Omega_K(R)$, by the universal property of the latter:

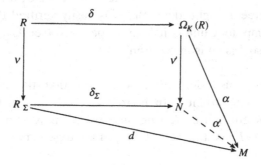

By the universal property of $N = {}^{\otimes}\Omega_K(R)^{\otimes}$, α can be factored by ν' to give a homomorphism $\alpha'\colon N \to M$, and we have to show that $d = \delta_\Sigma \alpha'$. By construction we have $\nu d = \delta\alpha = \delta\nu'\alpha' = \nu\delta_\Sigma\alpha'$, and since ν is an epimorphism we conclude that $d = \delta_\Sigma\alpha'$. This shows δ_Σ to be the desired universal derivation. ∎

Under suitable assumptions we can strengthen this result to a criterion for localization in terms of the universal derivation bimodule:

THEOREM 5.8.10. *Let R be a right hereditary K-ring, where K is a field. Then a ring homomorphism* $\varphi\colon R \to F$ *to a field F is a localization if and only if* φ *induces an isomorphism*

$$\delta\varphi\colon {}^{\otimes}\Omega_K(R)^{\otimes} \cong \Omega_K(F).$$

Proof. If F is a localization of R, the result follows by Th. 8.9. For the converse assume $\delta\varphi$ to be an isomorphism. Then every derivation of F over K is uniquely determined by its restriction to the image of R, i.e. $\Omega_R(F) = 0$, so φ is an epimorphism, by Cor. 8.3. Its singular kernel is a prime matrix ideal and if its complement in $\mathfrak{M}(R)$ is Σ, then R_Σ is a local ring and we have a commutative diagram as shown. As localization of a right hereditary ring R_Σ is again right hereditary, by Th. 7.4, so all ideals of R_Σ are free, as right ideals in a right hereditary local ring.

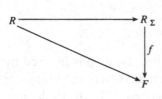

Now the above diagram yields the commutative diagram shown, where the top row is an isomorphism by Th. 8.9 and the slanting arrow is an isomorphism by hypothesis. Hence the vertical arrow is an isomorphism, and so the kernel of the homomorphism $f\colon R_\Sigma \to F$ is idempotent, by Th. 8.5; it is a proper ideal

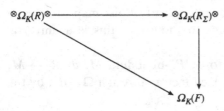

of R_Σ and is free as right ideal. But it is easily verified that a free ideal cannot be idempotent unless it is improper or 0 (see Ex. 7, 1.6). Hence $\ker f = 0$ and so f is an isomorphism. ∎

We shall now use derivations to prove that the number of free generators is an invariant both in free algebras and in free fields. Let $T = K_k\langle X \rangle$ be the tensor K-ring on a set X; if K is a field, then it is determined by T as the set of all units, together with 0. The free generating set X is of course not unique, e.g. $x \in X$ may be replaced by $x + a$, where $a \in K$, but as we shall now show, its cardinal is unique.

THEOREM 5.8.11. *Let K be a field which is a k-algebra and let X, Y be any sets. If $K_k\langle X \rangle \cong K_k\langle Y \rangle$, then $|X| = |Y|$.*

Proof. Let α be any retraction from $T = K_k\langle X \rangle$ to K; such a map is obtained for example by mapping X to 0. Now consider the K-bimodule D of all (α, α)-derivations of T in K. Such a derivation is a K-linear map determined by the images of X, which may be any prescribed elements of K; hence we have $D \cong K^{(X)}$, similarly $D \cong K^{(Y)}$, and by the invariance of the dimension of a vector space it follows that $|X| = |Y|$, as we wished to show. ∎

In order to study free fields, we first establish a property of subfields of free fields. We shall take $D\langle X \rangle$ to mean $D_D\langle X \rangle$, thus X centralizes D, but D need not be commutative.

PROPOSITION 5.8.12. *Let $E \subset D$ be fields and X a set. Then the subfield of $D\langle\!\langle X \rangle\!\rangle$ generated by X over E is naturally isomorphic to $E\langle\!\langle X \rangle\!\rangle$.*

Proof. Since $D\langle\!\langle X \rangle\!\rangle$ is a localization of $D\langle X \rangle$, we have by Th. 8.9,

$$\Omega_D(D\langle\!\langle X \rangle\!\rangle) \cong {}^{\otimes}\Omega_D(D\langle X \rangle)^{\otimes}.$$

Now by Prop. 8.7, $\Omega_D(D\langle X \rangle)$ is generated by the elements x^δ ($x \in X$) subject to

$$x^\delta a = ax^\delta \text{ for all } x \in X, a \in D.$$

Hence $\Omega_D(D\langle\!\langle X \rangle\!\rangle)$ is the free $D\langle\!\langle X \rangle\!\rangle$-bimodule on the generators x^δ, centralizing D. Denote by L the subfield generated by X over E and consider the L-bimodule generated by the x^δ in $\Omega_D(D\langle\!\langle X \rangle\!\rangle)$. It is the image of $\Omega_E(L)$ in $\Omega_D(D\langle\!\langle X \rangle\!\rangle)$ under the homomorphism $L \to D\langle\!\langle X \rangle\!\rangle$, and so is the free L-bimodule on the E-centralizing generators x^δ. Since $\Omega_E(E\langle\!\langle X \rangle\!\rangle)$ is the free bimodule on the E-centralizing generators x^δ, we see that the natural homomorphism $E\langle X \rangle \to L$ induces an isomorphism $\Omega_E(L) \cong {}^{\otimes}\Omega_E(E\langle X \rangle)^{\otimes}$. Now $E\langle X \rangle$, being a fir, is hereditary, so we can apply Th. 8.9 to conclude that $L \cong E\langle\!\langle X \rangle\!\rangle$. ∎

Our aim is to show that for any set X, the cardinal $|X|$ is an invariant of the free field $k\langle\!\langle X \rangle\!\rangle$. For infinite sets X this follows from a result in universal algebra which we quote without proof (Prop. 1.4.4 of A.3). In the finite case we shall actually find a slightly more general result, analogous to weak finiteness, or the Hopf property for groups. Namely we shall show that if $|X| = n$, then any set of n elements generating $k\langle\!\langle X\rangle\!\rangle$ is a free generating set. It is no harder to prove this result for the more general case of universal fields of fractions of tensor rings, although we need an extra hypothesis at this stage. For any k-algebra R we define its *multiplication algebra* as $R^\circ \otimes_k R$, where R° denotes the opposite (anti-isomorphism) of R. We note that R may be considered as an R-bimodule in a natural way, using the left and right multiplications, hence it is an $R^\circ \otimes_k R$-module.

THEOREM 5.8.13. *Let D be a skew field which is a k-algebra and consider $F = D_k\langle\!\langle X\rangle\!\rangle$, where $X = \{x_1, \ldots, x_n\}$. Further, assume that its multiplication algebra $F^\circ \otimes_k F$ is weakly finite. Then any D-ring epimorphism*

$$f: D_k\langle X\rangle \to F = D_k\langle\!\langle X\rangle\!\rangle \tag{7}$$

extends to an automorphism of F.

Proof. The universal derivation bimodule $\Omega_D(D_k\langle X\rangle)$ is the free $D_k\langle X\rangle$-bimodule on the generators x_i^δ, hence the universal derivation bimodule $\Omega_D(F)$ is the free F-bimodule on the x_i^δ, by Th. 8.9. Since (7) is an epimorphism, by hypothesis, the elements $x_i f$ generate F as D-field and it follows that the elements $(x_i f)^\delta$-generate $\Omega_D(F)$ as F-bimodule, i.e. as right $(F^\circ \otimes F)$-module, but this is a free right $(F^\circ \otimes F)$-module of rank n, hence by weak finiteness, these elements form a free generating set. Thus the map (7) induces an isomorphism $^\otimes\Omega_D(D_k\langle X\rangle)^\otimes \to \Omega_D(F)$. By Th. 8.10, (7) is a localization, so f can be extended to the universal field of fractions of $D_k\langle X\rangle$ to yield an automorphism of F. ∎

Taking $D = k$, we obtain

COROLLARY 5.8.14. *For any commutative field k, any generating set of n elements of $k\langle\!\langle x_1, \ldots, x_n\rangle\!\rangle$ is a free generating set; in particular, any two free generating sets of a free field have the same number of elements.*

Proof. The first part follows from Th. 8.13, because the multiplication algebra of $k\langle\!\langle X\rangle\!\rangle$ is weakly finite: writing $k\langle\!\langle X\rangle\!\rangle^\circ$ as E, we can express $k\langle\!\langle X\rangle\!\rangle^\circ \otimes_k k\langle\!\langle X\rangle\!\rangle$ as $E \otimes_k k\langle\!\langle X\rangle\!\rangle$ and this can be embedded in $E\langle\!\langle X\rangle\!\rangle$. Now assume that $k\langle\!\langle X\rangle\!\rangle \cong k\langle\!\langle Y\rangle\!\rangle$; if one of X, Y is infinite, then so is the other and both have the same cardinal, by Prop. 1.4.4 of A.3. So we may assume that X, Y are both finite. Applying the first part to the map $k\langle Y\rangle \to k\langle\!\langle X\rangle\!\rangle$ giving rise to the isomorphism, we find that $|Y| \geqslant |X|$. By symmetry we also have $|X| \geqslant |Y|$, hence $|Y| = |X|$, as claimed. ∎

Exercises

1. Let R, A, B be K-rings and $\alpha\colon R \to A$, $\beta\colon R \to B$ be homomorphisms. Show that the universal (α, β)-derivation bimodule for R is $A \otimes_R \Omega_K(R) \otimes_R B$.

2. Given ring monomorphisms $A \to R$, $R \to S$, obtain the exact sequence $0 \to \mathrm{Der}_R(S, M) \to \mathrm{Der}_A(S, M) \to \mathrm{Der}_A(R, M)$ for any S-module M. Deduce the exact sequence $\Omega_A(R) \to \Omega_A(S) \to \Omega_R(S) \to 0$.

3. Let $T = R_K\langle U\rangle = R \underset{K}{*} K\langle U\rangle$ be the tensor ring on a K-bimodule U. Show that $\Omega_K(T) = {}^{\otimes}\Omega_K(R)^{\otimes} \oplus {}^{\otimes}U'^{\otimes}$, where $U' = R \otimes U \otimes R$ and the exponent $\otimes$ refers to the map $R \to T$.

4°. (A. H. Schofield) Let F be a field with a central subfield k. Is the multiplication algebra of F, $F^\circ \otimes_k F$, weakly finite?

5. (A. H. Schofield) Let $E_1, \ldots, E_n$ be finite-dimensional division algebras over k and denote the least common multiple of the degrees $[E_i\colon k]$ by m; further write $\underset{k}{\bigcirc} E_i = F$. (i) Show that E_i can be embedded in $\mathfrak{M}_m(k)$, and deduce that F can be embedded in an appropriate localization of $\underset{k}{*}\mathfrak{M}_m(k)$ with n factors $\mathfrak{M}_m(k)$. (ii) Show that F can be embedded in a localization of $\mathfrak{M}_m(k) \underset{k}{*} k(y)$ by embedding E_r in $y^{-r}\mathfrak{M}_m(k)y^r$ $(r = 1, \ldots, n)$. (iii) Show that $F_k\langle\!\langle X\rangle\!\rangle$ can be embedded in $\mathfrak{M}_m(k\langle\!\langle Z\rangle\!\rangle)$, where $Z = \{z_{ijx}|i, j = 1, \ldots, n, x \in X \cup \{y\}\}$.

6. (A. H. Schofield) Let E be a division algebra with $[E\colon k] < \infty$ and write $F = E_k\langle\!\langle X\rangle\!\rangle$. Using Ex. 5, show that $F^\circ \otimes_k F$ is weakly finite.

7. (W. Dicks) Let D be a field which is a k-algebra and write $F = D_k\langle\!\langle x\rangle\!\rangle$ for the free field on an indeterminate x. Verify that $\Omega_k(F)$ is the free F-bimodule on x as k-centralizing set: $\Omega_k(F) = {}^{\otimes}(D \otimes_k D)^{\otimes}$. Show that there can be no equation $x = \sum[a_i, b_i]$ $(a_i, b_i \in F)$ in F. (Hint. Apply first $\delta\colon x \mapsto 1 \otimes 1$ and then reverse multiplication, $p \otimes q \mapsto qp$, to obtain a contradiction. Makar-Limanov [89] has constructed a field where every element can be written as a sum of at most two commutators, and for some elements two are really needed.)

5.9 Field extensions with different left and right degrees

To construct a field extension with different left and right degrees we shall first examine a type of binomial extension introduced in 3.5. Given a positive integer $n > 1$ and a primitive nth root of 1, ω say, in our ground field k, let us take a field E with an endomorphism α and an α-derivation δ such that

$$\delta\alpha = \omega\alpha\delta, \tag{1}$$

and construct fields K and L as in Th. 3.5.4. We then have an extension of right degree n and its left degree will be greater than n provided that $K^\alpha \neq K$. More generally, if $[K:K^\alpha]_{\mathrm{L}} = \infty$, then $[L:K]_{\mathrm{L}} = \infty$, but some care is needed here: it is *not* enough to take $[E:E^\alpha]_{\mathrm{L}} = \infty$, for whatever α is, we shall have $K^\alpha = K$ if $\delta = 0$, because α is then an inner automorphism of L, with inverse $c \mapsto tct^{-1}$. Likewise we have $[L:K]_{\mathrm{L}} = [L:K]_{\mathrm{R}}$ whenever K is commutative, by Prop. 3.1.4, or when L/K is Galois, by Cor. 3.3.8. Further, we cannot take E to be commutative (say, generated over k by commuting indeterminates), or even of finite degree over k, because then δ would be inner, by Th. 2.1.3 and Prop. 2.1.4, and so could be reduced to zero.

To construct our example, we take any commutative field k with a primitive nth root of 1, ω say, and let Λ be any set. When the characteristic p of k divides n, this is taken to mean that ω is a primitive mth root of 1, where $n = mp^r$, $p \nmid m$. We form the free algebra $R = k\langle x_{\lambda ij}\rangle$ on a family of indeterminates indexed by $\Lambda \times \mathbf{N}^2$ and write $E = k \langle\!\langle x_{\lambda ij}\rangle\!\rangle$ for its universal field of fractions. On R we have an endomorphism α defined by

$$x_{\lambda ij}^\alpha = \omega^i x_{\lambda i j+1}.$$

If β is the endomorphism of R defined by

$$x_{\lambda ij}^\beta = \begin{cases} \omega^{-i}x_{\lambda i j-1} & \text{if } j > 1, \\ 0 & \text{if } j = 1, \end{cases}$$

then $\alpha\beta = 1$ (read from left to right), hence α is a retraction, and so an honest map, by Prop. 4.5.1. It follows that α extends to an endomorphism of E, again denoted by α.

Next we define an α-derivation on R by

$$x_{\lambda ij}^\delta = x_{\lambda i+1 j}.$$

This extends to an α-derivation of E, still denoted by δ. Consider the (α, α^2)-derivation $\delta\alpha - \omega\alpha\delta$ on E; this is easily seen to vanish on each generator $x_{\lambda ij}$, hence it is zero on R and so it is zero on E. Thus we have $\delta\alpha = \omega\alpha\delta$ on E. We now form $L = E(t; \alpha, \delta)$ and $K = E(t^n; \alpha^n, \delta^n)$ as in Th. 3.5.4; by that result, L/K is a binomial extension of right degree n. To show that the left degree is $> n$ it is enough to prove that $K^\alpha \neq K$. This can be done quite easily whatever Λ, e.g. we could take Λ to consist of one element. But we are then left with the task of finding whether

$[K:K^\alpha]_L$ is finite or infinite. It is almost certainly infinite, but this does not seem easy to show when $|\Lambda| = 1$, whereas it becomes easy for infinite Λ.

For any $\mu \in \Lambda$ denote by E_μ the subfield of E generated over k by all $x_{\lambda ij}$ such that $j > 1$, or $j = 1$ and $\lambda \neq \mu$; thus we take all xs except $x_{\mu i1}$ ($i \in \mathbf{N}$). It follows that $E_\mu \supseteq E^\alpha$ for all μ, and $E_\mu(t) \supseteq L^\alpha \supset K^\alpha$. We claim that $x_{\lambda 11} \in E_\mu(t)$ if and only if $\lambda \neq \mu$. Assuming this for the moment, we see that the $x_{\lambda 11}$ are left linearly independent over K^α, for if $\sum a_\lambda x_{\lambda 11} = 0$, where $a_\lambda \in K^\alpha$ and some $a_\mu \neq 0$, then we could express $x_{\mu 11}$ in terms of the $x_{\lambda 11}$, $\lambda \neq \mu$, and so $x_{\mu 11} \in E_\mu(t)$, which contradicts our assumption.

That $x_{\lambda 11} \in E_\mu(t)$ for $\lambda \neq \mu$ is clear from the definition. To show that $x_{\mu 11} \notin E_\mu(t)$, let us write $A = E_\mu[t]$ (for a fixed μ) and observe that for any $a \in E$, $at = ta^\alpha + a^\delta$, hence

$$at \equiv a^\delta \pmod{A},$$

so by induction on r, $at^r \equiv a^{\delta^r} \pmod{A}$.

If $x_{\mu 11} \in E_\mu(t)$, we would have $x_{\mu 11} = fg^{-1}$, where $f, g \in A$; thus $x_{\mu 11}g \equiv f \equiv 0 \pmod{A}$, and if $g = \sum t^i b_i$, where $b_i \in E_\mu$, then

$$0 \equiv \sum x_{\mu 11} t^i b_i \equiv \sum x_{\mu 11}^{\delta^i} b_i \pmod{A}$$

$$\equiv \sum x_{\mu i+1 1} b_i \pmod{A}.$$

Here we have multiplied a congruence mod A by elements of A, which is permissible. Thus we have

$$\sum x_{\mu i+1 1} b_i + b = 0, \quad \text{where } b \in A, b_i \in E_\mu. \tag{2}$$

Now we have $E[t] = E \oplus tE[t]$, as a direct sum of E-spaces, so we may in (2) equate terms of degree 0 in t and so obtain an equation

$$\sum x_{\mu i+1 1} b_i + b_0 = 0, \quad \text{where } b_i, b_0 \in E_\mu,$$

and not all b_i vanish. But this is impossible, because the $x_{\mu i1}$ are clearly linearly independent over E_μ. This proves that $x_{\mu 11} \notin E_\mu(t)$ and it follows that $[L:K]_L \geq |\Lambda|$. Thus we have proved most of

THEOREM 5.9.1. *Given any two cardinal numbers λ, μ of which at least one is infinite, both greater than 1, there is a field extension L/K of left degree λ and right degree μ, and of prescribed characteristic.*

Proof. For finite μ this has just been proved. When both λ, μ are infinite and $\lambda \geq \mu$ say, we take any extension of left and right degree μ and follow it by an extension of left degree λ and right degree 2. The resulting extension has left degree $\lambda\mu = \lambda$ and right degree $2\mu = \mu$. ∎

There remains the case where the degrees on both sides are finite but different. We shall present Schofield's example, by proving

THEOREM 5.9.2. *Let $F \subseteq E$ be an extension of fields and r, s any integers greater than 1. Then there exists an extension field $\bar{E}$ of E with a subfield $\bar{F}$ such that $\bar{F} \cap E = F$ and*

$$[\bar{E}:\bar{F}]_\mathrm{R} = r, \ [\bar{E}:\bar{F}]_\mathrm{L} = s. \tag{3}$$

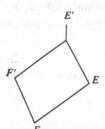

The proof will be in a number of stages. Suppose first that E' is any extension field of E with a subfield F' such that $F' \cap E = F$. Then $EF' = \{\sum x_i y_i | x_i \in E, \ y_i \in F'\}$ is an (E, F')-bimodule and the natural homomorphism $E \otimes_F F' \to EF'$ shows that

$$[EF':F'] \leqslant [E \otimes F':F'] = [E:F]_\mathrm{R}.$$

Here and in the rest of this section generally, tensor products of two fields are taken over their intersection. Our aim will be to construct, for the given extension E/F, an extension E'/F' such that $[EF':F']$ has a preassigned value, while $F' \otimes E \cong F'E$. We state the precise result needed as

PROPOSITION 5.9.3. *Let E/F be a field extension and r an integer such that $1 < r \leqslant [E:F]_\mathrm{R}$. Then there exist a field E' containing E and a subfield F' of E' such that*
  (i) *$F' \cap E = F$ and $F'E \cong F' \otimes E$ by the natural homomorphism;*
  (ii) *$[EF':F'] = r$ and any r right F-independent elements of E form a basis of EF' as right F'-space.*

We remark that since EF' is a subspace of E' as right F'-space, it follows that $[E':F']_\mathrm{R} \geqslant r$.

Before proving this proposition let us show how to deduce Th. 9.2 from it. We write $F_0 = F$ and let E_0 be an extension of E such that $[E_0:F_0]_\mathrm{R} \geqslant r$, $[E_0:F_0]_\mathrm{L} \geqslant s$, e.g. by taking $E_0 = E(t)$. We then define an extension E_n/F_n recursively, using Prop. 9.3 at the odd stage and its left–right dual at the even stage. In detail, given an extension $F_{n-1} \subseteq E_{n-1}$ with $[E_{n-1}:F_{n-1}]_\mathrm{R} \geqslant r$, $[E_{n-1}:F_{n-1}]_\mathrm{L} \geqslant s$, there exists an extension $F_n \subseteq E_n$ such that $E_{n-1} \subseteq E_n$, $F_n \cap E_{n-1} = F_{n-1}$, and further, (i) for odd n, $F_n E_{n-1} \cong F_n \otimes E_{n-1}$, while any r right F_{n-1}-independent elements of E_{n-1} form a basis of $E_{n-1} F_n$ as right F_n-space, (ii) for even n, $E_{n-1}F_n \cong E_{n-1} \otimes F_n$, while any s left F_{n-1}-independent elements of E_{n-1} form a basis of $F_n E_{n-1}$ as left F_n-space.

Given an extension $F \subseteq E$ as in Th. 9.2, we take $F_0 \subset E_0$ as above and suppose that $u_1, \ldots, u_r \in E_0$ are right F_0-independent and $v_1, \ldots, v_s \in E_0$ are left F_0-independent. If $F_{n-1} \subseteq E_{n-1}$ are already constructed, with $E_{n-1} \supseteq E_0$ such that $u_1, \ldots, u_r$ are right F_{n-1}-independent and $v_1, \ldots, v_s$ left F_{n-1}-independent, we construct $F_n \subseteq E_n$ as above. When n is odd, we have $F_n \cap E_{n-1} = F_{n-1}$ and $F_n E_{n-1} \cong F_n \otimes E_{n-1}$, hence $v_1, \ldots, v_s$ are left F_n-independent; further, $[E_n:F_n]_R \geqslant [E_{n-1}F_n:F_n] = r$ and $u_1, \ldots, u_r$ form a basis of $E_{n-1}F_n$ as right F_n-space. When n is even, $u_1, \ldots, u_r$ are right F_n-independent and $v_1, \ldots, v_s$ form a basis of $F_n E_{n-1}$ as left F_n-space.

Now put $\bar{E} = \bigcup E_n$, $\bar{F} = \bigcup F_n$; then $\bar{F} \subset \bar{E}$ is a field extension and the set $\{u_1, \ldots, u_r\}$ is a right $\bar{F}$-basis for $\bar{E}$, for it is linearly independent over $\bar{F}$ and spans $E_{2m}F_{2m+1}$ over F_{2m+1}, for all m; hence it spans $\bar{E}\bar{F}$ over $\bar{F}$. Similarly $v_1, \ldots, v_s$ is a left $\bar{F}$-basis for $\bar{E}$. Thus (3) holds; moreover, $F \subset \bar{F}$, $E \subset \bar{E}$ and any element of $\bar{F} \cap E$ lies in $F_n \cap E_0$ for some n, hence $\bar{F} \cap E = F$, and Th. 9.2 is proved.

It remains to prove Prop. 9.3, by constructing F' and E'. Suppose for a moment that E' and F' exist, satisfying the conditions of Prop. 9.3. Then $M = EF'$ is an (E, F')-bimodule generated by 1, which for clarity we shall denote by u. Since $E \cap F' = F$, we will have

For any $a \in E$, $b \in F'$, $au = ub$ if and only if $a = b \in F$.   (4)

Given any element x of M, we define its *left normalizer* in E as

$$\{a \in E \,|\, ax \in xF'\}.$$

Then (4) shows that the left normalizer of u is precisely F; in fact (4) asserts a little more: it also tells us that F is the precise centralizer of u (in E and F'). Our task is firstly to construct a field extension F' of F and an (E, F')-bimodule M with these properties, and then to find a field E' containing E and F' to satisfy (i) and (ii) of Prop. 9.3.

If there is an (E, F')-bimodule M such that $[M:F'] = r$, then there is an embedding of E in $\mathfrak{M}_r(F')$ induced by the left action of E on M; so we look for a suitable ring T such that E embeds in $\mathfrak{M}_r(T)$. Here T is restricted to be an F-ring and we want our bimodule to have a single F-centralizing generator, so we put

$$T = \mathfrak{W}_r(E; F),$$

where $\mathfrak{W}$ is the matrix reduction functor of 1.7. By Cor. 7.7, T is again a fir. If its universal field of fractions is denoted by F', we have

$$\mathfrak{M}_r(F) \underset{F}{*} E = \mathfrak{M}_r(T) \subseteq \mathfrak{M}_r(F'), \qquad (5)$$

and if $\{e_{ij}\}$ is the set of matrix units centralizing F, then $\mathfrak{M}_r(F')e_{11}$ is an (E, F')-bimodule with F-centralizing generator e_{11}. Moreover, in $\mathfrak{M}_r(T)$ we have $E \cap F' = F$, for from (5) we see that the elements of F' centralize the e_{ij} and the only elements of E centralizing the e_{ij} are those of F. We shall need to show that linear independence is preserved in passing from F to F':

LEMMA 5.9.4. *Given a field extension* $F \subset E$ *and elements* $u_1, \ldots,$ $u_r \in E$ *that are right F-independent, put* $T = \mathfrak{W}_r(E; F)$ *and let F' be the universal field of fractions of the fir T. Then the elements* $u_1 e_{11}, \ldots, u_r e_{11}$ *form a right F'-basis of* $\mathfrak{M}_r(F')e_{11}$.

Proof. For simplicity we shall put $P = \mathfrak{M}_r(F)$ and

$$C = \mathfrak{M}_r(T) = P \underset{F}{*} E.$$

Since $\mathfrak{M}_r(F')e_{11}$ has dimension r as right F'-space, it will be enough to show that the $u_i e_{11}$ are right F'-independent. We may assume without loss of generality that $u_1 = 1$ if 1 is in the F-span of the u_i. If the $u_i e_{11}$ are right F'-dependent, then, bearing in mind that they are columns of elements of F', we see that the matrix formed by these columns is singular over F' and hence non-full over T. Thus we have

$$\sum u_i e_{1i} = yz, \text{ where } y, z \in C \text{ and } ye_{rr} = 0. \qquad (6)$$

We choose a right F-basis for P of the form $\{1, e_{ij}\}$ for $(i, j) \neq (r, r)$ and label the e_{ij} as a_λ ($\lambda = 1, \ldots, r^2 - 1$), and we extend $u_1, \ldots, u_r$ to a right F-basis of E which we label as $1, b_\mu$ ($\mu = 1, \ldots$). We take the a_λ to be well-ordered, as well as the b_μ, and then well-order the set $\{a_\lambda, b_\mu\}$ so that $a_\lambda < b_\mu$ for all λ, μ. The words on this set $\{a_\lambda, b_\mu\}$ with as and bs alternating (including the empty word 1) form a right F-basis for C, which we shall take to be ordered by length and lexicographically, with 1 as first element.

For any $c \in C$ we denote by $\bar{c}$ the leading term, i.e. the maximal element in its support. We can also express c as a right P-linear combination of words whose last letter comes from $\{b_\mu\}$; the words occurring in this expression for c will be called the P-support of c and for any such word q there is a right P-linear map $c_q: C \to P$ which consists in taking the right-hand coefficient of q in c.

Assume that y in (6) has been chosen so that its P-support is minimal in the lexicographic ordering of finite descending sequences of words. Now consider the P-module yP; we may choose a right F-basis $\{y_k\}$ of yP such that the leading terms all have F-coefficients 1 and are distinct. If an element of yP had a leading term ending in some b_μ, it would generate a free right P-module, but this cannot be so, because $ye_{rr} = 0$, hence yP is a proper homomorphic image of P and so its dimension as right F-space is $< r^2$.

We note further that no leading term $\bar{y}_i$ can be an initial segment of another, $\bar{y}_j$. For suppose we have $\bar{y}_1 = \bar{y}_2 p$, where p is a non-empty word with initial factor b_μ. Let $\bar{y}_1 = q$, $c_q(y) = t$ and write $y_2 = ya$; then $ta = c_q(y)a = c_q(y_2) = 0$, because $\bar{y}_2$ is shorter than $\bar{y}_1$. So we have

$$(y - yapt)(z + aptz) = yz;$$

hence we may in (6) replace y by $y - yapt$ and z by $z + aptz$, which removes q from the P-support of y and replaces it by lower terms. But this contradicts the minimality of the P-support of y, and it proves our claim.

Now any element of yC can be written as $yc = \sum y_k f_k$, where $\bar{f}_k$ does not begin with a term a_λ. Since the leading term of y_k has a coefficient 1 and this term ends in a factor a_λ, we have $\overline{y_k f_k} = \bar{y}_k \bar{f}_k$. If two of these terms were the same, say $\overline{y_1 f_1} = \overline{y_2 f_2}$, then one of $\bar{y}_1, \bar{y}_2$ would be an initial segment of the other, but we have seen that this cannot happen, so all the $\overline{y_k f_k}$ are distinct.

By (6), we have

$$u_i e_{1j} = \sum_h u_h e_{1h} e_{ij} \in yC, \quad \text{for } i, j = 1, \ldots, r;$$

hence $u_i e_{1j} = \bar{y}_k q$ for some word q with initial factor not in P. Now $u_i e_{1j}$ is a word of length 2 with last factor in P. The only ways to split it into two factors of which the second does not start with a factor in P are $1 \cdot u_i e_{1j}$ and $u_i e_{1j} \cdot 1$. If $\bar{y}_k = 1$, then 1 is the leading term of some member of yP, hence yP contains an element of F, so $yP = P$ and hence $[yP:F] = r^2$, in contradiction with the fact that $ye_{rr} = 0$. Hence $\bar{y}_k = u_i e_{1j}$, so yP contains r^2 right F-independent elements, which leads to a contradiction as before. Therefore the $u_i e_{1j}$ are right F'-independent and so they form a right F'-basis, as claimed. ∎

Let us write M for the (E, F')-bimodule $\mathfrak{M}_r(F')e_{11}$ and u for its generator e_{11}. Our next aim is to construct a field E' containing F' and E such that the conditions of Prop. 9.3 are satisfied. We remark that none of these conditions limits the size of E' in any way: if we have found a field E' to satisfy all these conditions, then any field containing E' will also satisfy them.

Let $D = E \underset{F}{\circ} F'$ be the field coproduct of E and F' over F, and with $M = \mathfrak{M}_r(F')e_{11}$ form the D-bimodule

$$^\otimes M^\otimes = D \otimes_E M \otimes_{F'} D.$$

The tensor ring $R = D_F \langle {}^\otimes M^\otimes \rangle$ is a fir (see FR. 2.6), and so it has a universal field of fractions, which we shall denote by E'. In terms of the natural grading on R, the generator u of M is of degree 1 and hence it is an atom in R. Moreover, the left normalizer of u in R is $F + uR$, for by expressing the elements of R in terms of a right D-basis of ${}^\otimes M^\otimes$ including u, we see that this normalizer lies in $D + uR$ and DuF' $\cong D \otimes_E M$; hence if $c \in D$ satisfies $cu = uc'$, then $c \in F'$ and since the centralizer of u in F' is F, we find that $c \in F$.

Let us write $E^u = u^{-1}Eu$; then $E^u F' = u^{-1}EuF'$, and this (E^u, F')-bimodule is isomorphic to M as (E, F')-bimodule, as we see by multiplying by u on the left. Our next task is to determine the structure of $F'E^u$ as (F', E)-bimodule, and we shall do so in quite general circumstances.

LEMMA 5.9.5. *Let R be a fir with universal field of fractions U, and let v be an atom in R with left normalizer N in R. Then the mapping given by multiplication*

$$R \otimes_N (v^{-1}R) \to Rv^{-1}R \tag{7}$$

is an isomorphism; likewise for the mapping induced by multiplication

$$(R/Rv) \otimes_K (v^{-1}R/R) \to Rv^{-1}R/R, \tag{8}$$

where $K = N/Rv$.

Proof. We note that in $R \otimes_N v^{-1}R$ the two natural embeddings of R coincide, since $r \otimes 1 = r \otimes vv^{-1} = rv \otimes v^{-1} = 1 \otimes rvv^{-1} = 1 \otimes r$.

It is clear that both mappings (7) and (8) are surjective. If (8) is not injective, then there is a relation

$$p_1 v^{-1} q_1 + \ldots + p_n v^{-1} q_n = r \in R, \tag{9}$$

which is not a consequence of the relations in $R \otimes v^{-1}R$; a corresponding statement holds for (7), with r replaced by 0 in (9), so it is enough to consider (9). Let us take a relation (9) for which n is minimal. Writing $p = (p_1 \ldots p_n)$, $q = (q_1 \ldots q_n)^T$, we can express (9) as $p(v^{-1}I_n)q = r$, and by elementary row and column operations we have

$$\begin{pmatrix} I & 0 \\ 0 & r - p(v^{-1}I)q \end{pmatrix} \to \begin{pmatrix} I & 0 \\ p & r - p(v^{-1}I)q \end{pmatrix} \to \begin{pmatrix} I & (v^{-1}I)q \\ p & r \end{pmatrix}$$

$$\to \begin{pmatrix} vI & q \\ p & r \end{pmatrix}.$$

Thus we obtain a linearized matrix which is singular over U and hence non-full over R:

$$\begin{pmatrix} vI & q \\ p & r \end{pmatrix} = \begin{pmatrix} B \\ b \end{pmatrix} (C \quad c), \tag{10}$$

where B, C are $n \times n$ matrices over R, b is a row and c a column. By the partition lemma we can modify the right-hand side so that B, C are lower triangular. Writing $B = (b_{ij})$, $C = (c_{ij})$, we then have $b_{ii}c_{ii} = v$ and since v is an atom, one of b_{ii}, c_{ii} must be a unit; by a further adjustment we may assume that one of b_{ii}, c_{ii} is 1 and the other is v. Now $b_{11}c_1 = q_1$, and if $b_{11} = v$, then $p_1 v^{-1} q_1 = p_1 c_1$ and we can shorten (9) to $\sum_2^n p_i v^{-1} q_i = r - p_1 c_1$. Hence we may assume that $b_{11} = 1$. Let k be the first index for which $b_{kk} = v$; then $(b_{ij})_{i,j<k}$ is an invertible matrix, so by a modification we can reduce it to the unit matrix, while $(c_{ij})_{i,j<k}$ becomes vI_{k-1}, and $c_i = q_i$ for $i < k$. Now consider the (k, j)-entry of the product BC, where $j < k$; this is the kth row of B multiplied by the jth column of C:

$$b_{kj} v + v c_{kj} = 0.$$

Comparing $(k, n + 1)$-entries in (10), we find

$$\sum_{j<k} b_{kj} q_j + v c_k = q_k.$$

Hence

$$v^{-1} q_k = \sum v^{-1} b_{kj} q_j + c_k = -\sum c_{kj} v^{-1} q_j + c_k,$$

and when this is inserted into (9), that relation is again shortened. There remains the case $B = I$, $C = vI$. Then $c_{nn} = v$, so $b_n v = p_n$ and we can again shorten our relation. Hence every relation (9) is a consequence of the relations in $R \otimes v^{-1}R$; it follows that (8) is an isomorphism and similarly for (7). ∎

Applied to our particular construction, this result shows that for $R = D_F \langle {}^\otimes M^\otimes \rangle$ the submodule $F'E^u$ is isomorphic as (F', E^u)-bimodule to $F' \otimes_F E^u$, by a mapping which sends 1 to $1 \otimes 1$, for as we saw, u is an atom with left normalizer $F + uR$. Since E^u embeds in $u^{-1}R/R$ by the map $y^u \mapsto u^{-1} y \pmod R$ and F' embeds in R/Ru by mapping x to x $\pmod R$, the isomorphism (8) shows that the image of $F'E^u$ in $Ru^{-1}R/R$ is isomorphic to $F' \otimes_F E^u$ in a natural way. In detail, denoting residue classes by $[\,]$, we find that the mapping

$$F' \otimes_F E^u \to (R/uR) \otimes_K (u^{-1}R/R) \to U/R,$$

$$x \otimes y^u \mapsto [x] \otimes [u^{-1}y] \mapsto [xu^{-1}y]$$

is injective. Hence so is the map $F' \otimes_F E^u \to U$ given by $x \otimes y^u \mapsto xu^{-1}y$. Following this map by right multiplication by u, we see that the map $x \otimes y^u \mapsto xy^u$ is injective, as claimed. Together with Lemma 9.4 this completes the proof of Prop. 9.3. ∎

Exercises

1. Verify that in Th. 9.1, $K^\alpha \neq K$ when $|\Lambda| = 1$.

2°. Show that in the conditions of Th. 9.1, $[L:K]_L = \infty$ if $|\Lambda| = 1$.

3. (Schofield) Let L/K be any field extension such that $[L:K]_R = [L:K]_L < \infty$. If $u_1, \ldots, u_r \in L$ are left and right linearly independent over K but do not form a left or right K-basis, show that $U = \sum u_i K$ and $V = \sum K u_i$ are proper subgroups of L^+; hence find $u_{r+1} \in L$ not left or right linearly dependent on $u_1, \ldots, u_r$. Deduce that there is a right K-basis for L which is also a left K-basis. More generally, if $[L:K]_R \leqslant [L:K]_L$, show that there is a right K-basis which can be enlarged to a left K-basis.

5.10 Coproducts of quadratic field extensions

We have seen in 5.3 that any coproduct of fields is a fir; it is natural to ask when this fir is an Ore domain and hence principal. This turns out to be the case precisely when there are two factors, each a quadratic extension. We shall see that in this special case the coproduct has a form resembling a skew polynomial ring, but we begin by establishing the conditions for a coproduct to be Ore.

THEOREM 5.10.1. *Let (E_λ) be a family of fields that are K-rings, where K is a field, and $E_\lambda \neq K$. If the coproduct $R = \underset{K}{*} E_\lambda$ is a right Ore domain, then there are at most two factors and $[E_\lambda:K]_R \overset{K}{=} 2$ ($\lambda = 1, 2$).*

Proof. Assume first that there are just two factors E_1, E_2 and that $[E_1:K]_R > 2$. Then there exist $x, y \in E_1$ such that 1, x, y are right linearly independent over K. Take $z \in E_2 \backslash K$ and put $a = xz + zx$, $b = yz + zy$; then it follows by Cor. 1.4 that a, b are non-units in R. If $aR \cap bR \neq 0$, then since R is a fir, $aR + bR$ is principal; thus there exist $d, a_1, b_1, p, q \in R$ such that

$$d = ap + bq, \tag{1}$$

$$a = da_1, b = db_1. \tag{2}$$

Here d may be taken to be right reduced in R, i.e. not right associated to an element of lower height. By (2) and Th. 1.3,

$$h(d) + h(a_1) - 1 \leqslant h(da_1) = h(a) = 2,$$

hence $h(d) \leqslant 3 - h(a_1)$, so d has height at most 2. If $h(d) = 2$, then $h(a_1) \leqslant 1$, by Cor. 1.4. If $h(a_1) = 1$, then a_1 and d are left and right λ-pure for some λ, so $a_1 \in E_\lambda$ and $d = aa_1^{-1}$, but then $2 = h(d) = h(aa_1^{-1}) = 3$, by Th. 1.3, a contradiction. Hence if $h(d) = 2$, then $h(a_1) = 0$, i.e. $a_1 \in K$; similarly $b_1 \in K$ and writing $\gamma = b_1^{-1}a_1$, we have $a = b\gamma$, i.e.

$$xz + zx = yz\gamma + zy\gamma.$$

Equating pure terms of type (2, 1), we find

$$zx - zy\gamma \equiv 0 \ (\mathrm{mod} \ H_1).$$

Hence $x - y\gamma \in H_0$, which contradicts the right linear independence $(\mathrm{mod} \ H_0)$ of x and y. It follows that $h(d) < 2$. Now $h(ap) \geqslant 2$, so by (1), $h(ap) = h(bq) = N$, say, where $N \geqslant 2$ and $h(p) = h(q) = N - 2$. Rewriting (2) as a congruence, we find

$$xzp + zxp + yzq + zyq \equiv 0 \ (\mathrm{mod} \ H_{N-1}),$$

or, taking pure components separately,

$$xzp + yzq \equiv z(xp + yq) \equiv 0 \ (\mathrm{mod} \ H_{N-1}). \qquad (3)$$

Now p, q are of height $N - 2$, so xp, $yq \in H_{N-1}^{1\cdot}$, and since $z \in E_2 \backslash K$, the second congruence (3) yields

$$xp + yq \equiv 0 \ (\mathrm{mod} \ H_{N-2}).$$

By the right linear independence of x, y $(\mathrm{mod} \ H_0)$ we find

$$p \equiv q \equiv 0 \ (\mathrm{mod} \ H_{N-2}^{1\cdot}).$$

Thus p, q are left 1-pure of height $N - 2$, hence zp, zq are left 2-pure of height $N - 1$. Now we obtain from the first congruence (3),

$$zp \equiv zq \equiv 0 \ (\mathrm{mod} \ H_{N-2}),$$

and this contradicts the relations $h(zp) = h(zq) = N - 1$. Thus a and b can have no common right multiple other than zero and this shows R not to be right Ore.

If there are more than two factors, let E_1 be the first factor and E_2 the coproduct of all the others. Then $R = E_1 \underset{K}{*} E_2$ and $[E_2:K]_R > 2$, so the previous argument can be applied to complete the proof. ∎

It remains to investigate the case of two extensions of right degree 2. We take our extensions to be E_1, E_2 over K and again write $R = E_1 \underset{K}{*} E_2$. Each E_i is pseudo-linear and we may take it to be generated by

u_i with defining and commutation relations

$$au_i = u_i a^{\alpha_i} + a^{\delta_i} \quad \text{for all } a \in K, \tag{4}$$

$$u_i^2 + u_i \lambda_i + \mu_i = 0 \quad \text{for some } \lambda_i, \mu_i \in K. \tag{5}$$

We shall show that this coproduct is right Ore and obtain a normal form for its elements.

THEOREM 5.10.2. *Let E_i/K (i = 1, 2) be two quadratic extensions of a field K, generated by u_i subject to (4), (5), and write $R = E_1 \underset{K}{*} E_2$. Then R is a principal right ideal domain and if $t = u_1 - u_2$, then every element of R can be expressed uniquely in the form of a finite sum*

$$\sum t^r c_r, \quad \text{where } c_r \in E_1. \tag{6}$$

Moreover, t satisfies the commutation relations

$$at = ta^{\alpha_2} + u_1 a^* + a' \quad \text{for all } a \in K, \tag{7}$$

where

$$a^* = a^{\alpha_1} - a^{\alpha_2}, \quad a' = a^{\delta_1} - a^{\delta_2}; \tag{8}$$

$$u_1 t = t^2 - t(u_1 + \lambda_2) - u_1 \lambda - \mu, \tag{9}$$

where

$$\lambda = \lambda_1 - \lambda_2, \quad \mu = \mu_1 - \mu_2. \tag{10}$$

Proof. We begin by deriving the commutation formulae (7), (9) for t. For any $a \in K$ we have

$$\begin{aligned}
at &= a(u_1 - u_2) \\
&= u_1 a^{\alpha_1} + a^{\delta_1} - u_2 a^{\alpha_2} - a^{\delta_2} \\
&= ta^{\alpha_2} + u_1(a^{\alpha_1} - a^{\alpha_2}) + a^{\delta_1} - a^{\delta_2} \\
&= ta^{\alpha_2} + u_1 a^* + a',
\end{aligned}$$

and this is (7). To prove (9), we have from (5),

$$(u_1 - t)^2 + (u_1 - t)\lambda_2 + \mu_2 = 0.$$

We expand the square and use (5) to obtain

$$-u_1 \lambda_1 - \mu_1 - u_1 t - tu_1 + t^2 + u_1 \lambda_2 - t\lambda_2 + \mu_2 = 0.$$

Taking $u_1 t$ to the other side and using the abbreviations (10), we obtain (9). Thus (7) and (9) are established. They show that the set of all polynomials (6) form a subring R' of R; clearly R' contains E_1, and since every element of E_2 has the form

$$u_2 a + b = -ta + u_1 a + b,$$

for some $a, b \in K$, it follows that R' also contains E_2 and so $R' = R$, because R is generated by E_1 and E_2. Thus every element of R has the form (6). To prove uniqueness, assume that

$$c_0 + tc_1 + \ldots + t^n c_n = 0, \quad \text{where } c_i \in E_1, c_n \neq 0.$$

Multiplying by c_n^{-1} on the right, we obtain a relation

$$b_0 + tb_1 + \ldots + t^{n-1} b_{n-1} + t^n = 0, \quad \text{where } b_i \in E_1. \tag{11}$$

Now t is left and right reduced of height 1, so $h(t^r b_r) \leqslant r + 1$, and $h(t^{n-1}(b_{n-1} + t)) = n$, but by (11),

$$n = h(t^{n-1}(b_{n-1} + t)) = h(b_0 + tb_1 + \ldots + t^{n-2} b_{n-2}) \leqslant n - 1,$$

and so we have reached a contradiction. Hence the powers of t are right linearly independent over E_1 and (6) is unique.

It remains to show that R is right Ore; in fact it is no harder to show that R is right principal and this follows as in the polynomial case, but extra care is necessary on account of the commutation formulae (7), (9). Given any non-zero right ideal $\mathfrak{a}$ of R, we take an element f in $\mathfrak{a}$ of least degree in t, chosen monic without loss of generality:

$$f = t^n + t^{n-1} c_1 + \ldots + c_n, \quad \text{where } c_i \in E_1.$$

This polynomial f is unique, for if $\mathfrak{a}$ contained another monic polynomial g of degree n, the difference $f - g$ would be a non-zero polynomial in $\mathfrak{a}$ of lower degree. Consider fR, the right ideal generated by f. Since $fR \subseteq \mathfrak{a}$, f is also the unique monic polynomial of least degree in fR. We claim that for any $N > n$, fR contains a monic polynomial of degree N. Let P_r be the set of all elements of R of degree at most r in t; clearly this is a right E_1-space. We have

$$ft = t^{n+1} + t^{n-1} c_1 t + t^{n-2} c_2 t + \ldots + c_n t.$$

Hence by (7) and (9),

$$ft \equiv t^{n+1} + t^{n-1} c_1 t \pmod{P_n}. \tag{12}$$

We write c_1 as $c_1 = u_1 a + b$, where $a, b \in K$; then by (5), (7) and (9),

$$u_1 a t = u_1 t a^{\alpha_2} + u_1^2 a^* + u_1 a' = t^2 a^{\alpha_2} + g,$$

where g is a linear polynomial in t. Similarly bt is a linear polynomial in t, by (7). Inserting these expressions in (12), we find

$$ft \equiv t^{n+1}(1 + a^{\alpha_2}) \pmod{P_n}. \tag{13}$$

If $1 + a^{\alpha_2} = 0$ (i.e. $a = -1$), then ft is a polynomial of degree at most n and since $ft \in fR$, we must have $ft = fb$ for some $b \in E_1$, but this would mean that $t = b$, a contradiction. Hence $1 + a^{\alpha_2} \neq 0$ and now (13) shows that fR contains a polynomial of degree $n + 1$; therefore it also contains a monic polynomial of degree $n + 1$.

Now let $N > n + 1$ and assume that fR contains a monic polynomial of degree m for each m satisfying $n \leqslant m < N$; we have to find a monic polynomial of degree N in fR. Let g be in fR and monic of degree $N - 1$. Since $N - 2 \geqslant n$, we may reduce the coefficient of t^{N-2} in g to zero by subtracting a suitable polynomial in fR. If $g = t^{N-1} + t^{N-3}b + \ldots$, then

$$gt = t^N + t^{N-3}bt + \ldots$$
$$= t^N + t^{N-1}b + \ldots,$$

by (9), so gt is a monic polynomial of degree N in fR. It follows by induction that fR contains a monic polynomial of degree N for all $N \geqslant n$. Now any element h of R can by subtraction of suitable right E_1-multiples of the polynomials just found be brought to the form $h = h_1 + h_2$, where $h_1 \in fR$ and $h_2 \in P_{n-1}$. In particular, if $h \in \mathfrak{a}$, then $h_2 \in \mathfrak{a} \cap P_{n-1} = 0$ and so $h = h_1 \in fR$, and this shows that $\mathfrak{a} = fR$. Thus R is a principal right ideal domain, and in particular it is right Ore. ∎

Whether the coproduct R is also left Ore will naturally depend on the left degrees of the E_i over K. Let us briefly consider the case of commutative fields; we shall see that the field coproduct of two commutative field extensions is a quaternion algebra, i.e. a 4-dimensional algebra over its centre.

THEOREM 5.10.3. *Let F_1, F_2 be two commutative quadratic field extensions of k, given as $F_i = k(u_i)$, where as before,*

$$u_i^2 + u_i\lambda_i + \mu_i = 0 \quad \text{for some } \lambda_i, \mu_i \in k, \tag{5}$$

and put $t = u_1 - u_2$, $\lambda = \lambda_1 - \lambda_2$. Then their field coproduct $E = F_1 \underset{k}{\circ} F_2$ is a quaternion algebra with centre $k(u)$, where

$$u = t^2 + \lambda t. \tag{14}$$

Proof. We have seen that every element of the coproduct $P = F_1 \underset{k}{*} F_2$ can be written as a polynomial in $t = u_1 - u_2$ with coefficients in F_1:

$$\sum t^r c_r, \quad \text{where } c_r \in F_1. \tag{6}$$

The commutation formulae now read

$$\alpha t = t\alpha \text{ for all } \alpha \in k, \tag{7}$$

$$u_1 t = t^2 - t(u_1 + \lambda_2) - u_1\lambda - \mu, \tag{9}$$

where as before, $\lambda = \lambda_1 - \lambda_2$, $\mu = \mu_1 - \mu_2$. Multiplying (9) by t on the right and using (9) to simplify the result, we obtain

$$u_1 t^2 = t^2(u_1 - \lambda) + t(2u_1 + \lambda_2)\lambda + (u_1\lambda + \mu)\lambda.$$

Hence

$$u_1(t^2 + \lambda t) = (t^2 + \lambda t)u_1. \tag{15}$$

Let us put $u = t^2 + \lambda t$; by (15), u commutes with u_1 and it also commutes with t, and hence with $u_2 = u_1 - t$. Thus u lies in the centre of P. If the field of fractions of P is denoted by E, then E is a quadratic extension of $k(t)$, by (5) and (9), which in turn is a quadratic extension of $k(u)$, so E is 4-dimensional over $k(u)$. Since $u_1 u_2 \neq u_2 u_1$, E is non-commutative and so is a quaternion algebra with centre $k(u)$. ∎

From the proof it is easy to deduce the form taken by specializations of the field coproduct:

COROLLARY 5.10.4. *Let L be a field which is a k-algebra, generated over k by two elements u_1, u_2 both quadratic over k. Then L is either commutative or a quaternion algebra.*

Proof. Clearly L is a homomorphic image of $P = k(u_1) \underset{k}{*} k(u_2)$ and since L is a field, it is a proper homomorphic image. Now P is 4-dimensional over the central subring $k[u]$, where $u = t^2 + \lambda t$ as in the proof of Th. 10.3. Let v_1, v_2, v_3, v_4 be the image in L of a basis of P over $k[u]$ and let C be the image of $k[u]$. No finite-dimensional extension of $k[u]$ is a field, so C is a proper homomorphic image of $k[u]$ and hence is a field. Further, L is spanned by the v_i over the central subfield C and hence is either commutative or a quaternion algebra over C, because $[L:C]$ is a perfect square ≤ 4. ∎

Let us return to the skew case and examine more closely the coproduct of two quadratic extensions that are isomorphic. We may take them to be E_1, E_2 where E_i is generated over K by u_i satisfying

$$au_i = u_i a^\alpha + a^\delta \text{ for all } a \in K, \tag{16}$$

$$u_i^2 + u_i\lambda + \mu = 0. \tag{17}$$

This coproduct $P = E_1 \underset{K}{*} E_2$ has an automorphism σ of order 2 which consists in interchanging u_1 and u_2 and leaving K fixed. The commutation formulae for $t = u_1 - u_2$ now reduce to

$$at = ta^\alpha \text{ for all } a \in K, \tag{18}$$

and $u_1 t = t^2 - t(u_1 + \lambda)$, which can be written

$$u_1 t = -t(u_2 + \lambda). \tag{19}$$

Now we know from Cor. 3.6.2 that α may be extended to an endomorphism α_i of E_i by putting

$$u_i^{\alpha_i} = -(u_i + \lambda).$$

Hence (19) may be written in the form

$$u_1 t = t u_1^{\alpha_1 \sigma} = t u_1^{\sigma \alpha_2}.$$

Thus the map $\alpha_1 \sigma = \sigma \alpha_2$ agrees with the inner automorphism of P induced by t. Since $\sigma^2 = 1$, we have $\alpha_2 = \sigma \alpha_1 \sigma$ and $\alpha_1^2 = \alpha_1 \sigma \alpha_1 \sigma = \sigma \alpha_2 \sigma \alpha_2$, and this is just the inner automorphism induced by t^2.

The field of fractions L of P may be constructed as follows. Form the skew polynomial ring $E_1[u; \alpha_1^2]$, let L_0 be its field of fractions and extend the endomorphism α_1 of E_1 to L_0 by putting $u^{\alpha_1} = u$. Then L is the quadratic extension of L_0 generated by t satisfying the equation $t^2 = u$. Since every quadratic extension is pseudo-linear, we have a commutation formula

$$ft = tf^\beta + f^\Delta \text{ for all } f \in L_0,$$

where β is an endomorphism and Δ a β-derivation of L_0. For the special case where $f \in E_1$, say $f = u_1 a + b$ $(a, b \in K)$ we have

$$(u_1 a + b)t = t(u_1 a + b) + u a^\alpha,$$

as is easily verified, while of course $ut = tu$. It follows that $\beta = \alpha$ and Δ is defined by

$$(u_1 a + b)^\Delta = u a^\alpha, \quad u^\Delta = 0.$$

The result may be stated as

THEOREM 5.10.5. *Let E/K be a quadratic extension with endomorphism α and α-derivation δ, and denote the extension of α to E by α_1. Then the field coproduct of two copies of E over K is the quadratic extension of the function field $K(u; \alpha_1^2)$ generated by t subject to $t^2 = u$.* ∎

Exercises

1. Let E/K be a quadratic extension and $L = E \underset{K}{\circ} E$ the field coproduct of two copies of E over K, with bases 1, u_i ($i = 1, 2$). By expressing the elements of L as Laurent series in $t = u_1 - u_2$ determine the centre of L.

2. Using Th. 10.3, determine the quaternion algebras that can be expressed as field coproducts of two quadratic extensions.

3. Let L/K be a finite extension generated by an element u such that $I(u)$ maps K into itself. Show that $I(u)$ has finite inner order on K and that L/K can be obtained as a central extension followed by a binomial extension.

4. Let E/K be a quadratic extension with endomorphism α and α-derivation δ, and let L be a field which is a homomorphic image of $E \underset{K}{*} E$. Show that $[L:K]_R = 2$ unless α has finite inner order.

5. (Schofield [85]) Show that for any field E, $\mathfrak{M}_2(E) \underset{E \times E}{*} \mathfrak{M}_2(E) \cong \mathfrak{M}_2(E[t, t^{-1}])$ for a central indeterminate t. (Hint. Take the matrix units of the first factor to be e_{ij} and $E \times E$ embedded along the diagonal; now show that the matrix units of the second factor may be taken to be e_{11}, e_{22}, f_{12}, f_{21} and show that $f_{12} = te_{12}$, $f_{21} = t^{-1}e_{21}$ for some t centralizing the first factor.)

6. Let D be a quaternion algebra over k with splitting field F. Use Ex. 1 to show that $(\mathfrak{M}_2(k) \underset{E}{*} D) \otimes_k E \cong \mathfrak{M}_2(E[t, t^{-1}])$ for any commutative extension E of k and by localization deduce that $(\mathfrak{M}_2(k) \underset{E}{\circ} D) \otimes_k E \cong \mathfrak{M}_2(E(t))$. Assuming uniqueness of the minimal splitting field of D, deduce that the centre of $\mathfrak{M}_2(k) \underset{E}{\circ} D$ is F. (Schofield [85] shows that every function field of genus 0 can occur as centre of a coproduct in this way.)

Notes and comments

Coproducts of groups, usually called free products with amalgamated subgroup, occur naturally in algebraic topology, where the Seifert–van Kampen theorem (proved in 1933) expresses the fundamental group of a union of two spaces as a free product of groups with an amalgamated subgroup. Schreier [27] gave a topological proof that subgroups of free groups are free, and an algebraic proof of the existence of free products of groups (Th. 1.1), which was later simplified by van der Waerden [48]. It is also possible to define a notion of free product of groups with different amalgamated subgroups, where we have a family of groups (G_i) and subgroups (H_{ij}) of G_i such that $H_{ij} \cong H_{ji}$. We can form the coproduct of the G_i amalgamating H_{ij} and H_{ji}, but this coproduct will in general be neither faithful nor separating, see B. H. Neumann [54] and the references given there.

The coproduct of rings was introduced (as 'free product') by Cohn [59], where the coproduct of a family of K-rings was shown to be faithful and separating whenever all the factors are faithfully flat as left and right K-modules. This condition holds in particular when K is a field and all the factors are faithful K-rings, a case treated in more detail in Cohn [60], where a form of the weak algorithm was established when all the factors are fields, and this was used to prove that the coproduct is then (in today's terminology) a semifir; Th. 1.3 and Cor. 1.4 were also proved there. The method used was adapted and greatly generalized by Bergman [74], to prove the results of 5.2 and 5.3, culminating in Theorems 3.5, 3.8, 3.9. These sections follow essentially Bergman's paper, with some simplifications.

The consequences 4.1–4 for tensor rings in 5.4 first appeared in *Skew Field Constructions*, Lemma 4.7 is taken from Cohn [60]; the rest of 5.4 is new. The HNN-extension for fields, Th. 5.1, and the matrix form Th. 5.5 are taken from Cohn [73]. Th. 5.8 and the results leading up to it first appeared in Cohn [73']. The embedding theorem 6.4 first appeared in Cohn [58] with a direct proof; the present proof is new. A (not necessarily associative) algebra over a field can always be embedded in a non-associative division ring; this was shown by B. H. Neumann [51]. Bokut' [81] showed that every ring which is a k-algebra can be embedded in a 1-simple ring, again a k-algebra.

The results of 5.7 are largely taken from Bergman [74']. Part of Th. 7.6 (the $(n-1)$-fir property) was proved in direct fashion by Cohn [69], as a kind of analogue to 'small cancellation theory' in groups. The interpretation of the universal derivation bimodule as kernel of the multiplication map (Th. 8.2) is due to Eilenberg, while much of the rest of 5.8 is folklore (cf. also Bergman and Dicks [75, 78], where Th. 8.9 is proved). The invariance of the generating number of free fields (Th. 8.13) was proved by Schofield [85]; the corresponding result for free algebras appears in FR, 0.11.

The question whether left and right degrees of a field extension are equal was first raised by E. Artin, though there seems to be no written record and it was not listed among the problems in the preface to his collected works (E. Artin [65]). The first example of a quadratic field extension of infinite left degree (Cohn [61"]) was a construction based on the embedding theorem of 2.6 (Th. 2.6.3, see Cohn [61']); pseudo-linear extensions of right degree $n \geqslant 2$ and infinite left degree were constructed in Cohn [66']. Field extensions of different finite left and right degrees (Th. 9.2) were first constructed by Schofield [85'], whose presentation we follow here. In the text we referred to FR for the fact that the tensor ring $D_F \langle {}^{\otimes}M^{\otimes} \rangle$ is a fir (p. 266). In fact this is a semifir by Th. 4.5; the fir-property is only needed in the proof of Lemma 9.5 to show that elements have complete factorizations, but this follows easily by a direct argument using the degree.

The results on coproducts that are right Ore in 5.10 are taken from Cohn [60, 61″]. Using results from Cohn [85] (cf. also FR, Th. 7.8.4, p. 441) it can be shown that the centre of the field coproduct $E \underset{k}{\circ} F$ lies in K unless E, F are both at most quadratic over K. When E, F are both commutative quadratic extensions of K, then the centre is a rational function field $K(t)$. In general the centre of the field coproduct can be any function field of genus 0 (Schofield [85]; see also Ex. 6 of 5.10).

6

General skew fields

In any variety of algebras such as groups or rings the members can be defined by presentations in terms of generators and defining relations. Fields do not form a variety, as we saw in 1.2, and they do not have presentations in this sense. In fact the usual mode of construction of commutative fields is quite different: one takes a rational function field over some ground field and then makes an algebraic extension. For skew fields this method works only in very special cases, but there is an analogue of a presentation, in terms of matrices that become singular. Now it is necessary to verify in each case that the outcome is a field and this forms the subject of 6.1. At this stage free fields form a natural topic, but first we need to prepare some tools: In 6.2 we prove the specialization lemma, which generalizes the density principle of the commutative theory: A non-zero polynomial over an infinite (commutative) field assumes non-zero values. The analogue for skew fields states that a non-zero element of a free field (with infinite centre and of infinite degree over it) can be specialized to a non-zero element in the field.

The elements of the free field are described in terms of matrices over the tensor ring and it is therefore of interest to provide a normal form for these matrices. This is done in 6.3 and a corresponding 'normal form' for fractions is derived.

Section 6.4 is devoted to free fields. The relation between free fields over different base fields is described and some surprising properties are revealed, such as irreducible polynomials over a field D which become reducible on adjoining an indeterminate to D.

One commutative concept which has (so far) no satisfactory analogue in the general case is the algebraic closure of a commutative field. Some properties of the algebraic closure are shared by existentially closed fields, which form the topic of 6.5. These fields are less manageable, thus they are not axiomatizable (i.e. they cannot be defined by elementary sentences), unlike the algebraically closed commutative fields, but they also have properties not shared by the commutative fields, for example, being member of a given subfield can in an existentially closed field be expressed as an elementary sentence.

As in the case of groups one can formulate the word problem for presentations of fields, and we shall find in 6.6 that free fields have a solvable word problem (subject to a condition on the ground field), while examples given here show that there are also fields with unsolvable word problem.

The final section 6.7 uses the compactness theorem of logic to establish the claim made in 1.2 that the class of integral domains cannot be defined by a finite set of elementary sentences.

6.1 Presentations of skew fields

In the commutative case field extensions are very simple to describe. Any finitely generated commutative field may be formed by taking a purely transcendental extension of the ground field $K = k(x_1, \ldots, x_n)$ and making a finite number of simple adjunctions; each of the latter consists in adjoining a solution of an equation $f = 0$, where f is a polynomial in a single variable over K. In particular this shows that any simple extension of infinite degree is free. No such simple statement holds in the general case, in fact we shall need to define what we mean by a free extension (corresponding to purely transcendental extension in the commutative case).

To reach the appropriate definition, consider a finitely generated extension $E = D(\alpha_1, \ldots, \alpha_n)$ of a skew field D. As before we shall take all our fields to be k-algebras, where k is a commutative field. This represents no loss of generality (in fact a gain): if k is not present, we can take the prime subfield to play the role of the ground field. Given E as above, we have a D-ring epimorphism:

$$D_k\langle X \rangle \to E, X = \{x_1, \ldots, x_n\}, x_i \mapsto \alpha_i \ (i = 1, \ldots, n). \quad (1)$$

Here $D_k\langle X \rangle$ is the tensor D-ring on X introduced in 1.6. Let $\mathcal{P}$ be the singular kernel of (1); since E is an epic $D_k\langle X \rangle$-field, being the field generated by the α_i over D, it is determined up to isomorphism by $\mathcal{P}$. If Φ is any set of matrices generating $\mathcal{P}$ as matrix ideal, then E is already determined by Φ and we shall write

$$E = D_k \langle\!\langle X; \Phi \rangle\!\rangle, \quad (2)$$

and call this a *presentation* of E. In particular, we call the α_i *free over* D

if the presentation can be chosen with $\Phi = \varnothing$; this just means that (1) is an honest map, i.e. $\mathscr{P}$ consists precisely of all non-full matrices over $D_k \langle X \rangle$. From Th. 5.4.5 we know that $D_k \langle X \rangle$ is a semifir (in fact, $D_k \langle X \rangle$ is even a fir, see FR, 2.4), so the set of all non-full matrices is a prime matrix ideal and the corresponding universal field of fractions is the free D-field on X, $D_k \langle\!\langle X \rangle\!\rangle$.

Let E be any field with the presentation (2); we shall say that E is *finitely generated* over D if X may be taken to be finite; E is *finitely related* if Φ may be chosen finite, and E is *finitely presented* when X and Φ can both be chosen finite. As for groups we have the following decomposition for finitely related fields:

THEOREM 6.1.1. *Any finitely related field can be expressed as the field coproduct of a finitely presented and a free field.*

Proof. Let $E = D_k \langle\!\langle X; \Phi \rangle\!\rangle$, where Φ is finite. Then the subset X' of members of X occurring in matrices from Φ is finite. If X'' denotes the complement of X' in X, we clearly have $E = E' \underset{D}{\circ} E''$, where $E' = D_k \langle\!\langle X'; \Phi \rangle\!\rangle$ and $E'' = D_k \langle\!\langle X'' \rangle\!\rangle$. Here E' is finitely presented and E'' is free. ∎

In the special case when E has finite (left or right) degree over D, the above construction can be simplified a little. In that case (1) is surjective, not merely epic. Moreover, instead of taking the free algebra, we can incorporate the commutation relations as follows. Suppose that $[E:D]_R = n$ and let $u_1, \ldots, u_n$ be a right D-basis of E. Since E is a D-bimodule, we have the equations

$$\alpha u_j = \sum u_i \rho_{ij}(\alpha) \quad \text{for all } \alpha \in D, \tag{3}$$

where the map $\alpha \mapsto (\rho_{ij}(\alpha))$ is a homomorphism from D to $\mathfrak{M}_n(D)$. Let M be the right D-space on $u_1, \ldots, u_n$ as basis; by the equations (3), M becomes a pointed D-bimodule, as we see by changing our basis of E so that $u_1 = 1$. Let T be the filtered tensor ring on this bimodule, constructed as in 5.4. Th. 5.4.5 can be adapted to show that this ring is a semifir. Now E is obtained as a homomorphic image of T, so we need to look for ideals in T which as right D-spaces have finite codimension, in fact the kernel of (1) in this case is a complement of M in T. But it is not at all clear how this would help in the classification of extensions of finite degree.

Let us return to the general case. Given any set X and any set Φ of square matrices over $D_k \langle X \rangle$, we may ask: When does there exist a field with the presentation

$$D_k \langle\!\langle X; \Phi \rangle\!\rangle ? \tag{4}$$

Writing (Φ) for the matrix ideal of $D_k \langle X \rangle$ generated by Φ, we have two possibilities:

(i) (Φ) is improper. This means that there is an equation

$$I = C_1 \nabla C_2 \nabla \dots \nabla C_r, \tag{5}$$

where each C_i is either non-full or a diagonal sum of a matrix in Φ and another matrix, and the sum on the right of (5) is suitably bracketed. Then there is no field (4), in fact there is no field over which all the matrices of Φ become singular. Here there is no solution because we do not allow the one-element set as a field.

(ii) (Φ) is proper. This means that no equation (5) is possible. Now there is always a field over which the matrices of Φ become singular, possibly more than one. The different such fields correspond to the prime matrix ideals containing (Φ), and there is a universal one among them precisely when the radical $\sqrt{(\Phi)}$ is prime. In particular, this is so when (Φ) itself is prime, and that will be the only case in which the notation (4) will be used.

In practice most of the presentations we shall meet are given by equations rather than singularities, but the latter are important in theoretical considerations, e.g. when we want to prove that an extension is free we must check that there are no matrix singularities. We shall return to this question in Ch. 8.

A special case occurs when E (and hence also D) is of finite degree over k. In that case the singularity of a matrix can be expressed by the vanishing of a norm and hence it will be enough to consider equations. Only in the case of infinite extensions are the matrix singularities really necessary.

We conclude with some examples. Although the notation (4) is used, a justification is needed in each case that the matrix ideal generated is indeed prime.

1. The *Hilbert field*

$$H = K \langle\!\langle u, v; uv = vu^2 \rangle\!\rangle. \tag{6}$$

To construct this field we take a field K and form the rational function field $K(u)$ in a central indeterminate u; on $K(u)$ we have an endomorphism $\alpha: f(u) \mapsto f(u^2)$. Hence we can form the skew polynomial ring $K(u)[v; \alpha]$ and its field of fractions $H = K(u)(v; \alpha)$, which is clearly given by the presentation (6).

We remark that if the centre of K is C, then the Hilbert field H again has centre C and $[H:C] = \infty$. For the proof we write $u^{1/2^i} = v^i u v^{-i}$, so that

$$(u^{1/2^i})^{2^i} = v^{-i}(v^i u v^{-i})v^i = u.$$

If we put $E = K(u, u^{1/2}, u^{1/4}, \ldots)$, then on E α is an automorphism and $H = E(v; \alpha) \subset E((v; \alpha))$. It is clear that α has infinite inner order on E and fixed field K, hence the centre of H is C (Th. 2.2.10). Moreover, 1, u, u^2, ... are linearly independent over C, so $[H:C] = \infty$ (for another proof see Prop. 2.3.5).

2. The *Weyl field*

$$W = K \langle\!\langle x, y; xy - yx = 1 \rangle\!\rangle. \tag{7}$$

To form W we again take the rational function field $K(y)$ and on it define a derivation δ by the rule $y^\delta = 1$. This extends to a unique derivation δ on $K(y)$, which is just (formally) differentiation with respect to y. The skew polynomial ring $A(K) = K(y)[x; 1, \delta]$ is called the *Weyl algebra* and its field of fractions is the Weyl field W with the presentation (7).

When K has centre C and characteristic zero, we can again show that W has centre C and is of infinite degree over it. For on writing $z = x^{-1}$, we can embed W in the field of formal Laurent series in z (see 2.3); the commutation rule now reads $yz = z(y + z)$. We take an element f in the centre of W and write it as a Laurent series in $z: f = \sum z^i a_i$. By conjugating with z we see that $a_i(y) = a_i(y + z)$, hence a_i is independent of y, hence so is f. By symmetry f is independent of x and so $f \in K$; since f centralizes K, it must belong to C, hence C is the centre of K. It is again clear that 1, x, x^2, ... are linearly independent over C, so $[W:C] = \infty$.

3. Let $X = \{x_0, x_1, \ldots, x_4\}$ and consider the presentation

$$D = k \langle\!\langle X; A - x_0 \rangle\!\rangle, \text{ where } A = \begin{pmatrix} x_1 & x_2 \\ x_3 & x_4 \end{pmatrix}. \tag{8}$$

The fact that $A - x_0$ is singular may be expressed by the rational relation

$$x_2 - (x_1 - x_0)x_3^{-1}(x_4 - x_0) = 0.$$

As this relation shows, D may be regarded as the free field on x_0 and any three of x_1, x_2, x_3, x_4. In Ch. 8 we shall see that D may also be obtained as an extension of the free field on x_1, x_2, x_3, x_4.

Exercises

1. Show that the Weyl algebra on x and y over a commutative field k of characteristic p has centre $k(x^p, y^p)$ and its degree over its centre is p^2.

2. Show that for a commutative field k of characteristic 0 and an integer $r > 1$,

the field $k \langle u, v; vu = ruv \rangle$ is infinite-dimensional over its centre k. What happens in finite characteristic?

3. Show that $E \underset{F}{\circ} (F \underset{K}{\circ} G) \cong E \underset{K}{\circ} G$.

6.2 The specialization lemma

In commutative field theory the density principle is of importance; it states that any polynomial over an infinite field which is not identically zero assumes a non-zero value for some value of the argument. In the general theory there is an analogue, Amitsur's theorem on generalized polynomial identities, but for our purposes the result is more usefully expressed in terms of matrices. As we shall need Amitsur's theorem in the proof, we shall state it in the form required here.

If A is a k-algebra, then by a *generalized polynomial identity* (GPI) in a set X of variables over A one understands a non-zero element p of $A_k \langle X \rangle$ which vanishes under all maps $X \to A$. We shall only need the special case of Amitsur's theorem for skew fields, but strengthened by a restriction on the set of values that may be substituted. The precise form is as follows:

GPI-THEOREM. *Let E be a field, Σ a multiplicative subset of E and C its centralizer in E. If for all $a \in E^\times$, $Ca\Sigma$ is infinite-dimensional as left C-space, then any non-zero multilinear element p of $E_C \langle X \rangle$ has a non-zero value for some choice of values of X in Σ.*

For a proof, see A.3, Th. 9.8.3, p. 391 (see also FR, 5.9). The restriction on p to be multilinear was necessary here, because Σ may not be a subspace. However, if we take a subfield D for Σ, the multilinearity need not be assumed. In that case the condition $[CaD:C] = \infty$ follows if we assume (i) $[E:C] = \infty$ and (ii) $E = CD$. For in that case each $a \in E^\times$ has the form $a = \sum c_i d_i$, where $c_i \in C$, $d_i \in D$ and $c_i d_i \neq 0$ for some i. hence $CaD = CD = E$, and so $[CaD:C] = [E:C] = \infty$. Thus we obtain the following form of the GPI-theorem:

THEOREM 6.A. *Let $D \subseteq E$ be a field extension, denote the centralizer of D in E by C and assume that (i) $[E:C] = \infty$ and (ii) $E = CD$. Then any non-zero element of $E_C \langle X \rangle$ has a non-zero value for some choice of values of X in D.* ∎

We shall also need some auxiliary results. In the first place there is the inertia lemma for semifirs, which is concerned with pulling down factorizations from Laurent series to power series. Given a ring A and a subring B, we shall say that B is *finitely inert* in A, if for any matrix $M \in {}^r B^s$, such that over A we have $M = PQ$, where P is $r \times n$ and Q is $n \times s$, there exists $U \in \mathbf{GL}_n(A)$ such that PU^{-1}, UQ have all their entries in B.

LEMMA 6.2.1 (Inertia lemma). *Let R be a semifir. Then for a central indeterminate t, the ring $R[\![t]\!]$ of formal power series is finitely inert in $R(\!(t)\!)$, the ring of formal Laurent series.*

Proof. We write $A = R(\!(t)\!)$, $B = R[\![t]\!]$ and indicate the natural homomorphism $B \to R$, obtained by putting $t = 0$, by a bar. Given $M \in {}^r B^s$, suppose that over A we have

$$M = PQ, \text{ where } P \text{ is } r \times n \text{ and } Q \text{ is } n \times s. \tag{1}$$

We may suppose $P, Q \neq 0$, since otherwise the conclusion holds trivially. Over A every non-zero matrix T can be written uniquely as $t^v T'$, where $v \in \mathbf{Z}$, T' has entries in B and $\bar{T}' \neq 0$. Let $P = t^\mu P'$, $Q = t^v Q'$, where $\bar{P}', \bar{Q}' \neq 0$. Writing $\mu + v = -\lambda$, and dropping the dashes, we can now rewrite (1) in the form

$$M = t^{-\lambda} PQ, \text{ where } P \in {}^r B^n, Q \in {}^n B^s \text{ and } \bar{P}, \bar{Q} \neq 0. \tag{2}$$

If $\lambda \leq 0$, the conclusion follows; if $\lambda > 0$, we have $\bar{P}\bar{Q} = 0$. Then, since R is a semifir, we can find a matrix $U \in \mathbf{GL}_n(R)$ trivializing this relation, and on replacing P, Q by PU^{-1}, UQ we find that all the columns in $\bar{P}$ after the first h are 0 and the first h rows of $\bar{Q}$ are 0. If we multiply P on the right by $V = tI_h \oplus I_{n-h}$ and Q on the left by V^{-1}, then P becomes divisible by t, while Q has all its entries still in B. We can now cancel a factor t in (2) and so replace λ by $\lambda - 1$; after λ steps we obtain the same equation (2) for M but with $\lambda = 0$, and this proves the finite inertia. ∎

There is a stronger form, total inertia, which will not be needed here (see FR, 2.9), but we note the next result, which is proved by total inertia in FR; below we give a direct proof.

We remark that any matrix over a tensor ring $D_K \langle X \rangle$ is stably associated to a linear matrix, by the process of linearization by enlargement (also called Higman's trick, see Higman [40]). We fix an entry of our matrix in which a product occurs, say $f + ab$; for simplicity we take it to be in the bottom right-hand corner. Now take a diagonal sum with 1 and apply elementary transformations. Writing only the last two entries in the last two rows, we have

$$f + ab \rightarrow \begin{pmatrix} f + ab & 0 \\ 0 & 1 \end{pmatrix} \rightarrow \begin{pmatrix} f + ab & a \\ 0 & 1 \end{pmatrix} \rightarrow \begin{pmatrix} f & a \\ -b & 1 \end{pmatrix}.$$

In this way any matrix over $D_K\langle X \rangle$ can be reduced to a matrix which is linear in X.

PROPOSITION 6.2.2. *Let D be a field and K a subfield of D. Then for any finite set X, the natural homomorphism $D_K\langle X \rangle \rightarrow D_K\langle\langle X \rangle\rangle$ is honest.*

Proof. We have to show that any full matrix over $R = D_K\langle X \rangle$ remains full over $\overline{R} = D_K\langle\langle X \rangle\rangle$; by linearization it is stably associated to a matrix of the form $A = A_0 + \sum x_i A_i$, where $x_i \in X$ and A_0, A_i are matrices over D. Enlarging X if necessary, we may suppose that it contains an element x_0 not occurring in A. Now consider $x_0 A$; this is linear homogeneous in the variables $x_0, x_0 x_i$ ($i > 0$), which again form a free set, and it will be enough to show that $x_0 A$ remains full over $\overline{R}$ (which contains the power series ring on $x_0, x_0 x_i$). Thus we may assume A to be homogeneous of degree 1. Suppose that A is not full over $\overline{R}$ and take a rank factorization over $\overline{R}$:

$$A = BC, \tag{3}$$

where B is $n \times r$ say, C is $r \times n$ and $r < n$. Write $B = \sum B_i$, $C = \sum C_i$, where B_i, C_i are the homogeneous components of degree i. Then $B_0 C_0 = 0$, and B_0, C_0 are matrices over D, hence on replacing B_0 by an associate, we may assume

$$B_0 = \begin{pmatrix} I & 0 \\ 0 & 0 \end{pmatrix},$$

where I is $s \times s$, say. Since $B_0 C_0 = 0$, the first s rows of C_0 are now 0, and on replacing C_0 by an associate (which leaves B_0 unaffected), we find that

$$C_0 = \begin{pmatrix} 0 & 0 \\ 0 & I \end{pmatrix},$$

where I is $t \times t$ and since C_0 has r rows, $s + t \leq r$. Equating homogeneous components in (3), we have $A = B_0 C_1 + B_1 C_0$ or on writing B_1, C_1 in block form to conform with C_0, B_0 respectively:

$$A = \begin{pmatrix} C_1' & C_1'' \\ 0 & 0 \end{pmatrix} + \begin{pmatrix} 0 & B_1' \\ 0 & B_1'' \end{pmatrix} = \begin{pmatrix} C_1' & B_1' + C_1'' \\ 0 & B_1'' \end{pmatrix}.$$

Thus A has an $(n - s) \times (n - t)$ block of zeros. But

$$n - s + n - t = 2n - s - t \geqslant 2n - r > n,$$

so A is hollow; by Prop. 4.5.4 it is not full, which is a contradiction. ∎

To avoid complications we have taken X to be finite, as that is the only case needed, but with a little care the result can be proved generally.

Next we come to a property of inner ranks. As we saw in Prop. 1.4.3, over a weakly finite ring every invertible matrix is full; we shall be interested in matrices with an invertible submatrix:

LEMMA 6.2.3. *Let R be a weakly finite ring and consider a partitioned matrix over R:*

$$A = \begin{pmatrix} A_1 & A_2 \\ A_3 & A_4 \end{pmatrix}, \quad where\ A_1\ is\ r \times r.$$

If A_1 is invertible, then $rA \geqslant r$, with equality if and only if

$$A_3 A_1^{-1} A_2 = A_4.$$

Proof. Since the inner rank is unchanged by elementary transformations, we can make the following changes without affecting the inner rank:

$$A \to \begin{pmatrix} I & A_1^{-1}A_2 \\ A_3 & A_4 \end{pmatrix} \to \begin{pmatrix} I & A_1^{-1}A_2 \\ 0 & A_4 - A_3 A_1^{-1} A_2 \end{pmatrix} \to \begin{pmatrix} I & 0 \\ 0 & A_4 - A_3 A_1^{-1} A_2 \end{pmatrix}.$$

If $rA = s$, say, this last matrix can be written as

$$PQ = \begin{pmatrix} P' \\ P'' \end{pmatrix} (Q' \quad Q''),$$

where P' is $r \times s$ and Q' is $s \times r$. Thus $I = P'Q'$ and by weak finiteness I is full, so $r \leqslant s$. Thus $rA \geqslant r$; when equality holds, we have $Q'P' = I$, again by weak finiteness, but $P'Q'' = 0$, hence $Q'' = 0$ and similarly $P'' = 0$. Hence $A_4 - A_3 A_1^{-1} A_2 = 0$ and the conclusion follows; now the converse is clear. ∎

We shall also need a weak form of the specialization lemma, referring to a central indeterminate.

LEMMA 6.2.4. *Let D be a field which is a k-algebra over an infinite field k, and consider the polynomial ring $D[t]$ in a central indeterminate, with field of fractions $D(t)$. If $A = A(t)$ is a matrix over $D[t]$, then the rank of A over $D(t)$ is the maximum of the ranks of $A(\alpha)$, $\alpha \in k$, and this maximum is assumed for all but a finite number of values of α.*

Proof. By the diagonal reduction of matrices over the principal ideal domain $D[t]$ (see 10.5 of A.1 or 8.1 of FR) we can find invertible matrices P, Q over $D[t]$ such that

$$PAQ = \text{diag}(p_1, \ldots, p_r, 0, \ldots, 0), \text{ where } p_i \in D[t]. \qquad (4)$$

The product of the non-zero p_i gives us a polynomial f whose zeros in k are the only points of k at which $A = A(t)$ falls short of its maximum rank, and by the density principle this cannot happen for more than $\deg f$ values. ∎

We now come to the main result of this section:

THEOREM 6.2.5 (Specialization lemma). *Let D be a field which is a k-algebra over an infinite field k. Let E be an extension field of D, C the centralizer of D in E and assume that $E = CD$ and $[E:C] = \infty$. Then any full matrix over $E_C\langle X \rangle$ (for any set X) is invertible for some choice of values of X in D.*

Proof. Let $A = A(x)$ be a full $n \times n$ matrix over $E_C\langle X \rangle$ and denote by r the maximum of the ranks of $A(a)$ as a ranges over D^X. We have to show that $r = n$, so assume that $r < n$. Since only finitely many xs occur in A, we may take X finite. By a translation $x \mapsto x + a$ $(x \in X, a \in D)$ we may assume that the maximum rank is attained for $x = 0$; further, by interchanges of rows and of columns we may take the $r \times r$ principal minor of $A(0)$ to be invertible. Thus if

$$A(x) = \begin{pmatrix} B_1(x) & B_2(x) \\ B_3(x) & B_4(x) \end{pmatrix},$$

where B_1 is $r \times r$, then $B_1(0)$ is invertible. Given $a \in D^X$ and any $\lambda \in k$, we have $rA(\lambda a) \leq r$, hence by Lemma 2.4, the rank of $A(ta)$ over $E(t)$ is at most r, and the same holds over $E((t))$. Now $B_1(ta)$ is a polynomial in t with matrix coefficients and constant term $B_1(0)$, a unit, hence $B_1(ta)$ is invertible over the power series ring $E[\![t]\!]$. By Lemma 2.3, the equation

$$B_4(ta) = B_3(ta) B_1(ta)^{-1} B_2(ta) \qquad (5)$$

holds over $E[\![t]\!]$ for all $a \in D^X$. Therefore the matrix

$$B_4(tx) - B_3(tx) B_1(tx)^{-1} B_2(tx) \qquad (6)$$

vanishes when the elements of X are replaced by any values in D. Now (6) may be regarded as a power series in t whose coefficients are matrices over $E_C\langle X \rangle$, so the entries of these matrices are generalized polynomial identities or identically 0; hence by the GPI-theorem 6.A the expression

(6) vanishes as a matrix over $E_C\langle X\rangle[\![t]\!]$; putting $t = 1$, we obtain a matrix $A(x)$ over $E_C\langle\!\langle X\rangle\!\rangle$ of inner rank r, and since $r < n$, it is non-full. But this contradicts the fact that $A(x)$ is full over $E_C\langle X\rangle$ and the inclusion in $E_C\langle\!\langle X\rangle\!\rangle$ is honest. This shows that $r = n$ and so $A(a)$ is invertible for some $a \in D^X$. ∎

Taking $E = D$, $k = C$, we obtain an important special case:

COROLLARY 6.2.6. *Let D be a field with infinite centre C and suppose that $[D:C] = \infty$. Then every full matrix over $D_C\langle X\rangle$ is invertible for some choice of values of X in D.* ∎

For the specialization lemma to hold it is clearly necessary that C should be the precise centralizer of D. For C must centralize D to make the substitution possible, and if C were smaller, say there existed α in the centralizer of D but not in C, then $\alpha x - x\alpha$ would be non-zero and hence full, as 1×1 matrix, even though it vanishes for all values of x in D. Secondly, if $[E:C]$ were finite, there would be non-trivial identities over E, so again the lemma does not hold. On the other hand, it is not known whether it is essential to assume an infinite ground field.

One way in which the specialization lemma can be used is to specialize the variables in $D_k\langle\!\langle X\rangle\!\rangle$ so as to preserve certain elements. If E is a field containing D (and still a k-algebra), then any map $\alpha: X \to E$ can be extended to a homomorphism, again written α, of $D_k\langle X\rangle$ into E; if $\mathcal{P} = \text{Ker}\,\alpha$ is the singular kernel, then α can be extended to the localization of $D_k\langle X\rangle$ at $\mathcal{P}$. This yields the following result:

PROPOSITION 6.2.7. *Let $D \subseteq E$ be an extension of fields such that the centre k of E is contained in D, and assume that k is infinite and $[E:k] = \infty$. Then for any finite set of elements $p_1, \ldots, p_r$ of $D_k\langle\!\langle X\rangle\!\rangle$ there is a map $\alpha: X \to E$ such that the induced homomorphism α of $D_k\langle X\rangle$ into E is defined on $p_1, \ldots, p_r$. In particular, the intersection of all the singular kernels of the homomorphisms of $D_k\langle X\rangle$ into E is the set of all non-full matrices over $D_k\langle X\rangle$.*

Proof. By Prop. 5.4.3 we have a natural embedding $D_k\langle\!\langle X\rangle\!\rangle \to E_k\langle\!\langle X\rangle\!\rangle$, hence an honest homomorphism $D_k\langle X\rangle \to E_k\langle X\rangle$. Now let A_i be the denominator of p_i in some representation (see 4.2); then A_i is full over $D_k\langle X\rangle$, hence also over $E_k\langle X\rangle$, and so is $A = A_1 \oplus \ldots \oplus A_r$. By the specialization lemma (Th. 2.5) we can find $\alpha: X \to E$ such that A^α is invertible over E, hence each A_i^α is invertible over E and it follows that

α is defined on p_i for $i = 1, \ldots, r$. The last part follows because for each full matrix A we can find a singular kernel of a map to E which excludes A, so the intersection of these singular kernels is just the set of all non-full matrices. ∎

As a further application we prove a form of the normal basis theorem for Galois extensions. Let L/K be a finite Galois extension with Galois group G, and denote by G_0 the subgroup of inner automorphisms. Write $(G:G_0) = r$, denote the centre of L by C and recall that the algebra associated with G (see 3.3) is defined as

$$A = \{a \in L | a = 0 \text{ or } I(a) \in G\}.$$

It is clear that A is a C-algebra and if $[A:C] = s$, then $[L:K] = |G|_{\text{red}} = rs = n$, say, by Th. 3.3.7. Choose a transversal $\sigma_1, \ldots, \sigma_r$ of G_0 in G and a C-basis $v_1, \ldots, v_s$ of A; the set of rs elements $\sigma_i I(v_j)$ will be called a *reduced automorphism set* of L/K. By Cor. 3.3.3, the rs elements $\sigma_i I(v_j)$ form a right A-basis for the right A-space spanned by all the $\sigma I(a)$ ($\sigma \in G, a \in A$), for this corollary shows them to be linearly independent, and if $\sigma \in G$, then $\sigma = \sigma_i I(a)$, where $a = \sum \lambda_j v_j$ ($\lambda_j \in C$); hence for all $x \in L$,

$$x^\sigma = x^{\sigma_i} I(a) = a x^{\sigma_i} a^{-1} = \sum \lambda_j v_j x^{\sigma_i} a^{-1} = \sum v_j x^{\sigma_i} v_j^{-1} b_j,$$

where $b_j = \lambda_j v_j a^{-1} \in A$. It follows that any element σ of G can be expressed in the form

$$\sigma = \sigma_i I(a) = \sum_j \sigma_i I(v_j) \quad \text{where } b_j \in A \text{ and } i \text{ depends on } \sigma. \tag{7}$$

As in the commutative case we have the following criterion for a set to be a K-basis of L:

LEMMA 6.2.8. *Let L/K be a Galois extension of degree n with reduced automorphism set $\theta_1, \ldots, \theta_n$. Then a given set $\{u_1, \ldots, u_n\}$ of n elements of L forms a left K-basis of L if and only if the matrix $P = (p_{ij})$, where $p_{ij} = u_i^{\theta_j}$, is non-singular.*

Proof. The n elements $u_1, \ldots, u_n$ form a basis if and only if they are linearly independent over K. If there is a dependence relation $\sum a_i u_i = 0$, where the $a_i \in K$ are not all 0, then $\sum a_i u_i^{\theta_j} = 0$ for all j, so P is then singular. Conversely, assume that the u_i form a basis. By Dedekind's lemma (Th. 3.3.2), the θ_j are linearly independent, thus for any $x \in L$,

$$\sum x^{\theta_j} c_j = 0 \quad (c_j \in L) \Rightarrow c_1 = \ldots = c_n = 0. \tag{8}$$

Now $x = \sum a_i u_i$ for suitable $a_i \in K$, hence (8) is equivalent to

$$\sum u_i^{\theta_j} c_j = 0 \Rightarrow c_j = 0, \; j = 1, \ldots, n,$$

and this shows P to be non-singular. ∎

For our purpose we need to transform the specialization lemma as follows:

LEMMA 6.2.9. *Let L/K be a Galois extension of degree n with reduced automorphism set $\theta_1, \ldots, \theta_n$, and denote the centralizer of K in L by H. Further, assume that L, K are k-algebras, where the field k is infinite and $L = HK$, $[L:H] = \infty$. If $A = A(x_1, \ldots, x_n)$ is a full matrix over $L_H\langle x_1, \ldots, x_n \rangle$, then $A(u^{\theta_1}, \ldots, u^{\theta_n})$ is non-singular for some $u \in L$.*

Proof. Let $u_1, \ldots, u_n$ be a left K-basis of L and define a matrix B by

$$B(y_1, \ldots, y_n) = A\left(\sum y_i u_i^{\theta_1}, \ldots, \sum y_i u_i^{\theta_n}\right). \tag{9}$$

Suppose that $A(u^{\theta_1}, \ldots, u^{\theta_n})$ is singular for all $u \in L$; the lemma will follow if we show that $A(x_1, \ldots, x_n)$ is non-full. Now by (9), under the hypothesis stated, $B(a_1, \ldots, a_n)$ is singular for all $a_i \in K$, so by the specialization lemma (Th. 2.5), $B(y_1, \ldots, y_n)$ is not full over $L_H\langle y_1, \ldots, y_n \rangle$; but the equations

$$x_j = \sum y_i u_i^{\theta_j} \tag{10}$$

can be solved for the y_i because the matrix $(u_i^{\theta_j})$ is non-singular, by Lemma 2.8. Hence the matrix $A(x_1, \ldots, x_n) = B(y_1, \ldots, y_n)$ is non-full, as we wished to show. ∎

We now come to the normal basis theorem. We recall that a *normal basis* of a Galois extension of commutative fields is a basis consisting of the conjugates of a single element, and the existence of such a normal basis is well known in that case (see e.g. A.3, Th. 5.7.5, p. 205).

THEOREM 6.2.10. *Let L/K be a finite Galois extension and assume that L, K are algebras over an infinite field k, and that $L = HK$, $[L:H] = \infty$, where H is the centralizer of K in L. Then L/K has a normal basis.*

Proof. Let $\theta_1, \ldots, \theta_n$ be a reduced automorphism set for L/K and let $A(G)$ be the associated algebra (G being the group of L/K). From Lemma 2.8 we know that the elements $u_1, \ldots, u_n \in L$ form a basis if and only if the matrix $(u_i^{\theta_j})$ is non-singular, so we have to find $u \in L$ such that the matrix $(u^{\theta_i\theta_j})$ is non-singular. By (7) we have

$$\theta_i\theta_j = \sum \theta_h b_{hij}, \quad \text{where } b_{hij} \in A(G). \tag{11}$$

Let us consider the matrix $A = (a_{ij})$ over $L_H\langle x_1, \ldots, x_n\rangle$, where

$$a_{ij} = \sum x_h b_{hij}. \tag{12}$$

We can partition A into blocks corresponding to the transversal $\sigma_1, \ldots,$ σ_r of G_0 in G that was used for the set $\{\theta_i\}$. Thus in block form we have $A = (A_{\lambda\mu})$, where $\lambda, \mu = 1, \ldots, r$. If θ_i, θ_j are represented by σ_λ, σ_μ respectively and $\sigma_\lambda\sigma_\mu \equiv \sigma_v \pmod{G_0}$, then in (11), $b_{hij} = 0$ unless θ_h is represented by σ_v. Hence when A is written in block form, it is a *monomial matrix*, i.e. each row has just one non-zero block $A_{\lambda\mu}$ and likewise each column. To determine when A is singular we can therefore confine our attention to a single block.

We claim that $A(1 \ 0 \ \ldots \ 0) = (b_{1ij})$ is non-singular. For if it were singular, then in some row every block would be singular, i.e. there would exist a suffix i and elements $c_j \in A(G)$, not all 0, such that

$$\sum b_{1ij}c_j = 0, \tag{13}$$

and here all the θ_j for which $b_{1ij} \neq 0$ belong to the same σ_μ. Hence

$$\sum \theta_j c_j = \theta = \sigma_\mu I(v), \quad \text{where } v \in A(G).$$

Now the relation $(ab)^\theta = a^\theta b^\theta$, where $a, b \in L$, can be written as an operator equation, $b\theta = \theta b^\theta$, and by (11),

$$\theta_i\theta = \sum \theta_i\theta_j c_j = \sum \theta_h b_{hij}c_j.$$

This is independent of θ_1 by (13), so for any $a_i \in L$,

$$\sum \theta_i a_i \theta = \sum \theta_i \theta a_i^\theta = \sum_{h=2}^n \theta_h b_{hi} a_i^\theta, \text{ where } b_{hi} = \sum b_{hij}c_j.$$

This means that θ is a semilinear transformation of the right L-space spanned by $\theta_1, \ldots, \theta_n$, whose image is a proper subspace. But θ as an automorphism of L is invertible, so we have a contradiction, and this shows that (b_{1ij}) is non-singular. Similarly for the other blocks, hence $A = A(x_1, \ldots, x_n)$ must be full over $L_H\langle x_1, \ldots, x_n\rangle$. By Lemma 2.9 there exists $u \in L$ such that $A(u^{\theta_1}, \ldots, u^{\theta_n})$ is non-singular; this is the matrix (p_{ij}), where

$$p_{ij} = \sum u^{\theta_h} b_{hij}.$$

But this is just $u^{\theta_i\theta_j}$, by (11), and it is what we had to show. ∎

Exercises

1. Find the modifications needed to prove Prop. 2.2 in the case when X is infinite (distinguish the case where the degrees of $x \in X$ are bounded).

2. Find a counter-example for Lemma 2.3 when R is not weakly finite.

3. The GPI-theorem may be stated more generally as follows: Let E be a field, H a subring and K its centralizer. If for all $a \in E^\times$ the left K-space KaH is infinite-dimensional over K, then any non-zero element of $E_K\langle X \rangle$ has a non-zero value for some choice of values of X in H (see Cohn [89″]). Use it to show that if H is a semifir with universal field of fractions D such that the centre C of D is infinite, and H contains C and is of infinite degree over C, then any full matrix over $H_C\langle X \rangle$ is full for some choice of values of X in H.

4. Let k be any commutative field and $U = k \langle\!\langle X \rangle\!\rangle$ the free field on a set X over k. Show that no element of $U \backslash k$ can be algebraic over k. (Hint. Put $F = k(t)$ and let E be the Hilbert field on F; if $u \in U \backslash k$, fix $\lambda \in k$, take a specialization such that $u \neq \lambda$ and use the fact that k is relatively algebraically closed in E, see Ex. 7 of 2.3.)

5. Let K be a skew field whose multiplicative group is simple; verify that its centre is $\mathbf{F}_2$. Show that there exists such a field of any infinite cardinal. (Hint. Form $U = \mathbf{F}_2 \langle\!\langle X \rangle\!\rangle$ with a set X of the given cardinal, apply Cor. 5.5.2 to embed it in a homogeneous field and use Ex. 4.)

6.3 Normal forms for matrices over a tensor ring

When we are given a specialization φ of a free field U, to determine whether φ is defined at an element p of U we need to examine the image under φ of a denominator for p, as is made clear by the proof of Prop. 2.7. This makes it of interest to reduce the denominator to the simplest possible form; we shall accomplish this objective by finding a form for matrices over the tensor ring which is unique up to stable association, with a minimal representative in its class, unique up to association.

Let E be any field and D a subfield and consider the tensor E-ring $R = E_D\langle X \rangle$ on a set X, and its universal field of fractions $U = E_D \langle\!\langle X \rangle\!\rangle$. We are interested in full matrices over R, occurring as denominators of elements of U, but to begin with we take any matrix over R, not necessarily square. As a first simplification we shall reduce our matrix to be linear in X. By linearization by enlargement, any $m \times n$ matrix over $E_D\langle X \rangle$ can be reduced to the form

$$A = A_0 + \sum_1^d A_i' x_i A_i'', \tag{1}$$

where the x_i are distinct elements of X and A_0, A_i', A_i'' are matrices with entries in E. In the first place such a reduction gives r_1 terms $a_\nu' x_1 a_\nu''$, say, where the a_ν' are columns and the a_ν'' are rows; we combine the a_ν' to an $m \times r_1$ matrix A_1' and the a_ν'' to an $r_1 \times n$ matrix A_1'', and similarly for the other variables. Thus in (1), A_0 is $m \times n$, A_i' is $m \times r_i$ and A_i'' is $r_i \times n$. We may further assume that the columns of A_i' are linearly independent over D and likewise for the rows of A_i'', for when this is not so, we can make a transformation which will reduce r_i.

Any matrix A over $E_D\langle X \rangle$ is called *linear* if it has the form (1), where A_0, A_i', A_i'' have entries in E; if the r_i are minimal, A is said to be *reduced*. From what has been said it follows that when A is reduced, the columns of each A_i' are right linearly independent and the rows of each A_i'' are left linearly independent over D. The matrix A of the form (1) is called *right monic* if it is linear, reduced and A_1', $\ldots$, A_d' are right comaximal, i.e. $(A_1', \ldots, A_d')$ has a right inverse; A is *left monic* if it is linear reduced and A_1'', $\ldots$, A_d'' are left comaximal, and a *monic* matrix is one which is left and right monic.

In a right monic matrix A the terms of highest degree are left regular. For assume that A has the form (1) and suppose that

$$P\left(\sum A_i' x_i A_i'' \right) = 0.$$

Since the rows of each A_i'' are left linearly independent, we can equate the left cofactors of x_i and obtain $PA_i' = 0$. Now by hypothesis there exists a $\sum r_i \times m$ matrix T such that $(A_1', \ldots, A_d')T = I$, hence $P = P(A_1', \ldots, A_d')T = 0$. It follows that A itself is left regular.

The following lemma represents the analogue of the weak algorithm for matrices. For any matrix A we shall denote by $\bar{A}$ the terms of highest degree in A, and write $d(A)$ for the degree of A in the x_i.

LEMMA 6.3.1. *Let* $R = E_D\langle X \rangle$ *be a tensor* E-*ring, and consider a linear matrix* A *over* R *such that* $\bar{A}$ *is right regular, and a right monic matrix* B *over* R. *If* P, Q *are matrices over* R *such that*

$$PB = AQ, \tag{2}$$

then there exists a matrix C *over* R *such that*

$$P = AC + P_0, \quad Q = CB + Q_0 \quad \text{and } P_0 B = AQ_0, \tag{3}$$

where P_0, Q_0 *are matrices over* E.

Proof. Equating highest terms in (2), we obtain

$$\bar{P}\bar{B} = \bar{A}\bar{Q}. \tag{4}$$

If $d(Q) = 0$, then deg $P = 0$ by (4), because B is right monic and so $\bar{B}$ is left regular; so the result follows in this case. We may now assume that $d(Q) \geqslant 1$ and write $\bar{Q} = \sum F_i x_i Q_i$, where F_i is over R and Q_i over E. If $\bar{B}$ in reduced form is $\bar{B} = \sum B_i' x_i B_i''$, then (4) becomes

$$\sum \bar{P} B_i' x_i B_i'' = \sum \bar{A} F_i x_i Q_i. \tag{5}$$

By taking the number of columns in F_1 to be minimal we may also assume that these columns are right linearly independent over D. If we now equate the terms with last factor x_1 in (5), we obtain

$$\bar{P} B_1' x_1 B_1'' = \bar{A} F_1 x_1 Q_1.$$

Since B is reduced, the rows of B_1'' are left linearly independent over D; hence we have $Q_1 = V_1 B_1''$ for some matrix V_1 over D and so

$$\bar{P} B_1' = \bar{A} F_1 V_1.$$

Similarly, equating terms with x_i as last factor we obtain a corresponding equation for $\bar{P} B_i'$ and so find that

$$\bar{P}(B_1', \ldots, B_d') = \bar{A}(F_1 V_1, \ldots, F_d V_d).$$

Let T be a right inverse of $(B_1', \ldots, B_d')$ and put $C = (F_1 V_1, \ldots, F_d V_d)T$. Then T is a matrix over E and we have

$$\bar{P} = \bar{P}(B_1', \ldots, B_d')T = \bar{A}C. \tag{6}$$

Now by (2)

$$(P - AC)B = A(Q - CB);$$

this is an equation of the same form as (2), but $P - AC$ has lower degree than P, by (6), and since $\bar{A}$ is right regular, $Q - CB$ has lower degree than Q, so the result follows by induction on $d(Q)$. ∎

We now have the following normal form theorem for matrices over a tensor ring:

THEOREM 6.3.2. *Let $R = E_D\langle X \rangle$ be a tensor E-ring. Any matrix over R is stably associated to a matrix $A \oplus 0$, where A is monic. Moreover, if $A \oplus 0$, $A' \oplus 0$ are two matrices that are stably associated and such that A, A' are monic, then the numbers of rows of A, A' agree, as do those of $A \oplus 0$, $A' \oplus 0$, and likewise for columns, and there exist invertible matrices P, Q over E such that $PA' = AQ$.*

A linear matrix stably associated to a matrix T is also called a *linear companion* of T.

Proof. By the linearization process we reach a linear companion for our matrix, of the form $A \oplus 0$, where A is an $m \times n$ matrix of the reduced form (1). Suppose that $m + n$ is minimal for the given matrix; we claim that A is then monic. For if $(A'_1, \ldots, A'_d)$ has no right inverse, then its rank is less than m and so by elementary row operations its last row can be reduced to zero. If now the last row of A_0 is also 0, we can reduce m by 1, contradicting the minimality; this amounts to writing A as $B \oplus N$, where N is the 1×0 zero matrix. Hence the last row of A_0 is not 0, and by further row operations over R and column operations over E we find that A is associated to $A' \oplus 1$, where A' is again linear. So A is stably associated to A', but this again contradicts the minimality of $m + n$. Hence A is right monic; by symmetry it is also left monic, so the existence of a monic form (1) is established. We remark that if A is invertible, it is stably associated to the 0×0 matrix.

Suppose now that A, A' are both monic and $A \oplus 0$, $A' \oplus 0$ are stably associated; then $F \oplus 0$ and $F' \oplus 0$ are associated, where $F = A \oplus I_r$, $F' = A' \oplus I_s$ for suitable r and s. Thus there exist invertible matrices U, V such that for appropriate partitioning,

$$\begin{pmatrix} U_1 & U_2 \\ U_3 & U_4 \end{pmatrix}\begin{pmatrix} F & 0 \\ 0 & 0 \end{pmatrix} = \begin{pmatrix} F' & 0 \\ 0 & 0 \end{pmatrix}\begin{pmatrix} V_1 & V_2 \\ V_3 & V_4 \end{pmatrix}. \tag{7}$$

Equating (2, 1)-blocks, we find that $U_3 F = 0$, while the (1, 2)-blocks show that $F'V_2 = 0$. Now A, A' are regular, hence so are F, F' and therefore $U_3 = 0 = V_2$. Let F be $m \times n$, F' $m' \times n'$ and $F \oplus 0$, $F' \oplus 0$ both $t \times u$. Then U_3 is $(t - m') \times m$, and since U is full, we must have $t - m' + m \leq t$, i.e. $m \leq m'$ (by Prop. 4.5.4). Similarly V_2 is $n' \times (u - n)$, so $n' + u - n \leq u$, and hence $n' \leq n$. By symmetry we have $m' \leq m$, $n \leq n'$, and so F and F' are both $m \times n$. It follows that U_1, U_4, V_1, V_4 are all square and are invertible, since this is true of U, V. Moreover, $U_1 F = F'V_1$, so F and F' are associated and A, A' are stably associated. By Th. 1.5.1 we thus have a comaximal relation $AB' = BA'$, and there exist matrices C, D, C', D' over R such that

$$\begin{pmatrix} A & B \\ C & D \end{pmatrix} \text{ and } \begin{pmatrix} D' & -B' \\ -C' & A' \end{pmatrix} \text{ are mutually inverse.} \tag{8}$$

Further, by Lemma 3.1 there exists P such that

$$B = AP + B_0, \quad B' = PA' + B'_0,$$

where B_0, B'_0 are over E. Hence on multiplying the matrices (8) by $\begin{pmatrix} I & -P \\ 0 & I \end{pmatrix}$ on the right and by $\begin{pmatrix} I & P \\ 0 & I \end{pmatrix}$ on the left, respectively, we obtain a pair of mutually inverse matrices

$$\begin{pmatrix} A & B_0 \\ C & D_1 \end{pmatrix} \text{ and } \begin{pmatrix} D_1' & -B_0' \\ -C' & A' \end{pmatrix}.$$

We also have $A'C = C'A$, so by another application of Lemma 3.1 we obtain a pair of inverse matrices

$$\begin{pmatrix} A & B_0 \\ C_0 & D_2 \end{pmatrix} \text{ and } \begin{pmatrix} D_2' & -B_0' \\ -C_0' & A' \end{pmatrix},$$

where C_0, C_0' are over E. Thus we have the equation

$$AD_2' - B_0C_0' = I.$$

Equating highest terms we find $\sum A_i' x_i A_i'' \overline{D_2'} = 0$. Taking right cofactors of x_i we have $A_i'' \overline{D_2'} = 0$, and since $(A_1'', \ldots, A_d'')^{\mathrm{T}}$ is right regular, we conclude that $\overline{D_2'} = 0$; hence $D_2' = 0$ and so $-B_0C_0' = I$. Thus B_0 has a right inverse over E, and by symmetry B_0' has a left inverse. Hence, denoting the index of a matrix A by $i(A)$, we have $i(B) \geqslant 0 \geqslant i(B')$; since A, A' are stably associated, they have the same index, hence so do B, B' and therefore $i(B) = i(B') = 0$. It follows that B_0, B_0' are invertible over E, and we have $AB_0' = B_0A'$, as claimed. ∎

The proof shows that once we have reached a linear form for our matrix, if the latter is right regular, we can achieve left monic form by elementary row operations over E and column operations over R, and similarly on the other side. Thus we obtain

COROLLARY 6.3.3. *If A is a right regular linear matrix over $R = E_D\langle X \rangle$, then there exists an invertible matrix P over E and an invertible matrix T over R such that $PAT = B \oplus I$, where B is left monic. Moreover, the entries of P and the coefficients in the entries of T can be chosen to lie in the subfield of E generated by the coefficients in the entries of A.* ∎

Next we derive a bound on the degrees of the factors in a matrix factorization. For any matrix $C = (c_{ij})$ over $R = E_D\langle X \rangle$ we write again $d(C) = \max \{d(c_{ij})\}$, where d denotes the degree. In the proof we shall need the weak algorithm. We recall from Ch. 2 of FR that a family (a_i) of elements of R is called *right d-dependent* if some $a_i = 0$ or there exist $b_i \in R$, almost all 0, such that

$$d\left(\sum a_i b_i\right) < \max \{d(a_i) + d(b_i)\}.$$

An element a is said to be *right d-dependent on* (a_i) if $a = 0$ or if there exist $b_i \in R$, almost all 0, such that

$$d\left(a - \sum a_i b_i\right) < d(a), \qquad d(a_i) + d(b_i) \le d(a).$$

Now the *weak algorithm* for the tensor ring R asserts that in any right d-dependent family some element is right d-dependent on the elements of lower degree in the family. As a consequence any right ideal of R has a right d-independent basis (FR, Th. 2.2.4); here this result will only be needed for finitely generated ideals.

LEMMA 6.3.4. *Let* $R = E_D\langle X \rangle$ *be the tensor E-ring over D on X and let* $C \in {}^m R^n$. *If*

$$C = AB \tag{9}$$

is a factorization of C, where A is $m \times r$ and right regular and B is $r \times n$ and left regular, then there is an invertible matrix U_0 over R such that $d(AU_0) \le d(C)$.

Proof. Write $C = (c_{ij})$, $A = (a_{i\lambda})$, $B = (b_{\lambda j})$ and consider the tensor ring $S = E_D\langle Z \rangle$, where $Z = X \cup \{y_1, \ldots, y_m\}$ with new variables y_i. The embedding $R \to S$ is honest, since R is a retract of S. Consider the right ideal $\mathfrak{a}$ of S generated by $p_\lambda = \sum y_i a_{i\lambda}$ $(\lambda = 1, \ldots, r)$. Since A is right regular, $\mathfrak{a}$ is free on the p_λ. By the weak algorithm we can find a right d-independent basis $q_1, \ldots, q_r$ of $\mathfrak{a}$ with

$$d(q_1) \le \ldots \le d(q_r).$$

Since (p_λ) and (q_λ) are two bases of $\mathfrak{a}$, we have

$$p_\lambda = \sum y_i a_{i\lambda}, \qquad q_\lambda = \sum y_i a'_{i\lambda},$$

where A and $A' = (a'_{i\lambda})$ are right associated, and we have $C = AB = A'B'$, where $B' = (b'_{\lambda j})$ is left associated to B. We claim that $d(q_r) \le 1 + d(C)$; for if $d(q_r) > 1 + d(C)$, suppose further that $b'_{rj} \ne 0$ for some j. Since $f_j = \sum y_i c_{ij}$ can be written as $f_j = \sum q_\lambda b'_{\lambda j}$, we have

$$d\left(\sum q_\lambda b'_{\lambda j}\right) = d(f_j) \le 1 + d(C) < d(q_r) \le d(q_r) + d(b'_{rj}).$$

This shows that the q_λ are right d-dependent, a contradiction, unless $b'_{rj} = 0$ for all j, which in turn contradicts the fact that B' like B is left regular. This then shows that $d(q_r) \le 1 + d(C)$, and therefore $d(A') \le d(C)$. Now $A' = AU$ for an invertible matrix U over S. Setting $y_i = 0$, we obtain a homomorphism from S to R which maps U to an invertible matrix U_0 and does not raise the degree, hence $d(AU_0) \le d(C)$. ∎

We can now achieve our aim, of proving a factorization theorem for linear matrices over tensor rings. A matrix factorization PQ is said to be *proper* if P has no right inverse and Q has no left inverse. If a matrix with a proper factorization PQ is left regular, it follows that P has no left inverse, either. For if $P'P = I$, then

$$(PP' - I)PQ = P(P'P - I)Q = 0,$$

hence $PP' = I$, but this contradicts the fact that P has no right inverse.

THEOREM 6.3.5. *Let $R = E_D\langle X \rangle$ and consider a linear matrix C over R which is left regular. If*

$$C = FG \tag{10}$$

is a proper factorization of C, where F is right and G is left regular, then there exist invertible matrices U, W over E and V over R such that

$$UFV = \begin{pmatrix} A & 0 \\ 0 & I \end{pmatrix}, \quad V^{-1}GW = \begin{pmatrix} I & 0 \\ P & B \end{pmatrix}, \quad UCW = \begin{pmatrix} A & 0 \\ P & B \end{pmatrix}.$$

Proof. Let (10) be the given factorization of C; here F is regular, because it is right regular and C is left regular. Suppose that F is $m \times r$ and G is $r \times n$. By Lemma 3.4 we can make an internal modification such that F has degree at most 1. Since a regular matrix over E is a unit, F must have degree exactly 1. By Cor. 3.3 there exist $U \in \mathbf{GL}_m(E)$, $V \in \mathbf{GL}_r(R)$ such that

$$UFV = A \oplus I, \tag{11}$$

where A is left monic $(m - s) \times (r - s)$. Hence by suitably partitioning $V^{-1}G$, we can rewrite (10) as

$$UC = \begin{pmatrix} A & 0 \\ 0 & I \end{pmatrix}\begin{pmatrix} G' \\ G'' \end{pmatrix}.$$

Now UC is linear, hence G'' is linear and G' has degree 0, since otherwise the leading term of G' would be in the right annihilator of $\bar{A}$, contradicting the fact that A is left monic (and so $\bar{A}$ is right regular). The matrix G' has $t = r - s$ rows and is left regular (because G is left regular), hence it has a right inverse over E, i.e. there exists $W \in \mathbf{GL}_n(E)$ such that $G'W = (I_t, 0)$. It follows that

$$V^{-1}GW = \begin{pmatrix} G' \\ G'' \end{pmatrix}W = \begin{pmatrix} I & 0 \\ P & B \end{pmatrix}, \quad UCW = \begin{pmatrix} A & 0 \\ 0 & I \end{pmatrix}\begin{pmatrix} I & 0 \\ P & B \end{pmatrix},$$

and this, together with (11) is the required form. ∎

We shall mainly need the following consequence. We recall that an $n \times n$ matrix is called *hollow* if it has an $r \times s$ block of zeros, where $r + s > n$. As we saw in Prop. 4.5.4, such a matrix cannot be full.

COROLLARY 6.3.6. *A linear square matrix over* $R = E_D\langle X \rangle$ *which is not full is associated over E to a linear hollow matrix.*

Proof. Let $C \in R_n$, $C = FG$, where F is $n \times r$, G is $r \times n$ and $r < n$. We may assume C left regular since otherwise it is left associated to a matrix with a zero row (and so hollow). By Th. 3.5 there exist $U, W \in \mathbf{GL}_n(E)$ such that

$$UCW = \begin{pmatrix} A & 0 \\ P & B \end{pmatrix},$$

where A is $s \times (r + s - n)$ and B is $(n - s) \times (2n - r - s)$. Hence the zero block is $s \times (2n - r - s)$ and $2n - r - s + s = 2n - r > n$. ∎

These results allow us to express denominators for the free field in monic form:

THEOREM 6.3.7. *Let* $R = E_D\langle X \rangle$ *be the tensor ring on X and* $U = E_D \langle\!\langle X \rangle\!\rangle$ *its universal field of fractions. Then every element of U can be described by a reduced block with a universal denominator,*

$$\begin{pmatrix} u & c \\ A & v \end{pmatrix} \tag{12}$$

with a monic matrix A, determined up to unit factors over E.

Proof. Let $f \in U$ and suppose f is represented by the block (12). Here A may be replaced by any matrix stably associated to it. For if P, Q are invertible matrices of the same order as A, then the block

$$\begin{pmatrix} uQ & c \\ PAQ & Pv \end{pmatrix}$$

defines the same element as (12), and so does

$$\begin{pmatrix} u & 0 & c \\ A & 0 & v \\ 0 & I & 0 \end{pmatrix}, \tag{13}$$

while for $A = B \oplus I$, we can partition u, v accordingly and reduce (12) as follows:

$$
\begin{pmatrix} u' & u'' & c \\ B & 0 & v' \\ 0 & I & v'' \end{pmatrix} \rightarrow \begin{pmatrix} u' & 0 & c - u''v'' \\ B & 0 & v' \\ 0 & I & v'' \end{pmatrix} \rightarrow \begin{pmatrix} u' & 0 & c - u''v'' \\ B & 0 & v' \\ 0 & I & 0 \end{pmatrix}
$$

$$
\rightarrow \begin{pmatrix} u' & c - u''v'' \\ B & v' \end{pmatrix},
$$

where the last step is inverse to the step passing from (12) to (13). Thus by appropriate operations we can transform a block for $f \in U$ to another block of this form in which A is monic, using Th. 3.2, and hence reduced. Since R is a semifir, we can use Th. 4.7.1 to conclude that A is determined up to stable association and by Th. 3.2, if A, A' are monic matrices in two blocks representing f, then

$$
PA' = AQ, \tag{14}
$$

for invertible matrices P, Q over E. By Th. 4.7.3 the denominator is universal. ∎

By using the specialization lemma we obtain a representation of elements of U by power series:

THEOREM 6.3.8. *Let K be a field with centre C such that C and $[K:C]$ are infinite. Then every element of $K_C \langle\!\langle x_1, \ldots, x_d \rangle\!\rangle$ can be represented as a power series in $x_i - \alpha_i$ for suitable $\alpha_i \in K$.*

Proof. Let $f \in U$, the universal field of fractions, and take a block (12) representing A, with A in monic form, say

$$
A = A_0 + \sum A_i' x_i A_i''. \tag{15}
$$

By the specialization lemma A is non-singular for some values α_i of x_i in K. Replacing x_i by $y_i = x_i - \alpha_i$, we obtain a representation (15) where A is non-singular for $y_i = 0$ and by passing to an associate, we may take A in the form $I - B$, where B is homogeneous of degree 1 in $Y = \{y_i\}$. Now over $K_C \langle\!\langle Y \rangle\!\rangle$ f takes the form

$$
f = c - uA^{-1}v = c - \sum uB^v v. \quad \blacksquare
$$

Exercises

(In Ex. 1–3, $R = E_D \langle X \rangle$ denotes the tensor ring, as in the text.)

1. Show that a linear homogeneous left regular matrix is right monic.

2. Use Th. 3.5 to show that a monic full matrix A over R is an atom if and only if

A is not associated over E to a matrix in block triangular form. Obtain a bound on the length of factorizations of monic full $n \times n$ matrices over R in terms of n.

3. Give an example of a monic matrix over R which is not full.

6.4 Free fields

Our main concern in this section is to investigate the relation between free fields with related base fields. But we begin by making use of the specialization lemma to prove the existence of free fields without using the theory of Ch. 4.

THEOREM 6.4.1. *Let E be a field with centre C and consider the tensor ring $R = E_C\langle X \rangle$ on any set X. Then there exists an epic R-field U with an honest embedding $R \to U$, hence U is the free field $E_C \{\!\!\{ X \}\!\!\}$.*

Proof. We shall prove the result by finding an honest homomorphism from R to an ultrapower of E. Suppose first that $[E:C] = \infty$ and that C is infinite, and consider the mapping

$$R = E_C\langle X \rangle \to E^{E^X}, \qquad (1)$$

where $p \in R$ is mapped to (p_f), with $p_f = p(xf)$, for $f \in E^X$. With each square matrix A over R we associate a subset $\mathcal{D}(A)$ of E^X, its *singularity support*, defined by

$$\mathcal{D}(A) = \{ f \in E^X | A(xf) \text{ is invertible over } E \}.$$

By Cor. 2.6 $\mathcal{D}(A) \neq \varnothing$ for any full matrix A, while for a non-full matrix $\mathcal{D}(A) = \varnothing$. For any invertible matrices P, Q over E, $P \oplus Q$ is again invertible, so $A(x) \oplus B(x)$ becomes singular precisely when $A(x)$ or $B(x)$ becomes singular, hence we have

$$\mathcal{D}(A) \cap \mathcal{D}(B) = \mathcal{D}(A \oplus B).$$

This shows that the family of sets $\mathcal{D}(A)$, where A ranges over all full matrices over R, is closed under finite intersections. Hence this family is contained in an ultrafilter $\mathcal{F}$ on E^X (see 1.2), and the homomorphism (1) gives rise to a homomorphism to an ultrapower:

$$R \to E^{E^X}/\mathcal{F}. \qquad (2)$$

As an ultrapower of E the right-hand side is again a field, and every full matrix A over R is invertible on $\mathcal{D}(A)$ and so is invertible in the ultrapower. Thus (2) is an honest homomorphism, so the subfield generated by the image of R is the desired epic R-field U.

There remains the case where C or $[E:C]$ is finite. We take indeterminates r, s, t and define $E_1 = E(r)$, $E_2 = E_1(s)$. On E_2 we have the automorphism α: $f(s) \mapsto f(rs)$, of infinite order, with fixed field E_1. We now form the skew function field $F = E_2(t; \alpha)$; by Th. 2.2.10 the centre of F is the centre of E_1, namely $C(r)$. This is infinite and $[F:C(r)] = \infty$, because the powers s^n are clearly linearly independent. The natural homomorphism $E_C\langle X \rangle \to F_{C(r)}\langle X \rangle = F^*_{E_1} E_{1_{C(r)}}\langle X \rangle$ is honest, by Prop. 5.4.4 and Lemma 5.4.2. We now construct the free field $F_{C(r)} \langle\!\langle X \rangle\!\rangle$ by the first part of the proof, and this is a field over which every full matrix over $E_C\langle X \rangle$ becomes invertible. ∎

Let us return to $E_C\langle X \rangle$ and suppose now that C is a central subfield of E. Given any subfield D of E, let us write $k = D \cap C$; then we have a natural homomorphism of tensor rings

$$D_k\langle X \rangle \to E_C\langle X \rangle, \tag{3}$$

and this map is injective, for the subring of $E_C\langle X \rangle$ generated by D and X is determined by the defining relations $ax = xa$ for all $x \in X$, $a \in D \cap C = k$, and hence is isomorphic to $D_k\langle X \rangle$. We shall be interested to know when there is a corresponding homomorphism of free fields:

$$D_k \langle\!\langle X \rangle\!\rangle \to E_C \langle\!\langle X \rangle\!\rangle. \tag{4}$$

Clearly this is so if and only if the map (3) is honest, but that is not always the case. We shall find that it is so precisely when C and D are *linearly disjoint* in E over k, i.e. when the natural homomorphism

$$C \otimes_k D \to E$$

is injective. But we shall need some preliminary results.

LEMMA 6.4.2. *Let R be a k-algebra and E/k any field extension (possibly skew) of finite degree. If R is a right Ore domain and $R_E = R \otimes_k E$ is an integral domain, then R_E is again a right Ore domain.*

Proof. Let K be the field of fractions of R and consider K_E; this is of finite degree over K, because $[K_E:K] = [E:k]$, so if we can prove that it is a domain, it must be a field. Now every element of K_E has the form $p = \sum \lambda_i \otimes a_i b^{-1}$, where $\lambda_i \in E$, $a_i \in R$, $b \in R^\times$. Hence $p = ub^{-1}$, where $u = \sum \lambda_i \otimes a_i \in R_E$. We conclude that $R^\times$ is a right Ore set in R_E: given $u \in R$, $b \in R^\times$, we have $b^{-1}u = u_1 b_1^{-1}$ for some $u_1 \in R_E$, $b_1 \in R^\times$, hence $ub_1 = bu_1$ and clearly $u_1 \neq 0$ if $u \neq 0$. It follows that K_E is a domain: given ub^{-1}, $vc^{-1} \neq 0$ in K_E, we have $b^{-1}v = v_1 b_1^{-1}$ say, hence

$ub^{-1}vc^{-1} = uv_1b_1^{-1}c^{-1} = uv_1(cb_1)^{-1}$, which is not 0 because $uv_1 \neq 0$. Thus K_E is a domain, hence a field, and so R_E is a right Ore domain. ∎

Next we examine the effect of a scalar extension of the coefficient field of a tensor ring:

LEMMA 6.4.3. *Let D be a field which is a k-algebra and let C/k be a commutative field extension. If $D \otimes_k C$ is a domain, then (i) $D \otimes_k C$ is an Ore domain and (ii) we have*

$$(D \otimes_k C)_C\langle X \rangle \cong D_k\langle X \rangle \otimes_k C. \tag{5}$$

Proof. To prove (i) it is enough to take the case where C is finitely generated over k, because the Ore condition involves only finitely many elements of C. Thus let C be a finite algebraic extension of $C_0 = k(T)$, where $T = \{t_1, \ldots, t_r\}$ is a finite set of indeterminates. By the Hilbert basis theorem the ring $D \otimes k[T] = D[T]$ is Noetherian, and hence an Ore domain, and so is its localization $D \otimes_k C_0$. Now $D \otimes_k C = (D \otimes_k C_0) \otimes_{C_0} C$ is an extension of finite degree, hence an Ore extension by Lemma 4.2.

To prove (ii) we note that we have a k-bilinear map from $D_k\langle X \rangle$ and C to $(D \otimes_k C)_C\langle X \rangle$, hence a homomorphism from right to left in (5), and a $(D \otimes_k C)$-ring homomorphism from left to right, and these two maps are clearly mutually inverse. ∎

Next we show that irreducibility is preserved by suitable extensions of the ground field.

PROPOSITION 6.4.4. *Let D be a field which is a k-algebra and let C/k be a simple algebraic (commutative) field extension whose generator has the minimal polynomial f. If f is irreducible over D, then it remains so over $T = D_k\langle\!\langle X \rangle\!\rangle$ and $T_C = T \otimes_k C$ is the universal field of fractions of $(D \otimes_k C)_C\langle X \rangle$.*

Proof. We have $C \cong k[t]/(f)$, hence $D \otimes_k C \cong D[t]/(f)$, where f is central and irreducible over D, by hypothesis, hence $D \otimes_k C$ is a field E, say. If we can show that the tensor ring $E_C\langle X \rangle$ has the universal field of fractions T_C, then f must be irreducible over T and the proof will be complete.

By Lemma 4.3 we have the isomorphism

$$E_C\langle X \rangle \cong D_k\langle X \rangle \otimes_k C, \tag{6}$$

so the right-hand side is a semifir. Now we have the commutative square

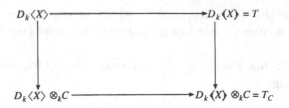

which shows that T_C can be obtained from $D_k\langle X\rangle \otimes C$ by inverting all full matrices over $D_k\langle X\rangle$ (it is not obvious a priori that these matrices remain full over $D_k\langle X\rangle \otimes C$). Now $E_C\langle\!\!\langle X\rangle\!\!\rangle$ is formed from $E_C\langle X\rangle = D_k\langle X\rangle \otimes C$ by inverting all full matrices, so we have a homomorphism

$$T_C \to E_C\langle\!\!\langle X\rangle\!\!\rangle, \tag{7}$$

which consists in inverting all full matrices. By Th. 5.5.11, $T = D_k\langle\!\!\langle X\rangle\!\!\rangle$ has the centre k, so T_C is simple (by Cor. 7.1.3 of A.3), and hence (7) is injective. Further, $[T_C : T] = [C : k] < \infty$, so T_C is Artinian and therefore a matrix ring over a field (see A.2, Th. 5.3.2, p. 174). But by (7), T_C is embedded in a field, so T_C is itself a field, and it follows that $T_C \cong E_C\langle\!\!\langle X\rangle\!\!\rangle$, as we had to show. ∎

We shall also need a criterion for a field to split under a commutative field extension.

LEMMA 6.4.5. *Let E be a field which is a k-algebra and let $C = k(\alpha)$ be a simple algebraic (commutative) extension of k, generated by an element α with minimal polynomial f over k. Then $E \otimes_k C$ is Artinian and moreover, (i) $E \otimes_k C$ is simple if and only if f is irreducible over the centre of E, (ii) $E \otimes_k C$ is a field if and only if f is irreducible over E.*

Proof. It is clear that $E \otimes C$ is Artinian, as E-ring of finite degree. Now we have

$$E \otimes_k C \cong E[t]/(f),$$

and the ideals of $E[t]$ are generated by invariant elements of $E[t]$, which, up to unit factors, are the monic polynomials with coefficients in the centre, Z say, of E (by Prop. 2.2.2). Thus $E[t]/(f)$ is simple precisely when f is irreducible over Z, i.e. (i). Suppose now that f is irreducible over Z; then it is an I-atom and by Th. 1.5.4 we have $E[t]/(f) \cong \mathfrak{M}_n(D)$, where D is a field and n is the length of f in a factorization over $E[t]$. Hence $E \otimes C$ is a field if and only if $n = 1$, i.e. f is irreducible over E. ∎

We now come to the main result of this section, giving conditions for an embedding of free fields:

THEOREM 6.4.6. *Let E be a field with a central subfield C, let D be a subfield of E and put $k = D \cap C$. Then for any set X there is a natural embedding*

$$D_k \langle\!\langle X \rangle\!\rangle \to E_C \langle\!\langle X \rangle\!\rangle, \tag{8}$$

if and only if D and C are linearly disjoint in E over k.

Proof. Suppose that there is a D-ring homomorphism (8) which is the identity on X. Then we have a k-bilinear map of $D_k \langle\!\langle X \rangle\!\rangle$ and C into $E_C \langle\!\langle X \rangle\!\rangle$ and hence a homomorphism

$$D_k \langle\!\langle X \rangle\!\rangle \otimes_k C \to E_C \langle\!\langle X \rangle\!\rangle. \tag{9}$$

Since $D_k \langle\!\langle X \rangle\!\rangle$ has centre k, the left-hand side of (9) is simple (Cor. 7.1.3 of A.3) and so (9) is an embedding. The restriction to $D \otimes_k C$ gives an embedding of the latter in E and this shows D and C to be linearly disjoint in E over k.

Conversely, if D and C are linearly disjoint in E over k, then $D \otimes C$ is an integral domain, as subring of E, so it is an Ore domain, by Lemma 4.2. Writing F for its field of fractions, we claim that there is a natural embedding

$$D_k \langle\!\langle X \rangle\!\rangle \otimes_k C \to F_C \langle\!\langle X \rangle\!\rangle. \tag{10}$$

It will be enough to show that there is a natural homomorphism

$$D_k \langle\!\langle X \rangle\!\rangle \to F_C \langle\!\langle X \rangle\!\rangle, \tag{11}$$

for then we have a bilinear map of $D_k \langle\!\langle X \rangle\!\rangle$ and C into $F_C \langle\!\langle X \rangle\!\rangle$, which gives rise to a homomorphism (10), and this must be an embedding, because the left-hand side of (10) is simple (by Cor. 7.1.3 of A.3). In the special case $C = k(t)$, $F = D(t)$ the result follows by Prop. 5.4.4, and by repetition the result follows when C is any purely transcendental extension of k; if C is a simple algebraic extension of k, say $C = k(\alpha)$, where α has the minimal polynomial f over k and $D \otimes C$ is an integral domain, then f remains irreducible over D (by Lemma 4.5) and hence remains irreducible over $D_k \langle\!\langle X \rangle\!\rangle$, by Prop. 4.4, so $D_k \langle\!\langle X \rangle\!\rangle \otimes C$ is a field. By Prop. 4.4 it is the universal field of fractions of $(D \otimes C)_C \langle X \rangle$ and $F = D \otimes C$, so (10) is an isomorphism in this case. Now suppose that the result holds for $k \subseteq C_1$ with $F_1 = D \otimes C_1$ and C/C_1 is a simple algebraic extension. Then we have maps $D_k \langle\!\langle X \rangle\!\rangle \to (D \otimes C_1)_{C_1} \langle\!\langle X \rangle\!\rangle$ and $(D \otimes_k C_1) \otimes_{C_1} \langle\!\langle X \rangle\!\rangle \to ((D \otimes C_1) \otimes C)_C \langle\!\langle X \rangle\!\rangle$ and by combining them and noting that $(D \otimes_k C_1) \otimes_{C_1} C = D \otimes C$, we obtain the map (11). Thus (11) holds generally, provided only that $D \otimes C$ is an integral domain.

Finally, since $D \otimes C$ is an Ore domain embedded in E, we have an embedding of F in E and by Prop. 5.4.3,

$$E_C \langle\!\langle X \rangle\!\rangle \cong E \underset{F}{\circ} F_C \langle\!\langle X \rangle\!\rangle,$$

therefore $F_C \langle\!\langle X \rangle\!\rangle$ is embedded in $E_C \langle\!\langle X \rangle\!\rangle$ and this together with (11) proves the result. ∎

As a first consequence we note the following strengthening of the specialization lemma:

THEOREM 6.4.7. *Let D be a field which is a k-algebra and let E be a field containing D as subfield, such that the centre C of E is infinite, $[E:C] = \infty$ and D, C are linearly disjoint in E over k. Then every finite set of elements of $D_k \langle\!\langle X \rangle\!\rangle$ is defined for some choice of values of X in E.*

For we have an embedding of $D_k \langle\!\langle X \rangle\!\rangle$ in $E_C \langle\!\langle X \rangle\!\rangle$ and we can apply the specialization lemma to the latter. ∎

This result may be expressed by saying that E satisfies no rational identities, a theorem first proved by Amitsur [66].

We return to polynomials over a field and briefly consider the relation between reducibility over a field and over its centre. Let D be a field with a central subfield k. In Prop. 4.4 we saw that a polynomial over k which is irreducible over D remains so over $D_k \langle\!\langle X \rangle\!\rangle$. For polynomials over D this need no longer hold, as we shall see below. Let us define a field extension D/k, where k is central in D, to be *normal* if every $f \in k[t]$ which is irreducible over k splits over D into factors all of the same degree. When D is a commutative algebraic field extension of k, this reduces to the usual notion (see A.2, 3.2, p. 72); in general it is probably stronger, though no examples are known (see Ex. 2°). As a first result we note that every field is normal over its centre.

PROPOSITION 6.4.8. *Any field is normal over its centre.*

Proof. Let D be a field and C its centre. If $f \in C[t]$ is irreducible over C, then f is an I-atom in $D[t]$, hence by Th. 1.5.4, all its atomic factors are similar and so are of the same degree. ∎

For non-normal extensions we can find irreducible polynomials which become reducible after a free adjunction:

THEOREM 6.4.9. *Let D be a field with a central subfield k. If D/k is not normal, then there exist irreducible polynomials over D which become reducible over $D_k \langle\!\langle X \rangle\!\rangle$, where $X \neq \varnothing$.*

Proof. By hypothesis there is an irreducible polynomial f over k with a complete factorization over D, $f = p_1 \ldots p_r$, in which the p_i are not all of the same degree. If we factorize each p_i over $D_k \langle\!\langle X \rangle\!\rangle$, we get a factorization $f = q_1 \ldots q_n$ in which each q_j divides some p_i. Now $D_k \langle\!\langle X \rangle\!\rangle$ is a field with centre k, hence normal over k and so all the q_j have the same degree. This means that some p_i splits into at least two factors over $D_k \langle\!\langle X \rangle\!\rangle$, but by construction p_i is irreducible over D. ∎

As an example of an irreducible polynomial over D which has a zero in $D_k \langle\!\langle X \rangle\!\rangle$, let us take an irreducible polynomial f of degree $n > 2$ in $k[t]$, such that over D we have

$$f = (t - \alpha)g, \tag{12}$$

where g is irreducible over D. Such a field may be obtained by taking a field k with a commutative extension E such that E/k is the splitting field of a separable polynomial f with symmetric group of degree n and taking D to be the subfield of E obtained by adjoining a single zero α of f. Then $[D:k] = n$ and over D, f has the form (12). By Th. 4.9, g, which is irreducible over D, splits into linear factors over $D_k \langle\!\langle X \rangle\!\rangle$.

A zero of g can be found as follows. Let $x \in X$ and put $c = x\alpha - \alpha x$, $\beta = c^{-1}\alpha c$. If $g = \sum t^i b_i$, then since $f = (t - \alpha)\sum t^i b_i = \sum t^i(b_{i-1} - \alpha b_i)$, it follows that $b_{i-1} - \alpha b_i \in k$. Now we have

$$cb_i = [x, \alpha]b_i = [x, \alpha b_i] + \alpha[b_i, x]$$

$$= [x, \alpha b_i - b_{i-1}] + \alpha[b_i, x] - [b_{i-1}, x],$$

and here the first term on the right vanishes, because $b_{i-1} - \alpha b_i \in k$. Hence we find

$$cg(\beta) = \sum c\beta^i b_i = \sum \alpha^i c b_i$$

$$= \sum \alpha^{i+1}[b_i, x] - \sum \alpha^i[b_{i-1}, x] = 0,$$

since the last two sums cancel each other. It follows that $g(\beta) = 0$. The other factors of f can be found similarly from conjugates of α, though these factors need not of course commute with each other.

Exercises

1. Let k be a commutative field, f an irreducible polynomial over k and E a (commutative) splitting field for f over k. Show that all zeros of f are conjugate over $E_k \langle\!\langle x \rangle\!\rangle$.

$2°$. Let E be a field and D a subfield such that any irreducible polynomial over D which has a linear factor over E splits completely. Is E/D normal?

3. Let D be a field with a central subfield k. Show that $D \otimes E$ is an integral domain for every simple algebraic extension E of k if and only if every irreducible polynomial over k remains irreducible over D. Show that this always holds provided that $D \otimes_k k^a$ is a field, where k^a is the algebraic closure of k. Prove the converse of this assertion when k is perfect and give a counter-example for an imperfect k.

4. Give an example of a field D with subfields E, F such that E, F are linearly disjoint (over $E \cap F$), but F, E are not.

5. Construct a field E which is a k-algebra (for a given commutative field k) such that every element of $E \backslash k$ is a multiplicative commutator. Show also that such a field cannot be finitely generated.

6. (Amitsur [66], Hua [50]). Use Th. 4.7 to show that for a skew field D with infinite centre, $D^{\times}$ satisfies no group identity, and deduce that $D^{\times}$ cannot be soluble.

6.5 Existentially closed skew fields

Let k be a commutative field. By an *algebraic closure* of k one usually understands a field k^a with the properties:

(i) k^a *is algebraic over k,*

(ii) *every polynomial equation over k^a has a solution in k^a.*

It is well known (see e.g. A.2, 3.3) that every commutative field has an algebraic closure and that the latter is unique up to isomorphism, though not necessarily a unique isomorphism. Thus the correspondence $k \mapsto k^a$ is not a functor, in fact the different isomorphisms of k^a over k form the subject of Galois theory. When one tries to perform an analogous construction for skew fields one soon finds that it is impossible to combine (i) and (ii); in fact (i) is rather restrictive, so we give it up altogether and concentrate on (ii). Here it is convenient to separate two problems, namely (a) which equations are soluble (in some extension) and (b) whether every soluble equation has a solution in the closure. So far only partial results are available on (a); we shall return to it later, in Ch. 8, and for the moment concentrate on (b).

The assertion that an equation $f(x_1, \ldots, x_n) = 0$ has a solution can be expressed as follows:

$$\exists a_1, \ldots, a_n\, f(a_1, \ldots, a_n) = 0.$$

Any sentence of the form $\exists a_1, \ldots, a_n\, P(a_1, \ldots, a_n)$, where P is an expression obtained from equations by negation, conjunction and disjunction, is called an *existential sentence*, more precisely an *elementary existential sentence*, because only element variables are quantified.

By an *existentially closed field over k*, *EC-field* for short, we understand a field D which is a k-algebra, such that any existential sentence with constants from D, which holds in some extension of D, already holds in D itself. Such a sentence can always be expressed as a finite conjunction of finite disjunctions of basic formulae, a *basic formula* being of the form $f = g$ or its negation, where f, g are polynomials in $x_1, \ldots, x_n$ over D, i.e. elements of $D_k\langle x_1, \ldots, x_n \rangle$. In the case of fields all existential sentences can actually be reduced to equations:

PROPOSITION 6.5.1. *Any existential sentence over a field D is equivalent to a finite set of equations*

$$\exists x_1, \ldots, x_n \, (f_1 = 0 \wedge \ldots \wedge f_r = 0). \tag{1}$$

Proof. Any basic formula $f = g$ can be written as $f - g = 0$, and $f \neq g$ can be expressed as $(f - g)y = 1$, where y is a new variable. A disjunction $g_1 = 0 \vee g_2 = 0 \vee \ldots \vee g_m = 0$ is equivalent to $g_1 g_2 \ldots g_m = 0$; thus our existential sentence can be written as a conjunction of a finite number of equations $f_1 = 0 \wedge f_2 = 0 \wedge \ldots \wedge f_r = 0$, and by prefixing an existential quantifier we obtain (1). ∎

By this result a field D is existentially closed if and only if every finite system of equations which is consistent (i.e. has a solution in some extension field) has a solution in D itself. As a trivial example, k itself is existentially closed over k precisely when k is algebraically closed, but for $D \supset k$ it is possible for D to be existentially closed even when k is not algebraically closed. In fact we shall find in Th. 5.3 that every field D can be embedded in an EC-field, but the latter will not be unique in any way.

We know from Ch. 4 that we shall need to consider, besides the vanishing of elements, also the singularity of matrices, but fortunately one can be reduced to the other. On the one hand, equations represent the special case of 1×1 matrices; on the other hand, if $A = (a_{ij})$ is an $n \times n$ matrix, let us write sing (A) for the existential sentence

$$\exists u_1, \ldots, u_n v_1, \ldots, v_n$$

$$\left(\sum a_{1j} u_j = 0 \wedge \ldots \wedge \sum a_{nj} u_j = 0 \wedge (1 - u_1 v_1) \ldots (1 - u_n v_n) = 0 \right),$$

and nonsing (A) for the sentence

$$\exists x_{ij}(i, j = 1, \ldots, n) \left(\sum a_{iv} x_{vj} = \delta_{ij}, i, j = 1, \ldots, n \right).$$

It is clear that sing (A) asserts that (over a field) A is singular and

$$\mathrm{nonsing}\,(A) \Leftrightarrow \neg\,\mathrm{sing}\,(A).$$

It is now an easy matter to obtain criteria for existential closure in terms of the singularity of matrices:

PROPOSITION 6.5.2. *A field D over k is an EC-field if and only if every finite set of matrices over $D_k\langle X\rangle$ which all become singular for a certain set of values in some extension field of D already become singular for some set of values in D.*

Proof. The conditions for existential closure concern the vanishing of a finite set of elements, i.e. the singularity of 1×1 matrices, and so are a special case of the matrix condition, which is therefore sufficient. Conversely, when D is existentially closed and we are given matrices A_1, ..., A_r which become singular in some extension, then $\mathrm{sing}\,(A_1) \wedge \ldots \wedge \mathrm{sing}\,(A_r)$ is consistent and hence has a solution in D. ∎

Let D be a field with k as central subfield. To find an EC-field containing D one might first construct an extension D_1 in which a given finite consistent system of equations over D has a solution, and repeat this process infinitely often. However there is no guarantee that the EC-field so obtained will contain solutions of every finite consistent system over D. For this to hold we need to be assured that any two consistent systems over D are jointly consistent. This follows from the existence of field coproducts: if Φ_1, Φ_2 are two consistent systems of equations over D, say Φ_i has a solution in K_i, then any field L containing both K_1 and K_2 will contain a solution of $\Phi_1 \cup \Phi_2$; here we can take the field coproduct $K_1 \underset{D}{\circ} K_2$ for L. More generally, a class of algebras is said to possess the *amalgamation property* if any two extensions of an algebra A are contained in some algebra C. Thus the class of fields has the amalgamation property; an example of a class not possessing the amalgamation property is the class of formally real fields (see 9.6).

To construct an EC-field extension of D we take the family $\{C_\lambda\}$ of all finite consistent systems of equations over D and for each λ take an extension E_λ of D in which C_λ has a solution. If we put $D_1 = \underset{D}{\circ} E_\lambda$, then every finite consistent set of equations over D has a solution in D_1. By repeating this process we obtain a tower

$$D \subset D_1 \subset D_2 \subset \ldots ,$$

whose union L is again a field, of the same cardinal as D, or countable if D was finite. Any finite consistent set of equations over L has its coefficients in some D_i and so has a solution in D_{i+1}, hence also in L. Thus L is an EC-field and we have proved

THEOREM 6.5.3. *Let D be any field with k as central subfield. Then there exists an EC-field L containing D, in which every finite consistent set of equations over D has a solution. When D is infinite, L can be chosen to have the same cardinal as D, while for finite D, L may be taken countable.* ■

The EC-field constructed here is not in any way unique; even a minimal EC-field containing a given field D need not be unique up to isomorphism, as will become clear later on. Further, it will no longer be possible to find an EC-field that is algebraic (in any sense) over D.

Th. 5.3 is actually a special case of a result in universal algebra. A general algebraic system A is said to be *existentially closed* (EC) if every existential sentence which holds in an extension of A already holds in A itself. For any class Σ of algebraic systems which is *inductive*, i.e. closed under isomorphisms and unions of ascending chains of Σ-systems, it can be shown that EC Σ-systems exist and in fact any Σ-system is contained in an EC Σ-system, although there may not be a least EC Σ-system containing it (see e.g. UA, IX. 3, p. 327). Applied to fields, this just gives Th. 5.3.

Sometimes a stronger version of closure is needed, where the above property holds for *all* sentences, not merely existential ones. We shall not need this stronger form, and therefore just state the results without proofs (which may be found e.g. in UA, Ch. IX).

Let $\mathscr{A}$ be an inductive class of algebras (of some sort). A homomorphism $f: A \to B$ between $\mathscr{A}$-algebras is said to be *elementary* if for every sentence $P(x)$ which holds in A, $P(xf)$ holds in B. Now a *forcing companion*(*) of $\mathscr{A}$ is defined as a subclass $\mathscr{C}$ of $\mathscr{A}$ such that

(F.1) *Any $\mathscr{A}$-algebra can be embedded in a $\mathscr{C}$-algebra,*

(F.2) *Any inclusion $C_1 \subseteq C_2$ between $\mathscr{C}$-algebras is elementary,*

(F.3) *$\mathscr{C}$ is maximal subject to (F.1, 2).*

It can be shown that every inductive class has a unique forcing companion (see UA, Th. IX. 4.3, p. 330). This applies in particular to skew fields. Here we also have the amalgamation property, but an important difference between the commutative and the non-commutative case is that algebraically closed commutative fields are axiomatizable (we can write down a set of elementary sentences asserting that all equations have solutions), whereas the corresponding statement for EC-fields is false. This follows from the fact that the class of EC-fields is not closed under ultrapowers (see Hirschfeld and Wheeler [75] and Ex. 10 below).

(*) More precisely, this is an infinite forcing companion, but we shall not have occasion to meet others.

Although EC-fields do not share all the good properties of algebraically closed fields, they have certain new features not present in the commutative case. For example, the property of being transcendental over the ground field can now be expressed by an elementary sentence:

$$\text{transc}\,(x)\colon \exists y, z\; (xy \neq yx \wedge x^2 y = yx^2 \wedge xz = zx^2 \wedge z \neq 0). \quad (2)$$

This sentence, due to Wheeler, states that there is an element y commuting with x^2 but not with x, hence $k(x^2) \subset k(x)$; secondly x is conjugate to x^2, so $k(x) \cong k(x^2)$, in particular, $[k(x):k] = [k(x^2):k]$, and since $k(x^2) \subset k(x)$, the degree must be infinite. Conversely, when x is transcendental over k, we can take an HNN-extension E in which x is conjugate to x^2 and form the field coproduct of E with $k(x^2, y)$ over $k(x^2)$, to obtain an extension in which (2) is satisfied. Hence (2) holds in the EC-field. Wheeler has generalized (2) to find (for each $n \geq 1$) an elementary formula $\text{transc}_n\,(x_1, \ldots, x_n)$ expressing the fact that $x_1, \ldots, x_n$ commute pairwise and are algebraically independent over k (see Ex. 2).

To describe EC-fields in more detail we shall need a result that is useful in forming constructions. We recall that an EC-field can be embedded in a homogeneous extension, by Cor. 5.5.2, and so it is itself homogeneous.

LEMMA 6.5.4 (Zig-zag lemma). *If K, L are two EC-fields, countably generated over k, then $K \cong L$ if and only if K and L have the same family of finitely generated subfields.*

Proof. Clearly the condition is necessary. Conversely, suppose that K, L are countably generated EC-fields with the same finitely generated subfields. Write $K = k(a_1, a_2, \ldots)$, $L = k(b_1, b_2, \ldots)$; we shall construct finitely generated subfields K_n, L_n of K and L respectively such that (i) $K_n \subseteq K_{n+1}$, $L_n \subseteq L_{n+1}$, (ii) $K_n \supseteq k(a_1, \ldots, a_n)$, $L_n \supseteq k(b_1, \ldots, b_n)$, (iii) there is an isomorphism between K_{n+1} and L_{n+1} extending a given isomorphism between K_n and L_n. Since $K = \bigcup K_n$, $L = \bigcup L_n$ by (ii), it will follow that $K \cong L$, by taking the common extension of the isomorphisms in (iii).

Put $K_0 = L_0 = k$; if K_n, L_n have been defined, with an isomorphism $\varphi_n \colon K_n \to L_n$ and we put $K'_n = K_n(a_{n+1})$, then K'_n is finitely generated, hence isomorphic to a subfield L'_n of L containing an isomorphic copy of L_n. Since L is homogeneous, there is an inner automorphism of L mapping L'_n to a subfield L''_n containing L_n in such a way that the restriction to K_n is φ_n. Let $\varphi'_n \colon K'_n \to L''_n$ be the isomorphism so obtained, put $L_{n+1} = L''_n(b_{n+1})$ and find an isomorphic copy of L_{n+1} in K. It will contain a subfield isomorphic to K'_n and by applying a suitable inner automorphism of K we obtain an isomorphism of L_{n+1} with a subfield K_{n+1}, say, of K such that its restriction to L_n'' is $(\varphi_n')^{-1}$. Now K_{n+1} and L_{n+1} satisfy (i)–(iii) and the result follows by induction. ∎

We now turn to applications, and begin by showing that in an EC-field, belonging to a finitely generated subfield can be expressed by an elementary sentence. For any subset S of a field D we shall write $\mathscr{C}(S)$ for the centralizer of S in D.

PROPOSITION 6.5.5. *Let D be an EC-field over k. Then for any a_1, . . . , a_r, $b \in K$,*

$$b \in k(a_1, \ldots, a_r) \Leftrightarrow \mathscr{C}(b) \supseteq \mathscr{C}(a_1, \ldots, a_r).$$

This means that the formula '$b \in k(a_1, \ldots, a_r)$', not at first sight elementary (and in fact not so in the commutative case), can be expressed as an elementary sentence in an EC-field:

$$\forall x \ (a_i x = x a_i \ (i = 1, \ldots, r) \Rightarrow bx = xb). \tag{3}$$

Proof. Put $E = k(a_1, \ldots, a_r)$; if $b \in E$, then (3) clearly holds, while for $b \notin E$, (3) is false in $D \underset{E}{\circ} E(x)$ and hence also in D, because the latter is existentially closed (and the negation of (3) is an existential sentence). ∎

Taking $r = 0$, we obtain a result which is well known in the special case when k is the prime subfield:

COROLLARY 6.5.6. *The centre of an EC-field over k is k.* ∎

EC-fields are in some way analogous to algebraically closed groups, which have been studied by B. H. Neumann [73]; the next result corresponds to a property proved by Neumann for groups:

PROPOSITION 6.5.7. *An EC-field cannot be finitely generated or finitely related.*

Proof. Let D be an EC-field over k. Given $a_1, \ldots, a_r \in D$, the sentence

$$\exists x, y \ (a_i x = x a_i \ (i = 1, \ldots, r) \land xy \neq yx)$$

holds in $D(x) \underset{k}{\circ} k(y)$, hence it holds in D itself, and by Prop. 5.5 this means that D contains an element $y \notin k(a_1, \ldots, a_r)$; therefore D cannot be finitely generated. Now suppose that D is finitely related; then it can be expressed as a field coproduct of a finitely presented field E and a free field F, and here F cannot be finitely generated, by the first part. If x is a free generator, it occurs in no relation, so the sentence $\exists y(x = y^2)$ is not satisfied in D, though it is clearly consistent, but this contradicts the fact that D is an EC-field. Hence D cannot be finitely related. ∎

As we saw in 5.5, there are continuum-many non-isomorphic finitely generated fields, hence no countable EC-field can contain them all, thus there are no countable *universal* EC-fields. However, it is possible to construct a countable EC-field containing all finitely presented fields: we enumerate all finitely presented fields $D_1, D_2, \ldots$ over k, form their field coproduct over k and take a countable EC-field containing this co-product, which exists by Th. 5.3. The result is a countable EC-field containing each finitely presented field over k; such an EC-field is sometimes called *semiuniversal*.

Any EC-field has proper EC-subfields, thus there are no minimal EC-fields, as the next result shows.

THEOREM 6.5.8. *Let D be an EC-field over k and c any element of D. Then the centralizer C of c in D is an EC-field over $k(c)$.*

Proof. It is clear that $k(c)$ is contained in the centre of C. Now let

$$f_1 = f_2 = \ldots = f_r = 0 \tag{4}$$

be any finite set of equations in $x_1, \ldots, x_n$ over C which has a solution in some extension of C over $k(c)$. This means that the solution $(x_1, \ldots, x_n)$ of (4) also satisfies

$$x_1 c - c x_1 = \ldots = x_n c - c x_n = 0. \tag{5}$$

Hence the equations (4), (5) are consistent and so have a solution in D. By (5) this means that we have found a solution of (4) in C, so C is an EC-field over $k(c)$, as claimed. ∎

By taking intersections we can get EC-fields over k itself. Such a construction can also be obtained in a more straightforward fashion:

THEOREM 6.5.9. *Let D be an EC-field over k and let $a \in D$ be transcendental over k. Then there exists $b \in D$ such that*

$$ba = a^2 b \neq 0. \tag{6}$$

If E is the centralizer of such a pair a, b in D, then E is again an EC-field over k and the inclusion $E \subset D$ is an elementary embedding.

Proof. Since a and a^2 are both transcendental over k, the system (6) has a solution in an HNN-extension and hence in D itself. Given a, b satisfying (6), let E be their centralizer in D and consider a consistent system of equations

$$f_1 = \ldots = f_r = 0, \tag{7}$$

in the variables $x_1, \ldots, x_n$ over E. Since D is an EC-field, this system has a solution in D. Let $c_1, \ldots, c_s \in E$ be the coefficients occurring in (7) and consider the system consisting of (7) and

$$x_i y = y x_i, \ x_i z = z x_i \ (i = 1, \ldots, n), \tag{8}$$

$$c_j y = y c_j, \ c_j z = z c_j, \ zy = y^2 z \neq 0 \ (j = 1, \ldots, s). \tag{9}$$

This system is consistent: we form first $D(y)$ and with the endomorphism $\sigma: f(y) \mapsto f(y^2)$ form $K(y)(z; \sigma) = H$; this is just the Hilbert field already encountered in 6.1. It shows (7), (8) and (9) to be consistent and so to have a solution in D itself, which we again denote by x_i, y, z. Now the mapping $y \mapsto a$, $z \mapsto b$ defines an isomorphism

$$k(c_1, \ldots, c_s, y, z) \cong k(c_1, \ldots, c_s, a, b), \tag{10}$$

for the left-hand side is the field of fractions of L, the skew Laurent polynomial ring in z, z^{-1} over $k(c_1, \ldots, c_s, y)$; this ring L is simple and has the corresponding ring in a, b, b^{-1} as homomorphic image; by the simplicity we have an isomorphism, which yields the isomorphism (10). By homogeneity there exists $t \in D$ such that $c_j t = t c_j$, $yt = ta$, $zt = tb \neq 0$. If we put $t^{-1} x_i t = x_i'$, then $x_i' \in E$ and x_i' is a solution of (7) in E. This shows E to be an EC-field.

To prove that the inclusion $E \subset D$ is an elementary embedding, we need only show that every finitely generated subfield of D can be embedded in E. Let $c_1, \ldots, c_s \in D$ and consider the system (8), (9); this system is consistent and so has a solution in D. Since $k(y, z) \cong k(a, b)$ with $y \mapsto a$, $z \mapsto b$, it follows that there exists $t \in K$ such that $yt = ta$, $zt = tb \neq 0$. If we put $t^{-1} c_j t = c_j'$, then $c_j' \in E$ and $k(c_1, \ldots, c_s) \cong k(c_1', \ldots, c_s')$, but the latter is a subfield of E and so the result follows. ∎

When D itself is countable, it follows from the zig-zag lemma 5.4 that $E \cong D$ and we obtain

COROLLARY 6.5.10. *If D is a countable EC-field over k and a, $b \in D$ are transcendental elements such that $ba = a^2 b \neq 0$, then D is isomorphic over k to the centralizer of a, b in D. Thus every countable EC-field has a proper subfield isomorphic to itself.* ∎

An important and useful result due to Wheeler states that every countable EC-field has outer automorphisms:

THEOREM 6.5.11. *Every countable EC-field has $2^{\aleph_0}$ distinct automorphisms, and hence has outer automorphisms.*

Proof. Let D be an EC-field, generated over k by $a_1, a_2, \ldots$, where the a_i are so chosen that $a_n \notin k(a_1, \ldots, a_{n-1})$; this is possible, because D is not finitely generated, by Prop. 5.7. By Prop. 5.5 there exists b_n in D commuting with $a_1, \ldots, a_{n-1}$ but not with a_n. Let β_n be the inner automorphism induced by b_n and consider the formal product

$$\alpha = \beta_1^{\varepsilon_1} \beta_2^{\varepsilon_2} \ldots,$$

for a given choice of exponents $\varepsilon_i = 0, 1$. We claim that α defines an automorphism of D. Its effect on $k(a_1, \ldots, a_n)$ is

$$\beta_1^{\varepsilon_1} \ldots \beta_n^{\varepsilon_n},$$

for when $i > n$, then β_i leaves $k(a_1, \ldots, a_n)$ elementwise fixed. Thus it is an endomorphism which is in fact invertible since each β_i is. Since the ε_i are independent and each choice gives a different automorphism, we have indeed $2^{\aleph_0}$ distinct automorphisms; of course there cannot be more than this number since D is countable. But D has only countably many inner automorphisms, so there are outer automorphisms. ∎

This proof is of course highly non-constructive; since EC-fields themselves are not given in any very explicit form, there seems little hope of actually finding a particular outer automorphism.

An important but difficult question is: Which fields are embeddable in finitely presented fields? It would be interesting if some analogue of Higman's theorem could be established. This asserts that a finitely generated group is embeddable in a finitely presented group if and only if it can be recursively presented (Higman [61]).

Exercises

1. For a commutative field F of prime characteristic p, the *perfect closure* is defined as a p-radical (= purely inseparable) extension which is perfect. Show that the perfect closure $\hat{F}$ of F is unique up to a unique isomorphism, and so defines a functor $F \mapsto \hat{F}$.

2. (W. H. Wheeler) Define a sentence $\mathrm{transc}_n (x_1, \ldots, x_n)$ expressing that $x_1, \ldots, x_n$ are commuting independent indeterminates, in terms of

$transc_{n-1}(x_2, \ldots, x_n)$, $transc_1(x_1)$ and the fact that x_1 commutes with the x_i but $x_1 \notin k(x_1^2, x_2, \ldots, x_n)$.

3. (W. H. Wheeler) Show that every EC-field contains a commutative algebraically closed subfield of infinite transcendence degree over the ground field.

4°. Does every EC-field over k contain a free field (of rank > 1) over k?

5. Show that the class of commutative fields has the amalgamation property.

6. Prove an analogue of Th. 5.9 in which (6) is replaced by $ba = \lambda ab$, where $\lambda \neq 0$, $\lambda^n \neq 1$ for all $n > 0$.

7. Let K be a field with a presentation $K = k\langle X; \Phi \rangle$, where Φ is a set of square matrices. This presentation is called *absolute*, if K is the only field generated, as field and k-algebra, by X and in which all the matrices of Φ become singular. Show that the given presentation is absolute if and only if the matrix ideal generated by Φ is maximal.

8. Show that every EC-field contains a copy of every finitely absolutely presented field and an epimorphic image under a local homomorphism of every finitely presented field.

9. (Boffa and Van Praag [72]) Show that in an EC-field over a perfect ground field, $transc(x)$ can be described by the sentence: $\exists y \, (xy - yx = 1)$.

10. (Boffa and Van Praag [72]) Let K be an EC-field over a perfect ground field k and (a_n) a sequence of algebraic elements in K of unbounded degrees. Show that in an ultrapower L of K the element $(a_1, a_2, \ldots)$ is transcendental but does not satisfy the criterion of Ex. 9. Deduce that the class of EC-fields does not admit ultrapowers. (A similar result was obtained by Wheeler [72], using (2).)

11. Let K be an EC-field with an automorphism α and an α-derivation δ and consider $R = K[t; \alpha, \delta]$. Show that a polynomial in R is totally unbounded if and only if its linear companion is totally transcendental. Deduce that two totally unbounded polynomials are similar if and only if they have the same degree.

12°. Let E be a free field on a countable set over a ground field k. Are any two countable EC-fields containing E as subfield isomorphic?

6.6 The word problem for skew fields

The *word problem* in a variety of algebras, e.g. groups, is the problem of deciding, for a given presentation of a group, when two expressions represent the same element. This is often a highly non-trivial problem; if there is an algorithm for deciding this question in a finite number of steps, the given presentation is said to have a *solvable* word problem. In the case of skew fields we again have a presentation, as explained in 6.1, and we can ask the same question, but the word problem is now a relative one. Generally we have a coefficient field K, itself a k-algebra, and we need to know how K is given. It may be that K itself is given by a presentation with solvable word problem, and the algorithm which achieves this is then incorporated in the algorithm to be constructed; or more generally, we merely postulate that certain questions about K can be answered in a finite number of steps and use this fact to construct a relative algorithm. A process or construction which can be carried out in a finite number of steps is often called *effective*.

We shall give one example of a field with solvable word problem and one with unsolvable word problem. Our first task will be to show that free fields have a solvable word problem; of the two alternatives described above we shall take the second, thus our solution will not depend on the precise algorithm in K, but merely on the fact that it exists. In fact it is not enough to assume that K has a solvable word problem; we need to assume that K is dependable over its centre: Given a field K which is a k-algebra, we shall call K *dependable* over k if there is an algorithm which for each finite family of expressions for elements of K, in a finite number of steps either leads to a linear dependence relation between the elements over k, or shows them to be linearly independent over k.

When K is dependable over k, then K has a solvable word problem (relative to k), as we see by testing one-element sets for linear dependence. Let K have centre C; our task will be to solve the word problem for the free field $K_C \langle\!\langle X \rangle\!\rangle$. Here it will be necessary to assume K dependable over C; this assumption is indispensable, for we shall see that it holds whenever $K_C \langle\!\langle X \rangle\!\rangle$ has a solvable word problem.

There is another difficulty which needs to be briefly discussed. As observed earlier, we need to deal with *expressions* of elements in a field and our problem will be to decide when such an expression represents the zero element. But in forming an expression we may need to invert non-zero elements, therefore we need to solve the word problem already in order to form meaningful expressions. This problem could be overcome by allowing formal expressions such as $(a - a)^{-1}$, but we shall be able to bypass it altogether: instead of building up rational expressions step by step, we can obtain them in a single step by solving suitable matrix equations, as we saw in 4.2.

We begin by recalling some relevant definitions from logic. Let $\mathbf{N}$ be the set of natural numbers, as usual. A subset S of $\mathbf{N}$ is said to be *recursive* if for each $n \in \mathbf{N}$ there is an algorithm for deciding in a finite number of steps whether or not $n \in S$. The set S is called *recursively enumerable* (r.e.) if there is a function f defined on N whose range of values is all of S and such that for each $n \in N$, $f(n)$ can be computed in a finite number of steps. It is clear that every recursive set is r.e., but there exist r.e. sets that are not recursive; we omit the proof (which uses Cantor's diagonal argument) and refer to H. Rogers Jr [67] for details. However the following easy consequence of the definitions helps to elucidate these concepts.

PROPOSITION 6.6.1. *A subset S of N is recursive if and only if both S and its complement S' in N are recursively enumerable.*

Proof. If S is recursive, then so is S', and so both are r.e. If both S and S' are r.e., let f, g be the functions with ranges S and S' respectively. Then every natural number r is a value of f or g, so it will occur in the sequence $f(1)$, $g(1)$, $f(2)$, $g(2)$, ... ; hence after a finite number of steps we see whether r is in S or in S'. ∎

More generally, these concepts may be applied to any countable set, whose elements are effectively enumerated.

Now we have the following reduction theorem:

THEOREM 6.6.2. *Let R be a semifir and U its universal field of fractions. Then the word problem for U can be solved if the set of full matrices over R is recursive.*

Of course this condition can only hold in a countable ring.

Proof. Any element p of U is obtained as component of the solution of a matrix equation

$$Au = 0,$$

and $p = 0$ if and only if its numerator A^0 is non-full over R. By hypothesis there is an algorithm to decide whether A^0 is full or not, and this provides the answer to our question. ∎

We note that it is enough to assume that the set of full matrices over R is r.e., because its complement, the set of all non-full matrices, is always r.e. in a countable ring.

We shall also need a property of matrices, which generalizes the well known fact that an $n \times n$ nilpotent matrix A over a field satisfies the equation $A^n = 0$.

LEMMA 6.6.3. *Let P be an $n \times n$ matrix, A a matrix with n columns and B a matrix with n rows over a field K. If*

$$AP^v B = 0 \quad for \; v = 0, 1, \ldots, n - 1, \tag{1}$$

then $AP^v B = 0$ for all $v \geq 0$.

Proof. Let $^n K$ be the right K-space of columns with n components. The columns of B span a subspace V_0 of $^n K$, while the columns annihilated by the rows of A form a subspace W of $^n K$, and since $AB = 0$ by hypothesis, we have $V_0 \subseteq W$. Regarding P as an endomorphism of $^n K$, we may define a subspace V_v of $^n K$ for $v > 0$ recursively by the equations

$$V_v = V_{v-1} + PV_{v-1}.$$

Thus $V_v = V_0 + PV_0 + \ldots + P^v V_0$ and it follows that

$$V_0 \subseteq V_1 \subseteq \ldots \subseteq V_{n-1}. \tag{2}$$

Moreover, by (1), $V_v \subseteq W$ for $v = 0, 1, \ldots, n - 1$. Now if $A = 0$ or $B = 0$, then there is nothing to prove. Otherwise $V_0 \neq 0$, $W \neq {}^n K$ and we must have equality at some point in (2), since $\dim V_{n-1} \leq n - 1$. Suppose that $V_{k-1} = V_k$ for some $k < n$; then $PV_{k-1} \subseteq V_{k-1}$, hence $PV_k \subseteq PV_{k-1} \subseteq V_k$ and so $V_{k+1} = V_k + PV_k = V_k$, so the sequence is stationary from V_k onwards. We conclude that $V_v \subseteq W$ for all v, i.e. $AP^v B = 0$ for all v, as we had to show. ■

To solve the word problem for a free field, we must be able to decide when a matrix over a tensor ring is full. This is accomplished by the next result, which may be regarded as a constructive form of the specialization lemma.

PROPOSITION 6.6.4. *Let K be a field, dependable over its centre C. Assume further that C and $[K:C]$ are infinite. Then for any full matrix over the tensor ring $F = K_C \langle X \rangle$ there is an algorithm for finding a set of values α of X in K such that $A(\alpha)$ is invertible.*

Proof. Since being full or invertible is unaffected by stable association, we can limit ourselves to the case where A is linear, by passing to a linear companion. Thus we may assume that

$$A = A_0 + A_1, \tag{3}$$

where A_i is homogeneous of degree i in the xs. Thus A_0 has entries in K; if A is not full, then it will remain non-full when the xs are replaced by 0, so A_0 must then be singular. Hence if A_0 is non-singular, then A is necessarily full.

We may therefore assume that A_0 is singular, of rank $r < N$, say, where N is the order of A; we shall use induction of N. By a diagonal reduction over K (which leaves the fullness of A unaffected), we can reduce A_0 to the form $I \oplus 0$; this is an effective procedure because K is dependable over C. If we partition A_1 accordingly, we have

$$A = \begin{pmatrix} I - P & Q \\ R & S \end{pmatrix},$$

where P, Q, R, S are homogeneous of degree 1 (and the sign of P is chosen for convenience in what follows). Now pass to the completion $\hat{F} = K_C\langle\langle X \rangle\rangle$; by inertia (Prop. 2.2) A is full over F if and only if it is full over $\hat{F}$. The matrix $I - P$ is invertible over $\hat{F}$, so by Lemma 2.3 the inner rank of A is at least r, with equality precisely when

$$S - R(I - P)^{-1}Q = 0. \tag{4}$$

To check whether (4) holds, we have to verify that for each $v = 0, 1, \ldots$ the terms of degree v in (4) vanish. Now $S - R(I - P)^{-1}Q = S - \sum RP^v Q$, and equating terms of a given degree we find that (4) is equivalent to

$$S = 0, \quad RP^v Q = 0, \quad v = 0, 1, \ldots; \tag{5}$$

by Lemma 6.3 it is enough to take $v = 0, 1, \ldots, N - 1$ in (5). This provides us with an algorithm for determining whether (4) holds. When this equation holds, A is non-full, by Lemma 2.3. Otherwise, writing $X = \{x_\lambda\}$, we can by the specialization lemma (Cor. 2.6) specialize x_λ to values $\alpha_\lambda \in K$ such that $I - P$ remains non-singular (using induction on N) and $S - R(I - P)^{-1}Q$ is non-zero. Translating back to A, we find that by specializing x_λ to α_λ we obtain a matrix of rank $> r$. We now replace x_λ by $x_\lambda + \alpha_\lambda$ in A and start again from (3). This time we have a matrix A_0 over K of rank $> r$. By repeating this process a finite number of times (at most N times, where N is the order of A), we either find values of X in K for which A is non-singular, or find that A is non-full. ∎

Prop. 6.4 provides an algorithm for deciding when a matrix A over $K_C\langle X \rangle$ is full, in case C and $[K:C]$ are infinite, and so it shows the set of full matrices to be recursive in this case. When C is finite, or even when it is not 'constructively' infinite in the way described earlier, we can reach the conclusion as follows. Given a field K with centre C, let $K' = K(t)$ be the rational function field in a central indeterminate t. By Prop. 2.1.5, the centre of K' is $C' = C(t)$. We claim that K' is dependable over C'

whenever K is dependable over C. For let $u_1, \ldots, u_n \in K'$ and write these elements as functions in t with a common denominator: $u_i = f_i g^{-1}$, where $f_i, g \in K[t]$. Clearly it will be enough to test $f_1, \ldots, f_n$ for linear dependence over C', and we may take the fs to be numbered so that $\deg f_1 \geqslant \deg f_2 \geqslant \ldots$. Consider the leading coefficients of $f_1, \ldots, f_n$; if they are linearly independent over C, then the fs are linearly independent over C'. Otherwise we can find $i, 1 \leqslant i \leqslant n, \alpha_{i+1}, \ldots, \alpha_n \in C$ and non-negative integers $v_{i+1}, \ldots, v_n$ such that $f_i' = f_i - \sum_{i+1}^{n} f_j \alpha_j t^{v_j}$ has lower degree than f_i. Now the linear dependence over C' of $f_1, \ldots, f_n$ is equivalent to that of $f_1, \ldots, f_{i-1}, f_i', f_{i+1}, \ldots, f_n$, and here the sum of the degrees is smaller. Now the conclusion follows by induction on the sum of the degrees of the fs.

When C is finite but $[K:C]$ is infinite, we can thus by adjoining t obtain K', C' such that C' is infinite in the constructive sense, e.g. we can take $1, t, t^2, \ldots$; moreover, $[K':C'] = [K:C]$. It follows that the set of full matrices over $K_{C'}'\langle X \rangle$ is recursive, and since the natural map $K_C\langle X \rangle \to K_{C'}'\langle X \rangle$ is honest, by Prop. 5.4.4, we conclude that the set of full matrices over $K_C\langle X \rangle$ is recursive.

When $[K:C]$ is finite but C is infinite, we again adjoin another central indeterminate u and obtain $K' = K(u)$, $C' = C(u)$. On K' we have the endomorphism $\sigma: f(u) \mapsto f(u^2)$. Now form the skew polynomial ring $K'[v; \sigma]$ and its field of fractions K''. This is again the Hilbert field, its centre is C and by Th. 4.6 the inclusion $K_C\langle X \rangle \to K_C''\langle X \rangle$ is honest. This reduces the problem again to the first case. Finally, when C and $[K:C]$ are both finite, we carry out the last two reductions in succession. We have thus proved

PROPOSITION 6.6.5. *Let K be a field, dependable over its centre C. Then for any set X, the set of full matrices over the tensor ring $K_C\langle X \rangle$ is recursive.* ∎

As we have seen, the recursiveness of the set of full matrices is equivalent to the solvability of the word problem; it only remains to establish the connexion with dependability; we shall prove the following somewhat more general result:

LEMMA 6.6.6. *Let K be a field with central subfield k. If for every finite set Y the word problem for the free field $K_k \langle\!\langle Y \rangle\!\rangle$ is solvable, then the free K-field $K_k \langle\!\langle X \rangle\!\rangle$ on any set X is dependable over k.*

Proof. If X is any set and $Y \subseteq X$, then the natural map $K_k\langle Y \rangle \to K_k\langle X \rangle$ has a retraction and so is honest (Prop. 4.5.1). Now let

X be an infinite set and consider a matrix A over $K_k\langle X\rangle$. Since A involves only finitely many elements from X, A is defined over a finite subset Y of X; by hypothesis we can determine whether A is full over $K_k\langle Y\rangle$ and hence we can do the same over $K_k\langle X\rangle$. So we may take X infinite in what follows.

Let $U = K_k \langle\!\langle X\rangle\!\rangle$ be the free field on an infinite set X; given $u_1, \ldots, u_n \in U$, we have to determine whether the us are linearly dependent over k. We shall use induction on n, the case $n = 1$ being essentially the word problem for U. We may assume that $u_1 \neq 0$, and hence on dividing by u_1 we may suppose that $u_1 = 1$. Only finitely many elements of X occur in $u_2, \ldots, u_n$, so we can find another element in X, y say. We write $u_i' = u_i y - y u_i$ and check whether $u_2', \ldots, u_n'$ are linearly dependent over k. If so, let $\sum_2^n u_i' \alpha_i = 0$, where $\alpha_2, \ldots, \alpha_n$ are in k and not all zero; hence $u = \sum_2^n u_i \alpha_i$ satisfies $yu = uy$. Since u does not involve y, it follows that we can specialize y to any element of U, hence u lies in the centre of U, so $u \in k$, say $u = \alpha$. Now $1 \cdot \alpha - \sum_2^n u_i \alpha_i = 0$ is a dependence relation over k. Conversely, if there is a dependence relation $\sum_1^n u_i \alpha_i = 0$, where not all the α_i vanish, then not all of $\alpha_2, \ldots, \alpha_n$ can vanish, because $u_1 = 1 \neq 0$, and so $\sum_2^n u_i' \alpha_i = 0$ is a dependence relation between $u_2', \ldots, u_n'$. Now the result follows by induction on n. ∎

This lemma, taken together with the earlier remarks, shows that for any field K dependable over its centre C, the free field $K_C\langle\!\langle X\rangle\!\rangle$ on any set X is dependable over C. Conversely, if $K_C\langle\!\langle X\rangle\!\rangle$ with an infinite set X is dependable over C, then K is dependable over C. In fact it is enough if the hypothesis holds for $K_C\langle\!\langle x, y\rangle\!\rangle$, for we can transform any word, using the infinite alphabet $x_i = y^{-i} x y^i$. Thus we have established

THEOREM 6.6.7. *Let K be a field, dependable over its centre C. Then the free K-field on any set X over C, $K_C\langle\!\langle X\rangle\!\rangle$, has a solvable word problem and is again dependable over C. Conversely, if the word problem in $K_C\langle\!\langle X\rangle\!\rangle$, where $|X| \geqslant 2$, is solvable, then K is dependable over C.* ∎

By taking $K = C$ we obtain a special case, where no hypothesis on dependability is needed:

COROLLARY 6.6.8. *Let k be any commutative field with solvable word problem. Then for any set X, the free field $k\langle\!\langle X\rangle\!\rangle$ has a solvable word problem.* ∎

We now give an example (due to Macintyre) of a field with unsolvable word problem. The idea is to take a finitely presented group with unsolvable word problem and use these relations in the group algebra of the free group. We first show how to encode a given group presentation in a direct product of free groups.

LEMMA 6.6.9. *Let F_x be the free group on the family $\{x_\lambda | \lambda \in \Lambda\}$ and F_y the free group on $\{y_\lambda | \lambda \in \Lambda\}$. In the direct product $F_x \times F_y$ let H be the subgroup generated by the elements $x_\lambda y_\lambda$ ($\lambda \in \Lambda$) and a subset U of F_x. Then $H \cap F_x$ is the normal subgroup of F_x generated by U.*

Proof. Let us denote by N the normal subgroup of F_x generated by U. Since the xs and ys commute in the direct product, we have for $u \in U$, $x_\lambda^{-1} u x_\lambda = (x_\lambda y_\lambda)^{-1} u (x_\lambda y_\lambda) \in H$; it follows that $N \subseteq H$ and so $N \subseteq H \cap F_x$ and we have to establish equality here.

Consider the natural homomorphism

$$f: F_x \times F_y \to (F_x/N) \times F_y, \tag{6}$$

which maps U to 1. If $w \in H \cap F_x$, then since $uf = 1$ for $u \in U$, it follows that wf is a product of the $(x_\lambda y_\lambda)f$, and since the xs and ys commute, we can write it as

$$wf = [v(x)v(y)]f = v(xf)v(yf),$$

for some group word v. Since $w \in F_x$, $wf \in F_x/N$ and so $v(yf) = 1$, but the $y_\lambda f$ are free, so $v = 1$ and $wf = 1$, hence $w \in \ker f = N$. ∎

Let F_x, F_y, H, N be as above and consider $G = F_x \times F_y$. This group can be ordered: we order the factors as in 2.4 and then take the lexicographic order on G. Hence we can form the power series field $K = k((G))$. The power series with support in H form a subfield L; we take a family of copies of K indexed by $\mathbf{Z}$ and form their field coproduct amalgamating L. The result is a semifir with universal field of fractions D, say. If σ denotes the shift automorphism in D, we can form the skew polynomial ring $D[t; \sigma]$ and its field of fractions $D(t; \sigma)$. It turns out that we can describe N as the centralizer of t in F_x:

LEMMA 6.6.10. *With the above notation, let $w \in F_x$, where $F_x \subseteq G \subseteq K = K_0 \subseteq D$; then $w \in N$ if and only if $tw = wt$ in $D(t; \sigma)$.*

Proof. If $w \in N$, then $w \in H$ by Lemma 6.9, hence $w \in L$ and so $wt = tw$. Conversely, when $tw = wt$, then w is fixed under σ and so lies in the fixed field of σ, i.e. $w \in L$. But L consists of all power series with support in H, hence $w \in H \cap F_x = N$, by Lemma 6.9. ∎

Now let A be a finitely presented group with unsolvable word problem, say

$$A = \langle x_1, \ldots, x_n; u_1 = \ldots = u_m = 1 \rangle, \tag{7}$$

where $u_1, \ldots, u_m$ are words in the xs. We shall construct a finitely presented field whose word problem incorporates that of A. Let k be a commutative field and put

$$M = k \langle\!\!\langle x_1, \ldots, x_n, y_1, \ldots, y_n, t; \Phi \rangle\!\!\rangle,$$

where Φ consists of the following equations:

$$x_i y_j = y_j x_i, \; x_i y_i t = t x_i y_i \; (i, j = 1, \ldots, n), \; u_\mu t = t u_\mu \; (\mu = 1, \ldots, m). \tag{8}$$

To see that this definition is meaningful, let $P = P_x$ be the free field over k on $x_1, \ldots, x_n$ and form the field coproduct of the $P(y_i)$ over P. The resulting field Q has $F_x \times F_y$ naturally embedded in it; in fact Q is the universal field of fractions of the group algebra of $F_x \times F_y$ over k. In Q consider the subfield generated by H over k, $k(H)$ say, and let S be the field coproduct of copies of Q indexed by $\mathbf{Z}$ amalgamating $k(H)$. With the shift automorphism σ in S we can form the skew function field $T = S(t; \sigma)$; from its construction this is just M (see Lemma 5.5.6). By the universality of T we have a specialization from T to $D(t; \sigma)$. We claim that

$$\text{for any } w \in F_x, \, w \in N \Leftrightarrow tw = wt \text{ in } T. \tag{9}$$

Clearly if $w \in N$, then $tw = wt$; conversely when $wt = tw$ in T, then this also holds in $D(t; \sigma)$, hence $w \in N$ by Lemma 6.10.

Now (9) shows that the word problem in M $(= T)$ is unsolvable because this is the case for $A = F_x/N$.

It only remains to find an example of a finitely presented group with unsolvable word problem. Such examples are given by Lyndon and Schupp [77], to whom we refer for details. Perhaps the simplest way (described there) is to take a r.e. but non-recursive set S of integers and form

$$G = \langle a, b, c, d; a^{-i} b a^i = c^{-i} d c^i, i \in S \rangle.$$

This presentation has unsolvable word problem since $a^{-n} b a^n = c^{-n} d c^n$ holds if and only if $n \in S$ and S is not recursive. Since S is r.e., G can be embedded in a finitely presented group H (by Higman's theorem quoted in 6.5), and so H has unsolvable word problem.

Exercises

1. Let R be any ring and K an epic R-field. Show that the word problem in K can be solved if the singular kernel is recursive.

2. (G. M. Bergman) Let k be a countable commutative field. Show that the Hilbert field $H = k \langle u, v; uv = vu^2 \rangle$ is again countable. Use the specialization lemma to enumerate all the full matrices over $k \langle X \rangle$ and hence solve the word problem for this field.

3°. Let K be a field with a central subfield k, such that K is dependable over k. Find an extension E of K with exact centre k and dependable over k. Deduce that $K_k \langle X \rangle$ has a solvable word problem.

4°. Let K be a field with central subfield k. If $K_k \langle x \rangle$ for a single indeterminate x has a solvable word problem, is K necessarily dependable over k?

5°. Is the conjugacy problem in $k \langle X \rangle$, i.e. deciding when two elements are conjugate, solvable?

6. Show that any EC-field has unsolvable word problem.

7. Verify the assertion after Lemma 6.10 that Q is the universal field of fractions of the group algebra of $F_x \times F_y$ over k.

6.7 The class of rings embeddable in fields

In 1.2 we saw on general grounds that the class of rings embeddable in fields can be characterized by a set of quasi-identities, together with the condition excluding zero-divisors. An explicit form of these quasi-identities was provided by Th. 4.4.5, which tells us that a ring R can be embedded in a field if and only if $R \neq 0$ and no diagonal matrix with non-zero diagonal elements can be written as a determinantal sum of non-full matrices. Let us call this the *embedding condition*. To express this condition in the form of quasi-identities we need to separate out the integral domain condition:

PROPOSITION 6.7.1. *A ring R can be embedded in a field if and only if R is an integral domain and no scalar matrix can be written as a determinantal sum of non-full matrices.*

Proof. These conditions are certainly necessary: that R must be an integral domain is of course well known; it also follows from the observation that if $ab = 0$, then

$$\begin{pmatrix} a & 0 \\ 0 & b \end{pmatrix} = \begin{pmatrix} a & 0 \\ 1 & b \end{pmatrix} \nabla \begin{pmatrix} 0 & 0 \\ -1 & b \end{pmatrix} = \begin{pmatrix} a \\ 1 \end{pmatrix}(1 \quad b) \nabla \begin{pmatrix} 0 \\ 1 \end{pmatrix}(-1 \quad b).$$

Hence by the embedding condition, $a = 0$ or $b = 0$. Conversely, when the conditions of the proposition hold, suppose that we have an equation

$$D = A_1 \nabla \ldots \nabla A_r, \tag{1}$$

where D is diagonal and $A_1, \ldots, A_r$ are non-full. If $D = \operatorname{diag}(d_1, \ldots, d_n)$, we multiply (1) by $\operatorname{diag}(1, d_1, d_1 d_2, \ldots, d_1 \ldots d_{n-1})$ on the left and by $\operatorname{diag}(d_2 \ldots d_n, d_3 \ldots d_n, \ldots, d_n, 1)$ on the right. Then we have the scalar matrix $d_1 \ldots d_n I$ expressed as a determinantal sum of non-full matrices. By hypothesis it follows that $d_1 \ldots d_n = 0$, hence $d_i = 0$ for some i. This proves the sufficiency of the conditions. ∎

The conditions of Prop. 7.1, apart from the integral domain condition, can be expressed as quasi-identities, for we have

$$cI = P_1 Q_1 \nabla \ldots \nabla P_r Q_r \Rightarrow c = 0, \tag{2}$$

where the P_i are $n \times (n-1)$ and the Q_i are $(n-1) \times n$ and it is understood that each determinantal sum refers to a particular column and a system of bracketing is given. Moreover, we use the convention that the right-hand side is undefined unless at each stage the matrices to be operated on agree in all but the columns that are to be added.

Thus we now have an explicit set of quasi-identities that are necessary and sufficient for the embeddability of an integral domain in a field. Of course we shall need (2) for all n and all r, so we have an infinite set of quasi-identities. So far there is nothing to tell us whether this infinite set is perhaps equivalent to a finite set of quasi-identities, or indeed any finite set of elementary sentences. We shall now show that this is not the case. For the proof we shall need the compactness theorem of logic, which states that a set of elementary sentences is consistent (i.e. has a model) if and only if every finite subset is consistent. If for each set Σ of sentences $\mathcal{P}(\Sigma)$ denotes the class of its models, the space of all models may be regarded as a topological space, whose closed sets are finite unions of sets of the form $\mathcal{P}(\Sigma)$. For any infinite set Σ of sentences we clearly have

$$\mathcal{P}(\Sigma) = \bigcap \mathcal{P}(\Sigma_f), \tag{3}$$

where Σ_f ranges over the finite subsets of Σ. Now the compactness theorem may be interpreted as saying that the model space is compact, for it asserts that the directed system $\mathcal{P}(\Sigma_f)$ of non-empty closed sets has the non-empty intersection $\mathcal{P}(\Sigma)$ (see Malcev [73], IV. 8, p. 163, or UA V.5, p. 213 or also Ex. 5 below).

We shall need two results about rings:

(I) *Every semifir is embeddable in a field.*

(II) *For every $n \geqslant 1$, there exists an n-fir which is not embeddable in a field.*

(I) was proved in Cor. 4.5.9, while (II) follows by taking the universal non-IBN ring of type $(n + 1, n + 2)$, see 5.7.

We remark that the class of n-firs, for any $n \geqslant 1$, may be defined by a single elementary sentence, namely $P_1 \wedge \ldots \wedge P_n$, where

$$P_n: 1 \neq 0 \wedge \forall x_1, \ldots, x_n, y_1, \ldots, y_n \left(\sum x_i y_i = 0 \right) \Rightarrow$$

$$\exists p_{ij}, q_{ij} \ (i, j = 1, \ldots, n)$$

$$\left(\sum p_{ij} q_{jk} = \sum q_{ij} p_{jk} = \delta_{ik} \wedge \left(\sum x_i p_{i1} = 0 \vee \sum q_{1j} y_j = 0 \right) \right).$$

If $\mathcal{F}_n$ denotes the class of n-firs and $\mathcal{F}$ the class of semifirs, then we clearly have

$$\mathcal{F}_1 \supseteq \mathcal{F}_2 \supseteq \ldots, \qquad \bigcap \mathcal{F}_n = \mathcal{F}. \tag{4}$$

Suppose now that there is a finite set of elementary sentences which is necessary and sufficient for an integral domain to be embeddable in a field. On replacing these sentences by their conjunction, we obtain a single sentence Q, say. Then for any $n \geqslant 1$, the sentences $\neg Q, P_1, P_2, \ldots, P_n$ have a model in the n-fir not embeddable in a field, by (II). By the compactness theorem, the set $\{ \neg Q, P_1, P_2, \ldots \}$ has a model; this is a semifir not embeddable in a field, and it contradicts (I). Thus we have a contradiction and it proves

THEOREM 6.7.2. *The class of integral domains embeddable in a field can be characterized by a set of quasi-identities, but it cannot be defined by a finite set of elementary sentences.* ∎

Exercises

1. Show that inclusions in (4) are all proper.

2. Show that the class of all rings with IBN cannot be defined by a finite set of elementary sentences.

3. Show that the class of rings with a universal field of fractions cannot be defined by a finite set of elementary sentences.

4. Show that the class of rings R with an R-field cannot be defined by a finite set

of elementary sentences. (Hint. Find an elementary sentence describing 2×2 matrix rings over n-firs.)

5. Let Σ be a set of elementary sentences such that each finite subset Σ_0 is consistent. Show that the sets $\mathscr{P}(\Sigma_0)$ of models of Σ_0 possess the finite intersection property and by forming ultraproducts obtain a model for Σ, and so prove the compactness theorem.

Notes and comments

The notion of a presentation of a field (which depends essentially on the theory of matrix ideals given in 4.4) was developed in Cohn [75] to obtain a form of the Nullstellensatz (see Ch. 8). The specialization lemma was first proved by Cohn [72″] (where the proof used important suggestions by G. M. Bergman). Since the appearance of *Skew Field Constructions* the proof has been further simplified, to be independent of the inertia theorem. The normal basis theorem 2.10 is taken from Cohn [80].

The normal form of 6.3 in the special case $E = D = k$ was described by M. L. Roberts [84], while Lemma 3.4 and Th. 3.5 for this case are due to A. H. Schofield (see FR, 5.8); the present generalization is new. For applications to a normal form in the power series representation see Cohn and Reutenauer [94]. The existence proof for free fields (Th. 4.1) is taken from A.3, Th. 11.3.3, p. 446. Th. 4.6 on relations between free fields and the work leading up to it were presented in *Skew Field Constructions*, albeit with some errors, which were corrected in Cohn [82′]. Prop. 4.8, Th. 4.9 and the example following it are taken from Cohn and Dicks [80].

No really satisfactory notion of 'algebraic closure' has so far been found for skew fields; some possibilities will be discussed in Ch. 8, but if we turn to logic we find two versions of 'completeness'. On the one hand there are the EC-fields; their construction is relatively straightforward and they have various useful properties, which are discussed in 6.5 (the concept was developed in the 1950s, see A. Robinson [63]; the applications to fields in 6.5 are taken from Cohn [75]). On the other hand there is the smaller class of 'generic fields', obtained as the (infinite) forcing companion of the class of fields. For an inductive theory where the forcing companion is axiomatizable, there is also a model companion, but this is not the case for the theory of fields, as was noted by Sabbagh [71] (cf. Macintyre [77] and Ex. 10 of 6.5). This follows because the class of generic fields does not admit ultraproducts (in this respect skew fields behave like groups). We have not defined generic fields in the text, as they have so far no immediate application. For a fuller account see Macintyre [77] or UA, Ch. IX. Th. 5.11 is taken from Wheeler [72], with a proof from Cohn [75].

The word problem for groups was first formulated by Max Dehn in 1911, in connexion with group presentations arising in topology. It has been solved in many special cases; the first examples of finitely presented groups with unsolvable word problem are due to P. S. Novikov [55] and W. W. Boone [59]. Th. 6.7, giving conditions for the word problem for free fields to be solvable, was proved by Cohn [73″], in response to a question by Macintyre. The first example of a field with unsolvable word problem was given by Macintyre [73], our account follows another example by Macintyre; Lemma 6.9 used in the proof is due to K. A. Mikhailova [58]. The proof of Th. 7.2, showing that integral domains embeddable in fields cannot be defined by finitely many elementary sentences, is taken from Cohn [74]. It is modelled on similar proofs by Robinson [63].

It may be of interest to recall the various classes of rings considered by Bokut' [81] (cf. also FR, p. 486f.). Let $\mathfrak{D}_0$ be the class of integral domains, $\mathfrak{D}_1$ the class of rings R such that $R^\times$ is embeddable in a group, $\mathfrak{D}_2$ the class of rings such that the universal $R^\times$-inverting map is injective and $\mathfrak{E}$ the class of fields. Then

$$\mathfrak{D}_0 \supset \mathfrak{D}_1 \supset \mathfrak{D}_2 \supset \mathfrak{E},$$

where the inclusions are obvious (though not the fact that they are proper). B. L. v. d. Waerden [30] had asked whether $\mathfrak{D}_0 = \mathfrak{E}$ and this was answered by Malcev [37] by showing that $\mathfrak{D}_0 \neq \mathfrak{D}_1$. At about this time Malcev asked whether $\mathfrak{D}_1 = \mathfrak{E}$; the answer came 30 years later from three people working independently: Bowtell [67] and Klein [67] gave examples showing that $\mathfrak{D}_2 \neq \mathfrak{E}$, while Bokut' [67, 69] proved that $\mathfrak{D}_1 \neq \mathfrak{D}_2$. His proofs are quite long and have not been simplified. The examples of Bowtell and Klein can be verified using the methods of 5.7. Valitskas [87] uses defining relations expressed in terms of the blocks of 4.3 to construct a ring R which is in $\mathfrak{D}_1$ but not in $\mathfrak{D}_2$. By the same method he finds a ring whose adjoint semigroup (with multiplication $a \circ b = a + b + ab$) is embeddable in a group, but which is not embeddable in a radical ring.

7

Rational relations and rational identities

The specialization lemma in one of its forms (Prop. 6.2.7) states that in a field of infinite degree over its centre, itself infinite, there are no rational identities, and the proof depended on Amitsur's GPI-theorem. In 7.1 we again take up GPIs and examine their relation with ordinary polynomial identities.

The functional approach leads in 7.2 to another treatment of rational identities in fields, and the rational topology, a topic to which we shall return in Ch. 8. To study rational identities we need, besides the free field, the generic division algebras of different PI-degrees. They are introduced in 7.3; the specializations between them are described there and are illustrated in 7.4.

The rest of the chapter is devoted to an exposition of Bergman's theory of specializations. The basic notions of *rational meet* and *support relation* are explained in 7.5 and in 7.6 we see how they are realized in generic division algebras. Finally in 7.7 examples of the different support relations are given, showing the totally different behaviour in the non-commutative case.

7.1 Polynomial identities

Every ring satisfies certain identities such as the associative law: $(xy)z = x(yz)$. In a field the situation is less simple; we have rational identities like $xx^{-1} = 1$ or $(xy)^{-1} = y^{-1}x^{-1}$, but here it is necessary to restrict x and y to be different from zero. In order to discuss rational identities over a field it is helpful first to summarize the situation for rings.

Let k be a commutative field and $F = k\langle X \rangle$ the free k-algebra on a set $X = \{x_1, x_2, \ldots\}$. Any k-algebra A is said to *satisfy* the polynomial identity

$$p(x) = 0, \tag{1}$$

if p is an element of F which vanishes for all values of the xs in A. If A satisfies a *non-trivial* identity, i.e. this is true for some $p \neq 0$, it is called a *PI-algebra*. The basic result on PI-algebras is

KAPLANSKY'S PI-THEOREM (Kaplansky [48]). *Let R be a primitive PI-algebra with a polynomial identity of degree d. Then R is a simple algebra of finite degree n^2 over its centre, where $n \leqslant d/2$.*

For a proof see Jacobson [56], p. 226, Herstein [68], p. 157 or A.3, Th. 10.4.6, p. 401. We recall that a ring is (left) *primitive* if it has a faithful simple (left) module.

In (1) the coefficients were restricted to lie in the centre; without this restriction the result clearly fails to hold. E.g. in the matrix ring $\mathfrak{M}_n(k)$ over k we have the identity

$$e_{11}xe_{11}ye_{11} - e_{11}ye_{11}xe_{11} = 0,$$

which has degree 2, but the degrees of the algebras satisfying it are unbounded. Of course this is not an identity of the form (1), but a generalized polynomial identity (GPI), as defined in 6.2. The existence of such a generalized identity again limits the algebra, as Amitsur has shown. For reference we state his theorem (of which a special case was used in 6.2):

AMITSUR'S GPI-THEOREM (Amitsur [65]). *A primitive ring R satisfies a generalized polynomial identity if and only if it is isomorphic to a dense ring of linear transformations in a vector space over a field of finite degree over its centre, and R contains a linear transformation of finite rank.*

A proof can be found in Herstein [76], p. 31.

The connexion between generalized and ordinary polynomial identities may be described as follows. Let D be a simple algebra of finite dimension n over its centre k; then k is a field, n is a perfect square, say $[D:k] = n = d^2$ and so for $m \geqslant 1$, $[D^m:k] = mn$ (see A.3, Ch. 7). In terms of a k-basis $u_1, \ldots, u_n$ for D we can write the elements of the tensor D-ring $F = D_k\langle x_1, \ldots, x_m\rangle$ as linear combinations (over k) of monomials $u_{i_0}x_{j_1}u_{i_1} \ldots x_{j_r}u_{i_r}$. There is a pairing

$$F \times D^m \to D, \quad (f, a) \mapsto f(a),$$

where $f(a) = f(a_1, \ldots, a_m)$ arises from $f(x_1, \ldots, x_m)$ on replacing x_j by $a_j \in D$. If we fix $f \in F$, we have a mapping $D^m \to D$ and if we fix $a \in D^m$, we get a mapping $F \to D$; clearly the latter mapping is a D-ring homomorphism. Writing $\mathrm{Hom}(F, D)$ for the set of all D-ring homomorphisms, we can state the result as follows:

LEMMA 7.1.1. *If D is a k-algebra and $F = D_k\langle x_1, \ldots, x_m \rangle$, then*

$$\mathrm{Hom}(F, D) \cong D^m.$$

Explicitly we have $\varphi \mapsto (x_1\varphi, \ldots, x_m\varphi)$.

The proof is almost immediate: we have seen that $a \in D^m$ defines a homomorphism and conversely, each homomorphism φ provides an element $(x_1\varphi, \ldots, x_m\varphi)$ of D^m. ∎

This result expresses the left adjoint property of F, viz. the set $\mathrm{Hom}(F, D)$ corresponds to the set of mappings from X to D, i.e. D^X, which is D^m, because $|X| = m$.

In a similar way we get a mapping θ, the *evaluation mapping*

$$\theta: F \to D^{D^m}, \quad f \mapsto \bar{f}, \tag{2}$$

where $\bar{f}: D^m \to D$ is the function on D^m defined by f. Now (2) is also a D-ring homomorphism if we regard the functions from D^m to D as a ring under pointwise operations; this amounts to treating the right-hand side of (2) as a product of rings. The image of F under θ is written $\bar{F}$; it is the ring of polynomial functions in m variables on D, and the kernel of θ is just the set of all generalized polynomial identities in m variables over D.

By identifying D^m with k^{mn} via a k-basis of D, we may view the ring of polynomial functions $k^{mn} (= D^m) \to k$ as a central k-subalgebra G of D^{D^m}; clearly G does not depend on the choice of k-basis of D. Since the canonical map $k^{D^m} \otimes_k D \to D^{D^m}$ is injective, the subring C of D^{D^m} generated by D and G is of the form $G \otimes_k D$. If, moreover, k is infinite, then G is just the k-algebra of polynomials in mn commuting indeterminates, so C is the D-ring of polynomials in mn central indeterminates. Another way of expressing C is as the image of the tensor ring:

THEOREM 7.1.2. *Let D be a central simple k-algebra of dimension n. Then $F = D_k\langle x_1, \ldots, x_m \rangle$ may be expressed as the tensor ring on mn D-centralizing indeterminates and $C = \bar{F}$ is the image of the evaluation map θ.*

Proof. We may regard F as the tensor D-ring on the D-bimodule $(D \otimes_k D)^m$; since D is central simple, the map

$$\varphi: D \otimes_k D \to \text{End}_k(D) \cong D^n, \quad \text{where } (a \otimes b)\varphi: x \mapsto axb, \qquad (3)$$

is injective, and hence is a D-bimodule isomorphism, by the density theorem (A.3, Th. 7.1.1, p. 259). It follows that F is the tensor D-ring on mn D-centralizing indeterminates.

Now fix a k-basis $u_1, \ldots, u_n$ of D and consider the dual k-basis u_1^*, $\ldots, u_n^*$ in $\text{Hom}_k(D, k) \subseteq \text{End}_k(D)$. For each $\mu = 1, \ldots, n$ there exists $v_\mu = \sum a_{\mu\lambda} \otimes u_\lambda \in D \otimes D$ mapping onto u_μ^*, i.e. such that $v_\mu \varphi = u_\mu^*$. If in $F = D_k\langle x_1, \ldots, x_m \rangle$ we take $v_{i\mu} = \sum a_{\mu\lambda} x_i u_\lambda$, then $F = D\langle v_{i\mu} \rangle$ is the free D-ring on the mn D-centralizing indeterminates $v_{i\mu}$, $i = 1, \ldots, m$, $\mu = 1, \ldots, n$. If we put $\xi_{i\mu} = v_{i\mu}\theta: (\sum b_{1\lambda}u_\lambda, \ldots, \sum b_{m\lambda}u_\lambda) \mapsto b_{i\mu}$, then it is clear that $\bar{F}$ is the D-ring generated by the $\xi_{i\mu}$. Now G is by definition the k-algebra generated by the $\xi_{i\mu}$, hence we conclude that $C = GD = \bar{F}$, as claimed. ∎

If we examine the role played by θ, we obtain the following explicit expression for it:

THEOREM 7.1.3. *Let D be a central simple k-algebra of degree n, where k is infinite. Then the evaluation map θ can be expressed in the form*

$$F = D\langle v_{i\mu} \rangle \to D[\xi_{i\mu}] = C \subseteq D^{D^m},$$

$$\text{where } v_{i\mu} \mapsto \xi_{i\mu} \ (i = 1, \ldots, m, \ \mu = 1, \ldots, n).$$

Hence the kernel of θ is generated by the commutators of pairs of the $v_{i\mu}$. ∎

Exercises

1. Prove Th. 1.2 by tensoring D with a splitting field K, so as to reduce D to the form K_m.

2. In $\mathfrak{M}_n(A)$ denote the matrix units by e_{ij}. Verify that the product of e_{11}, e_{12}, e_{22}, $\ldots, e_{n-1 n}, e_{nn}$ in any order other than the given one is zero. Deduce that for any non-zero k-algebra A, the matrix ring $\mathfrak{M}_n(A)$ satisfies no polynomial identity of degree $< 2n$. (This is the staircase lemma, see A.3, p. 379. When A is commutative, $\mathfrak{M}_n(A)$ satisfies an identity of degree $2n$, viz. the standard identity $S_{2n} = 0$, by the Amitsur–Levitzki theorem, A.3, p. 378.)

3. Let D be a central simple k-algebra with k-basis $u_1, \ldots, u_n$ and put $X = \{x_1, \ldots, x_n\}$, $\Xi = \{\xi_{\lambda i} | \lambda = 1, \ldots, m, \ i = 1, \ldots, n\}$. Show that the map $D \otimes k\langle X \rangle \to D \otimes k[\Xi]$ defined by $x_\lambda \mapsto \sum \xi_{\lambda i} u_i$ is surjective. Is it injective?

7.2 Rational identities

The basic result on rational identities, again due to Amitsur [66], states that there are no non-trivial rational identities over a skew field which is of infinite degree over its centre and whose centre is infinite. But it is now more tricky to decide what constitutes a 'non-trivial' identity. Here are some 'trivial' ones:

$$(x + y)^{-1} - y^{-1}(x^{-1} + y^{-1})^{-1}x^{-1} = 0,$$

$$[x^{-1} + (y^{-1} - x)^{-1}]^{-1} - x + xyx = 0 \text{ (Hua's identity)},$$

$$[x - (1 + y)^{-1}x(1 + y)][(1 + y)^{-1}x(1 + y) - y^{-1}xy]^{-1} - y = 0.$$

Amitsur's result is implicit in Th. 6.4.7; we shall now sketch another proof, due to Bergman [70].

Our first task is to find a means of expressing rational functions. Let D be a field with centre k; we can form the polynomial ring $D[t]$ in a central indeterminate t and its field of fractions $D(t)$. Any $\varphi \in D(t)$ has the form $\varphi = fg^{-1}$, where $f, g \in D[t]$, and we can set $t = \alpha$ if $\alpha \in k$ is such that $g(\alpha) \neq 0$; then $\varphi(\alpha)$ will be defined. For a given φ we can choose f, g in $\varphi = fg^{-1}$ to be coprime, and then f, g will not both vanish for any $\alpha \in k$. Since we only had to avoid the zeros of g in defining $\varphi(\alpha)$ we see that φ is defined at all but finitely many points of k. More generally, the same reasoning applies to any commutative unique factorization domain R and its field of fractions U, with an epic R-field playing the role of k, but once we give up unique factorization, the situation changes. For example, consider the ring of polynomials in commuting variables x, y, z, t over k subject to the relation $xt = yz$ (this is the coordinate ring of a quadric, the simplest non-UFD). Here we have $x/y = z/t$, but there is no representation of this fraction which can be used for all specializations: x/y fails if we put $x = y = 0$, while z/t fails for $z = t = 0$.

In the case of several non-central variables the role of $D(t)$ can be taken by the free field, but that notion will not be needed in our first construction. What we shall do is to build up formal expressions in x_1, ..., x_m using $+$, $-$, $\times$, $\div$ and elements of D, and called *rational expressions*. The expressions will be defined on a subset of D^m or more generally, of E^m, where E is a D-field. Strictly speaking, our 'expressions' should be called 'generalized rational expressions', to emphasize that the coefficients need not commute with the variables, but we shall omit the qualifying adjective and often briefly refer to 'expressions'.

Let X be any set; by an *X-ring* we shall understand a ring R with a mapping $\alpha: X \to R$, and we sometimes write (R, α) to emphasize this mapping. More generally, for any field D an *(X, D)-ring* is a D-ring R

with a mapping $\alpha: X \to R$. If R is a field, we speak of an X-*field* or an (X, D)-*field*; the latter is essentially the same as a $D_k\langle X \rangle$-field (with k understood as ground field) in our previous terminology. An (X, D)-field is again called *epic* if it is generated, as D-field, by the image of X.

Given $X = \{x_1, \ldots, x_m\}$ and a field D, we write $\Re(X; D)$ for the free abstract algebra on X with constants D and operations $\{-_1, (\ )_1^{-1}, +_2, \times_2\}$, where the subscript indicates the arity of the operation. For each expression there is a unique way of building it up since no relations are imposed, thus e.g. $(x - x)^{-1}$ is defined. In contrast to 7.1 we now have a *partial* mapping

$$\Re(X; D) \times E^m \to E, \tag{1}$$

for any D-field E. Thus any map $\alpha: X \to E$ defines a map $\bar{\alpha}$ of a subset of $\Re(X; D)$ into E by the following rules:

  (i) if $a \in D$, then $a\bar{\alpha} = a$,
  (ii) if $a = x_i$, then $x_i\bar{\alpha} = x_i\alpha$,
  (iii) if $a = -b$ or $b + c$ or bc and $b\bar{\alpha}, c\bar{\alpha}$ are defined, in E, then
       $a\bar{\alpha} = -b\bar{\alpha}$ or $b\bar{\alpha} + c\bar{\alpha}$ or $b\bar{\alpha} \cdot c\bar{\alpha}$,
  (iv) if $a = b^{-1}$ and $b\alpha$ is defined and non-zero, then $a\bar{\alpha} = (b\bar{\alpha})^{-1}$.

Since $\bar{\alpha}$ just extends α, we can safely omit the bar. We thus obtain the following simple condition for an expression in X over D to define an element of a given (X, D)-field:

PROPOSITION 7.2.1. *Let D be a field, X a set, (E, α) an (X, D)-field and $a \in \Re(X; D)$. Then $a\alpha$ is undefined if and only if a has a subexpression b^{-1}, where $b\alpha = 0$.* ∎

With each $f \in \Re(X; D)$ we associate its *domain* $\mathrm{dom}\, f$, a subset of D^m consisting of the points at which f is defined; more generally we shall consider $\mathrm{dom}\, f$ in E^m, where E is a D-field. If $\mathrm{dom}\, f \neq \varnothing$, f is called *non-degenerate* on E. Taking $m = 1$, we obtain an expression $f(t)$ in a single variable t; if we regard t as a central variable, we may thus consider $f(t)$ as an element of $D(t)$. As such, it will (by Lemma 6.2.4) be defined for all but finitely many values of t in the centre of D (assumed infinite).

LEMMA 7.2.2. *Let D be a field which is a k-algebra, where k is an infinite field, and let E be a D-field with centre k. If $f, g \in \Re(X; D)$ are expressions that are non-degenerate on E, then $\mathrm{dom}\, f \cap \mathrm{dom}\, g \neq \varnothing$.*

Proof. Let $p \in \mathrm{dom}\, f$, $q \in \mathrm{dom}\, g$, write $r = tp + (1 - t)q$ and consider $f(r), g(r) \in E(t)$. For $t = 1$, $f(r)$ is defined and for $t = 0$, $g(r)$ is defined, hence each is defined for all but finitely many values of t in k and so for some value t_0 of t both are defined. ∎

Given $f, g \in \mathfrak{R}(X; D)$, let us put $f \sim g$ if f, g are non-degenerate (on a given E) and f, g have the same value at each point of dom $f \cap$ dom g. This is clearly an equivalence; reflexivity and symmetry are obvious and transitivity follows by Lemma 2.2. If f, g are non-degenerate, then so are $f + g, f - g, fg$, and they depend only on the equivalence classes of f, g, not on f, g themselves. Further, if $f \not\sim 0$, then f^{-1} is defined. Each equivalence class of expressions is called a *rational function*, and as we have just seen, they form a field:

THEOREM 7.2.3. *Let D be a field which is a k-algebra, where k is an infinite commutative field, and let E be a D-field with centre k. Then the rational functions from E^m to E with coefficients in D form a field $D_k(X; E)$.* ∎

When E is commutative, this field $D_k(X; E)$ reduces to the familiar function field $D(X)$ and is independent of E. In that case any element of $D(X)$ can be written as a quotient of two coprime polynomials, and this expression is essentially unique. The dependence on E in the general case will be examined below. Now there is no such convenient normal form for the elements of $D_k(X; E)$. But in any case each element of $D_k\langle X \rangle$ defines an element of $D_k(X; E)$, thus $D_k(X; E)$ is a $D_k\langle X \rangle$-field. It may not be epic, but when it is (e.g. when $E = D$), then $D_k(X; E)$ is a specialization of $D_k \langle\!\langle X \rangle\!\rangle$, and in any case it contains such a specialization as subfield. By a *generalized rational identity* in $D_k(X; E)$ we understand an element of $D_k \langle\!\langle X \rangle\!\rangle$ at which this specialization either is undefined or is defined and equal to zero.

The domains of functions form a basis for the open sets of a topology on E^m, the *rational topology* on E^m; this is in general distinct from the Zariski topology, a polynomial topology which is usually coarser (see 8.7). The closed sets in the rational topology are of the form

$$\mathcal{V}(P) = \{p \in E^m | f(p) = 0 \text{ or undefined, for all } f \in P\},$$

$$\text{where } P \subseteq D_k(X; E).$$

A subset S of E^m is called *irreducible* if it is non-empty and not the union of two closed proper subsets. Equivalently, the intersection of two non-empty open subsets of S is non-empty. Thus Lemma 2.2 states that E^m is irreducible in the rational topology when the centre of E is infinite.

A subset S of E^m is called *flat* if S contains with p and q also

$\alpha p + (1 - \alpha)q$ for infinitely many $\alpha \in k$. Of course a closed flat subset will then contain $\alpha p + (1 - \alpha)q$ for all $\alpha \in k$. Now the proof of Lemma 2.2 gives us

LEMMA 7.2.4. *If E is a field with infinite centre, then any non-empty flat subset of E^m is irreducible.* ∎

An example of a flat closed subset is the space S defined by

$$\sum a_{i\lambda}x_i b_{i\lambda} = c \quad (a_{i\lambda}, b_{i\lambda}, c \in D). \tag{2}$$

By Lemma 2.4, S is irreducible (if non-empty) and so, as in Th. 2.3, it yields a field $D_k(X; S)$ in $x_1, \ldots, x_m$ satisfying (2); it may be called the *function field* of the set defined by (2). In general it is not easy to decide whether a given set is irreducible, e.g. $x_1 x_2 - x_2 x_1 = 0$ for $E \supset D \supset k$. In the commutative case every closed set is a finite union of irreducible closed sets, but this need not hold in general.

It is clear that every polynomially closed set is rationally closed; we shall be interested in conditions for the converse to hold. This will be the case for flat sets, but first we shall need two general remarks:

(i) Let $S \subseteq E^m$ and p be a point not in $\bar{S}$, the closure of S; then there exists f defined at p but not anywhere on S. The degeneracy of f can only arise by inversion, so $f = g^{-1}$, where g is non-degenerate on S and is 0 at all points of S where defined, and $g(p) \neq 0$.

(ii) Any element of $D(t)$ defined at $t = 0$ can be expanded in a power series. If $f = a - tg$ say, then $f^{-1} = a^{-1}(1 - tga^{-1})^{-1} = \sum a^{-1}(tga^{-1})^n$, so we can build up any function in $D(t)$, provided that it is defined at $t = 0$.

We can now show that the closure of any flat set is polynomially closed, so for flat sets, rationally closed = polynomially closed.

LEMMA 7.2.5. *Let D be a field which is a k-algebra, where k is an infinite field, and let E be a D-field with centre k. If $S \subseteq D^m$ is flat, then its closure in E^m is polynomially closed.*

Proof. Let $p \notin \bar{S}$; we have to find a polynomial over D which is zero on S but not at p. We know that there is a rational function f, non-degenerate on S and $f = 0$ on S, but $f(p) \neq 0$. Let $q \in S$ be a point at which f is defined.

For any $x \in E^m$ consider $f((1 - t)q + tx)$; this is defined for $t = 0$, so it is a well-defined element of $E(t)$. If $x \in S$, f is zero by flatness, but for $x = p$ it is non-zero because it is non-zero for $t = 1$. In the power series expansion of $f((1 - t)q + tx)$, if we have to take the inverse of an expression $g(t)$, the constant term $g(0)$ is non-zero, because $f(q)$ is defined, and $g(0)$ does not involve the coordinates of x. Hence the expansion $f((1 - t)q + tx)$ has coefficients which are polynomials in x; their coefficients are in D, because $q \in S \subseteq D^m$. These polynomials are 0 on S, but at least one is non-zero at p, and this is the required polynomial. ∎

COROLLARY 7.2.6. *Let k, D, E be as before and assume that D and E satisfy the same generalized polynomial identities over k with coefficients in D. Then $D_k(X; E) \cong D_k(X; D)$, and for any D-subfield B of D, $B_k(X; E) \cong B_k(X; D)$.*

Proof. The rational closure of D^m in E^m is polynomially closed by Lemma 2.5, because D^m is flat. Since every GPI on D^m holds in E^m, the rational closure of D^m is E^m, i.e. D^m is dense in E^m. The rest follows because $B_k(X; E)$ is the subfield of $D_k(X; E)$ generated by B and X. ∎

If we bear in mind Amitsur's GPI-theorem, which for fields states that a field E of infinite degree over its centre satisfies no GPIs (except those stating that the centre of E commutes with the variables), then we obtain the following conclusion, where $E \supseteq D \supseteq B$ of Cor. 2.6 are replaced by $E' \supseteq E \supseteq D$:

PROPOSITION 7.2.7. *Let D be a field which is a k-algebra and let E, E' be D-fields with centres C, C' containing k, such that $E \subseteq E'$, $C \subseteq C'$ and $[E:C]$, $[E':C']$, C are infinite. Then for any set X,*

$$D_k(X; E) \cong D_k(X; E'). \quad \blacksquare$$

An extension of fields $E \subseteq E'$ is said to be *centralizing* if the centre of E' contains that of E; this just means that E' centralizes the centre of E (and we see that central extensions form a special class of centralizing extensions). We can now state Bergman's form of Amitsur's theorem on rational identities:

THEOREM 7.2.8. *Let D be a field. Then there exists a centralizing extension E of D with centre C, such that C and $[E:C]$ are infinite. For a given $m = 1, 2, \ldots$ all such fields E yield the same function field $D_k(x_1, \ldots, x_m; E)$ up to D-isomorphism.*

Proof. Write $X = \{x_1, \ldots, x_m\}$ and denote the centre of D by k. Then $D(t)$ has centre $k(t)$, and so does the Hilbert field H on $D(t)$. Moreover, H has infinite degree over $k(t)$ and $k(t)$ is infinite, so H is an extension of the required form. If E is any centralizing extension of D whose centre C is infinite and such that $[E:C] = \infty$, let E' be the Hilbert field on $E(t)$. Then E' is a centralizing extension of both E and H, so by Prop. 2.7, $D_k(X; E) \cong D_k(X; E') \cong D_k(X; H)$, therefore $D_k(X; E) \cong D_k(X; H)$, so $D_k(X; E)$ is independent of the choice of E. ∎

The result may be expressed by saying that for fields infinite over their centre, where the latter is infinite, there are no rational identities, a conclusion we already met in Th. 6.4.7.

When E is of finite degree over its centre, there are of course non-trivial identities, but Amitsur [66] has shown that they depend only on the degree (cf. also Bergman [70]). More precisely, if $[D:k] = n = d^2$ and E is any extension of D with infinite centre C containing k, where $[E:C] = (rd)^2$, then $D_k(X; E)$ depends only on D, d, r, $m = |X|$ and not on E. It can be shown that $D_k(X; E)$ has dimension $(rd)^2$ over its centre, hence these fields are different for different values of rd. Moreover, for $d_1 | d_2$ the field with d_1 is a specialization of that with d_2, see Bergman [70].

Exercises

1. Show that an integral domain which satisfies a polynomial identity is an Ore domain. (Hint. Use Prop. 1.6.6.)

2°. (Bergman) Show that the domain of definition of $f = (x^{-1} + y^{-1} + z^{-1})^{-1}$ is the intersection of dom (x^{-1}), dom (y^{-1}), dom (z^{-1}) and

$$\text{dom}\,(1 + xy^{-1} + xz^{-1})^{-1} \cup \text{dom}\,(yx^{-1} + 1 + yz^{-1})^{-1}$$

$$\cup \text{dom}\,(zx^{-1} + zy^{-1} + 1)^{-1},$$

and deduce that the set where f is undefined is irreducible. Is it polynomially closed?

3°. (Bergman) Is the set defined by $xy - yx = 0$ (for any $E \supseteq D \supseteq k$) irreducible, or at least a finite union of irreducible closed sets?

7.3 Specializations

We now examine how rational identities change under specialization. Over skew fields the situation is relatively straightforward. Consider a generalized rational expression $f(\xi_1, \ldots, \xi_m)$; we shall call f an *absolute generalized rational identity* (GRI) if in the free field $D_k \langle\!\langle x_1, \ldots, x_m \rangle\!\rangle$ the element $f(x_1, \ldots, x_m)$ is either undefined or zero. From the existence of universal denominators in $D_k\!\!<\!\!(X\}$ (Th. 6.3.7) it follows that any absolute GRI is in fact a GRI for all D-fields (understood as k-algebras). But rational identities can behave in unexpected ways in rings with few units. For example, in $\mathbf{Z}$ the identity $\xi - \xi^{-1} = 0$ holds, though it fails in homomorphic images such as $\mathbf{Z}/5\mathbf{Z}$. Even an algebra over a field, say a free k-algebra $k\langle X \rangle$ will satisfy the identity $\xi\eta^{-1} - \eta^{-1}\xi = 0$, though it has homomorphic images where this fails. Our aim will be to show that any absolute GRI over D is a GRI over any weakly finite D-ring. We begin by reducing the question to matrix form.

PROPOSITION 7.3.1. *Let D be a field which is a k-algebra. To every generalized rational expression f over D in the set of indeterminates $\{\xi_1, \ldots, \xi_r\}$ we can associate a matrix $A = A(x)$ over $D_k\langle X \rangle$ of index 1, such that*

(i) *at any point $a = (a_1, \ldots, a_r)$ over any weakly finite D-ring R which is a k-algebra, $f(a)$ is defined if and only if the denominator of $A(a)$ is invertible over R,*

(ii) *if the point $a \in R^r$ satisfies the equivalent conditions of (i), then $f(a)$ may be obtained as the last component u_∞ of the unique normalized solution $u = (1 \; u_* \; u_\infty)^\mathrm{T}$ of the equation $Au = 0$.*

Proof. We use induction on the complexity of f. If $f = \xi_i$ or $f \in D$, we can take $A = (f \; -1)$; both (i) and (ii) are clearly satisfied.

If $f = g + h$ and the matrices associated to g, h are B, C, then we associate to f the matrix

$$A = \begin{pmatrix} B_0 & 0 & -B_\infty & B_* & B_\infty \\ C_0 & C_* & C_\infty & 0 & 0 \end{pmatrix}.$$

When we evaluate this matrix at a point a of a weakly finite ring R, we see that its denominator will be invertible whenever those of B and C are; the converse also holds because all denominators are square and R is weakly finite. This proves the induction step for (i) in this case, and now (ii) also follows. If $f = gh$, we use

$$A = \begin{pmatrix} 0 & 0 & B_0 & B_* & B_\infty \\ C_0 & C_* & C_\infty & 0 & 0 \end{pmatrix},$$

with the same reasoning as before. When $g = -1$, this gives a matrix for $-h$.

There remains the case when $f = g^{-1}$. If the matrix associated to g is B, it would seem natural to take for f the matrix

$$A = (B_\infty \; B_* \; B_0).$$

This has almost the desired properties. For if $f(a)$ is defined, this means that $g(a)$ is defined and invertible and hence, by the induction hypothesis, $B(a)$ defines $g(a)$ and by Cramer's rule (Prop. 4.2.3), $B(a)$ will have an invertible numerator. Hence $A(a)$ then has an invertible denominator and so defines $f(a)$. However, the converse may not hold; namely if $g(a)$ itself is undefined, so that $f(a) = g(a)^{-1}$ also cannot be defined, the denominator of A may still be invertible. Thus for $f = g^{-1}$ to be defined it is necessary for g to be defined and invertible; this will be so if the denominator of A contains both the numerator and the denominator of B. To achieve this aim, we put

$$A = \begin{pmatrix} B_\infty & 0 & 0 & B_* & B_0 \\ 0 & B_* & B_\infty & 0 & 0 \end{pmatrix};$$

now it is easily checked that (i) and (ii) hold, and the result follows by induction. ∎

Since $D_k\langle X \rangle$ is a semifir, any square matrix over it is invertible over its universal field of fractions $D_k \langle\!\langle X \rangle\!\rangle$ if and only if it is full. Hence an expression f is an absolute GRI if and only if the corresponding matrix A constructed in Prop. 3.1 has either a non-full denominator (making $f(x)$ undefined) or a non-full numerator (making $f(x) = 0$, if it is defined). We can now achieve our aim announced earlier.

THEOREM 7.3.2. *Let D be a field which is a k-algebra and let R be a D-ring which is a k-algebra. Then every absolute generalized rational identity is a generalized rational identity for R if and only if R is weakly finite.*

Proof. Assume first that R is weakly finite; let f be an absolute GRI in $x_1, \ldots, x_r$ with associated matrix A, and take $a \in R^r$. Since $A(x)$ has a non-full numerator or denominator, the same is true of $A(a)$. Now a non-full matrix over a weakly finite ring cannot be invertible, so either (i) $A(a)$ has a non-invertible denominator, or (ii) $A(a)$ has an invertible denominator and a non-full numerator. By Prop. 3.1, in case (i) $f(a)$ is not defined; in case (ii) the numerator is non-full, so by Cramer's rule (Prop. 4.2.3) we find that $f(a) \oplus I$ is not full, say

$$\begin{pmatrix} f(a) & 0 \\ 0 & I \end{pmatrix} = \begin{pmatrix} p \\ P \end{pmatrix} (q \quad Q),$$

where P, Q are square. Thus $PQ = I$, $pQ = 0 = Pq$, $pq = f(a)$. Since R is weakly finite, $QP = I$, so $p = 0 = q$ and $f(a) = 0$, as we wished to show.

Now suppose that R is not weakly finite. Then R contains square matrices P, Q such that $PQ = I$ and $QP \neq I$. Let P, Q be $n \times n$ say; writing S, T for $n \times n$ matrices with indeterminate entries, let us consider the matrix equation

$$T(ST)^{-1}S - I = 0.$$

Written out in full, the left-hand side consists of n^2 expressions in the entries of S, T and $(ST)^{-1}$; thus they are rational expressions which are defined and equal to zero in the free field $D_k \langle\!\langle X \rangle\!\rangle$, so they are absolute GRIs, but not all of them hold when we set $S = P$, $T = Q$, though all are defined. ∎

To describe rational identities that are not absolute we shall need the notion of PI-degree. If A is a commutative ring, then $\mathfrak{M}_n(A)$ is of dimension n^2 over its centre as free A-module and it satisfies the standard identity of degree $2n$, by the Amitsur–Levitzki theorem (see A.3, Th. 9.5.8, p. 378):

$$S_{2n}(x_1, \ldots, x_{2n}) = \sum \operatorname{sgn} \sigma x_{1\sigma} x_{2\sigma} \ldots x_{(2n)\sigma} = 0,$$

where the sum is taken over all permutations σ of $(1, 2, \ldots, 2n)$ and sgn σ indicates the sign of σ. Let R be any PI-ring which is *prime* (i.e. the product of any two non-zero ideals is non-zero); by Posner's theorem it has a ring of fractions Q which is simple Artinian and satisfies the same polynomial identities as R (see A.3, Th. 10.7.6, p. 420, or Jacobson [75], p. 57). If Q is d^2-dimensional over its centre, then R satisfies $S_{2d} = 0$ and no standard identity of lower degree. We shall call d the *PI-degree* of R (and Q) and write $d = \text{PI-deg } R$. For a prime ring satisfying no polynomial identity the PI-degree is defined as ∞.

We shall also need the notion of a generic matrix ring. Let k be a commutative field and m, d two positive integers, and write $k[T]$ for the commutative polynomial ring over k in the family $T = \{x_{ij}^{\lambda}\}$ of md^2 commuting indeterminates, where $i, j = 1, \ldots, d$, $\lambda = 1, \ldots, m$. Let $k(T)$ be its field of fractions and consider the matrix ring

$$\mathfrak{M}_d(k[T]) \subseteq \mathfrak{M}_d(k(T)).$$

We have a canonical m-tuple of matrices $X_{\lambda} = (x_{ij}^{\lambda})$; the k-algebra generated by these m matrices is written $k\langle X \rangle_d$ and is called the *generic matrix ring* of order d. This may be regarded as the analogue of the matrix reduction functor for the category of commutative rings. Another way of describing it is as the free k-algebra on $X = \{X_{\lambda}\}$ in the variety of k-algebras generated by $d \times d$ matrix rings over commutative k-algebras. This ring $k\langle X \rangle_d$ is an Ore domain (see A.3, Prop. 9.7.2, p. 385); its field of fractions, written $k \Cleft X \Cright_d$, is called the *generic division algebra* of degree d; like $k \langle X \rangle_d$ it has PI-degree d, if $m > 1$. Of course, for $m = 1$, $k \langle X \rangle_d$ reduces to a polynomial ring in one variable; this case is of no interest here and we henceforth assume that $m > 1$.

Let (D, α) be any X-field and define its *domain* $\mathscr{E}(D)$ as the subset of $\mathscr{R}(X; D)$ on which α is defined. Let $\mathfrak{A}(D)$ be the subset of $\mathscr{E}(D)$ consisting of all functions which vanish for α. Any $f \in \mathfrak{A}(D) \cup (\mathscr{R}(X; D) \backslash \mathscr{E}(D))$ is called a *rational relation*, or k-rational relation if coefficients in k are allowed. Explicitly we have $f^\alpha = 0$, provided that f^α is defined. More precisely, a member of $\mathfrak{A}(D)$ is a *non-degenerate* relation, while a member of $\mathscr{R}(X; D) \backslash \mathscr{E}(D)$ is a *degenerate* relation. Now Amitsur's theorem on rational identities (Th. 2.8) may be expressed as follows: For any field D with infinite centre k and any set X, there is an X-field E over k such that the k-rational identities over D are the k-rational relations satisfied by X over E. Thus we can speak of E as the (relatively) *free* X-field for this set of identities. Moreover, the structure of E depends only on k, $|X|$ and the PI-degree of D:

If PI-deg $D = d$, then $E = k \Cleft X \Cright_d$ is the field of generic matrices,

If PI-deg $D = \infty$, then $E = k \Cleft X \Cright$ is the free k-field on X.

In particular, two k-fields satisfy the same rational identities if and only if they have the same PI-degree. Our aim is to describe the specializations between different generic matrix rings; here we shall need a theorem of Bergman and Small on PI-domains. We recall that a ring R with Jacobson radical $\mathfrak{J}$ is called *local* if $R/\mathfrak{J}$ is a field, and *matrix local* if $R/\mathfrak{J}$ is simple Artinian, i.e. a full matrix ring over a field.

THEOREM 7.3.3 (Bergman and Small [75]). (i) *If R is a prime PI-ring which is also local (or even matrix local) with maximal ideal* $\mathfrak{m}$, *then* PI-deg$(R/\mathfrak{m})$ *divides* PI-deg R.

(ii) *If $R_1 \subseteq R$ are PI-domains, then* PI-deg R_1 *divides* PI-deg R.

Since the result is somewhat peripheral, we shall sketch the proof of (ii) only. Let d, d_1 be the PI-degrees of R, R_1; they are also the PI-degrees of their fields of fractions Q, Q_1. We denote their centres by k, k_1; by enlarging Q_1 we may assume that $k_1 \supseteq k$. Now choose a maximal commutative subfield F_1 of Q_1 and enlarge F_1 to a maximal commutative subfield F of Q. Then $[F_1:k_1]$ divides $[F:k]$, and this means that $d_1|d$. ∎

With the help of this result we can derive Bergman's description of specializations between generic matrix rings.

THEOREM 7.3.4. *Let k be a commutative field, c, d positive integers and X a set with more than one element. Then the following conditions are equivalent:*

(a) $\mathscr{E}(k\langle\!\langle X\rangle\!\rangle_c) \subseteq \mathscr{E}(k\langle\!\langle X\rangle\!\rangle_d)$, *i.e. every rational identity in PI-degree d is also one for PI-degree c,*

(b) *there is an X-specialization $k\langle\!\langle X\rangle\!\rangle_d \to k\langle\!\langle X\rangle\!\rangle_c$,*

(c) *there is a surjective local homomorphism $D_d \to D_c$, where D_n is a division algebra over k of PI-degree n,*

(d) *c divides d.*

Proof. The implications (a) $\Rightarrow$ (b) $\Rightarrow$ (c) are clear. To prove that (c) $\Rightarrow$ (d), let $D'_d \subseteq D_d$ be a local ring with residue-class field D_c; then $c = $ PI-deg $D_c|$PI-deg $D'_d|$PI-deg D_d by Th. 3.3. To show that (d) $\Rightarrow$ (a), take an infinite k-field E. Since $c|d$, we can embed $\mathfrak{M}_c(E)$ in $\mathfrak{M}_d(E)$ by mapping α to $(\alpha \; \alpha \ldots \alpha)$. Then every rational identity in $\mathfrak{M}_d(E)$ holds in $\mathfrak{M}_c(E)$; but these identities are just the rational relations in $k\langle\!\langle X\rangle\!\rangle_d$, $k\langle\!\langle X\rangle\!\rangle_c$, hence $\mathscr{E}(k\langle\!\langle X\rangle\!\rangle_c) \subseteq \mathscr{E}(k\langle\!\langle X\rangle\!\rangle_d)$, and so (a) follows. ∎

Exercises

1. (Bergman) Let $X = \{x\}$, $D = Q(x)$, $D' = Q$, considered as X-field by $x \mapsto 0$. Find a relation which is defined and holds over D but is not defined over D'.

2. Show that a finite-dimensional division algebra satisfying the identity $[[x, y]^2, z] = 0$ must be a quaternion algebra. (Hint. Split the algebra by extending the centre and apply the staircase lemma (7.2, Ex. 1).)

7.4 A particular rational identity for matrices

As a consequence of Th. 3.4 there are rational identities holding in PI-degree 3 but not in PI-degree 2. We shall now describe a particular example of such an identity which was found by Bergman [76]. From the results of Bergman and Small [75] (see 7.3 above) it follows that there is no (x, y)-specialization

$$k\langle\!\langle x, y\rangle\!\rangle_3 \to k\langle\!\langle x, y\rangle\!\rangle_2.$$

So there must be a relation holding in PI-degree 3 but not 2, and we are looking for an explicit such relation. We shall need some preparatory lemmas; as usual, we put $[X, Y] = XY - YX$.

LEMMA 7.4.1. *Let C be a commutative ring. Then for any $X, Y \in \mathfrak{M}_3(C)$ we have*

$$[X, [X, Y]^2] = (\det[X, Y]) \cdot [X, [X, Y]^{-1}], \tag{1}$$

whenever $[X, Y]^{-1}$ *is defined. For* 2×2 *matrices the left-hand side of (1) is zero.*

Proof. Writing $Z = [X, Y]$, we have tr $Z = 0$, hence Z has the characteristic equation $Z^3 + pZ - q = 0$, where $p, q \in C$, in fact $q = \det Z$. When Z is invertible, we can write this as $Z^2 + p - qZ^{-1} = 0$, and applying $[X, -]$, we obtain $[X, Z^2] - q[X, Z^{-1}] = 0$, i.e. (1).

For 2×2 matrices the characteristic equation reduces to $Z^2 - q = 0$, and so $[X, Z^2] = 0$. ∎

Let us write $Y' = [X, Y]$; then we can express the conclusion of the lemma as

$$\frac{((Y')^2)'}{((Y')^{-1})'} = \begin{cases} 0 & \text{for } 2 \times 2 \text{ matrices,} \\ \det Y' & \text{for } 3 \times 3 \text{ matrices.} \end{cases} \tag{2}$$

Here we have used the convention of writing $u/v = \alpha$ if $u = \alpha v$ for a scalar α.

We shall need a second matrix identity. As before, C is a commutative ring and $Y' = [X, Y]$.

LEMMA 7.4.2. *Let* $X, Y \in \mathfrak{M}_3(C)$ *and write* Δ *for the discriminant of the characteristic polynomial of X. Then*

$$\det Y''' = \Delta \cdot \det Y'. \tag{3}$$

Proof. If C is an algebraically closed field, we can transform X to diagonal form whenever $\Delta \neq 0$, say $X = \text{diag}(\lambda_1, \lambda_2, \lambda_3)$. Then $\Delta = \delta^2$, where $\delta = (\lambda_1 - \lambda_2)(\lambda_2 - \lambda_3)(\lambda_3 - \lambda_1)$. Now an iterated commutator formed from $Y = (y_{ij})$ has the form

$$Y^{(n)} = \begin{pmatrix} 0 & (\lambda_1 - \lambda_2)^n y_{12} & (\lambda_1 - \lambda_3)^n y_{13} \\ (\lambda_2 - \lambda_1)^n y_{21} & 0 & (\lambda_2 - \lambda_3)^n y_{23} \\ (\lambda_3 - \lambda_1)^n y_{31} & (\lambda_3 - \lambda_2)^n y_{32} & 0 \end{pmatrix},$$

so its determinant is given by

$$\begin{aligned} \det Y^{(n)} &= (\lambda_1 - \lambda_2)^n (\lambda_2 - \lambda_3)^n (\lambda_3 - \lambda_1)^n y_{12} y_{23} y_{31} \\ &\quad + (\lambda_1 - \lambda_3)^n (\lambda_3 - \lambda_2)^n (\lambda_2 - \lambda_1)^n y_{13} y_{32} y_{21} \\ &= \delta^n (y_{12} y_{23} y_{31} + (-1)^n y_{13} y_{32} y_{21}). \end{aligned}$$

For $n = 1$ and 3 these expressions differ by a factor $\delta^2 = \Delta$, hence (3) follows. This proves (3) for an algebraically closed field whenever $\Delta \neq 0$; hence it holds generally. ∎

Using (2) and (3) we can write down rational identities for 3×3 matrices, but most of them will hold for 2×2 matrices too. What we need is a relation between determinants of commutators of 3×3 matrices which fails when these commutators are replaced by 0. For any matrices X and Y let us again write $Y' = [X, Y]$ and consider the expression

$$\det Y' \det Y''(\det (Y''^{-1})')(\det (Y'''^{-1})'). \tag{4}$$

Since $'$ is a derivation, we have $\det (Y^{-1})' = \det(-Y^{-1}Y'Y^{-1}) = (\det Y)^{-2} \cdot \det (-Y')$, so (4) becomes, on writing $y_n = \det Y^{(n)}$,

$$y_1 y_2 y_2^{-2} y_3 y_3^{-2} y_4 = y_1 y_2^{-1} y_3^{-1} y_4.$$

Now the identity of Lemma 4.2 can be stated as $y_3 = \Delta y_1$, hence we have

$$y_1 y_2^{-1} y_3^{-1} y_4 = y_1 y_2^{-1} \Delta^{-1} y_1^{-1} \Delta y_2 = 1. \tag{5}$$

Thus we obtain

THEOREM 7.4.3 (Bergman [76]). *Let k be a commutative field and n equal to 2 or 3. For $X, Y \in \mathfrak{M}_n(k)$ write $Y' = [X, Y]$, $\delta(Y) = (Y^2)'[(Y^{-1})']^{-1}$, so that by (2), $\delta(Y') = (n-2) \det Y'$. Then there are rational identities:*

$$\delta(Y')\delta(Y'')[(\delta(Y'')^{-1})'][(\delta(Y''')^{-1})'] = \begin{cases} 1 & \text{if } n = 3, \\ 0 & \text{if } n = 2. \end{cases} \tag{6}$$

Proof. By (4) and (5), the left-hand side of (6) is identically 1 when $n = 3$; for $n = 2$ the left-hand side is 0, if defined, so we need only find X and Y such that the left-hand side is defined. Let E be an extension of k with more than two elements and write S for the set of all matrices $\begin{pmatrix} 0 & a \\ b & 0 \end{pmatrix}$, where $a, b \in E^\times$ when $n = 2$, or

$$\begin{pmatrix} 0 & a & 0 \\ 0 & 0 & b \\ c & 0 & 0 \end{pmatrix}, \quad \begin{pmatrix} 0 & 0 & a \\ b & 0 & 0 \\ 0 & c & 0 \end{pmatrix}, \quad a, b, c \in E^\times, \quad \text{when } n = 3.$$

Then S consists of invertible matrices and is closed under inversion and commutation by diagonal matrices with distinct elements. If we choose Y in S and X diagonal with distinct entries, then all terms lie in S and so (6) is defined. ■

Exercise

Verify that every polynomial identity satisfied by $\mathfrak{M}_n(C)$, where C is a commutative ring, also holds for $\mathfrak{M}_r(C)$ for $r < n$. Examine where the proof breaks down for rational identities.

7.5 The rational meet of a family of X-rings

We shall now make a closer study of specializations, following Bergman [76]. We shall find that for skew fields they cannot be reduced to the situation involving only two fields, as in the commutative case. We shall be concerned with two basic notions: an *essential term* in a family of X-fields and the *support relation*.

Given rings $R_1 \subseteq R_2$, we shall say that R_1 is *rationally closed* in R_2 if the inclusion is a local homomorphism; thus for every element of R_1 which is invertible in R_2, the inverse also lies in R_1. The intersection of a family of rationally closed subrings is again rationally closed, so we can speak of the *rational closure* of a subset X of R, which is the least rationally closed subring of R containing X. If this is R itself, we shall call R a *strict X-ring*; e.g. $\mathbf{Q}(x, y)$ is a strict (x, y)-ring and so is $\mathbf{Z}[x, y, y^{-1}]$, but not $\mathbf{Z}[x, y, xy^{-1}]$. Generally, if Σ is the set of all matrices over $\mathbf{Z}\langle X \rangle$ which are mapped to invertible matrices over R, then the rational closure of X in R is contained in the Σ-rational closure of $\mathbf{Z}\langle X \rangle$, in the sense of Ch. 4, but the two may be distinct: if $x, y, u, v \in X$, then the entries of $\begin{pmatrix} x & y \\ u & v \end{pmatrix}^{-1}$ lie in the latter, but not generally in the former. We note that an epic $\mathbf{Z}\langle X \rangle$-field, briefly an epic X-field, is just a strict X-field.

A local homomorphism between X-fields $\varphi \colon D \to D'$ may be described as a partial homomorphism from D to D' whose graph is rationally closed in $D \times D'$; hence if there is any X-specialization at all, then the rational closure of X in $D \times D'$ is the unique least X-specialization. So there is at most one minimal X-specialization between two X-fields. Our aim is to study the rational closure of X in finite direct products; to do so we need to introduce the following basic concepts.

DEFINITION. Let $\{R_s\}_S$ be a family of strict X-rings. Then their *rational meet* $\bigwedge_S R_s$ is the rational closure of X in the product $\prod_S R_s$. When S is finite, say $S = \{1, \ldots, n\}$, we also write $R_1 \wedge \ldots \wedge R_n$ for $\bigwedge_S R_s$; for $n = 1$ this reduces to R_1 by the definition of strict X-ring.

The rational meet can also be viewed as the product in the category of strict X-rings. We note that the bigger S is, the smaller is $\bigwedge_S R_s$, in the sense that for $T \subseteq S$ we have a projection $p_{ST} \colon \bigwedge_S R_s \to \bigwedge_T R_s$. For example, whether $D_1 \wedge D_2$ is the graph of a specialization in one direction or the other depends on which projection maps are injective.

Our first task is to determine the domain and the zero-set of a rational meet:

LEMMA 7.5.1. *Let $\{R_s\}_S$ be any family of strict X-rings, for some set X. Then*

$$\mathscr{C}\left(\bigwedge_s R_s\right) = \bigcap_s \mathscr{C}(R_s), \quad \mathscr{L}\left(\bigwedge_s R_s\right) = \bigcap_s \mathscr{L}(R_s).$$

For $\bigwedge_s R_s$ is the set of all rational expressions evaluable in each R_s, modulo the relation of having equal values in each R_s: $f \sim g \Leftrightarrow f^{R_s} = g^{R_s}$ for all $s \in S \Leftrightarrow f - g \in \mathscr{L}(R_s)$ for all $s \in S$. ∎

Let $\{D_s\}_S$ be a finite family of epic X-fields and let $t \in S$. We shall call the index t (and also the field D_t) *essential* in S if

$$\bigcap_s \mathscr{C}(D_s) \cap \mathscr{L}(D_t) \not\subseteq \bigcup_{s \neq t} \mathscr{L}(D_s); \tag{1}$$

thus there exists f defined on all D_s and vanishing on D_t but on none of the others. Equivalently we have

$$\mathscr{C}(D_t) \not\supseteq \bigcap_{s \neq t} \mathscr{C}(D_s). \tag{2}$$

For when (2) holds, take $f \in \bigcap \mathscr{C}(D_s) \backslash \mathscr{C}(D_t)$; then f contains a sub-expression g^{-1} such that $g^{D_t} = 0$ but $g^{D_s} \neq 0$ for $s \neq t$. Conversely, given such g, we find that g^{-1} belongs to the right- but not the left-hand side of (2). Using this notion, we can say when the rational meet reduces to a direct product:

PROPOSITION 7.5.2. *Let X be a set and $\{D_s\}_S$ a finite family of epic X-fields. Then the following conditions are equivalent:*
(a) *Each s is essential in S,*
(b) *for each $s \in S$, there exists $e_s \in \bigcap_s \mathscr{C}(D_t)$ such that $e_s^{D_t} = \delta_{st}$,*
(c) $\bigwedge_s D_s = \prod_s D_s.$
Here δ_{st} in (b) is the Kronecker delta.

Proof. (a) $\Rightarrow$ (b). Choose f_s defined in all D_t and vanishing in D_s but not in D_t for $t \neq s$. Then the product $g_t = \prod_{s \neq t} f_s$ (in any order) vanishes on all Ds except D_t, so $e_s = g_s(\sum_t g_t)^{-1}$ satisfies the required condition.

(b) $\Rightarrow$ (c). By (b), $\bigwedge_s D_s$ contains a complete set of central idempotents e_s, which shows that $\bigwedge_s D_s = \prod_s R_s$ for some $R_s \subseteq D_s$. Now $\bigwedge D_s$ is rationally closed in $\prod D_s$, hence R_s is rationally closed in D_s and it contains X, so $R_s = D_s$.

(c) $\Rightarrow$ (a). Given $s \in S$, choose $g \in \bigcap \mathscr{C}(D_t)$ such that $g^{D_s} = 0$ but $g^{D_t} \neq 0$ for all $t \neq s$; this shows s to be essential in S. ∎

As an illustration, consider $D_1 \wedge D_2$; if $\mathscr{C}(D_1) \supseteq \mathscr{C}(D_2)$, then $D_1 \wedge D_2$ is a local ring, the graph of a specialization $D_1 \rightarrow D_2$. Similarly if

$\mathscr{E}(D_1) \subseteq \mathscr{E}(D_2)$, while if neither inclusion holds, then $D_1 \wedge D_2 = D_1 \times D_2$ by Prop. 5.2. For more than two factors we shall find that $\bigwedge D_s$ is a *semilocal* ring, i.e. a ring R which modulo its Jacobson radical $J(R)$ is semisimple (Artinian).

LEMMA 7.5.3. *Let $f: R \to R'$ be a ring homomorphism such that Rf rationally generates R'. Then f is local if and only if f is surjective and $\ker f \subseteq J(R)$.*

Proof. $\Rightarrow$. Rf is rationally closed because f is local and it rationally generates R', hence $Rf = R'$. If $af = 0$ for $a \in R$, then $1 + ax$ maps to 1, hence $1 + ax$ is a unit, for any $x \in R$, and this means that $a \in J(R)$.

$\Leftarrow$. If the conditions hold, take $a \in R$ such that af is a unit, say $af \cdot bf = 1$ for some $b \in R$. Then $ab = 1 + n$, where $n \in J(R)$, hence $ab(1 + n)^{-1} = 1$ and similarly $(1 + m)^{-1}ba = 1$ for some $m \in J(R)$, and so a is a unit. ∎

We note that the extra hypothesis (Rf rationally generates R') is only needed for the first part of the proof.

We can now prove a result which describes the structure of rational meets:

PROPOSITION 7.5.4. *Let X be a set, $\{D_s\}_S$ a finite family of epic X-fields, pairwise non-isomorphic as X-fields, and write U for the set of essential indices in S. Then for any subset T of S, with projection map*

$$p: \bigwedge_S D_s \to \bigwedge_T D_s, \tag{3}$$

the following conditions are equivalent:
  (a) *p is a local homomorphism,*
  (b) *p is surjective and $\ker p \subseteq J(\bigwedge_S D_s)$,*
  (c) *$\bigcap_S \mathscr{E}(D_s) = \bigcap_T \mathscr{E}(D_s)$,*
  (d) *$T \supseteq U$.*
Moreover, $\bigwedge_S D_s$ is a semilocal ring with residue-class ring $\bigwedge_U D_s = \prod_U D_s$:

$$\bigwedge_S D_s / J\left(\bigwedge_S D_s\right) \cong \prod_U D_s. \tag{4}$$

Proof. (a) $\Leftrightarrow$ (b) by Lemma 5.3 and (a) $\Leftrightarrow$ $\mathscr{E}(\bigwedge_S D_s) = \mathscr{E}(\bigwedge_T D_s)$, which is equivalent to (c), by Lemma 5.1. Now let V be a subset of S which is minimal subject to (a)–(c). By the minimality of V applied to (c) we see that V contains no inessential index, i.e. $V \subseteq U$; hence $\bigwedge_V D_s = \prod_V D_s$, by Prop. 5.2. This ring is semisimple and V satisfies (b), so $\bigwedge_S D_s / \mathbf{J}(\bigwedge_S D_s) \cong \bigwedge_V D_s = \prod_V D_s$, hence $\bigwedge_S D_s$ is semilocal, with residue class fields isomorphic to the D_s $(s \in V)$. But distinct Ds are non-isomorphic as X-fields, hence V is the unique minimal subset of S satisfying (a)–(c), so (a)–(c) are equivalent to $U \supseteq V$.

Now for any $t \in S$, t is inessential if and only if $S \backslash \{t\}$ satisfies (c), which holds precisely when $S \backslash \{t\} \supseteq V$ (because (c) is equivalent to $U \supseteq V$) and this holds if and only if $t \notin V$. Hence $V = U$ as claimed, and now (4) also follows. ∎

By (4) we have

Cated and some COROLLARY 7.5.5. *The set U of essential indices in S can also be characterized as the set of those $t \in S$ for which $p_{St}: \bigwedge_S D_s \to D_t$ is surjective.* ∎

Still assuming our family $\{D_s\}$ to consist of pairwise non-isomorphic epic X-fields, we can describe the complement of U, i.e. the set of all *inessential* indices, as the set of all $t \in S$ such that for $S' = S \backslash \{t\}$, (a) $p_{SS'}$ is a local homomorphism, (b) $p_{SS'}$ is surjective with kernel in the Jacobson radical, or equivalently, (c) $\bigcap_S \mathscr{E}(D_s) = \bigcap_{S'} \mathscr{E}(D_s)$. This follows from Prop. 5.4. Here (c) states essentially that $\mathrm{Ker}\,(D_t) \subseteq \bigcup_S \mathrm{Ker}\,(D_s)$. In the commutative case this can happen only when $\ker\,(D_t) \subseteq \ker\,(D_s)$ for some s, by the prime avoidance lemma (A.2, Lemma 9.8.10, p. 343, or Lemma 6.4 below):

COROLLARY 7.5.6. *In the case of a family $\{D_s\}_S$ of commutative (pairwise non-isomorphic) fields an index t is inessential in S if and only if D_t has some D_s $(s \neq t)$ as specialization.* ∎

For example, let $S = \{0, 1, 2\}$ and suppose that 0 is inessential in S; then for commutative fields, every relation holding in D_0 holds in D_1 or in D_2. For skew fields the result need not hold, i.e. there may be f_1, f_2 such that f_i vanishes in D_0 but not in D_i $(i = 1, 2)$. Now $f_1 + f_2$ would seem to be 0 in D_0 but not in D_1 or D_2; but in fact it need not be defined, for $f_1 \neq 0$ may hold 'degenerately' in D_1, if f_1 is not defined in D_1. This will become clear later.

We now come to the second basic notion, the support relation. We have seen (in Cor. 5.5) that $p_{St}: \bigwedge_S D_s \to D_t$ is surjective precisely when t is essential in S. Our next question is: When is p_{St} injective? It is answered by

PROPOSITION 7.5.7. *Let X be a set and $\{D_s\}_S$ a family of epic X-fields. Then for any $t \in S$ the following conditions are equivalent:*
  (a) $\mathfrak{X}(D_t) \cap \bigcap_S \mathfrak{E}(D_s) \subseteq \bigcap_S \mathfrak{X}(D_s),$
  (b) *any relation defined in each D_s and holding in D_t holds in each D_s,*
  (c) $p_{St}: \bigwedge_S D_s \to D_t$ *is injective.*
Note that by (c) there is a local homomorphism $D_t \to \prod_S D_s$.

Proof. An expression e in $\mathfrak{R}(X; Z)$ represents an element in ker p_{St} if it is in the left-hand side of (a) and it represents 0 if it is in the right-hand side of (a); this just expresses (b), and the equivalence with (c) is now clear. ∎

When these conditions hold we shall say that t *supports* S, or also: D_t supports $\{D_s\}_S$. More generally, if $t \notin S$, we say that t *supports* S if it supports $S \cup \{t\}$ in the above sense.

To gain an understanding of the support relation we begin by proving some trivial facts:

PROPOSITION 7.5.8. *Let X be a set and $\{D_s\}_S$ a family of epic X-fields. Then*
  (i) *Given $t \in S$, $U \subseteq S$, if t supports U and $D_t \not\equiv D_u$ for some $u \in U$, then t is inessential for $U \cup \{t\}$,*
  (ii) *If t supports U, then it supports $U \cup \{t\}$,*
  (iii) *If t supports S_i ($i \in I$), then it supports $\bigcup_I S_i$,*
  (iv) *If t supports U and for each $u \in U$, u supports a non-empty set S_u, then t supports $\bigcup_U S_u$.*

Proof. To say that 't is inessential for S' means 'any relation defined in all D_s and D_t and holding in D_t also holds in *some* D_s, $s \neq t$', while 't supports S' means 'any relation defined in all D_s and D_t and holding in D_t holds in *all* D_s'. Now (i) is clear and (ii) also follows. To prove (iii), let f be defined in D_t and D_s ($s \in S_i$) and $f = 0$ in D_t; then $f = 0$ in all D_s, $s \in S_i$, so t supports $\bigcup S_i$. (iv) Let f be defined in D_t and D_v, where $v \in S_u$, for all $u \in U$. If $f = 0$ in D_t, then $f = 0$ in D_u ($u \in U$) by hypothesis, hence $f = 0$ in D_v ($v \in S_u$), so t supports $\bigcup_U S_u$. ∎

We remark that if S' is a subset of S such that $\bigwedge_S D_s \to \bigwedge_{S'} D_s$ is surjective, then any t which supports S also supports S', but in general this need not be so. In the commutative case the support relation still simplifies:

COROLLARY 7.5.9. *If all the D_s are commutative and t supports S, then* *either* $S = \emptyset$ *or* D_t *specializes to some* D_s $(s \in S)$. *More precisely: t supports* $\{s\}$ *if and only if there is a specialization* $D_t \to D_s$.

For if t supports S and $S \neq \emptyset$, then either $t \in S$ or t is inessential for $S \cup \{t\}$; in the latter case there exists $s \neq t$ in S such that D_s is a specialization of D_t. ∎

To clarify the relation between support and essential set we have the following lemma. We note that by (i) above, a supporting index is a special kind of inessential index.

LEMMA 7.5.10. *Let* X *be a set,* $\{D_s\}_S$ *a finite family of pairwise non-isomorphic epic* X-*fields and* D_t *an epic* X-*field. Then the following conditions are equivalent:*
  (a) $S \cup \{t\}$ *is a minimal set in which t is inessential,*
  (b) S *is a minimal non-empty set supported by t.*

Proof. Let us write (a_0), (b_0) for (a), (b) without the minimality clause. Then (b) $\Rightarrow$ (a_0) by (i) of Prop. 5.8, provided that S contains an element $s \neq t$. To prove that (a) $\Rightarrow$ (b_0), let us write $S' = S \cup \{t\}$. We know by hypothesis that S is minimal subject to $\bigcap_S \mathscr{E}(D_s) = \bigcap_{S'} \mathscr{E}(D_s)$. By Prop. 5.4, S is the set of essential indices in S', hence the projection $p_{S'S}$ is surjective. For any function f let us abbreviate f^{D_u} as f_u. If t does *not* support S, there exist $a \in \bigwedge_{S'} D_s$ and $u \in S$ such that $a_t = 0$ but $a_u \neq 0$. Since the map $\bigwedge_{S'} D_s \to \prod_S D_s$ is surjective, there exists $b \in \bigwedge_{S'} D_s$ such that $b_u = a_u^{-1}$, $b_s = 0$ for all $s \neq u, t$, where $s \in S$. Then $e = ab$ is in $\bigwedge_{S'} D_s$ and has value 1 in D_u and 0 everywhere else, for $b_s = 0$ for $s \neq t$ and $a_t = 0$. Thus e is a central idempotent and so $\bigwedge_{S'} D_s = R \times D_u$ for some ring R. Now write $T = S\backslash\{u\}$, $T' = T \cup \{t\}$; then $R \subseteq \prod_{T'} D_s$ and R is rationally generated by X and is rationally closed, hence $R = \bigwedge_{T'} D_s$. Further, $p_{S'S}$ is a local homomorphism, hence so is $p_{T'T}$ (where we have to factor by D_u), therefore by Prop. 5.4, T includes all the essential indices in S, which contradicts the minimality of S. So D_t supports $\{D_s\}_S$ and (b_0) follows.

We thus have (a) $\Rightarrow$ (b_0) and (b) $\Rightarrow$ (a_0); in an obvious terminology, if S is a minimal (a)-set, it is a (b)-set. Now let S be a minimal (a)-set and take a minimal (b)-subset S_1 of S; this is also an (a)-set contained in S, hence $S_1 = S$, i.e. S was a minimal (b)-set. Thus (a) $\Rightarrow$ (b) and similarly (b) $\Rightarrow$ (a). ∎

COROLLARY 7.5.11. *Let* X, $\{D_s\}_S$ *be as in Lemma 5.10 and let* $t \in S$. *Then* $\mathscr{E}(D_t) \supseteq \bigcap_S \mathscr{E}(D_s)$ *if and only if* D_t *supports some non-empty subfamily of* $\{D_s\}_S$.

For the left-hand side expresses the fact that t is inessential in $S \cup \{t\}$. Now pick $S_0 \subseteq S$ minimal with this property and apply the lemma to reach the desired conclusion. ∎

The relation

$$t \text{ supports } S \tag{5}$$

may be called *trivial* if S is $\varnothing$ or $\{t\}$; thus (5) will be called *non-trivial* if $S \neq \varnothing, \{t\}$.

COROLLARY 7.5.12. *For $\{D_s\}_S$ as before, each $s \in S$ is essential in S if and only if there are no non-trivial support relations in S.* ∎

The essential relations are determined by the minimal essential relations, but there is no corresponding statement for support relations. However, Cor. 5.12 shows that essential relations are determined by the support relations.

Let us call a set S *essential* if each member is essential in it. Thus Cor. 5.12 states that S is essential if and only if there are no non-trivial support relations in it. For essential sets the support relation can be described as follows.

PROPOSITION 7.5.13. *Write $S' = S \cup \{t\}$ and let $\{D_s\}_{S'}$ be a finite family of pairwise non-isomorphic epic X-fields. Then the following conditions are equivalent:*
  (a) *t supports S and S is essential,*
  (b) *$\bigwedge_{S'} D_s$ is a semilocal ring contained in D_t (via the projection map) with residue-class fields D_s,*
  (c) *there exists a semilocal X-ring $R \subseteq D_t$ with residue-class fields D_s $(s \in S)$,*
  (d) *$\mathfrak{L}(D_t) \cap \bigcap_S \mathfrak{C}(D_s) \subseteq \bigcap_S \mathfrak{L}(D_s)$ and no $\mathfrak{C}(D_s)$ contains the intersection of all the others.*

Proof. (a) $\Rightarrow$ (b) follows by Prop. 5.4 and Prop. 5.7; (b) $\Rightarrow$ (c) $\Rightarrow$ (d) holds trivially and (d) $\Rightarrow$ (a) is also clear. ∎

COROLLARY 7.5.14. *Let $\{D_s\}_S$ be a finite family of pairwise non-isomorphic epic X-fields, $t \in S$ and suppose that t supports S and U is the subset of all essential indices in S. Further write $U' = U \cup \{t\}$. Then the projection map $\bigwedge_S D_s \to \bigwedge_{U'} D_s$ is an isomorphism and t supports U.* ∎

Exercises

1. Verify that any non-empty family of strict X-rings has an essential index.

2. Give an example of an infinite family of epic X-fields which is not essential, though every finite subfamily is.

7.6 The support relation on generic division algebras

We have seen that for two epic X-fields D_1, D_2 the rational meet $D_1 \wedge D_2$ is either a local ring – namely when one of D_1, D_2 is a specialization of the other, and a field precisely when $D_1 \cong D_2$ – or the full direct product $D_1 \times D_2$. For three factors D_1, D_2, D_3 there are many more possibilities, e.g. the subrings $D_i \wedge D_j \subseteq D_i \times D_j$ ($1 \leqslant i < j \leqslant 3$) may each be the full direct product and yet the rational meet $D_1 \wedge D_2 \wedge D_3 \subseteq D_1 \times D_2 \times D_3$ could be a semilocal ring which embeds in D_1 and has two residue class fields D_2, D_3. This happens when every rational relation satisfied in D_1 is satisfied either in D_2 or in D_3 but neither D_2 nor D_3 accounts for all such relations. In order to see what possibilities can be realized we shall take the generic division algebras $k \langle X \rangle_n$ and describe all possible support relations in this case, using the work of Bergman and Small [75] (and still following Bergman [76]). We shall need Th. 6.8 of that paper, which for our purpose may be stated as follows:

THEOREM 7.A. *Let R be a prime PI-ring and $\mathfrak{p}_0$ a prime ideal of R. Then the integer* (PI-deg R − PI-deg$(R/\mathfrak{p}_0)$) *can be written as a sum of integers* PI-deg$(R/\mathfrak{p})$ *(allowing repetitions), where $\mathfrak{p}$ ranges over the maximal ideals of R.*

We omit the rather lengthy proof (see Bergman and Small [75]).

Let us say that an integer n *supports* a set M of positive integers if for each $m \in M$, $n - m$ lies in the additive monoid generated by the elements of M. Clearly M must then be a subset of $\{1, 2, \ldots, n\}$. The Bergman–Small theorem shows the truth of the following:

If R is any prime PI-ring, then PI-deg R supports the set

$$\{\text{PI-deg}\,(R/\mathfrak{p})|\mathfrak{p} \text{ prime in } R\}.$$

In what follows, X will be fixed, with more than one element, so that $k \langle X \rangle_n$ has PI-degree n. We shall write $E(n) = \mathscr{E}(k \langle X \rangle_n)$, $Z(n) = \mathscr{Z}(k \langle X \rangle_n)$ for brevity.

THEOREM 7.6.1 (Bergman [76]). *Let n be a positive integer and M a finite non-empty set of positive integers. Then the following conditions are equivalent:*

(a) $k\langle X\rangle_n$ *supports* $\{k\langle X\rangle_m | m \in M\}$,

(b) $Z(n) \cap \bigcap_M E(m) \subseteq \bigcap_M Z(m)$,

(c) *the projection map* $p_{M'\{n\}} : \bigwedge_{M'} k\langle X\rangle_r \to k\langle X\rangle_n$ *is injective, where* $M' = M \cup \{n\}$,

(d) *there exists a prime PI-ring R of PI-degree n such that the set of PI-degrees of the residue-class rings of R at the maximal ideals is precisely M, briefly,* $\{\text{PI-deg}(R/\max)\} = M$,

(e) *n supports M.*

Proof. (a)–(c) are equivalent by Prop. 5.7. Now we have two variants of (d), one weaker (d−) and one stronger (d+):

(d−) There is a prime ring R of PI-degree n such that

$$\{\text{PI-deg}(R/\text{prime})\} \supseteq M \supseteq \{\text{PI-deg}(R/\max)\},$$

(d+) there is a semilocal prime ring R of PI-degree n such that every non-zero prime ideal is maximal and $\{\text{PI-deg}(R/\max)\} = M$, and every residue-class field is infinite.

We complete the proof by showing that (c) $\Rightarrow$ (d−) $\Rightarrow$ (e) $\Rightarrow$ (d+) $\Rightarrow$ (b). Note that each condition implies that $M \subseteq \{1, 2, \ldots, n\}$.

(c) $\Rightarrow$ (d−). Put $R = \bigwedge_{M'} k\langle X\rangle_r$, where $M' = M \cup \{n\}$. By Prop. 5.4, the residue class rings at the maximal ideals are among the $k\langle X\rangle_m$ ($m \in M$), for n itself cannot occur by (c) and Prop. 5.8. Now for each $m \in M$, $\ker(R \to k\langle X\rangle_m) = \mathfrak{p}$ is prime and PI-deg$(R/\mathfrak{p}) = m$, so (d−) holds.

(d−) $\Rightarrow$ (e) is just the Bergman–Small theorem 7.A quoted earlier.

(e) $\Rightarrow$ (d+). By (e) we can write $n = m(1, 1) + \ldots + m(1, r_1) = \ldots = m(s, 1) + \ldots + m(s, r_s)$, $m(i, j) \in M$, where $s \geq 1$, $r_i \geq 1$ and each $m \in M$ occurs as some $m(i, j)$. Let A be a commutative k-algebra which is a semilocal principal ideal domain with just s non-zero prime ideals $\mathfrak{J}_1, \ldots, \mathfrak{J}_s$ each with infinite residue-class field $K_i = A/\mathfrak{J}_i$ (e.g. let $K \supseteq k$ be an infinite field extension and take a suitable localization of $K[t]$). Then we have $A/\mathbf{J}(A) = \prod_i K_i$, hence

$$\mathfrak{M}_n(A)/\mathbf{J}(\mathfrak{M}_n(A)) \cong \prod_i \mathfrak{M}_n(K_i).$$

Now for each $i = 1, \ldots, s$, $\mathfrak{M}_n(K_i)$ has a block diagonal subring isomorphic to

$$\mathfrak{M}_{m(i,1)}(K_i) \times \ldots \times \mathfrak{M}_{m(i,r_i)}(K_i) = L_i, \text{ say.}$$

Hence

$$Q = \prod_i L_i \subseteq \prod_i \mathfrak{M}_n(K_i) \cong \mathfrak{M}_n(A)/\mathbf{J}(\mathfrak{M}_n(A)), \tag{1}$$

where Q as a direct product of simple Artinian rings is semisimple. If R is the inverse image of Q in $\mathfrak{M}_n(A)$ by the isomorphism (1), then $\mathbf{J}(R) = \mathbf{J}(\mathfrak{M}_n(A))$, hence $R/\mathbf{J}(R) \cong Q$. Since $R/\mathbf{J}(R)$ is semisimple (Artinian), it follows that R is semilocal and PI-deg $(R/\text{max}) = M$. Let $\mathfrak{P}$ be any prime ideal in R and put $\mathfrak{p} = \mathfrak{P} \cap A$; then $\mathfrak{p}$ is prime in A, so $\mathfrak{p}$ is 0 or some $\mathfrak{J}_i$. Suppose that $\mathfrak{p} = 0$; then $A \subseteq R/\mathfrak{P}$. Since $A + \mathfrak{M}_n(\mathbf{J}(A)) \subseteq R \subseteq \mathfrak{M}_n(A)$, we have, on writing F for the field of fractions of A,

$$R_{A^\times} = \mathfrak{M}_n(A)_{A^\times} = \mathfrak{M}_n(F),$$

because A is a domain and $\mathbf{J}(A) \neq 0$. Hence $R_{A^\times}$ is simple with 0 as the only prime, so $\mathfrak{P}$ must be 0. If $\mathfrak{p} = \mathfrak{J}_i$, then $K_i = A/\mathfrak{J}_i \subseteq R/\mathfrak{P}$ and since R is a finitely generated R-module, $R/\mathfrak{P}$ is a finitely generated K_i-module, hence Artinian. It is also prime, hence simple, and so $\mathfrak{P}$ was maximal. Thus R satisfies (d+).

(d+) $\Rightarrow$ (b). Assume that e lies in the left-hand side of (b), thus $e = 0$ is a rational identity holding in PI-degree n and not degenerate in PI-degree m for any $m \in M$; we have to show that $e = 0$ holds in each PI-degree $m \in M$. Let R be as in (d+); this means that for each non-zero prime ideal $\mathfrak{P}$ of R there is given a map $\alpha_\mathfrak{P}: X \to R/\mathfrak{P}$ such that $e^{\alpha_\mathfrak{P}}$ is defined in $R/\mathfrak{P}$; we have to show that all the $e^{\alpha_\mathfrak{P}}$ are zero. Since R is semilocal, by the Chinese remainder theorem there exists $\alpha: X \to R$ inducing all the $\alpha_\mathfrak{P}$. Now e^α can be evaluated $(\text{mod } \mathfrak{P})$ for all maximal $\mathfrak{P}$, hence it can be evaluated in R. Since $e \in Z(n)$, we have $e^\alpha = 0$ and so $e^{\alpha_\mathfrak{P}} = 0$, as claimed. ∎

To give an illustration, we have $5 = 2 + 3$. Let A be a local principal ideal domain with maximal ideal $\mathfrak{J}$; then $\mathfrak{M}_5(A)$ contains the subring

$$R = \begin{pmatrix} A_2 & {}^2\mathfrak{J}^3 \\ {}^3\mathfrak{J}^2 & A_3 \end{pmatrix}$$

and we have the local homomorphism $\mathfrak{M}_5(A) \to \mathfrak{M}_2(K) \times \mathfrak{M}_3(K)$, where $K = A/\mathfrak{J}$. This gives rise to a specialization of fields, for when we replace $\mathfrak{M}_n(K)$ by the generic matrix ring, we get a field with the same identities as $\mathfrak{M}_n(K)$.

If we combine Th. 6.1 with Prop. 5.13, we obtain

COROLLARY 7.6.2. *For any integer n and set of integers M the following conditions are equivalent, where $M' = M \cup \{n\}$:*

(a) $p: \bigwedge_{M'} k \langle\!\langle X \rangle\!\rangle_r \to k \langle\!\langle X \rangle\!\rangle_n$ is injective, with residue-class fields $k \langle\!\langle X \rangle\!\rangle_m$ $(m \in M)$,

(b) $k \langle\!\langle X \rangle\!\rangle_n$ has a semilocal subring with residue-class fields $k \langle\!\langle X \rangle\!\rangle_m$ $(m \in M)$,

(c) n supports M, but no $m \in M$ supports any non-empty subset of $M \backslash \{m\}$. ∎

COROLLARY 7.6.3. *With n, M as before the following conditions are equivalent:*

(a) $k \langle\!\langle X \rangle\!\rangle_n$ *supports a non-empty subfamily of* $\{k \langle\!\langle X \rangle\!\rangle_m | m \in M\}$,

(b) *every rational identity holding in PI-degree n holds in PI-degree m for some $m \in M$, $Z(n) \cap \bigcap_M E(m) \subseteq \bigcup_M Z(m)$,*

(c) *there exists a prime PI-ring R of PI-degree n such that $\{\text{PI-deg}\,(R/\text{max})\} \subseteq M$,*

(d) *n supports a subset of M.* ∎

To describe the connexion between prime ideals and the support relation we shall need a couple of lemmas. The first is a form of the well known prime avoidance lemma:

LEMMA 7.6.4. *Let R be a ring, $\mathfrak{I}_1, \ldots, \mathfrak{I}_m$ any ideals in R and $\mathfrak{P}_1, \ldots, \mathfrak{P}_n$ any prime ideals such that $\mathfrak{I}_i \not\subseteq \mathfrak{P}_j$ for all i, j. Then*

$$\bigcap \mathfrak{I}_i \not\subseteq \bigcup \mathfrak{P}_j.$$

Proof. If $\mathfrak{P}_k \subseteq \mathfrak{P}_j$ for $k \neq j$, we can omit $\mathfrak{P}_k$. Since $\mathfrak{P}_j$ is prime, we then have $\mathfrak{I}_1 \ldots \mathfrak{I}_m \mathfrak{P}_1 \ldots \mathfrak{P}_{j-1} \mathfrak{P}_{j+1} \ldots \mathfrak{P}_n \not\subseteq \mathfrak{P}_j$. Choose a_j in the left- but not the right-hand side; then $a = a_1 + \ldots + a_n \in \mathfrak{I}_i$ for all i but $a \notin \mathfrak{P}_j$. ∎

In what follows, $\{D_s\}_S$ is a finite family of epic X-fields, $R = \bigwedge_S D_s$ and $\mathfrak{P}_s = \ker(R \to D_s)$. Thus $\mathfrak{P}_s = 0$ if and only if s supports S, and $\mathfrak{P}_s$ is maximal if and only if s is essential. We shall write $\text{Ess}\,(S)$ for the set of essential indices in S.

LEMMA 7.6.5. *Assume that the D_s $(s \in S)$ are pairwise non-isomorphic as X-fields and that $T \subseteq S$. Then the following conditions are equivalent:*

(a) $\bigwedge_S D_s \to \bigwedge_T D_s$ *is surjective,*

(b) *for each $t \in T$ and $s \in \text{Ess}\,(S)$, $\mathfrak{P}_t \subseteq \mathfrak{P}_s \Rightarrow s \in T$.*

When these conditions hold, then $\text{Ess}\,(T) = \text{Ess}\,(S) \cap T$.

Proof. (a) ⟹ (b). Put $R' = \bigwedge_T D_s$; then the projection $p: R \to R'$ is surjective, by hypothesis. If t, s are as in (b), then $\mathfrak{P}_s$ is a maximal ideal of

R containing $\mathfrak{P}_t$ and hence $\ker p$, so its image under p is a maximal ideal of R' with the same residue class field. But if R' has a residue-class field isomorphic to D_s, then $s \in \mathrm{Ess}\,(T) \subseteq T$, so (b) holds.

(b) $\Rightarrow$ (a). Since $\mathrm{im}\,p$ rationally generates R', it is enough to show that $\mathrm{im}\,p$ is rationally closed in R', i.e. the inclusion $\mathrm{im}\,p \subseteq R'$ is a local homomorphism; let $a \in R$ be such that ap is invertible in R'; then $a \notin \mathfrak{P}_t$ for all $t \in T$. Now consider those $s \in \mathrm{Ess}\,(S)$ for which $a \in \mathfrak{P}_s$; by (b), since $s \notin T$, we have $\mathfrak{P}_s \not\supseteq \mathfrak{P}_t$ for all $t \in T$. hence $\mathfrak{P}_s \not\supseteq \bigcap_T \mathfrak{P}_t = \ker p$. By Lemma 6.4 there exists $b \in R$ such that $b \in \ker p$ and for any $s \in \mathrm{Ess}\,(S)$, $b \notin \mathfrak{P}_s$ if and only if $a \in \mathfrak{P}_s$. But then $a + b$ lies in no maximal ideal of R and so is a unit, and $(a + b)p = ap$ is likewise a unit in $\mathrm{im}\,p$. ∎

We note that (a) shows that the residue-class rings of R' at maximal ideals are just the residue-class rings of R at the maximal ideals containing $\ker p$. We can now express the inclusion of prime ideals in terms of the support relation. We shall write $\mathrm{Supp}_S(t)$ for the maximal subset of S supported by t, i.e. the union of all the subsets supported by t.

THEOREM 7.6.6. *Let $\{D_s\}_S$ be a finite family of epic X-fields and put $R = \bigwedge_S D_s$, $\mathfrak{P}_s = \ker(R \to D_s)$. Then*

$$\mathrm{Supp}_S(u) = \{v \in S \mid \mathfrak{P}_u \subseteq \mathfrak{P}_v\}. \tag{2}$$

Proof. Isomorphic X-fields determine the same kernel in R, so we may without loss of generality take the D_s to be pairwise non-isomorphic. Fix $u \in S$ and let T be the right-hand side of (2), i.e. the set of all $s \in S$ for which $\mathfrak{P}_s \supseteq \mathfrak{P}_u$; then T satisfies (b) of Lemma 6.5, so the projection $R \to \bigwedge_T D_s$ is surjective. The kernel is $\bigcap_T \mathfrak{P}_t$, which contains $\mathfrak{P}_u$ by the definition of T; in fact since $u \in T$, we have $\bigcap_T \mathfrak{P}_s = \mathfrak{P}_u$. Hence the map $\bigwedge_T D_s \to D_u$ is injective, i.e. u supports T. It follows that $T \subseteq \mathrm{Supp}_S(u)$, but clearly also $\mathrm{Supp}_S(u) \subseteq T$, hence we have the equality (2). ∎

We remark that in Lemma 6.5, (b) just states that

$$T \supseteq \mathrm{Ess}\,(S) \cap \bigcup_T \mathrm{Supp}_S(t). \tag{3}$$

Hence we obtain

COROLLARY 7.6.7. *With the notation of Th. 6.6, if $T, T' \subseteq S$, then $\ker p_{ST} \subseteq \ker p_{ST'}$ if and only if $\bigcup_T \mathrm{Supp}_S(t) \supseteq \bigcup_{T'} \mathrm{Supp}_S(t)$. In particular,*
  (i) *p_{ST} is injective $\Leftrightarrow \bigcup_T \mathrm{Supp}_S(t) = S$,*
  (ii) *$\ker p_{ST} \subseteq J(R) \Leftrightarrow \bigcup_T \mathrm{Supp}_S(t) \supseteq \mathrm{Ess}\,(S)$.* ∎

In general $R = \bigwedge_S D_s$ will have prime ideals not of the form $\mathfrak{P}_s$; e.g. if C is a commutative local domain, X is a rational generating set and D_0, D_1 are the field of fractions and the residue-class field respectively, then $D_0 \wedge D_1 = C$, but C may have other primes (when C is not discrete).

We recall that Prop. 5.2 asserted that $\bigwedge_S D_s = \prod_S D_s$ if and only if S is essential. More generally we can now say that

$$\bigwedge_S D_s = \prod_i \left(\bigwedge_{S_i} D_s \right), \quad S = S_1 \cup \ldots \cup S_r,$$

whenever the S_i are disjoint support sets in S, i.e. if for any $t \in S_i$, $\text{Supp}_S(t) \subseteq S_i$.

We have seen that non-isomorphic X-fields may have the same kernels, eg. $k\langle x, y \rangle \subseteq k[t][x; \alpha_i]$, where $\alpha_i: f(t) \mapsto f(t^i)$. Here $y = xt$ and $tx = xt^i$ (see 2.1). The resulting embeddings of $k\langle x, y \rangle$ in D_i are distinct for $i = 2, 3, \ldots$ and none is a specialization of the others.

By contrast, if R is a right Ore X-domain, the epic R-fields can be determined by their kernels, e.g. $R = k\langle X \rangle_n$. If, moreover, D_t is commutative, then $\text{Supp}_S(t) = \{u \in S | t \text{ supports } \{u\}\}$, and this is just the set of u such that $D_t \to D_u$ is a specialization. For let us write $T = \text{Supp}_S(t)$ and put $C = \bigwedge_T D_s$; we have an injection $C \to D_t$, so C is a commutative integral domain with D_t as field of fractions. Let $u \in T$; then $C/\mathfrak{P}_u$ is an integral domain with field of fractions D_u. Hence the localization at $\mathfrak{P}_u$ is a local ring $L_u \subseteq D_t$ with residue class ring D_u, i.e. we have a specialization $D_t \to D_u$.

We conclude this section by showing how to express the notions of essentiality and support in terms of singular kernels. For any prime matrix ideal $\mathcal{P}$ we define $\mathcal{P}^\cdot$ as the set of matrices all of whose first order minors lie in $\mathcal{P}$. Clearly $\mathcal{P}^\cdot \subseteq \mathcal{P}$ and under a homomorphism into a field, if $\mathcal{P}$ represents the singular matrices, then $\mathcal{P}^\cdot$ represents the matrices of nullity at least two.

PROPOSITION 7.6.8. *Let X be a set, $\{D_s\}_S$ a finite family of epic X-fields and $t \in S$. Then the following conditions are equivalent, where $\mathcal{P}_s$ is the singular kernel of the map $k\langle X \rangle \to D_s$:*
  (a) *t is essential in S,*
  (b) *$\mathcal{E}(D_t) \not\supseteq \bigcap_{s \neq t} \mathcal{E}(D_s)$,*
  (c) *$\mathcal{P}_t \backslash \mathcal{P}_t^\cdot \not\subseteq \bigcup_{s \neq t} \mathcal{P}_s$,*
  (d) *$\mathcal{P}_t \not\subseteq \bigcup_{s \neq t} \mathcal{P}_s$.*

Proof. We saw that (a) $\Leftrightarrow$ (b) in 7.5. When (b) holds, there is a rational expression f which is defined in D_s for $s \neq t$ but not in D_t. Hence the denominator of f lies in $\mathcal{P}_t$ but not in $\mathcal{P}_t^\cdot$, nor in any $\mathcal{P}_s$ for $s \neq t$, so (c) holds. Clearly (c) $\Rightarrow$ (d) and when (d) holds, then we can find $A \in \mathcal{P}_t$,

$A \notin \mathcal{P}_s$ $(s \neq t)$. We may assume that A has been chosen to be of minimal order subject to these conditions. Further, we may assume A to be an atom, for if $A = BC$, where B, C are (square) non-units, then $B, C \notin \mathcal{P}_s$ for $s \neq t$ and either $B \in \mathcal{P}_t$ or $C \in \mathcal{P}_t$ because $\mathcal{P}_t$ is prime and now we can use induction on the degree. If we now consider the $n \times (n + 1)$ matrix $(c \quad A)$, where $c = (y \quad xy \quad \ldots \quad x^{n-1}y)^{\mathrm{T}}$, where $x, y \in X$, $x \neq y$, we obtain an admissible block for D_s $(s \neq t)$, so the rational expression f defined by this block is defined in D_s but not in D_t; for if f were defined in D_t and we take a reduced admissible block for f, its denominator B would be stably associated to A, but $A \in \mathcal{P}_t$, hence $B \in \mathcal{P}_t$ and this contradicts the fact that f is defined in D_t. ∎

We shall need an auxiliary result on ranks of matrices.

LEMMA 7.6.9. *Given any prime matrix ideals* $\mathcal{P}_1, \ldots, \mathcal{P}_r$ *in* $k\langle X \rangle$, *where* X *is infinite, any* $n \times n$ *matrix* $A \notin \bigcup \mathcal{P}_i$ *can be extended to an* $n \times (n + 1)$ *matrix which has rank* $n \bmod \mathcal{P}_i$ *for* $i = 1, \ldots, r$.

Proof. Write $A = (a_1 \quad \ldots \quad a_n)$; if a is another column, we put $A^* = (a_1 \quad \ldots \quad a_n \quad a)$ and we write $A^* \notin \mathcal{P}_i$ to indicate that A^* has rank $n \bmod \mathcal{P}_i$; this just means that the square matrix obtained by omitting some column of A^* is not in $\mathcal{P}_i$. We shall use induction on r; when $r = 1$, A has nullity $1 \bmod \mathcal{P}_1$ and we can make it non-singular by adjoining a suitable column. When $r > 1$, we can by the induction hypothesis adjoin a column to A to obtain an $n \times (n + 1)$ matrix A_i such that $A_i \notin \mathcal{P}_j$ $(j \neq i)$. If for some i, $A_i \notin \mathcal{P}_i$, this will show that $A_i \notin \bigcup \mathcal{P}_j$. Otherwise $A_i \in \mathcal{P}_i$ for all i and we form $A^* = (A \quad a)$, where $a = \sum a_i x_i$, with distinct elements $x_i \in X$ not occurring in any A_i. If $A^* \in \mathcal{P}_1$ say, then by specializing x_i to δ_{i2} we obtain A_2, but $A_2 \notin \mathcal{P}_1$; hence $A^* \notin \mathcal{P}_1$ and similarly $A^* \notin \mathcal{P}_i$, $i = 2, \ldots, r$. ∎

Now the support relation is described by

THEOREM 7.6.10. *Let* $\{D_s\}_S$ *be a family of epic* X-*fields and* $\mathcal{P}_s$ *the singular kernel of* D_s. *Then* D_t *supports* $\{D_s | s \in S\}$ *if and only if*

$$\mathcal{P}_t \subseteq \left(\bigcap \mathcal{P}_s \right) \cup \bigcup \mathcal{P}_s'. \tag{4}$$

In words: every matrix which becomes singular in D_t is either singular in each D_s or of nullity > 1 in some D_s.

Proof. Suppose that t supports S and let $A \in \mathcal{P}_t$, $A \notin \bigcup \mathcal{P}_s'$. By the lemma we can find a column a such that $(A \quad a) = A^* \notin \bigcup \mathcal{P}_s$. Hence the equation $A^*u = 0$ defines $u = (u_1 \ \ldots \ u_n \ u_0)$ up to a scalar multiple in any D_s. Since $A \in \mathscr{P}_t$, $u_0 = 0$ in D_t and so $u_0 = 0$ in all D_s; but this means that $A \in \mathscr{P}_s$ for all s, because u_0 is stably associated to its numerator. Hence (4) holds.

Conversely, assume (4) and let f be defined in all D_s and $f = 0$ in D_t. We can find a denominator for f, say A, with numerator A_1, and then $A_1 \in \mathscr{P}_t$, so either $A_1 \in \bigcap \mathscr{P}_s$, which means that $f = 0$ in all D_s or $A_1 \in \mathscr{P}_s'$ for some s. But then $A \in \mathscr{P}_s$ and this contradicts the fact that A was a denominator. ■

7.7 Examples of support relations

Before constructing examples let us summarize the properties of supports. This is most easily done by introducing the notion of an *abstract support system*, by which we understand a set S with a relation on $S \times \mathscr{P}(S)$, written $t \propto U$ and called the *support relation*, with the following properties:

(S.1) *If $t \in S$ and $U \subseteq S$, then $t \propto U \Leftrightarrow t \propto U \cup \{t\}$,*

(S.2) *If $t \propto S_i$ ($i \in I$), then $t \propto \bigcup_I S_i$,*

(S.3) *If $t \propto U$ and for each $u \in U$, $u \propto S_u \neq \varnothing$, then $t \propto \bigcup_U S_u$.*

If in (S.2) we take the index set I to be empty, the hypothesis is vacuous, hence $t \propto \varnothing$ and by (S.1), $t \propto \{t\}$ always holds. By Prop. 5.8 the support relation on any family of epic X-fields is a support relation in the above sense.

A special case of the support relation is that where

(S.4) $t \propto U \Leftrightarrow t \propto \{u\}$ *for all $u \in U$.*

The relation is then completely determined by all pairs t, u such that $t \propto \{u\}$ and if we write $t \leq u$ to indicate that $t \propto \{u\}$, we obtain a preordering of S. Conversely, every preorder on S leads to a support relation satisfying (S.4) in this way. Thus preorders may be regarded as a special case of support relations.

A support relation on S induces a support relation on any subset of S. If a support relation is such that

(S.5) $s \propto \{t\}$ *and $t \propto \{s\} \Rightarrow s = t$,*

the relation is said to be *separated*. For example, the separated preorders are just the partial orders. Note that this is *not* the same as $s \in \text{Supp}_S(t)$, $t \in \text{Supp}_S(s)$, which may well hold for distinct s, t in a separated support relation.

We now construct all possible separated support relations on a three-element set. There are ten in all, five of them orders (if we allow non-separated ones and do not identify isomorphic ones, we get 53 support systems, 29 of them preorders).

We list the ten below, the orders first, with rising arrows to indicate specializations. In the examples, the p_i are primes and $|X| > 1$, unless stated, while $G_n = k\langle X\rangle_n$.

1.

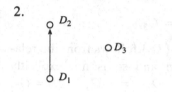

$\circ\, D_1 \qquad \circ\, D_2 \qquad \circ\, D_3$

Examples. (a) $X = \varnothing$, $D_i = \mathbf{Z}/p_i$ ($i = 1, 2, 3$). (b) $X = \{x\}$, $x \mapsto 1, 2, 3$ in $\mathbf{Q}$. (c) G_n, $n = 3, 4, 5$.

2.

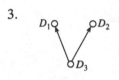

Here D_1 specializes to D_2 and $D_1 \wedge D_2 \wedge D_3 = (D_1 \wedge D_2) \times D_3$. (a) $X = \{x, y, z\}$, $D_1 = \mathbf{Q}(x, y)$, $z \mapsto 0$, $D_2 = \mathbf{Q}(x)$, $y, z \mapsto 0$, $D_3 = \mathbf{Q}(z)$, $x, y \mapsto 0$. (b) $D_1 = G_4$, $D_2 = G_2$, $D_3 = G_3$.

3.

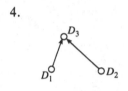

$R = D_1 \wedge D_2 \wedge D_3$ a semilocal domain $\subseteq (D_1 \wedge D_3) \cap (D_2 \wedge D_3) \subseteq D_3$. (a) $D_3 = \mathbf{Q}$, $D_i = \mathbf{Z}/p_i$ ($i = 1, 2$), (b) $D_1 = G_3$, $D_2 = G_2$, $D_3 = G_6$.

4.

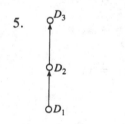

$R = D_1 \wedge D_2 \wedge D_3$ is a local ring with two minimal prime ideals, subdirect product of $D_1 \wedge D_3$ and $D_2 \wedge D_3$. (a) $D_1 = \mathbf{Q}(x)$, $y \mapsto 0$, $D_2 = \mathbf{Q}(y)$, $x \mapsto 0$, $D_3 = \mathbf{Q}$, $x, y \mapsto 0$. (b) $D_1 = \mathbf{Q}$, $x \mapsto 0$, $D_2 = \mathbf{Q}$, $x \mapsto p$, $D_3 = \mathbf{Z}/p$, $x \mapsto 0$.

5.

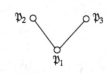

$R = D_1 \wedge D_2 \wedge D_3 = D_1 \wedge D_3$ (by Cor. 5.14). (a) $D_1 = \mathbf{Q}(x, y)$, $D_2 = \mathbf{Q}(x)$, $y \mapsto 0$, $D_3 = \mathbf{Q}$, $x, y \mapsto 0$. (b) $D_1 = \mathbf{Q}(x)$, $D_2 = \mathbf{Q}$, $x \mapsto 0$, $D_3 = \mathbf{Z}/p$.

In each case there were commutative examples. For the remaining support systems (non-orders) we have of course only non-commutative examples. In each case $s \propto T$ is indicated by drawing an arrow from s to a balloon enclosing T. We also indicate the partially ordered set of primes $\mathfrak{P}_i = \ker(\wedge\, D_j \to D_i)$; in each case the lowest prime is 0.

6.

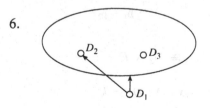

$$D_1 = G_5, \; D_2 = G_2, \; D_3 = G_3.$$

7.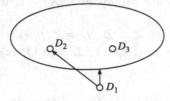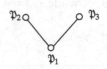

$$D_1 = G_8, \; D_2 = G_2, \; D_3 = G_3.$$

8.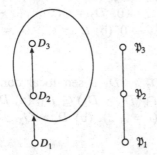

Here $D_1 \propto \{D_3\}$ follows from the relations shown and so is not explicitly indicated. $D_1 = G_3$, $D_2 = G_2$, $D_3 = G_1$.

9.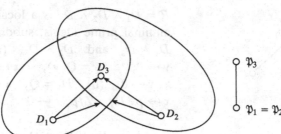

$$D_1 \wedge D_2 \wedge D_3 \cong D_1 \wedge D_3 \cong D_2 \wedge D_3.$$

$D_1 = k\langle\!\langle x, y; (x^{-1}y)^2 = yx^{-1}\rangle\!\rangle$, $D_2 = k\langle\!\langle x, y; (x^{-1}y)^3 = yx^{-1}\rangle\!\rangle$,
$D_3 = k$ (see 2.1).

10.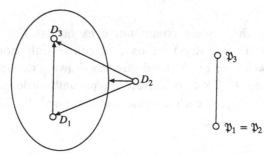

$$D_1 \wedge D_2 \wedge D_3 \cong D_1 \wedge D_3 \cong D_2 \wedge D_3.$$

$D_1 = k\langle\!\langle x, y\rangle\!\rangle$, $D_2 = k\langle\!\langle x, y; (x^{-1}y)^2 = yx^{-1}\rangle\!\rangle$, $D_3 = k$.

Exercises

1. Construct the support relations on a four-element set and illustrate them by field specializations. For example, draw a support diagram for G_n with $n = 2, 3, 4, 12$ and $n = 2, 3, 5, 12$.

2. (G. M. Bergman) Show that in a non-empty separated support system S there exists t with $\mathrm{Supp}_S(t) = \{t\}$.

3. (G. M. Bergman) Show that abstract support systems form a category, taking as morphisms the maps compatible with the support relation. Describe products and coproducts in this category.

4°. (G. M. Bergman) Show that in case 3 above, $D_1 \wedge D_2 \wedge D_3 = (D_1 \wedge D_2) \cap (D_1 \wedge D_3)$ when all fields are commutative. Does this hold generally?

5°. (G. M. Bergman) Can every abstract support system be realized by a family of X-fields?

Notes and Comments

The problem of polynomial identities was first considered in geometry, in the study of Desarguesian projective planes in which Pappus' theorem does not hold (and which are thus coordinatized by skew fields), see Dehn [22], Wagner [37]. The first general result was the PI-theorem of Kaplansky [48], followed by the GPI-theorem of Amitsur [65]. The presentation of 7.1 follows Procesi [68], who has now given a more general treatment using Azumaya algebras in Procesi [73]. The special case of Th. 1.2 when D is a field was proved by Gordon and Motzkin [65].

The functional approach of 7.2 follows Bergman [70]; 7.3–4 largely follow Bergman [76] and Bergman and Small [75], except for Prop. 3.1 and Th. 3.2, which are taken from Cohn [82]. In 7.5–7 we again follow Bergman [76]. For the equivalence (b) ⇔ (d) of Cor. 6.3 a brief proof using Azumaya algebras was recently given by L. Le Bruyn [93].

8

Equations and singularities

The theory of equations over skew fields is in many ways the least developed part of the subject. An element of an extension field whose powers are (right) linearly dependent over a subfield satisfies a rather special kind of equation (discussed in 3.4). More general equations are much less tractable, and in some ways it is more appropriate to consider singularities of matrices, a wider problem. Here we can limit ourselves to linear matrices, but even in this case there is as yet no comprehensive theory.

We begin by discussing the different possible notions of an algebraically closed skew field; although their existence in all cases has not yet been established, the relations between them are described in 8.1. We find that to solve equations we need to find singular eigenvalues of a matrix and this has so far been done only in special cases (discussed in 8.5). By contrast the similarity reduction to diagonal form requires the notion of left and right eigenvalues; it is shown in 8.2 that they always exist (in a suitable extension field) and 8.3 describes the reduction to normal form based on these eigenvalues. While this normal form (over an EC-field) is quite similar to the commutative case, there is no full analogue of the Cayley–Hamilton theorem (owing to the lack of a determinant function), although such a result exists for 'skew cyclic' matrices (i.e. matrices A such that $xI - A$ is stably associated to a 1×1 matrix) and is presented in 8.3.

In 8.4 various notions of 'algebraic' over a central subfield are defined and compared and earlier results are used to construct V-rings (rings over which all simple modules are injective) as central localizations of polynomial rings, following R. Resco, and to construct right Noetherian but not Artinian annihilator rings, following Faith and Menal. In 8.5 we return to the topic of equations and show how left and right eigenvalues can be used in certain cases to find singular eigenvalues and hence solve the equations considered in 3.4. Singular eigenvalues of a matrix A are also found in two special cases: (i) when A is 2×2 and (ii) when A has indeterminate entries.

In the remaining three sections we discuss the problems arising when one tries to do algebraic geometry over a skew field. The main difficulty is of course that the intuitive geometric picture is missing, but there are also many technical problems and these sections can only be regarded as a programme. It is possible to define affine and projective space as in the commutative case and to define the analogue of the Zariski topology, the rational topology (already encountered in 7.2), and a notion of specialization. This is done in 8.6 and illustrated in 8.7 by some examples of varieties. The final 8.8 brings a non-commutative analogue of the Nullstellensatz and describes the (as yet unsolved) problem of finding a satisfactory elimination procedure.

8.1 Algebraically closed skew fields

The first algebraically closed field we encounter is usually the field of complex numbers, but that it is algebraically closed depends on topological properties of the real numbers. For a truly algebraic construction we rely on a theorem of Kronecker which tells us that every polynomial equation of positive degree over a commutative field k has a solution in some extension field of k. One effect of this result has been to try to reduce any search for solutions to a single equation. For example, to find the eigenvalues of a matrix A we solve the equation $\det(xI - A) = 0$.

In the general case no such simple theorem exists (so far!) and in any case we do not have a good determinant function – the determinant introduced by Dieudonné [43] is not really a polynomial but a rational function – so the above reduction is not open to us. In fact we shall find it more profitable to go from scalar equations to matrices.

Our first problem is to write down the general equation in one variable x over a skew field D. We cannot allow x to be central if we want to be able to substitute non-central values of D, but some elements of D are bound to commute with x, such as $1, -1$ etc. and it is clear that these elements form a subfield k. Moreover, if $\alpha \in k$, so that $\alpha x = x\alpha$, then α must lie in the centre of D if arbitrary substitutions of x are to be allowed. Thus we have a field D which is a k-algebra, for a commutative field k, and a polynomial in x is an element p of the tensor D-ring $D_k\langle x \rangle$. Explicitly p has the form

$$a + b_1 x c_1 + \ldots + b_r x c_r + d_1 x e_1 x f_1 + \ldots + d_s x e_s x f_s + \ldots, \quad (1)$$

where $a, b_i \ldots \in D$. Thus even a polynomial of quite low degree can already have a complicated form, and the problem of finding a solution seems at first sight quite hopeless. A little light can be shed on the problem by linearizing our polynomial. As we saw in 6.2, every matrix over $D_k\langle X \rangle$, in particular every polynomial (considered as a 1×1 matrix), is stably associated to a linear matrix which may be taken in the form $A_0 + \sum A_i' x_i A_i''$ or also $A_0 + \sum x_i A_i$. Thus a general polynomial p in x over D is stably associated to $A + xB$, where $A, B \in D_n$ for some n, and instead of finding extension fields E where $p = 0$ has a root we can look for $\alpha \in E$ such that $A + \alpha B$ is singular.

Given any D-ring R, a square matrix A over R will be called *proper* if A becomes singular under some D-ring homomorphism of R into a field; otherwise A is said to be *improper*. The following result is an immediate consequence of the definitions:

PROPOSITION 8.1.1. *Let D be a field and R a D-ring. If a matrix A over R is invertible, or proper, then any matrix stably associated to A has the same property.* ■

To elucidate the relation between invertible and improper matrices, we have the following result:

PROPOSITION 8.1.2. *Let R be a D-ring, where D is a field. Then an invertible matrix over R is improper. When R is commutative, the converse holds: every improper matrix is invertible, but this does not hold generally.*

Proof. If A is invertible over R and $f : R \to E$ is a D-ring map into a D-field E, then Af is again invertible, hence A is then improper. When R is commutative and A is not invertible, then $\det A$ is also a non-unit and so is contained in a maximal ideal $\mathfrak{m}$ of R. The natural map $R \to R/\mathfrak{m}$ is clearly a D-ring map into a field and it maps $\det A$ to 0, so A becomes singular over $R/\mathfrak{m}$.

To find a counter-example we can limit ourselves to 1×1 matrices, thus we must find a non-unit of R which maps to a unit under any homomorphism into a field. Take a ring with no R-fields, such as $\mathfrak{M}_2(D)$; any element c of $\mathfrak{M}_2(D)$ which is not zero or a unit is improper but not invertible. ■

Of course this result does not exclude the possibility that the converse holds in some non-commutative rings. As we shall see below, it is of particular interest to know whether this holds for the tensor ring.

Our original problem was this: Does every polynomial p in $F = D_k\langle x \rangle$

which is a non-unit have a zero in some D-field E? We note that p has a zero α in E if and only if the D-ring map $F \to E$ defined by $x \mapsto \alpha$ maps p to zero. Thus the question now becomes: Is every non-invertible element of F proper? It is natural to subsume this under the more general form:

PROBLEM 8.1. *Is every non-invertible matrix over $D_k\langle x \rangle$ proper?*

We shall find that this is certainly not true without restriction on k and D, and find a positive answer in certain special cases, but the general problem is still open. As a first necessary condition we have the following result.

THEOREM 8.1.3. *Let D be a field with centre k. If every non-constant polynomial equation over D has a solution in some field extension of D, then k is relatively algebraically closed in D.*

Proof. Suppose that the conclusion fails to hold and let $a \in D$ be algebraic over k but not in k. We consider the *metro-equation*

$$ax - xa = 1; \tag{2}$$

by hypothesis it has a solution λ in an extension of D, so that $a\lambda - \lambda a = 1$. If f is the minimal polynomial for a over k, then by (2) we have

$$0 = f(a)\lambda - \lambda f(a) = f'(a),$$

where f' is the formal derivative of f (see 2.1). By the minimality of f it follows that $f' = 0$, so a is not separable over k; in particular, this shows that k must be separably closed in D. If a is p-radical (purely inseparable) over k, say $a^q \in k$, where $q = p^r$, then on writing $\delta(a)$ for the derivation $x \mapsto ax - xa$, we have $\delta(a)^q = \delta(a^q) = 0$. Since $a \notin k$, we have $\delta(a) \neq 0$, say $b\delta(a) \neq 0$ for some $b \in D$. Now the equation $x\delta(a)^{q-1} = b$ has a solution x_0 in some extension of D, but then $x_0\delta(a)^q = b\delta(a) \neq 0$, a contradiction. Hence $a \in k$ and k is relatively algebraically closed in D. ∎

Thus for Problem 1 to have a positive solution it is necessary for k to be relatively algebraically closed in D. We have already noted that every matrix $A(x)$ over $D_k\langle x \rangle$ has a stably associated linear form $xB + C$, where B, C are over D. If $A(0)$ is non-singular, then so is C, the result of putting $x = 0$ in $xB + C$, and so $xB + C$ is associated to $xBC^{-1} + I$. If B is non-singular, our matrix can be put in the form $xI - A$, where

$A = -CB^{-1}$. Any element of $D_k\langle x \rangle$ or more generally a matrix P whose linear companion can be put in the form $xI - A$ is said to be *non-singular at infinity*. For such a matrix the linear companion is unique up to conjugacy:

PROPOSITION 8.1.4. *Any matrix over $D_k\langle x \rangle$ which is non-singular at infinity has a linear companion $xI - A$, where A is unique up to conjugacy over k.*

Proof. If the given matrix is stably associated to $xI - A$ and to $xI - A'$, then $xI - A$ and $xI - A'$ are stably associated, hence by Th. 6.3.2, there are invertible matrices P, Q over D such that

$$P(xI - A) = (xI - A')Q.$$

Equating terms in x we find that $Px = xQ$, hence $P = Q$ and all entries of P commute with x and so lie in k, and $A' = PAP^{-1}$ as claimed. ∎

To give an example, let

$$p(x) = x^n + a_1 x^{n-1} + \ldots + a_n, \tag{3}$$

and write $p = a_n + p_1 x$; then the first step is

$$p \to \begin{pmatrix} x & -1 \\ a_n & p_1 \end{pmatrix}$$

where we have interchanged the rows. If we continue in this way, we obtain the matrix

$$\begin{pmatrix} x & -1 & 0 & 0 & \ldots & 0 \\ 0 & x & -1 & 0 & \ldots & 0 \\ & & \ldots & \ldots & \ldots & \\ 0 & 0 & \ldots & \ldots & & -1 \\ a_n & a_{n-1} & \ldots & \ldots & & x + a_1 \end{pmatrix}.$$

This is of the form $xI - A$, where $A \in D_n$, and this matrix or also A itself is usually known as the *companion matrix* of p.

We now come to define algebraically closed skew fields; here it is convenient to introduce several notions.

DEFINITION 1. A skew field K is called *characteristically algebraically closed* (CAC) if for each square matrix A over K there exists $\alpha \in K$ such that $\alpha I - A$ is singular. Any such α is called *singular eigenvalue* of A.

It is clear that for commutative fields this reduces to the usual definition of 'algebraically closed', for in that case 'singular eigenvalues' are just eigenvalues, and we can solve a polynomial equation by finding a singular eigenvalue for its companion matrix. The question whether every field has a CAC-extension is the *singular eigenvalue problem*.

The remaining definitions involve the centre of K and so have no strict analogue.

DEFINITION 2. A skew field K with centre C is called *polynomially algebraically closed* (PAC) if for every non-constant polynomial $p(x) \in K_C\langle x \rangle$ there exists $\alpha \in K$ such that $p(\alpha) = 0$.

DEFINITION 3. A skew field K with centre C is called *fully algebraically closed* (FAC) if every square matrix over $K_C\langle x \rangle$ which is not invertible becomes singular when x is replaced by a suitable element of K.

Our first task is to elucidate the connexion between these notions. It is clear that in Def. 3 we can limit ourselves to matrices that are linear in x; further, we may take the constant term to be non-singular and hence equal to I, without loss of generality. We begin by giving a criterion for $I - Ax$ to be invertible; it is no harder to do this for the case of several variables. A square matrix will be called 0-*triangular* if it has zeros on and below the main diagonal; if the matrix A over a ring R is such that $P^{-1}AP$ is 0-triangular, where P is an invertible matrix over a subring S of R, then A is said to be 0-*triangulable over S*.

THEOREM 8.1.5. *Let K be a field and C its centre. For any matrices $A_i \in \mathfrak{M}_n(K)$ ($i = 1, \ldots, r$) the following conditions are equivalent:*
  (a) $I - \sum A_i x_i$ *has an inverse over* $K_C\langle x_1, \ldots, x_r \rangle$,
  (b) $\sum A_i x_i$ *is nilpotent, as matrix over* $K_C\langle x_1, \ldots, x_r \rangle$,
  (c) $(\sum A_i x_i)^n = 0$,
  (d) $A_1, \ldots, A_r$ *are simultaneously 0-triangulable over C.*

Proof. (a) $\Rightarrow$ (b). Let us write $X = \{x_1, \ldots, x_r\}$. We have an embedding of $K_C\langle X \rangle$ in the formal power series ring $K_C \ll X \gg$ and in the latter, $(I - \sum A_i x_i)^{-1} = \sum_j (\sum_i A_i x_i)^j$. By uniqueness this must also be the inverse in $K_C\langle X \rangle$, which is possible only if $(\sum A_i x_i)^t = 0$ for some $t \geq 1$. Now (b) $\Rightarrow$ (c) follows because $K_C\langle X \rangle$ is a semifir, and so satisfies Klein's nilpotence condition (Prop. 1.6.7).

(c) $\Rightarrow$ (d). Write $T = \sum A_i x_i$ and let t be the least exponent of nilpotence, i.e. $T^t = 0$, $T^{t-1} \neq 0$. By equating left cofactors of degree 1 in $T^t = 0$ we obtain an equation $TB = 0$, where $B \neq 0$, $B \in \mathfrak{M}_n(K)$. We denote the columns of T by $T_1, \ldots, T_n$; if these columns are linearly independent over C, then $\sum T_j \otimes b_j \neq 0$ for any $b_j \in K$, not all 0; but this contradicts the equation $TB = 0$. Hence the T_j are linearly dependent over C, say $\sum T_j \beta_j = 0$, where $\beta_j \in C$ and not all vanish. Suppose that $\beta_1 \neq 0$ and write U for the matrix whose columns are $\beta = (\beta_1 \ \dots \ \beta_n)^T, e_2, \dots, e_n$. This matrix U has entries in C and $U^{-1}TU = T'$ has its first column 0, thus

$$T' = \begin{pmatrix} 0 & v \\ 0 & T^* \end{pmatrix},$$

where T^*, like T and T', is linear in the xs, say $T^* = \sum A_i^* x_i$. Since $T' = 0$, we have $T^{*\prime} = 0$, so by induction on n there exists $V \in \mathbf{GL}_n(C)$ such that $V^{-1}T^*V$ is 0-triangular. Write $P = U(1 \oplus V)$; then $P \in \mathbf{GL}_n(C)$ and $P^{-1}TP = \sum P^{-1}A_iPx_i$ is 0-triangular, therefore each $P^{-1}A_iP$ is 0-triangular, and so (d) holds. Now (d) $\Rightarrow$ (a) follows trivially. ∎

We can now describe the relations between the different forms of algebraic closure, or rather, algebraic closedness:

THEOREM 8.1.6. (i) *Every FAC field is PAC and CAC.*

(ii) *A field K with centre C is FAC if and only if every square matrix A over K has a non-zero singular eigenvalue in K unless A is 0-triangulable over C.*

(iii) *If a field K with centre C is FAC, then either $K = C$ and C is algebraically closed, or $[K : C] = \infty$ and every non-invertible square matrix over $K_C\langle x \rangle$ is proper.*

Proof. (i) follows because a non-constant polynomial is not invertible; neither is $xI - A$ and invertibility is preserved by stable association. To prove (ii), suppose that K is FAC and let A be a square matrix over K which is not 0-triangulable over C. Then $I - Ax$ is non-invertible, by Th. 1.5, so by FAC it becomes singular for some value of x in K, say $I - A\lambda$ is singular. Clearly $\lambda \neq 0$, hence λ^{-1} is a singular eigenvalue of A. Conversely, assume that every matrix over K either is 0-triangulable or has a non-zero singular eigenvalue, and consider any matrix over $K_C\langle x \rangle$ which is not invertible; without loss of generality we may replace it by its linear companion, $A + Bx$ say. If A is singular, we can put $x = 0$ to get a singular matrix. Otherwise we can replace the matrix by $A^{-1}(A + Bx) = I - Dx$, say. By hypothesis this is not invertible, so D is not 0-triangulable over C, and hence it has a singular eigenvalue $\delta \neq 0$. It follows that $I - D\delta^{-1}$ is singular, and this shows K to be FAC.

To establish (iii) it is enough by (ii) to show that the conditions hold if every square matrix that is not 0-triangulable over C has a non-zero singular eigenvalue. Suppose first that $K = C$; then every matrix A has a singular eigenvalue precisely when C is algebraically closed, and moreover, A has a non-zero singular eigenvalue unless it is 0-triangulable. Now suppose that $K \neq C$ and K is FAC. By (i) the metro-equation has a solution for all $a \in K \backslash C$, so by Th. 1.3, C is relatively algebraically closed in K, and in particular it follows that $[K : C] = \infty$. Moreover, any equation over C has a solution in K and hence in C, i.e. C is algebraically closed. Now take a square matrix A over $K_C\langle x \rangle$; we must show that A is either invertible or proper, and we may without loss of generality take it in linear form $A_0 + A_1 x$. If A_0 is singular, we obtain a singular matrix by putting $x = 0$; otherwise our matrix is associated to $I - Bx$. Now if B has a non-zero singular eigenvalue λ, then $I - B\lambda^{-1}$ is singular. The alternative, by (ii), is that B is 0-triangulable over C, say $B = P^{-1} B_0 P$, where B_0 is 0-triangular; hence

$$I - Bx = P^{-1}(I - B_0 x)P,$$

and $I - B_0 x$ has the inverse $I + B_0 x + \ldots + (B_0 x)^{n-1}$, if B_0 is $n \times n$. Thus $I - Bx$ and hence A is invertible, as we had to show. ∎

In the course of proving this result we have seen that the centre of a FAC field is always algebraically closed; it follows that a FAC field, if non-commutative, is of infinite degree over its centre. A CAC field can be of finite degree, but only in a very special case:

PROPOSITION 8.1.7. *Let K be a field with centre C and suppose that $1 < [K : C] < \infty$. If K is CAC, then C is real-closed and $[K : C] = 4$.*

Proof. We know that the dimension of K over C is a perfect square, say $[K : C] = m^2$, and any commutative subfield of K containing C has as dimension over C a divisor of m (see A.3, 7.1). We claim that C is perfect; for if not, then $p = \text{char } C$ is finite and there exists $a \in C \backslash C^p$. Choose r so large that $q = p^r \nmid m$. The equation $x^q = a$ has a root α in K and $x^q - a$ is irreducible over C (because $y^p - a$ is irreducible, by the choice of a), hence $[C(\alpha) : C] = q$, which is a contradiction. This shows C to be perfect.

Let F be a maximal commutative subfield of K; then $F \supseteq C$ and $[F : C] = m$. We claim that F is algebraically closed; for if not, then there is a finite extension E of F, $[E : F] = h > 1$. Since C is perfect, E/C is separable and so $E = C(\beta)$ for some $\beta \in E$. Let g be the minimal polynomial of β over C; then g has degree hm and is irreducible over C. By hypothesis g has a zero β_1 in K and $[C(\beta_1) : C] = hm$, which is again a contradiction. This shows F to be algebraically closed. Thus C has an extension of finite degree m which is algebraically closed. By a theorem of Artin (see A.2, p. 122), C is real-closed and $[F:C] = 2$, hence $[K:C] = 4$, as asserted. ∎

Exercises

1. Consider the matrix $N = \begin{pmatrix} i & j \\ -j & i \end{pmatrix}$ over the real quaternions. Verify that N is nilpotent and find a non-zero singular eigenvalue of N. Show also that $\begin{pmatrix} j & i \\ i & -j \end{pmatrix}$ is not nilpotent and has no singular eigenvalue except zero.

2. Find two 3×3 matrices A, B over a commutative field such that $\lambda A + \mu B$ is nilpotent for all λ, μ but such that A, B are not simultaneously 0-triangulable. (Hint. First take the field F_2; find a 0-triangulable matrix A and its transpose such that $\det(I + \lambda A + \mu A^{T}) = 1$.)

3°. Give an example of an improper non-invertible matrix over $k\langle x, y \rangle$.

4°. Prove (or disprove) that in a field K of infinite degree over its centre C, any finite-dimensional C-space U satisfies $U^{-1}U \neq K$, i.e. there exists $a \in K$ not of the form $u^{-1}v$, where $u, v \in U$.

5. Assuming a positive answer to Ex. 4°, show that if K is a FAC field with centre $C \neq K$, then for any set X, every matrix over $K_C\langle X \rangle$ which is non-invertible is proper.

6°. Let K be an infinite field. Show that for every square matrix A over K there exists $\alpha \in K$ such that $A - \alpha I$ is non-singular. (Clearly this is false if K is finite and it is easily proved if the centre of K is infinite; the general case is still open.)

8.2 Left and right eigenvalues of a matrix

In 8.1 we encountered singular eigenvalues of a matrix, but information on their existence is still very fragmentary. We now turn to another type of eigenvalue, whose existence can always be established, and which will allow us to effect a reduction of matrices to diagonal form (when possible). Let A be a square matrix over a field K and suppose that A is conjugate to a diagonal matrix $D = \text{diag}(\alpha_1, \ldots, \alpha_n)$. Thus there is a non-singular matrix U such that

$$AU = UD.$$

If we denote the columns of U by $u_1, \ldots, u_n$, this equation can also be written as

$$Au_i = u_i\alpha_i \quad (i = 1, \ldots, n).$$

This makes it clear that we have indeed an eigenvalue problem, but the α_i need not be singular eigenvalues of A, since they do not in general commute with the components of u_i.

For any square matrix A over a field K, an element α of K is called a *right eigenvalue* of A if there is a non-zero column vector u, called an *eigenvector* for α, such that

$$Au = u\alpha. \tag{1}$$

Similarly a *left eigenvalue* of A is an element $\beta \in K$ for which there exists a non-zero row vector v, an *eigenvector* for β, such that $vA = \beta v$. The set of all left and right eigenvalues of A is called the *spectrum* of A, spec A.

Let $c \in K^\times$; if $Au = u\alpha$, then $A \cdot uc = u\alpha c = uc \cdot c^{-1}\alpha c$. This shows that the right eigenvalues of A consist of complete conjugacy classes; similarly for left eigenvalues. If $P \in \mathbf{GL}_n(K)$, where n is the order of A, then $P^{-1}AP \cdot P^{-1}u = P^{-1}u\alpha$, hence α is also a right eigenvalue of $P^{-1}AP$. In other words, right (and left) eigenvalues are conjugacy invariants of A. For singular eigenvalues this is not the case; in fact, although the three notions clearly coincide for elements in the centre of K ('central eigenvalues'), there is in general no very close relation between them. Thus it is possible for a matrix to have a right but no left eigenvalue (see Ex. 2), but as we shall see in 8.3, over an EC-field the notions of left and right eigenvalue coincide.

In order to achieve a transformation to diagonal form one needs a basis of eigenvectors, and here one usually applies the well known result that eigenvectors for different eigenvalues are linearly independent. Over a skew field this takes the following form:

PROPOSITION 8.2.1. *For any square matrix A over a field K, eigenvectors belonging to inconjugate right eigenvalues are linearly independent. If α is a right and β a left eigenvalue of A and α, β are not conjugate, then the eigenvectors belonging to them are orthogonal, thus if $Au = u\alpha$, $vA = \beta v$, then $vu = 0$.*

Proof. Let $\alpha_1, \ldots, \alpha_r$ be right eigenvalues of A and $u_1, \ldots, u_r$ corresponding eigenvectors, and assume that the us are linearly dependent. By taking a minimal linearly dependent set, we may assume that

$$u_1 = \sum_2^r u_i\lambda_i, \quad \text{where } \lambda_i \in K.$$

By definition $u_i \neq 0$, hence by minimality $\lambda_i \neq 0$ and $r > 1$. Now $u_1\alpha_1 = Au_1 = \sum Au_i\lambda_i = \sum u_i\alpha_i\lambda_i$. Hence

$$\sum_{2}^{r} u_i(\lambda_i \alpha_1 - \alpha_i \lambda_i) = 0,$$

but $u_2, \ldots, u_r$ are linearly independent, so $\alpha_i = \lambda_i \alpha_1 \lambda_i^{-1}$ and the α_i are all conjugate.

Next, if $Au = u\alpha$, $vA = \beta v$, then $vAu = vu \cdot \alpha = \beta \cdot vu$, and if $vu \neq 0$, this would mean that α and β are conjugate. ∎

To describe the conditions for diagonalizability we shall need a lemma on the solution of linear equations over a bimodule.

LEMMA 8.2.2. *Let R, S be k-algebras and M an (R, S)-bimodule. Given $a \in R$, $b \in S$, assume that there is a polynomial f over k such that $f(a)$ is a unit while $f(b) = 0$. Then for any $m \in M$ the equation*

$$ax - xb = m \tag{2}$$

has a unique solution x in M.

Proof. If in $\mathrm{End}_k(M)$ we define $\lambda_a: x \mapsto ax$, $\rho_b: x \mapsto xb$, then (2) may be written

$$x(\lambda_a - \rho_b) = m. \tag{3}$$

We note that $\lambda_a \rho_b = \rho_b \lambda_a$ and by hypothesis $f(\lambda_a)$ is a unit and $f(\rho_b) = 0$. Now define a polynomial $\varphi(s, t)$ in the commuting variables s, t by

$$\varphi(s, t) = \frac{f(s) - f(t)}{s - t};$$

then

$$\varphi(\lambda_a, \rho_b)(\lambda_a - \rho_b) = (\lambda_a - \rho_b)\varphi(\lambda_a, \rho_b) = f(\lambda_a) - f(\rho_b) = f(\lambda_a).$$

Since $f(\lambda_a)$ is a unit, it follows that $\lambda_a - \rho_b$ has a two-sided inverse; hence (3) has a unique solution in M. ∎

The significance of the lemma lies in this: Given R, S, M as in the lemma, the set of all matrices $\begin{pmatrix} r & m \\ 0 & s \end{pmatrix}$, where $r \in R$, $s \in S$, $m \in M$, is a ring under the usual matrix multiplication, and (2) shows that

$$\begin{pmatrix} 1 & x \\ 0 & 1 \end{pmatrix}\begin{pmatrix} a & m \\ 0 & b \end{pmatrix} = \begin{pmatrix} a & 0 \\ 0 & b \end{pmatrix}\begin{pmatrix} 1 & x \\ 0 & 1 \end{pmatrix}, \tag{4}$$

thus $\begin{pmatrix} a & m \\ 0 & b \end{pmatrix}$ is conjugate to a 'diagonal' matrix, and the transformation is accomplished by the element x found by Lemma 2.2. We now show how this result can be used to provide a conjugacy reduction.

THEOREM 8.2.3. *Let K be any field and $A \in K_n$. Then $\operatorname{spec} A$ cannot contain more than n conjugacy classes, and when it consists of exactly n classes, all except at most one algebraic over the centre of K, then A is conjugate to a diagonal matrix, and in that case the left and right eigenvalues of A are the same up to conjugacy.*

Proof. We have seen that $\operatorname{spec} A$ consists of complete conjugacy classes. Let r be the number of classes containing right eigenvalues and s the number of the remaining classes in $\operatorname{spec} A$; by Prop. 2.1 the space spanned by the columns of eigenvectors for the right eigenvalues is at least r-dimensional, and the space of rows orthogonal to it is at least s-dimensional; hence $r + s \leqslant n$, and $r + s$ is just the total number of conjugacy classes in $\operatorname{spec} A$.

Assume now that $r + s = n$; let $\alpha_1, \ldots, \alpha_r$ be inconjugate right eigenvalues and $u_1, \ldots, u_r$ corresponding eigenvectors, while $\beta_1, \ldots, \beta_s$ are left eigenvalues not conjugate among themselves or to the αs, with corresponding eigenvectors $v_1, \ldots, v_s$. By Prop. 2.1 the us are right linearly independent, the vs are left linearly independent and $v_j u_i = 0$ for all i, j. Thus if we write U_1 for the $n \times r$ matrix consisting of the columns $u_1, \ldots, u_r$ and V_2 for the $s \times n$ matrix consisting of the rows $v_1, \ldots, v_s$, then $V_2 U_1 = 0$. Further, since the columns of U_1 are linearly independent, we can find an $r \times n$ matrix V_1 over K such that $V_1 U_1 = I$ and similarly there is an $n \times s$ matrix U_2 such that $V_2 U_2 = I$. We now put $U = (U_1 \quad U_2)$, $V = (V_1 \quad V_2)^{\mathrm{T}}$; then U, V are $n \times n$ matrices and

$$VU = \begin{pmatrix} V_1 \\ V_2 \end{pmatrix} (U_1 \quad U_2) = \begin{pmatrix} V_1 U_1 & V_1 U_2 \\ V_2 U_1 & V_2 U_2 \end{pmatrix} = \begin{pmatrix} I & W \\ 0 & I \end{pmatrix}.$$

The matrix on the right is clearly invertible and since fields are weakly finite (see 1.4), we have $U(VU)^{-1} = V^{-1}$, hence

$$AV^{-1} = A(U_1 \quad U_2 - U_1 W) = (u_1 \alpha_1 \quad \ldots \quad u_r \alpha_r \quad A(U_2 - U_1 W)),$$

$$VA = \begin{pmatrix} V_1 A \\ \beta_1 v_1 \\ \vdots \\ \beta_s v_s \end{pmatrix}.$$

It follows that $VAV^{-1} = \begin{pmatrix} \alpha & T \\ 0 & \beta \end{pmatrix}$, where $\alpha = \operatorname{diag}(\alpha_1, \ldots, \alpha_r)$, $\beta = \operatorname{diag}(\beta_1, \ldots, \beta_s)$ and $T \in {}'K^s$. Now all the αs and βs are inconjugate and all but at most one are algebraic over the centre C of K, hence their minimal equations are distinct (by Cor. 3.4.5). If only right or only left eigenvalues occur, we have reached diagonal form; otherwise let $\beta_1, \ldots,$ β_s be algebraic, say. Taking f to be the product of their minimal polynomials over C, we have $f(\beta) = 0$, while $f(\alpha)$ is a unit. By Lemma 2.2 we can find $X \in {}^r K^s$ such that $\alpha X - X\beta = T$ and on transforming our matrix by $\begin{pmatrix} I & X \\ 0 & I \end{pmatrix}$ we reach diagonal form. If this is $D = P^{-1}AP$, then the equation $AP = PD$ shows that the right eigenvalues are just the diagonal entries of D, while the equation $P^{-1}A = DP^{-1}$ shows the same for the left eigenvalues. ∎

The restriction on the eigenvalues, that there is to be at most one transcendental conjugacy class, is not as severe as appears at first sight, but is to be expected, since K can be extended so that all transcendental elements are conjugate (by Cor. 5.5.2).

Exercises

1. Show that every 2×2 matrix with a right eigenvalue has a left eigenvalue.

2. Show that a matrix has a right eigenvalue in a field K if and only if it is conjugate over K to a matrix in which there is a column with at most one non-zero element. Give an example of a matrix with a right but no left eigenvalue.

3°. Let $A \in K_n$; an *eigenspace* of A is an r-dimensional space V of columns in ${}^n K$ $(0 < r < n)$, such that the $n \times r$ matrix P formed from a basis of V satisfies $AP = PB$ for some $B \in K_r$. Verify that the 1-dimensional eigenspaces are the eigenvectors associated with right eigenvalues. Develop a reduction theory using eigenspaces.

4°. (G. M. Bergman) Let $A \in K_n$; an element α of K is called an *inner eigenvalue*, more precisely an *r-eigenvalue*, where $1 \le r \le n$, if ${}^n K$, regarded as a space on which A acts (on the left), has an A-invariant subspace W of dimension $r - 1$ and for some $u \notin W$, $Au \equiv u\alpha \pmod W$. Verify that right eigenvalues are 1-eigenvalues and left eigenvalues are n-eigenvalues. Develop a reduction theory using inner eigenvalues.

5. Let A be a Hermitian matrix over the quaternions. Show that the left and right eigenvalues of A are real, but the singular eigenvalues need not be real. If α is a singular eigenvalue of A with eigenvector u (i.e. $Au = \alpha u$), verify that $u^H(\alpha - \bar\alpha)u = 0$, where H indicates the Hermitian transpose.

6. Show that the singular eigenvalues of a skew Hermitian matrix over the

quaternions are pure (i.e. with zero real part), and the singular eigenvalues of a unitary matrix have norm 1. (Note that the equation $P = e^S$ relates a skew Hermitian matrix S to a unitary matrix P.)

7. Show that an $n \times n$ matrix over a field has at most n central eigenvalues. (Hint. Use Lemma 6.2.4.)

8.3 Normal forms for a single matrix over a skew field

As before let K be a field which is a k-algebra; our task is to find a canonical form for a square matrix under conjugacy. The results are not quite as precise as in the commutative case, but come very close, the main difficulty being the classification of polynomials over K, i.e. elements of $K[t]$. We shall need to make use of the reduction theory of matrices over a principal ideal domain, as given by Jacobson [43], Ch. 3 or FR, Ch. 8; the results needed will be stated with the appropriate reference.

Let $A \in K_n$; in 5.5 we called A *totally transcendental* over k if for any $f \in k[t]^\times$, $f(A)$ is non-singular. If there is non-zero polynomial f over k such that $f(A) = 0$, A is said to be *algebraic* over k. Of course when $K = k$ (the classical case) every matrix is algebraic, by the Cayley–Hamilton theorem. In general a matrix is neither algebraic nor totally transcendental, e.g. $\mathrm{diag}\,(\alpha, 1)$, where α is transcendental over k, but we have the following useful reduction.

PROPOSITION 8.3.1. *Let K be a field which is a k-algebra. Every square matrix A over K is conjugate to a matrix $A_0 \oplus A_1$, where A_0 is algebraic over k and A_1 is totally transcendental, and A_0, A_1 are unique up to conjugacy.*

Proof. Let A be $n \times n$ and consider $V = {}^nK$ as a $(K, k[t])$-bimodule, in which the action of t for a given K-basis $v_1, \ldots, v_n$ is given by

$$v_i t = \sum a_{ij} v_j, \quad \text{where } A = (a_{ij}). \tag{1}$$

Since $K \otimes k[t] = K[t]$, we may regard V as a left $K[t]$-module, generated by $v_1, \ldots, v_n$. Now $K[t]$ is a principal ideal domain, and every module over it has a unique torsion submodule with torsion-free quotient. Let V_0 be the torsion submodule of V; its quotient, being torsion-free and finitely generated, is free, so we can find a complement V_1 of V_0 in V:

$$V = V_0 \oplus V_1. \tag{2}$$

Using a basis adapted to the decomposition (2), we find that A takes the form $\begin{pmatrix} A_0 & A' \\ 0 & A_1 \end{pmatrix}$, where A_0 is algebraic and A_1 is totally transcendental.

Thus there is a polynomial f over k such that $f(A_0) = 0$ and $f(A_1)$ is non-singular. By Lemma 2.2 and the remark following it we can reduce A' to 0 and so obtain a conjugate of A in the form $A_0 \oplus A_1$. Now in (2), V_0 is unique while V_1 is unique up to isomorphism, hence A_0, A_1 are determined up to a change of basis in V, i.e. up to conjugacy. ∎

The totally transcendental part is in some ways simpler to deal with. For by suitably extending K, we obtain a field over which any two totally transcendental matrices of the same order are conjugate, by Th. 5.5.5. This means that every totally transcendental matrix is conjugate to scalar (not merely diagonal) form, where the scalar can be any transcendental element. Over K itself we cannot expect such a good normal form, but of course we can ensure it by assuming K to be an EC-field (see 6.5). We state the result as

THEOREM 8.3.2. *Let K be an existentially closed field over k. Then any totally transcendental matrix over K is conjugate to αI, where α is any transcendental element of K.* ∎

To describe the algebraic part, we recall the reduction theorem for matrices over a principal ideal domain. For any integral domain R, an element a is called a *total divisor* of b, in symbols $a\|b$, if there exists an invariant element c such that $Rb \subseteq Rc = cR \subseteq Ra$. We shall also use the notions of bounded elements and similarity as defined in 1.5.

REDUCTION THEOREM FOR MATRICES OVER PIDs. *Let R be a principal ideal domain and $C \in R_n$. Then there exist P, $Q \in \mathbf{GL}_n(R)$ such that*

$$PCQ = \mathrm{diag}\,(e_1, \ldots, e_r, 0^{n-r}), \text{ where } e_i\|e_{i+1}, e_r \neq 0. \qquad (3)$$

Moreover, the e_i are unique up to similarity. ∎

For a proof we refer to Jacobson [43], p. 43 or FR, Th. 8.1.1, p. 489. We apply the result by putting $R = K[t]$ and $C = tI - A$. Since C is non-singular, the zeros on the right of (3) do not occur and we have

$$P(tI - A)Q = \mathrm{diag}\,(\lambda_1, \ldots, \lambda_n), \lambda_i\|\lambda_{i+1}. \qquad (4)$$

The polynomials λ_i are called the *invariant factors* of A, and V as left R-module is isomorphic to the direct sum

$$R/R\lambda_1 \oplus \ldots \oplus R/R\lambda_n. \qquad (5)$$

We observe that this holds for any matrix A, algebraic or not. In fact we now see that A is algebraic if and only if λ_n divides a polynomial with coefficients in k. Let us take k to be the precise centre of K; then a polynomial over K is invariant if and only if it is associated to a polynomial over k (Prop. 2.2.2). It follows that A is algebraic if and only if λ_n is bounded in R.

To find when A is totally transcendental, suppose that λ_n has a bounded factor p, say. Then the R-module V has an element annihilated by p and hence by p^*, the least bound of p. Now $p^* = p^*(t)$ is invariant and $p^*(A)$ is singular, so A cannot be totally transcendental. Conversely, if A is not totally transcendental, then V has an element annihilated by an invariant polynomial, so some invariant factor λ_i has a factor which is bounded, and hence λ_n has a bounded factor. This proves most of

PROPOSITION 8.3.3. *Let K be a field with centre k and let $A \in K_n$ have invariant factors $\lambda_1, \ldots, \lambda_n$. Then (i) A is algebraic over k if and only if λ_n is bounded, (ii) A is totally transcendental over k if and only if λ_n is totally unbounded, and then $\lambda_1 = \ldots = \lambda_{n-1} = 1$.*

Only the last part still needs proof. Each λ_i ($i < n$) is a total divisor of λ_n, so there is an invariant element c such that $\lambda_i | c | \lambda_n$. But if λ_n is totally unbounded, the only invariant element dividing λ_n is 1, hence $\lambda_i = 1$ for $i = 1, \ldots, n - 1$. ∎

This result shows in particular that when A is totally transcendental, then the associated R-module (5) is cyclic. In that case A is called *cyclic*; for example, the linear companion of any polynomial is a cyclic matrix; more precisely, it is of the form $tI - A$, where A is cyclic. Prop. 3.3 has the following immediate consequence:

COROLLARY 8.3.4. *Any matrix over a field K which is totally trancendental over the centre of K is cyclic.* ∎

To obtain a normal form for algebraic matrices we need yet another decomposition of the R-module V. In (5) we had a decomposition into the *fewest* number of cyclic summands. We now look for a decomposition into the *largest* number of cyclic summands. Such a decomposition exists of course because V has finite composition-length, and it is unique by the Krull–Schmidt theorem. The precise result is as follows:

For any principal ideal domain R and any $a \in R^\times$ there exists a decomposition

$$R/Ra \cong R/Rq_1 \oplus \ldots \oplus R/Rq_k \oplus R/Ru, \tag{6}$$

where each q_i is a product of similar bounded atoms, while atoms in different q's are dissimilar, and u is totally unbounded. Moreover, q_i and u are unique up to similarity. ∎

For a proof see Jacobson [43], pp. 44f. or FR, Th. 6.4.5, p. 316. This process can again be applied to $tI - A$, with $R = K[t]$; then the q_i are called the *elementary divisors* of A. By (6) we can write A as a diagonal sum of terms corresponding to the different elementary divisors, together with a totally transcendental part, corresponding to R/Ru. We have already dealt with the totally transcendental part. Let now A be an algebraic matrix; then A is a diagonal sum of cyclic matrices, each with a single elementary divisor. We therefore consider a matrix with a single elementary divisor α; this is a product of similar atoms,

$$\alpha = p_1 \ldots p_s. \tag{7}$$

Each p_j has the same degree d say (as polynomial in t) and $sd = n$ is the order of A. The R-module associated to A is $V = R/R\alpha$; by definition it is cyclic, with generator v, say. Then $v, vt, \ldots, vt^{n-1}$ is a basis of V, for v, as generator of V, cannot be annihilated by a polynomial of degree less than n. We still have a basis if we take $v, vt, \ldots, vt^{d-1}, vp_s, vtp_s, \ldots, vp_{s-1}p_s, \ldots, vt^{d-1}p_2 \ldots p_s$. Relative to this basis t has the matrix

$$\begin{pmatrix} P_s & N & 0 & 0 & \ldots & 0 \\ 0 & P_{s-1} & N & 0 & \ldots & 0 \\ & \ldots & & \ldots & & \ldots \\ 0 & 0 & 0 & 0 & \ldots P_2 & N \\ 0 & 0 & 0 & 0 & \ldots & P_1 \end{pmatrix}$$

where P_i is the linear companion of p_i and $N = e_{d1}$ is a $d \times d$ matrix with 1 in the SW-corner and the rest zero.

This describes A completely and we obtain an expression much like the rational canonical form in the commutative case (see A.1, 11.4). However, unlike the latter, the above expression is not unique; in fact the p_i are determined only up to similarity and not every choice of ps in their similarity class is possible, thus the factors in (7) cannot be prescribed.

In order to obtain a reduction to triangular form and an analogue of the Jordan normal form we shall need to assume that every polynomial over k splits into linear factors over K. Even when this does not hold, we can enlarge K to a field L for which it holds by taking $L = K^{\circ}_k k^a$, where k^a is a (commutative) algebraic closure of k. Assuming K enlarged in this way, we find that all bounded atoms are linear, of the form $t - a$, up to a unit factor, and $t - a$ is similar to $t - a'$ precisely when a is conjugate to a', by Lemma 3.4.2.

For an analogue of the Jordan normal form we shall need a description of indecomposable bound polynomials, and here it is necessary to assume that k is perfect. We shall need Prop. 1.5.6 for a polynomial ring which tells us: (i) If K is a field with centre k, then in $K[t]$ a polynomial f which is indecomposable bounded has a bound of the form p^n, where p is an I-atom and hence is irreducible as polynomial over k. (ii) Two bounded indecomposable polynomials have the same bounds if and only if they are similar.

PROPOSITION 8.3.5. *Let K be a field with centre k. Assume that k is perfect and that there is a commutative field F, $k \subseteq F \subseteq K$, such that every polynomial over k splits into linear factors over F. Then every indecomposable bounded polynomial over K is similar to a polynomial of the form $(t - \alpha)^n$.*

Proof. Let $a \in K[t]$ be indecomposable bounded. By Prop. 1.5.6, its least bound has the form p^n, where p is an irreducible polynomial over k, which may be taken to be monic. Thus we have over F,

$$p = (t - \alpha_1) \ldots (t - \alpha_r), \quad \text{where } \alpha_i \in F,$$

and where the α_i are distinct, because k is perfect. We claim that $(t - \alpha_1)^n$ has the least bound p^n. For clearly $(t - \alpha_1)^n | p^n$, so the least bound is a factor of p^n, say p^m, where $m \leqslant n$. We now have

$$p^m = (t - \alpha_1)^n q = (t - \alpha_1)^m \prod_{i=2}^{r} (t - \alpha_i)^m.$$

By unique factorization in $F[t]$ we must have $n = m$ because the α_i are distinct, hence p^n is the least bound of $(t - \alpha_1)^n$. Clearly $(t - \alpha_1)^n$ is indecomposable, and it is similar to a because it has the same bounds. ∎

Let K be any field with perfect centre k, and assume that K contains an algebraic closure of k. If A is an algebraic matrix over K with a single elementary divisor, then the latter has the form $(t - \alpha)^n$, by the result just proved. It follows that A is conjugate to

$$\begin{pmatrix} \alpha & 1 & 0 & 0 & \ldots & 0 \\ 0 & \alpha & 1 & 0 & \ldots & 0 \\ & \ldots & \ldots & & \ldots & \\ 0 & 0 & \ldots & & \ldots & \alpha & 1 \\ 0 & 0 & \ldots & & \ldots & 0 & \alpha \end{pmatrix}. \tag{8}$$

When K is any field with perfect centre k, we can extend K so as to contain a commutative field F containing an algebraic closure of k as well as an element transcendental over k, e.g. by forming the field coproduct $K \underset{k}{\circ} F$ for a suitable F. By combining the above steps we then can for any matrix over K find a conjugate over F in Jordan normal form. We shall denote the $n \times n$ matrix (8) by $\mathbf{J}_n(\alpha)$ and call it a *Jordan matrix*. Then we can state our conclusions as follows:

THEOREM 8.3.6. *Let K be a field with perfect centre k and F a commutative field containing an algebraic closure of k as well as an element λ transcendental over k. Form the coproduct $K \underset{k}{\circ} F$ and take an extension L which is matrix-homogeneous with centre k. Then every matrix A over L has a conjugate of the form*

$$\mathbf{J}_{r_1}(\alpha_1) \oplus \ldots \oplus \mathbf{J}_{r_m}(\alpha_m) \oplus \lambda I_s,$$

where $\sum r_i + s = n$, the order of A. ∎

In particular, we see that over a suitable extension every matrix has a conjugate in triangular form. As in the commutative case we deduce

COROLLARY 8.3.7. *Every field K has an extension L (with the same centre) such that every matrix over L has left and right eigenvalues in L.* ∎

This result shows that in many cases there is a very close analogy to the commutative situation. However, there are significant differences; one result without a full analogue is the Cayley–Hamilton theorem. To find a replacement, let us take a field K with centre k and form the tensor ring $F = K_k\langle x \rangle$. Any element $p(x)$ of F has a linear companion, which may be taken in the form $xI - A$, if p is non-singular at infinity. Of course not every matrix of the form $xI - A$ is stably associated to a polynomial; this is not even true in the commutative case (but see Ex. 6). Let us call a matrix A *skew cyclic* if $xI - A$ is stably associated to a polynomial $p(x)$ over $K_k\langle x \rangle$, and call $p(x)$ the *invariant factor* of A. It is clear that every cyclic matrix is skew cyclic, but not conversely. Now we are dealing with $K_k\langle x \rangle$ instead of $K[x]$, so we no longer have a PID (unless $K = k$), but in exchange we can now substitute arbitrary elements of K for x. We have the following almost obvious relation between the singular eigenvalues of a skew cyclic matrix and the zeros of its invariant factor.

PROPOSITION 8.3.8. *Let A be a skew cyclic matrix over a field K, with invariant factor $p(x)$. Then the singular eigenvalues of A are precisely the roots of the equation $p(x) = 0$; in particular, for any $\alpha \in K$, the nullity of $\alpha I - A$ is at most 1.*

The result follows because $xI - A$ is stably associated to $p(x)$ and so for any value of α of x both expressions have the same nullity. ∎

It is natural next to replace x by a matrix, but some care is needed: if in $xI - A$, where $A \in K_n$, we replace x by $B \in K_m$ we obtain, not $B - A$, which is not even defined unless $m = n$, but $I_n \otimes B - A \otimes I_m$. In fact this substitution defines a homomorphism $K_n[x] \to K_{mn}$. We shall need an estimate for the rank of this matrix in a special case:

LEMMA 8.3.9. *Let K be a field and $A \in K_n$. Then $I_n \otimes A - A \otimes I_n$ has nullity at least n.*

Proof. Clearly it is equivalent to prove that the rank is at most $n^2 - n$. We number the rows and columns of $I \otimes A - A \otimes I$ by double subscripts and apply the elementary row and column operations $\text{row}_{ij} \to \text{row}_{ij} + \text{row}_{ji}$, $\text{col}_{ij} \to \text{col}_{ij} + \text{col}_{ji}$ $(i < j)$. In the resulting matrix the entry in row_{ij} $(i \leqslant j)$ and col_{rs} $(r \leqslant s)$ is 0. Hence after permuting non-zero rows and columns we obtain an $N \times N$ block of zeros, where $N = (n^2 + n)/2$. It follows that all non-zero entries are now confined to $(n^2 - n)/2$ rows and $(n^2 - n)/2$ columns, hence the rank is at most $n^2 - n$, as we had to show. ∎

With the help of this result we can prove a 'Cayley–Hamilton theorem' for skew cyclic matrices:

THEOREM 8.3.10. *Let K be a field with centre k and let A be a skew cyclic matrix over K, with invariant factor $p(x)$. Then $p(A) = 0$.*

Proof. By hypothesis, $xI - A$ is stably associated to $p(x)$ over $K_k\langle x \rangle$, and this still holds if we substitute A for x. Hence $p(A)$ is an $n \times n$ matrix, whose nullity is at least n, by the lemma; thus $p(A) = 0$. ∎

In conclusion we briefly note the case of a field of finite dimension over its centre. Take such a field K and let k be its centre; then $[K:k] = n^2$ is a perfect square and any maximal commutative subfield of K has degree n over k (see A.3, 7.1). Let F be such a field and $u_1, \ldots, u_n$ a left F-basis for K. Then for any $a \in K$ we have

$$u_i a = \sum \rho_{ij}(a) u_j, \quad \text{where } \rho_{ij}(a) \in F, \tag{9}$$

and it is easily verified that the mapping $a \mapsto (\rho_{ij}(a))$ is a k-homomorphism of K into F_n. Since F is commutative, left, right and singular eigenvalues coincide, but of course none need exist in F. Suppose however that $(\rho_{ij}(a))$ has an eigenvalue α in F; then there exist $\gamma_1, \ldots, \gamma_n$ in F, not all 0, such that

$$\sum \gamma_i \alpha u_i = \sum \gamma_i \rho_{ij}(a) u_j = \sum \gamma_i u_i a.$$

Writing $c = \sum \gamma_i u_i$, we have $\alpha c = ca$ and $c \neq 0$. Thus $(\rho_{ij}(a))$ has an eigenvalue in F if and only if a is conjugate to an element in F. By the Skolem–Noether theorem (Cor. 3.3.6) this is so whenever $k(a)$ is isomorphic to a subfield of F. All this still applies if a is an $r \times r$ matrix over K, if we interpret $(\rho_{ij}(a))$ as an $nr \times nr$ matrix over F. Thus a matrix A over K has a left eigenvalue in F if and only if its image under ρ (given by (9)) has an eigenvalue in F. In general this is not much help, because F need not be algebraically closed, but in the special case of the real quaternions F reduces to $\mathbf{C}$, the complex numbers. Applying the above remarks, and using the fact that $\mathbf{C}$ is algebraically closed, we see that every matrix over the real quaternions has left and right eigenvalues.

Exercises

1. (W. S. Sizer) Let K be a field with centre k. A matrix A over K, algebraic over k, is called *separable* over k if the bounds of its elementary divisors are separable over k. Show that a separable matrix is conjugate to a matrix with entries in a commutative field, and here separability cannot be omitted.

2. (W. S. Sizer) Show that a family of pairwise commuting matrices that are separable and diagonalizable can be simultaneously diagonalized.

3. Let K be a field with centre k. If no irreducible polynomial over k has repeated zeros in K, show that K has a central extension to a field with perfect centre and formulate a version of Prop. 3.5 for such fields.

4°. Examine how far the reduction theory of matrices can be carried out over a projective-free k-algebra.

5. (W. S. Sizer) Show that in Th. 3.6 one can further require that no two distinct elements α_i, α_j have the same minimal polynomial over k.

6°. Let $k[x]$ be a polynomial ring over a commutative algebraically closed field k. Show that a square matrix over $k[x]$ is cyclic if and only if it is stably associated to a 1×1 matrix. Show that for a square matrix A over k with distinct eigenvalues $xI - A$ is cyclic and find conditions on a general square matrix A over k for $xI - A$ to be cyclic.

8.4 Central localizations of polynomial rings

This section is something of a digression, in that its results are not primarily about fields, but they are closely related and in any case make use of the results proved earlier.

Let D be a skew field; then, as we saw in 2.1, the polynomial ring $D[t]$ is a principal ideal domain and so has a field of fractions, the rational function field $D(t)$, obtained as the localization of $D[t]$ at the set $D[t]^\times$. Sometimes it is of interest to single out a central subfield k of D and consider the localization at $k[t]^\times$. Since $D[t]$ may be expressed as $D \otimes_k k[t]$, its localization at $k[t]^\times$ is the ring

$$T = D \otimes_k k(t). \tag{1}$$

As localization of $D[t]$, T is again a PID. We shall call T the *central localization* of $D[t]$ over k. Its elements may be expressed as fractions fg^{-1} with denominator g in $k[t]^\times$. These localizations have a number of interesting features, which we shall examine here. We shall need to assume some elementary properties of injective modules. An *injective* module I may be defined by the property that any homomorphism $M \to I$ can be extended to any module containing M. Every module M has an 'injective hull' $\mathbf{E}(M)$ which is a minimal injective module containing M, and which may also be defined as a maximal essential extension of M, i.e. a module containing M such that every non-zero submodule intersects M non-trivially (thus M is 'large' in $\mathbf{E}(M)$, see A.3, Ch. 3). The results of this section will not be used later and so may be omitted without loss of continuity.

Let D be a field with a central subfield k; if $[D:k]$ is finite, then since $[T:k(t)] = [D:k]$, where T is the central localization as in (1), it follows that T is of finite degree over $k(t)$ and so is itself a field. This conclusion still holds under more general conditions. We recall that a k-algebra A is said to be *locally finite* over k if every finite subset of A generates a subalgebra which is of finite degree over k. A k-algebra A is called *matrix-algebraic* over k if every square matrix over A is algebraic over k. Both these conditions are, on the face of it, stronger than algebraicity:

THEOREM 8.4.1. *Let D be a field which is a k-algebra. Then of the following conditions, each implies the next:*
   (a) *D is locally finite over k,*
   (b) *D is matrix-algebraic over k,*
   (c) *D is algebraic over k.*

Proof. (a) $\Rightarrow$ (b). Let $A \in D_n$; then the subalgebra L of D generated by the entries of A has finite degree over k, so does L_n and it follows that A is algebraic over k. Now (b) $\Rightarrow$ (c) is clear, since (c) is a special case of (b). ∎

Algebraic k-algebras that are not locally finite have been constructed by Golod and Shafarevich [64] (cf. A.3, p. 58), but it is not known whether they are matrix-algebraic and no examples of fields are known satisfying (c) but not (b) or (b) but not (a).

We shall be particularly interested in matrix-algebraic fields, and therefore list some equivalent conditions, which also explain the connexion with central localization.

THEOREM 8.4.2. *Let D be a field which is a k-algebra. Then*

$$D \otimes_k k(t) \subseteq D(t), \tag{2}$$

and the following conditions are equivalent:
  (a) *D is matrix-algebraic over k,*
  (b) *equality holds in (2),*
  (c) *$D \otimes k(t)$ is a field,*
  (d) *every non-zero polynomial over D divides a polynomial over k.*
When k is the exact centre of D, (a)–(d) are equivalent to
  (e) *every non-zero polynomial over D is bounded.*

Proof. The inclusion (2) is clear. Clearly (b) $\Rightarrow$ (c) and when $D \otimes k(t)$ is a field, every non-zero polynomial over D has an inverse, so we have equality in (2) and this shows that (c) $\Rightarrow$ (b); now it is easily verified that (c) $\Leftrightarrow$ (d).

Suppose for a moment that k is the exact centre of D. A matrix A over D is algebraic over k if and only if every invariant factor of A is bounded. Since every polynomial can occur as invariant factor of some matrix, e.g. its companion matrix, D is matrix-algebraic if and only if every polynomial is bounded, i.e. (a) $\Leftrightarrow$ (d) $\Leftrightarrow$ (e). But this also means that every element of $D(t)$ can be written as a fraction with denominator in $k[t]$, which is the condition for equality in (2), thus (d) $\Leftrightarrow$ (b). This proves the equivalence of (a)–(e) when k is the exact centre of D.

Now let C be the centre of D; then $C \supseteq k$ and by what we have shown, D is matrix-algebraic over C if and only if

$$D \otimes_C C(t) = D(t).$$

Assume that D is matrix algebraic over k. Then D is matrix-algebraic over C and C is matrix-algebraic over k, hence every polynomial over D divides a polynomial over C, which in turn divides a polynomial over k, so equality holds in (2), thus (a) $\Rightarrow$ (b). Conversely, assume (b); then $D \otimes C(t) = D(t)$, hence D is matrix-algebraic over C, and (2) also shows that every polynomial over C divides a polynomial over k; applying this to $t - \alpha$ ($\alpha \in C$), we see that C is algebraic over k, and this shows that D is matrix-algebraic over k, which proves that (b) $\Rightarrow$ (a). ∎

For a closer study of central localizations let us define a *right V-ring* as a ring R for which every simple right R-module is injective. It is clear that over a V-ring every module of finite length is semisimple and hence again injective. We begin by listing some properties of V-rings. An integral domain which is a V-ring will also be called a *V-domain*. Further, we recall that a *cogenerator* for a ring R is an R-module C such that for every R-module M and $x \in M$, $x \neq 0$, there exists $f: M \to C$ such that $xf \neq 0$.

PROPOSITION 8.4.3. (i) *For any ring R the following are equivalent:*
  (a) *R is a right V-ring,*
  (b) *the product of all simple right R-modules is a cogenerator,*
  (c) *in every right R-module the intersection of all maximal submodules is 0.*
  (ii) *Any V-ring is semiprimitive.*
  (iii) *In a right V-ring every right ideal is idempotent.*
  (iv) *Any V-domain is a simple ring.*

Proof. (i) (a) $\Rightarrow$ (b). Let C be the product of all simple right R-modules. Since R is a right V-ring, C is injective; we have to show that it is a cogenerator. Given $x \in M$, $x \neq 0$, let N_0 be a maximal submodule not containing x and let N_1 be the submodule generated by N_0 and x. Then N_1/N_0 is simple, and so injective, therefore the natural homomorphism $N_1 \to N_1/N_0$ can be extended to a homomorphism $M \to N_1/N_0$. Thus we have a homomorphism $M \to C$ which does not annihilate x, and so (b) follows.

(b) $\Rightarrow$ (c). For each $x \neq 0$ in M there is a homomorphism to a simple module not annihilating x; its kernel is a maximal submodule of M not containing x, so the intersection of all maximal submodules of M is 0, and this proves (c).

(c) $\Rightarrow$ (a). Let M be a simple right R-module and let E be its injective hull; then E is an essential extension of M, but since M is simple, this means that every non-zero submodule of E contains M, so if $E \neq M$, the intersection of all maximal submodules contains M. This contradicts (c), hence $E = M$, M is injective and this shows R to be a V-ring, i.e. (a).

(ii) In a V-ring the Jacobson radical, as the intersection of all maximal right ideals, must be 0, by (i), so R is semiprimitive.

(iii) Let $\mathfrak{a}$ be a right ideal in a right V-ring R; the result will follow if we show that any maximal right ideal containing $\mathfrak{a}^2$ also contains $\mathfrak{a}$, for then by (i),

$$\mathfrak{a} = \bigcap \{\mathfrak{m} | \mathfrak{m} \text{ maximal} \supseteq \mathfrak{a}\} = \bigcap \{\mathfrak{m} | \mathfrak{m} \text{ maximal} \supseteq \mathfrak{a}^2\} = \mathfrak{a}^2.$$

Suppose then that $\mathfrak{m}$ is a maximal right ideal of R and that $\mathfrak{m} \supseteq \mathfrak{a}^2$ but $\mathfrak{m} \not\supseteq \mathfrak{a}$. Then $\mathfrak{m} + \mathfrak{a} = R$, hence $\mathfrak{m}\mathfrak{a} + \mathfrak{a}^2 = R\mathfrak{a}$, and so $\mathfrak{m} \supseteq \mathfrak{m}\mathfrak{a} + \mathfrak{a}^2 = R\mathfrak{a} \supseteq \mathfrak{a}$, which contradicts our hypothesis, and it proves our claim.

Let R be a V-domain and $a \in R^\times$. Then by (iii), $aRaR = aR$, hence $RaR = R$, so the ideal generated by any non-zero element is the whole ring and R is simple. ∎

Let D be a field which is a k-algebra and let $T = D \otimes k(t)$ be its central localization over k. For any $n \geqslant 1$, ^nD can be considered as a right D-module in the usual way, by right multiplication. Further, for any $n \times n$ matrix $A = (a_{ij})$ over D we can regard ^nD as a $D[t]$-module by choosing a basis $u_1, \ldots, u_n$ of ^nD and writing $u_i t = \sum a_{ij} u_j$, briefly,

$$ut = Au. \tag{3}$$

If A is totally transcendental over k, this action can be extended to $T = D \otimes k(t)$ and we obtain a natural T-module structure on ^nD, determined by the matrix A; this T-module will be denoted by $(^nD, A)$. Suppose that $(^nD, B)$ is a second such T-module, whose defining equation corresponding to (3) is

$$vt = Bv. \tag{4}$$

If these two T-modules are isomorphic, then

$$v = Pu \tag{5}$$

for some $P \in \mathbf{GL}_n(D)$, and (4) takes the form $Put = BPu$, i.e. $ut = P^{-1}BPu$. Now a comparison with (3) shows that

$$A = P^{-1}BP, \tag{6}$$

so A and B are conjugate over D. Conversely, if A and B are conjugate over D, say (6) holds for some $P \in \mathbf{GL}_n(D)$, then the equation (5) can be used to define an isomorphism

$$(^nD, A) \cong (^nD, B). \tag{7}$$

Thus the necessary and sufficient condition for (7) to hold is that A and B are conjugate over D.

If D is matrix-algebraic over k, then T is a field, by Th. 4.2, and there are no modules of the form $(^nD, A)$. We shall now show that in every other case, *every* T-module of finite length is of the form $(^nD, A)$ for some transcendental matrix A.

PROPOSITION 8.4.4. *Let D be a field which is a k-algebra, not matrix-algebraic over k. Then every T-module of finite length is of the* form $(^nD, A)$ for some $n \geqslant 0$ and some totally transcendental matrix $A \in D_n$.

Proof. Let L be the given T-module of finite length; for $L = 0$ there is nothing to prove, so we may assume that $L \neq 0$. We claim that L as D-space has finite dimension. Since T is a PID (as localization of $D[t]$), L is a direct sum of cyclic torsion modules, and it will be enough to consider the case where L is a cyclic torsion module, say $L = T/aT$; here $a \neq 0$, because T is not a field and so has infinite length as T-module. On multiplying up by the denominator, we may assume a to be a polynomial in t, of degree n, say; hence $[L:D] = n$. Now take a D-basis of L_D, $u_1, \ldots, u_n$ say. If $u_i t = \sum a_{ij} u_j$, then the matrix $A = (a_{ij})$ is totally transcendental over k, because for every $\phi \in k(t)^{\times}$ the action of ϕ on L is well-defined. Hence $L \cong (^nD, A)$, as we had to show. ∎

We now come to the main result of this section, due to Resco [87], which gives conditions for a central localization to be a V-ring.

THEOREM 8.4.5. *Let D be a field with centre k and denote by $T = D \otimes k(t)$ the central localization over k. Then T is a simple principal ideal domain. Further, T is a V-domain with a unique simple module up to isomorphism if and only if D is matrix-homogeneous.*

Proof. We have seen that T must be a PID. Any ideal is generated by an invariant polynomial f; but such a polynomial is associated to 1 over k and hence a unit; therefore T is simple.

If T is a field, it is also the unique simple T-module, and this is injective. Thus T is then a V-ring; moreover, D is matrix-algebraic over k and so the conclusion holds vacuously. So we may assume that T is not a field.

Assume that T is a V-domain with the unique simple module M. Then any T-module of finite length r is isomorphic to M^r. Let $A, B \in D_n$ be both totally transcendental and consider the T-modules $(^nD, A)$, $(^nD, B)$. If we write $l(L)$ for the length of a finitely generated torsion T-module L, then

$$[L_D:D] = l(L_T)[M_D:D], \qquad (8)$$

and here $[M:D]$ is finite, because T is not a field. It follows that $(^nD, A)$ and $(^nD, B)$ have the same length and so are isomorphic; hence A, B are conjugate, so D is matrix-homogeneous.

Conversely, assume that D is matrix-homogeneous. We claim that any two simple T-modules are isomorphic. Let M, N be simple T-modules;

by Prop. 4.4 there exist totally transcendental matrices $A \in D_m$ and $B \in D_n$ such that $M \cong ({}^mD, A)$, $N \cong ({}^nD, B)$. Now $\oplus^n A$ and $\oplus^m B$ are totally transcendental matrices, both of order mn, and hence they are conjugate and so $M^n \cong N^m$. It follows that $m = n$ and $M \cong N$, as claimed, by the Jordan–Hölder theorem.

It remains to show that the unique simple T-module M is injective. Let $M \cong ({}^mD, A)$; then for any T-module of finite length l we have $L \cong ({}^nD, B)$, where $n = lm$ and B is conjugate to $\otimes^n A$, hence $L \cong M^l$. It follows that M cannot have any essential extension, for if $M \subseteq N$, then any finitely generated submodule of N is a direct sum of copies of M; but any two non-zero submodules must both contain M, so we must have $N = M$. This shows M to be its own maximal essential extension, and hence injective. ∎

For example, if D is existentially closed over k, then D is matrix-homogeneous by Th. 5.5.5, so the central localization T of D over k is an example of a simple V-domain, not a field, with a single simple T-module.

Consider the special case of $({}^nD, A)$, where $n = 1$. In that case D itself is a T-module with the action

$$c \otimes t: x \mapsto axc, \text{ where } c \in D, \qquad (9)$$

for some element a of D, transcendental over k. We claim that the endomorphism ring $\mathrm{End}_T((D, a))$ is $\mathscr{C}_D(a)$, the centralizer of a in D. For if α is a T-module endomorphism, then α commutes with all right multiplications, $(xc)\alpha = (x\alpha)c$. Taking $x = 1$ and putting $1\alpha = b$, we find that $c\alpha = bc$, so α is left multiplication by an element b of D. Moreover, $b(ax) = a(bx)$, so $b \in \mathscr{C}_D(a)$, as claimed.

Suppose now that D is a countable existentially closed field over k and $a \in D$ is transcendental over k. Then by Cor. 6.5.10, D can be embedded as a k-algebra in $C = \mathscr{C}_D(a)$. Thus we have an embedding $f: D \to C$ and an embedding $g: k(t) \to C$ such that $t \mapsto a$; since C and $k(a)$ centralize each other, we have a homomorphism

$$f \otimes g: T = D \otimes k(t) \to \mathscr{C}_D(a) = \mathrm{End}_T((D, a)_T). \qquad (10)$$

Here the left-hand side is simple, by Th. 4.5, hence (10) is an embedding and this allows us to regard D as a T-bimodule. Since D is matrix-homogeneous, it follows by Th. 4.5 that T is a V-domain with a unique simple right module.

This construction can be used to answer a question about annihilator rings. In any ring R let us define the left and right annihilators of a subset X as

$$l(X) = \{a \in R \mid aX = 0\}, \quad r(X) = \{b \in R \mid Xb = 0\}.$$

We note that this is a Galois connexion; in particular, we have $lrl(X) = l(X)$, $rlr(X) = r(X)$. A ring R in which every right ideal has the form of a right annihilator, $r(X)$ for some $X \subseteq R$, is called a *right annihilator ring*. Since in any case $rlr(X) = r(X)$, a right annihilator ring is characterized by the property

$$\mathfrak{a} = rl(\mathfrak{a}) \text{ for any right ideal } \mathfrak{a} \text{ of } R.$$

The question has been raised whether every right Noetherian right annihilator ring is right Artinian (see Johns [77]); a counter-example was given by Faith and Menal [92] using V-domains. In order to describe it we shall need some properties of annihilator rings. We recall that the *right socle* of a ring is the sum of its minimal right ideals; it is clearly a two-sided ideal.

PROPOSITION 8.4.6 (Johns [77]). *Let R be a right annihilator ring, $\mathfrak{s}$ its socle and $\mathfrak{J}$ its Jacobson radical. If R is right Noetherian, then $\mathfrak{J}$ is nilpotent, $r(\mathfrak{J}) = l(\mathfrak{J}) = \mathfrak{s}$ and $\mathfrak{s}$ is a large right ideal.*

Proof. Since $\mathfrak{J} \supseteq \mathfrak{J}^2 \supseteq \ldots$ is a descending chain, we have the ascending chain $l(\mathfrak{J}) \subseteq l(\mathfrak{J}^2) \subseteq \ldots$, which must become stationary, say $l(\mathfrak{J}^m) = l(\mathfrak{J}^{m+1})$. Then $\mathfrak{J}^m = rl(\mathfrak{J}^m) = rl(\mathfrak{J}^{m+1}) = \mathfrak{J}^{m+1}$; thus $\mathfrak{J}^m = \mathfrak{J}^m\mathfrak{J}$ and by Nakayama's lemma, $\mathfrak{J}^m = 0$.

Next we show that $l(\mathfrak{J})$ is a large right ideal. Given $x \in R$, either $x\mathfrak{J} = 0$ or for some $n \geq 1$, $x\mathfrak{J}^n \neq 0$, $x\mathfrak{J}^{n+1} = 0$; in either case $xR \cap l(\mathfrak{J}) \neq 0$, and so $l(\mathfrak{J})$ is right large. Let us define $\mathfrak{z}$, the *right singular ideal* of R, as the set of all elements with a large right annihilator. Then $l^2(\mathfrak{J}) = ll(\mathfrak{J}) \subseteq \mathfrak{z}$ and since R is right Noetherian $\mathfrak{z}$ is a nil ideal (even nilpotent, see A.3, p. 371); hence $\mathfrak{z} \subseteq \mathfrak{J}$ and so $l^2(\mathfrak{J}) \subseteq \mathfrak{J}$. Thus we have a descending chain $\mathfrak{J} \supseteq l^2(\mathfrak{J}) \supseteq l^4(\mathfrak{J}) \supseteq \ldots$, and an ascending chain $l(\mathfrak{J}) \subseteq l^3(\mathfrak{J}) \subseteq \ldots$. The latter becomes stationary, say $l^m(\mathfrak{J}) = l^{m+2}(\mathfrak{J})$. Then $rl^m(\mathfrak{J}) = rl^{m+2}(\mathfrak{J})$, hence $l^{m-1}(\mathfrak{J}) = l^{m+1}(\mathfrak{J})$, and continuing in this way, we find that $l^2(\mathfrak{J}) = \mathfrak{J}$. Hence $r(\mathfrak{J}) = rl^2(\mathfrak{J}) = l(\mathfrak{J})$.

Let $\mathfrak{a}$ be a large right ideal of R; then $l(\mathfrak{a}) \subseteq \mathfrak{z} \subseteq \mathfrak{J}$, therefore $\mathfrak{a} = rl(\mathfrak{a}) \supseteq r(\mathfrak{z}) \supseteq r(\mathfrak{J})$. Since the socle $\mathfrak{s}$ is the intersection of all large right ideals of R (see e.g. A.3, p. 374), we have $\mathfrak{s} \supseteq r(\mathfrak{J})$. But $r(\mathfrak{J})$ itself is large, so $\mathfrak{s} = r(\mathfrak{J})$ as claimed. ∎

The next result provides us with a source of right annihilator rings.

LEMMA 8.4.7. *Let T be a right Noetherian right V-domain with a T-bimodule W which is the unique simple right T-module. Then the split* *null extension $R = T \oplus W$, defined as ring with the multiplication*

$$(t, w)(t', w') = (tt', tw' + wt'),$$

is a right Noetherian right annihilator ring.

Proof. Since T is right Noetherian and W_T is simple, it follows that R is right Noetherian. Moreover, T is simple by Prop. 4.3 (iv), hence the left annihilator of W, being a proper ideal, must be zero. Since W is simple as right T-module, $aW = W$ for all $a \in T^\times$, so $\mathfrak{J} = (0, W)$ is a minimal right ideal of R and it is large as right ideal.

Let $\mathfrak{a}$ be any non-zero right ideal of R; then $\mathfrak{a} \cap \mathfrak{J} \neq 0$, so $\mathfrak{a} \supseteq \mathfrak{J}$. Since T is a domain, $l(\mathfrak{a}) \subseteq l(\mathfrak{J}) = \mathfrak{J}$. Explicitly we have $\mathfrak{a} = (\mathfrak{a}_0, W)$, where $\mathfrak{a}_0$ is a right ideal of T and $l(\mathfrak{a}) = (0, W_0)$, where W_0 is the left annihilator of $\mathfrak{a}_0$ in W. We complete the proof by showing that $rl(\mathfrak{a}) = \mathfrak{a}$. Given $x \in T \backslash \mathfrak{a}_0$, we see as in the proof of Prop. 4.3 that there is a homomorphism $\varphi : R/\mathfrak{a}_0 \to W$ such that $x\varphi \neq 0$; thus $x\varphi = ux$ for some $u \in W$, and $u\mathfrak{a}_0 = 0$, so $u \in W_0$. Hence $ux \neq 0$ and this shows that $\mathfrak{a}_0$ is the exact right annihilator of W_0. Thus $\mathfrak{a} = (\mathfrak{a}_0, W) = rl(\mathfrak{a})$ and this shows R to be a right annihilator ring. ∎

Now take a countable existentially closed field D over k, let T be its central localization over k and consider D as T-bimodule, as in (10). By Th. 4.5, T is a V-domain with a unique simple right module, so we can apply Lemma 4.7 to form the split null extension $R = T \oplus D$, a right Noetherian right annihilator ring. But this ring is not right Artinian, for if it were, then T would be right Artinian and hence a field, which is not the case. Thus R is an example of a right Noetherian right annihilator ring which is not right Artinian.

Exercises

1. Show that in a right V-ring, for any two proper right ideals $\mathfrak{a}$, $\mathfrak{b}$, $\mathfrak{a} \cap \mathfrak{b} = \mathfrak{a}\mathfrak{b}$, hence the multiplication of proper ideals is commutative.

2. Show that every right V-ring is semiprime.

3. Show that a commutative ring is a V-ring if and only if it is von Neumann regular.

4. Show that a PID is a right V-ring if and only if every non-zero element is fully reducible, i.e. $cR = \bigcap p_i R$, where the p_i are atoms.

5. Show that a PID R is a right V-ring if and only if for any atoms p, q there exist $u, v \in R$ such that $up - qv = 1$.

6°. Is the non-algebraic locally finite algebra of Golod and Shafarevich matrix-algebraic?

7. Let D be a field not algebraic over its centre. Show that the polynomial ring $D[t]$ is primitive.

8. (H.-G. Quebbemann [79]) Let D be a field of characteristic 0, matrix-algebraic over its centre C. On $D[t]$ define $f' = df/dt$ as the formal derivative. Given $f \in D[t]$, if $f' = fg - gf$ for some $g \in D[t]$, show that f is an element of D commuting with all the coefficients of g. Deduce that for $a, b \in D$, if $ab \neq ba$, then $ab^n \neq b^n a$ for all $n \geqslant 1$.

8.5 The solution of equations over skew fields

We now return to the problem of solving equations over a skew field. In 8.1 we have seen how the problem is related to that of finding singular eigenvalues of a matrix; unfortunately there are no general methods so far for finding singular eigenvalues of a matrix. By contrast, left and right eigenvalues can always be found in a suitable extension, as we saw in 8.2, but this is not necessarily of help in solving equations. There are just some exceptions where this is the case, which we shall describe now. They are (i) equations with all coefficients on one side, (ii) 2×2 matrices and (iii) $n \times n$ matrices with indeterminates as entries.

In solving equations it is convenient to write the coefficients on the left; of course a similar result holds on the right.

THEOREM 8.5.1. *Let K be any field. Then any equation of degree $n > 0$,*

$$x^n + a_1 x^{n-1} + \ldots + a_n = 0 \quad (a_i \in K), \tag{1}$$

has a right root in some extension field of K; moreover, the right roots coincide with the right eigenvalues of the companion matrix.

Proof. The equation has the linear companion

$$A = \begin{pmatrix} 0 & 1 & 0 & 0 & \ldots & 0 & 0 \\ 0 & 0 & 1 & 0 & \ldots & 0 & 0 \\ & & \ldots & & \ldots & \ldots & \\ 0 & 0 & 0 & \ldots & \ldots & 0 & 1 \\ -a_n & -a_{n-1} & \ldots & & \ldots & -a_2 & -a_1 \end{pmatrix}. \tag{2}$$

In a suitable extension of K we can find a right eigenvalue of A:

$$Au = u\alpha. \tag{3}$$

In detail we have $u_1\alpha = u_2, u_2\alpha = u_3, \ldots, u_{n-1}\alpha = u_n$, and

$$u_n\alpha = -a_nu_1 - a_{n-1}u_2 - \ldots - a_1u_n.$$

If $u_1 = 0$, it follows that $u_2 = \ldots = u_n = 0$ and so $u = 0$. But by hypothesis, (3) has a solution with $u \neq 0$, so we may take $u_1 = 1$. Then $u_2 = \alpha, u_3 = \alpha^2, \ldots, u_n = \alpha^{n-1}$ and the last equation reads

$$\alpha^n + a_1\alpha^{n-1} + \ldots + a_n = 0.$$

Thus α is a right root of (1). Conversely, any right root α of (1) leads to an eigenvector $(1 \quad \alpha \quad \ldots \quad \alpha^{n-1})^{\mathrm{T}}$ corresponding to the right eigenvalue α. ∎

Since the right eigenvalues of A fall into at most n conjugacy classes, this provides another proof of Cor. 3.4.4, that the right roots of (1) lie in at most n conjugacy classes.

Next we turn to the singular eigenvalue problem for the special case of 2×2 matrices. We shall need to consider two special quadratic equations.

THEOREM 8.5.2. *Let K be any field. Given $a, b, c, d \in K$, $c \neq 0$, the equation*

$$x(cx + d) = ax + b \tag{4}$$

has a solution in some extension field of K.

Proof. Write $A = \begin{pmatrix} a & b \\ c & d \end{pmatrix}$ and consider the equation $Av = v\alpha$, which we know has a solution (in an extension of K), with a right eigenvalue α of A and $v \neq 0$. Writing $v = (x \quad y)^{\mathrm{T}}$, we have

$$ax + by = x\alpha,$$

$$cx + dy = y\alpha.$$

If $y = 0$, the second equation reduces to $cx = 0$, and so $x = 0$. But x, y cannot both vanish, so $y \neq 0$ and we can adjust α so that $y = 1$. If we now eliminate α, we obtain $ax + b = x(cx + d)$, i.e. (4). ∎

It remains to consider (4) when $c = 0$. Changing the notation slightly, we have

$$ax - xb = c. \tag{5}$$

If we replace x by $y = xp$, where $p \in K^\times$, then (5) becomes $ayp^{-1} - yp^{-1}b = c$, i.e.

$$ay - y \cdot p^{-1}bp = cp.$$

Thus in (5) we can either replace c by 1 (provided only that $c \neq 0$), or replace b by a conjugate. We also recall from 8.2 that (5) is equivalent to the matrix equation

$$\begin{pmatrix} 1 & x \\ 0 & 1 \end{pmatrix} \begin{pmatrix} a & c \\ 0 & b \end{pmatrix} = \begin{pmatrix} a & 0 \\ 0 & b \end{pmatrix} \begin{pmatrix} 1 & x \\ 0 & 1 \end{pmatrix}. \tag{6}$$

Let us put $A = \begin{pmatrix} a & c \\ 0 & b \end{pmatrix}$, $B = \begin{pmatrix} a & 0 \\ 0 & b \end{pmatrix}$; if a, b are both transcendental over k, then the matrices A, B are totally transcendental over K and by Th. 5.5.5 they are conjugate over a suitable extension of K, i.e. there exists a non-singular matrix P such that $PA = BP$. Writing $P = \begin{pmatrix} p & q \\ r & s \end{pmatrix}$, we thus have the equations

$$pa = ap, \quad pc + qb = aq,$$

$$ra = br, \quad rc + sb = bs.$$

Since P is invertible, p, r cannot both vanish, say $p \neq 0$. Then from the first two equations we find that

$$c + p^{-1}qb = p^{-1}aq = ap^{-1}q,$$

so $x = p^{-1}q$ is a solution of (5). Similarly, if $r \neq 0$, then $x = r^{-1}s$ is a solution. Thus (5) is solved when a, b are both transcendental.

If one of a, b is algebraic over k and the other transcendental, or both are algebraic but with different minimal equations over k, then (5) has a unique solution within K, by Lemma 2.2. There remains the case where a, b have the same minimal equation over k. Here we need a lemma, as well as two results from FR about elements in a PID: (i) If p, p' are two atoms with the same least bound q, then $pp'|q^2$ and q^2 is the least bound of pp' if and only if pp' is indecomposable; (FR, Prop. 6.4.9, p. 318). (ii) Two elements a, b satisfy a comaximal relation $ab = b'a'$ if and only if u, v exist such that $ua - bv = 1$ (FR, Lemma 3.4.3, p. 171).

LEMMA 8.5.3. *Let K be a field with centre C and let $a, b \in K$ have the same minimal polynomial μ over C. Then the equation*

$$ax - xb = 1 \tag{7}$$

has a solution in K (or indeed in any extension field with centre C) if and only if in the polynomial ring $K[t]$, $(t - b)(t - a)$ divides $\mu(t)$.

Proof. In the ring $R = K[t]$, $t - a$ and $t - b$ are similar bounded atoms, with the same bound μ. By the above remark (i), the product $p = (t - b)(t - a)$ divides μ^2 and μ^2 is the exact bound of p if and only if p is indecomposable, i.e. $p|\mu$ if and only if p is decomposable, which means that there is a comaximal relation

$$(t - b)(t - a) = hk, \quad \text{where } h, k \in R.$$

By (ii) above this is so precisely when $f, g \in R$ exist such that

$$f(t - b) - (t - a)g = 1. \tag{8}$$

By the division algorithm in R we can write $f = (t - a)f_1 + u$, where $f_1 \in R$, $u \in K$. Inserting this expression in (8) and simplifying, we get

$$u(t - b) - (t - a)v = 1,$$

where $v = g - f_1(t - b)$. By comparing terms of highest degree in t, we see that $v \in K$ and $v = u$. If we now equate constant terms, we find

$$au - ub = 1,$$

and this solves (7). Conversely, when (7) has a solution $x = u$, then (8) holds with $f = g = u$ and so $(t - b)(t - a)$ is decomposable, whence $(t - b)(t - a)|\mu$, as claimed. ∎

Let us now return to (5); as we have seen, for $c \neq 0$ this is equivalent to $ay - y \cdot cbc^{-1} = 1$. Now if a, b have the same minimal polynomial μ over k, we can enlarge K so that k becomes the exact centre, e.g. by taking $K_k\langle x \rangle$ with an indeterminate x. Then (5) has a solution if and only if $(t - cbc^{-1})(t - a)|\mu$, by Lemma 5.3.

Finally we ask when the solution is unique; by linearity (5) has a unique solution if and only if $ax = xb$ has only $x = 0$ as solution, i.e. when a, b are not conjugate. Summing up our results, we have the following description of the solution of (5):

THEOREM 8.5.4. *Let K be a field which is a k-algebra and consider the equation*

$$ax - xb = c, \quad \text{where } a, b, c \in K. \tag{5}$$

(i) *If a, b are both transcendental over k, (5) has infinitely many solutions over a suitable extension of K.*

(ii) *If one of a, b is algebraic over k and the other is either transcendental* or algebraic with a different minimal equation, then (5) has a unique solution in K or any extension of K.

(iii) If a, b have the same minimal polynomial μ over the centre k of K, then (5) has a solution in K (or in any extension of K) if and only if either $c = 0$ or $(t - cbc^{-1})(t - a)$ divides μ, or equivalently, if $(t - cbc^{-1})(t - a)$ is decomposable in $K[t]$. ■

We can now solve the singular eigenvalue problem for 2×2 matrices.

THEOREM 8.5.5. *Let K be any field. Then every 2×2 matrix over K has a singular eigenvalue in some extension of K.*

If K is a k-algebra, where k is relatively algebraically closed in K, then A has a non-zero singular eigenvalue unless A is 0-triangulable over k.

Proof. Let $A = \begin{pmatrix} a & b \\ c & d \end{pmatrix}$, where $a, b, c, d \in K$; we have to find x such that $xI - A$ is singular. If $c = 0$, we can take $x = a$. Otherwise, on replacing x by cx, we may take $c = 1$. Now

$$\begin{pmatrix} a - x & b \\ 1 & d - x \end{pmatrix} \text{ is associated to } \begin{pmatrix} a - x & b - (a - x)(d - x) \\ 1 & 0 \end{pmatrix},$$

and this is singular precisely when $(a - x)(d - x) = b$, i.e.

$$x^2 - xd - ax + (ad - b) = 0. \tag{9}$$

By Th. 5.2 this has a solution in some extension field.

Suppose now that A has no non-zero singular eigenvalue. If $c = 0$, this means that $a = d = 0$ and A is already 0-triangular. Otherwise we may again take $c = 1$ and reach the equation (9). If this has no non-zero solution, we must have $ad = b$, and on writing $y = x^{-1}$ we can bring it to the form

$$dy + ya = 1.$$

By hypothesis this has no solution in any extension of K, so by Th. 5.4, d and $-a$ are algebraic over k with the same minimal equation over k (if they had different minimal equations over k we could extend K to reduce the centre to k). But k is relatively algebraically closed in K, hence $a, d \in k$ and $a = -d$. Thus A takes the form

$$A = \begin{pmatrix} a & -a^2 \\ 1 & -a \end{pmatrix}, \quad \text{where } a \in k.$$

But this is clearly conjugate to $\begin{pmatrix} 0 & 1 \\ 0 & 0 \end{pmatrix}$ over k. ■

For general matrices the singular eigenvalue problem is still open, but there is another case where a solution can be found; paradoxically here the matrix has to be sufficiently general.

THEOREM 8.5.6. *Let $K_k \langle x_{ij} \rangle$ be the tensor ring on n^2 indeterminates x_{ij}* *$(i, j = 1, \ldots, n)$ over a field K and denote by L its universal field of* *fractions. Then the matrix $X = (x_{ij})$ has a singular eigenvalue in a suitable* *extension of L.*

Proof. We shall operate in $M = L((t))$, the field of Laurent series in a central indeterminate t. Put

$$I' = \begin{pmatrix} 0 & 0 \\ 0 & I_{n-1} \end{pmatrix}.$$

We first find a singular eigenvalue for $I' + tX$ in M; this means that we have to find λ such that $I' + tX - \lambda I$ is singular. For convenience we shall write

$$X = \begin{pmatrix} x_{11} & u \\ v & X_{11} \end{pmatrix},$$

and put $\lambda = t\lambda'$. We have

$$I' + tX - \lambda I = \begin{pmatrix} t(x_{11} - \lambda') & tu \\ tv & I + tX_{11} - t\lambda' I \end{pmatrix} = \begin{pmatrix} a & b \\ c & d \end{pmatrix},$$

say. This matrix will be singular if $a = bd^{-1}c$, i.e.

$$t(x_{11} - \lambda') = tu(I - t(\lambda'I - X_{11}))^{-1}tv$$

$$= tu(I + t(\lambda'I - X_{11}) + t^2(\lambda'I - X_{11})^2 + \ldots)tv.$$

If we write $\lambda' = \sum f_i t^i$, where $f \in L$, insert this series and equate powers of t, we can solve the resulting equations recursively for the f_i. Thus we have found a singular eigenvalue for $I' + tX$ in M.

We now embed L in M by the rule $X \mapsto I' + tX$, $a \mapsto a$ for $a \in K$. Since the coefficients of $I' + tX$ are indeterminates, this is indeed an embedding, and since the image of X has a singular eigenvalue, so does X itself. ∎

With a reasonable analogue of specialization one could use this result to solve the general singular eigenvalue problem, but it seems unlikely that a strong enough analogue of specialization exists in this generality (see 8.8).

Exercises

1. (I. Niven [41]) Using the remark at the end of 8.3, show that every equation (1) over the real quaternions $\mathbf{H}$ has a right root in $\mathbf{H}$.

2°. Can an $n \times n$ matrix over a field have more than n inconjugate singular eigenvalues?

3°. Let $L = K_k \langle\!\langle x \rangle\!\rangle$ and consider a matrix $A = (a_{ij})$ with $a_{11} = x$ and all other entries in K. Adapt the method of Th. 5.6 to show that A has a singular eigenvalue in some extension of L.

8.6 Specializations and the rational topology

So far we have been concerned with finding one solution of any equation (where possible), but in algebraic geometry one has to deal with the system of *all* solutions. Now we can define varieties as solution sets of systems of equations, but very little is known so far about such varieties. Here is a very simple example which already shows the difference between the commutative and the general case.

For any $c \in K$ one would expect the equation

$$cx - xc = 0 \tag{1}$$

to define a zero-dimensional variety, provided that c is not in the centre of K. But the solution of (1) is a subfield of K – the centralizer of c in K – whereas in the commutative case 0-dimensional varieties are finite sets of points.

Consider a field K which is a k-algebra and let E be a K-field. We shall consider affine n-space over E, E^n or $\mathbf{A}^n(E)$, whose points are described by n-tuples in E. Given $\alpha, \beta \in E^n$, we shall write $\alpha \underset{K}{\rightarrow} \beta$ and call β a *specialization* of α over K, if the map $\alpha_i \mapsto \beta_i$ defines a specialization in the sense of 7.3. In terms of singular kernels, if $\mathscr{P} = \mathrm{Ker}\,(x \mapsto \alpha)$, $\mathscr{Q} = \mathrm{Ker}\,(x \mapsto \beta)$, this means that $\mathscr{P} \subseteq \mathscr{Q}$, or more concretely, for any square matrix $A = A(x)$ over $K_k\langle X \rangle$, where $X = \{x_1, \ldots, x_n\}$,

$$\mathrm{sing}\,A(\alpha) \Rightarrow \mathrm{sing}\,A(\beta),$$

where $\mathrm{sing}\,A$ denotes the elementary sentence stating that A is singular (as in 6.5). Since every matrix over $K_k\langle X \rangle$ is stably associated to a linear matrix, it is enough to require this to hold for all linear matrices. Thus we have

THEOREM 8.6.1. *Let E/K be an extension of fields. Then for $\alpha, \beta \in E^n$ we have $\alpha \underset{K}{\rightarrow} \beta$ if and only if*

$$\text{sing}\left(A + \sum A_i\alpha_i\right) \Rightarrow \text{sing}\left(A + \sum A_i\beta_i\right),$$

for any square matrices $A, A_1, \ldots, A_n$ *of the same order over* K. ∎

We shall call $\alpha \in E^n$ *free* over K if the map $x \mapsto \alpha$ defines an isomorphism $K_k\langle X \rangle \cong K(\alpha)$. The following criterion is an immediate consequence of the definitions.

COROLLARY 8.6.2. *Given a field extension* E/K, *a point* $\alpha \in E^n$ *is free over* K *if and only if* $A + \sum A_i\alpha_i$ *is singular only when* $A + \sum A_i x_i$ *is non-full over* $K_k\langle X \rangle$. ∎

Th. 6.1 also makes it clear how specializations in projective space should be defined. Each point of the projective space $\mathbf{P}^n(E)$ is described by an $(n + 1)$-tuple $\xi = (\xi_0, \ldots, \xi_n) \neq 0$ and ξ, η represent the same point if and only if $\xi_i = \eta_i\lambda$ for some $\lambda \in E^\times$. At first sight it is not clear how specializations in projective space are to be defined; instead of polynomials we would have to consider rational functions in the x_i and (in contrast to the commutative case) there is no simple way of getting rid of the denominators. However, with Th. 6.1 in mind we can define $\xi \underset{K}{\rightarrow} \eta$ if and only if

$$\text{sing}\sum_0^n A_i\xi_i \Rightarrow \text{sing}\sum_0^n A_i\eta_i,$$

for any square matrices $A_0, \ldots, A_n$ of the same order over K. But for the moment we shall concentrate on affine spaces.

The criterion of Th. 6.1 can still be simplified if we are specializing to a point in K:

THEOREM 8.6.3. *Let* E/K *be a field extension and* $\alpha \in E^n, \lambda \in K^n$. *Then* $\alpha \underset{K}{\rightarrow} \lambda$ *if and only if*

$$\text{nonsing}\left(I - \sum A_i(\alpha_i - \lambda_i)\right) \tag{2}$$

for all square matrices $A_1, \ldots, A_n$ *of the same order over* K.

Proof. Assume that $\alpha \underset{K}{\rightarrow} \lambda$. If $I - A_i(x_i - \lambda_i)$ becomes singular for $x_i = \alpha_i$, then it must become singular for $x_i = \lambda_i$; but then we have I, which is non-singular, a contradiction. So (2) holds.

Conversely, when (2) holds, let $\mathcal{P}$ be the singular kernel of the map $x \mapsto \alpha$ from $K_k\langle X \rangle$ to $K(\alpha)$. We must show that under the map $x \mapsto \lambda$ every matrix of $\mathcal{P}$ becomes singular, and here it is enough to test linear matrices $A + \sum A_i x_i$. Thus let $A + \sum A_i\alpha_i$ be singular; we have to show that $C = A + \sum A_i \lambda_i$ is also singular; we note that C has entries in K. If C were non-singular, we could write

$$A + \sum A_i \alpha_i = A + \sum A_i \lambda_i + \sum A_i (\alpha_i - \lambda_i)$$

$$= C + \sum A_i (\alpha_i - \lambda_i)$$

$$= C\left(I - \sum B_i (\alpha_i - \lambda_i)\right), \quad \text{where } B_i = -C^{-1} A_i.$$

Here the left-hand side is singular by hypothesis and the right-hand side is non-singular by (2), a contradiction, which establishes the result. ∎

The following special case is of interest:

COROLLARY 8.6.4. *Let E/K be a field extension and $\alpha \in E^n$. Then α has every point of K^n as a specialization if and only if $I - \sum A_i (\alpha_i - \lambda_i)$ is non-singular for all A_i (of the same order) over K and all $\lambda_i \in K$.* ∎

Let us call a point α satisfying the conditions of this corollary *quasi-free*, thus α is *quasi-free* if $\alpha \underset{K}{\nrightarrow} \lambda$ for all $\lambda \in K^n$. It is clear that every free point is quasi-free, and the converse holds when the centre C of K is infinite and $[K:C] = \infty$, so that the specialization lemma can be applied. For then, if $A + \sum A_i x_i$ is full, there exists $\lambda \in K^n$ such that $A + \sum A_i \lambda_i$ is non-singular, hence $A + \sum A_i \alpha_i$ is also non-singular.

To restate this corollary in a more intuitive form we make a couple of definitions. A point $\alpha \in E^n$ is called an *inverse eigenvalue* of the sequence $A_1, \ldots, A_n$ of matrices if $I - \sum A_i \alpha_i$ is singular. Given an extension E/K, we can regard the vector space K^n as a subgroup of E_n; its cosets will be called the *levels* in E^n over K. Thus $\alpha, \beta \in E^n$ are on the same level precisely when $\alpha - \beta \in K^n$. Now Cor. 6.4 may be stated as

COROLLARY 8.6.4'. *Let E/K be a field extension. Then a point α of E^n is quasi-free if and only if its level contains no inverse eigenvalue of any sequence of matrices over K.* ∎

We can put this in another way by saying that if the level of α contains an inverse eigenvalue (of some sequence of matrices over K), then there is a point $\lambda \in K^n$ which is *not* a specialization of α.

Let $K \subseteq E \subseteq L$ be fields. Given $\alpha \in L^n$, we define the *locus* of α in E over K as the set of all specializations of α in E over K.

EXAMPLES. 1. If $\alpha \in K^n$, the locus of α in K is just the point α.

2. The locus of α in K is all of K^n precisely when α is quasi-free over K.

3. If a square matrix A over E has an inverse eigenvalue α in L but none in E, then the locus of α in E is empty. Of course this cannot happen when E is an EC-field.

Sometimes it will be convenient to define specializations to infinity. Thus we define $\alpha \underset{K}{\to} \infty$ for $\alpha \in E \supseteq K$ to mean: $\alpha \neq 0$ and $\alpha^{-1} \underset{K}{\to} 0$. By Th. 6.3, $\alpha^{-1} \underset{K}{\to} 0$ holds if and only if $I - A\alpha^{-1}$ is non-singular for all matrices A over K, thus $\alpha \underset{K}{\to} \infty$ precisely when α is not a singular eigenvalue of any matrix over K. An element α of E such that $\alpha \not\to \infty$ is also called *finite* over K.

Example 3 above means in detail that $I - A\alpha$ is singular for some $\alpha \in L$, but $I - A\lambda$ is non-singular for all $\lambda \in E$, so α has no specializations in E over K; in particular, A has no non-zero singular eigenvalue in E. If 0 is also not an eigenvalue, then A is non-singular; this means that α is finite over K, for if we had $\alpha \to \infty$, then by Th. 6.3, $I - C\alpha^{-1}$ would be non-singular for any C over K. But $I - A\alpha$ is singular and A is non-singular, hence $I - A^{-1}\alpha^{-1} = -A^{-1}(I - A\alpha)\alpha^{-1}$ is singular, which is a contradiction.

To illuminate the situation in the general case, let us first look at the case of commutative fields $E \supseteq K$. If $\alpha \in E$ is algebraic over K, but not in K, then α satisfies an equation

$$f(x) = 0, \tag{3}$$

over K. If (3) also has a root λ in K, we can replace f by a polynomial of lower degree which still has α as zero but not λ. In the general case it may not be possible to separate out the rational solutions in this way; those that always accompany α represent the locus. To give an example of this behaviour we shall construct a point α which has a specialization in K without itself being in K or free over K.

In our example the locus of α consists of precisely one point, which may be taken to be ∞. The condition at ∞ means that $I - A\alpha^{-1}$ is non-singular for all A, i.e. $\alpha I - A$ is non-singular, while the condition at a point $\lambda \in K$ means that $I - A(\alpha - \lambda)$ is singular for some A. Let k be a commutative field of characteristic 0, form the rational function field $F = k(t)$ and let K be a field with centre F and such that $[K:F] = \infty$ (by Prop. 2.3.5). Let α be a root of the equation

$$(x + 1)t - tx = 0, \tag{4}$$

in a suitable extension field of K, which exists by Th. 5.4. Putting $x = \lambda \in K$ in (4), we obtain $t = 0$, which is false; hence (4) has no roots in K, so α cannot be specialized to any element of K. So it only remains to show that $\alpha \underset{K}{\to} \infty$, and this will follow if we show that $\alpha I - A$ is non-singular for all matrices A over K. By (4) we have $t\alpha = (\alpha + 1)t$ and $At = tA$ for any matrix A over K, hence

$$t(\alpha I - A) = ((\alpha + 1)I - A)t.$$

Hence if $\alpha I - A$ is singular, then so is $(\alpha + 1)I - A$, and by induction, $(\alpha + \nu)I - A$ is singular for all $\nu \in \mathbf{N}$. Since K has characteristic 0, this means that $A - \alpha I$ has infinitely many central eigenvalues, which contradicts Th. 2.3. Thus $\alpha I - A$ cannot be singular and it follows that $\alpha \underset{K}{\not\to} \infty$.

If we extend K to a field E which contains a root $x = s$ of (4), then the solution of (4) has infinitely many specializations in E, viz. ∞ and s, and hence the whole level of s over K, but no point in K itself.

We have already met the rational topology in 7.2, but it seems more natural to define it in terms of matrices rather than rational equations. To do so we define for each matrix $A = A(x)$ over $K_k\langle X \rangle$ a subset of E^n, its *singularity support*, as

$$\mathfrak{D}(A) = \{\alpha \in E^n | \text{nonsing}\,(A(\alpha))\}.$$

It is clear that

$$\mathfrak{D}(I) = E^n, \quad \mathfrak{D}(A \oplus B) = \mathfrak{D}(A) \cap \mathfrak{D}(B). \tag{5}$$

Hence we get a topology on E^n by taking the $\mathfrak{D}(A)$ as a basis for the open sets, called the *rational K-topology* on E^n; this is easily seen to agree with the rational topology as defined in 7.2. It is finer than the polynomial topology, because singularities of matrices give us more sets than zeros of polynomials. Like the Zariski topology in the commutative case, it is not Hausdorff, but it is a T_0-topology, i.e. for each pair of distinct points one can be chosen which does not lie in some neighbourhood of the other. It is clear that when E satisfies the conditions of the specialization lemma, $\mathfrak{D}(A)$ is non-empty precisely when the matrix A is full. In that case E^n is irreducible, i.e. every non-empty open set is dense. For if $\mathfrak{D}(A), \mathfrak{D}(B) \neq \varnothing$, then A, B are full, hence so is $A \oplus B$ and so

$$\mathfrak{D}(A) \cap \mathfrak{D}(B) = \mathfrak{D}(A \oplus B) \neq \varnothing,$$

by (5).

We note that $\alpha \in E^n$ is free over K precisely when

$$A \text{ full} \Rightarrow \text{nonsing}\,(A(\alpha)), \tag{6}$$

by Cor. 6.2. This means that $\alpha \in \mathfrak{D}(A)$ unless $\mathfrak{D}(A) = \varnothing$, so all the non-empty open sets have a non-empty intersection, consisting of all the free points. In K^n itself there are of course no free points, because

$x_1 - \alpha_1$ fails to satisfy (6). Put differently, we can say that in the E-topology the non-empty open sets have empty intersection. This shows that the E-topology on E^n is in general finer than the K-topology.

The topology can also be used to describe the locus of a point:

THEOREM 8.6.5. *Let E/K be a field extension and $\alpha \in E^n$. Then the locus of α in K^n is the closure of α in the rational K-topology.*

Proof. Let $\lambda \in K^n$; we have $\lambda \in \overline{\{\alpha\}}$ if and only if α lies in every neighbourhood of λ, i.e. $\lambda \in \mathcal{D}(A) \Rightarrow \alpha \in \mathcal{D}(A)$ for all matrices A over K. But this just means: $\mathrm{sing}\,(A(\alpha)) \Rightarrow \mathrm{sing}\,(A(\lambda))$, i.e. $\alpha \underset{K}{\to} \lambda$. ∎

In commutative algebraic geometry there is a very satisfactory theory of dimension which can be built up using the notion of algebraic dependence, and it is natural to ask whether a similar theory exists in general. This seems not to be the case, and it may be of interest briefly to discuss the reasons for this failure.

We consider a field extension E/K, where E and K are k-algebras. A finite family of elements of E is said to be *algebraically dependent* over K if it is not quasi-free. By Cor. 6.4 this means that $\alpha_1, \ldots, \alpha_n$ are algebraically dependent over K if $I - \sum A_i(\alpha_i - \lambda_i)$ is singular for some matrices A_i over K and $\lambda_i \in K$. An element β of E is said to be *algebraically dependent* on a set S over K if there is a finite subset S' of S which is quasi-free but such that $S' \cup \{\beta\}$ is not quasi-free. We ask whether this notion satisfies the usual axioms for a dependence relation; unfortunately this is not the case. Let us briefly recall the definitions involved (see A.3, 1.4).

An abstract *dependence relation* on a set S associates with each finite subset X of S certain elements of S, said to be *dependent* on X, subject to the conditions:

(D.0) *If $X = \{x_1, \ldots, x_n\}$, then each x_i is dependent on X,*

(D.1) (Transitivity) *If z is dependent on $\{y_1, \ldots, y_m\}$ and each y_j is dependent on $X = \{x_1, \ldots, x_n\}$, then z is dependent on X,*

(D.2) (Exchange axiom) *If y is dependent on $\{x_1, \ldots, x_n\}$ but not on $\{x_2, \ldots, x_n\}$, then x_1 is dependent on $\{y, x_2, \ldots, x_n\}$.*

Linear dependence (in a vector space over a field) and algebraic dependence over a commutative field are familiar examples of dependence relations. The above notion of algebraic dependence clearly satisfies (D.0): Given $\alpha_1, \ldots, \alpha_n \in E$, $\alpha_1 \in K(\alpha_1)$ and so α_1 is dependent on $\{\alpha_1, \ldots, \alpha_n\}$.

We can also verify (D.2): Let $\alpha_1, \ldots, \alpha_n, \beta \in E$ and suppose that β is dependent on $\{\alpha_1, \ldots, \alpha_n\}$, but not on $\{\alpha_2, \ldots, \alpha_n\}$. Let S be a minimal subset of $\{\alpha_1, \ldots, \alpha_n\}$ such that S is independent but $S \cup \{\beta\}$ is dependent. Then S must contain α_1, say $S = \{\alpha_1, \ldots, \alpha_r\}$, by suitable renumbering of the αs. Now $\{\beta, \alpha_1, \ldots, \alpha_r\}$ is dependent but $\{\beta, \alpha_2, \ldots, \alpha_r\}$ is independent, hence α_1 is dependent on $\{\beta, \alpha_2, \ldots, \alpha_r\}$ and so (D.2) holds.

If (D.1) were true, we could conclude in the usual way that every extension field of K has a transcendence basis of a uniquely determined cardinal v (the 'transcendence degree') and any algebraically independent subset of E has at most v elements (see A.3, 1.4). But in Th. 5.5.8 we saw that every countably generated field can be embedded in a two-generator field over k. This shows that no transcendence degree can exist, so (D.1) does not hold in general. A look at the commutative case shows that what is needed here is a form of elimination. This again emphasizes the point that in the case of skew fields a general elimination procedure is lacking, but it might be worthwhile to examine special situations where elimination can be used.

Exercises

1. For a field extension E/K verify that the rational K-topology on E^n is a T_0-topology and give conditions for it to be a T_1-topology (i.e. every point is closed).

2. Define a *dependence relation* on a set X as a collection of finite subfamilies, the 'independent' subfamilies, such that (i) there are independent families, (ii) every subfamily of an independent family is independent, (iii) every independent family consists of distinct members. An element x of X is said to be *dependent* on a subset S of X if there is a finite subset S' of S such that S' is independent, but not $S' \cup \{x\}$. Verify that (D.0), (D.2) hold, and give an example where (D.1) fails.

3. Verify from the definition that all coordinates of a quasi-free point are distinct.

4. (J. Treur [89]) For any subset S of a field K and $c \in K$ define c to be *dependent* on S if there is a finite subset S' of S such that every polynomial f over K having all members of S' as zeros also has c as zero. Show that for every finite set S there is a unique monic polynomial of least degree having all members of S as zeros. Verify that this is a dependence relation in the sense of the text, i.e. that it satisfies (D.0–2). Show also that a polynomial of degree n cannot have more than n independent zeros, and that equality holds for a generating set of a finite Galois extension.

5. Let E be a field with an uncountable subfield K and suppose that E is finitely generated as K-ring. Show that no element of E is free over K. (Hint. Compare Ex. 16 of 3.4.)

8.7 Examples of singularities

To define a notion of variety over a skew field we need to look at
singularities of matrices. For any set $\mathscr{A}$ of matrices over $K_k\langle X \rangle$ we define
its *variety* as the set

$$\mathscr{V}(\mathscr{A}) = \mathscr{V}_E(\mathscr{A}) = \{\alpha \in E^n | \operatorname{sing}(A(\alpha)) \text{ for all } A \in \mathscr{A}\}.$$

When $\mathscr{A} = \{A\}$ consists of a single matrix, we shall also write $\mathscr{V}(A)$
instead of $\mathscr{V}(\{A\})$; this is just the complement of its singularity support.
To avoid trivialities we need to take A to be full. When k is the centre of
K and E satisfies the assumptions of the specialization lemma, the
fullness of A ensures that $\mathscr{V}(A)$ is a proper subset of E^n.

It is natural to begin with the simplest case, of varieties on the line. In
the commutative case these are just finite sets of points; by contrast, in
the general case the varieties can be of two kinds. We may have an
equation

$$x - a = 0, \tag{1}$$

satisfied by a single value, or an equation

$$ax - xb = c, \tag{2}$$

which has a solution of the form $x = x_0 + \lambda x_1$, where x_0 is a particular
solution of (2), x_1 is a particular solution of the associated homogeneous
equation

$$ax - xb = 0, \tag{3}$$

and λ ranges over $\mathscr{C}_E(a)$, the centralizer of a in E. Let us call the
solutions of (1) *point singularities* and those of (2) *ray singularities*.
Without attempting a precise definition at this stage we can say that the
variety of a matrix will in general contain both point and ray singularities
and it raises the following question:

PROBLEM 8.2. *Can the variety of a full matrix over $K_k\langle x \rangle$ always be
written as a finite union of point and ray singularities?*

To give an answer one will need to have a much more precise
knowledge of the variety of a matrix. Here our main handicap is that the
knowledge of a point in the variety does not permit a reduction of the
matrix to one of lower order, in the way that knowledge of a zero of a
polynomial allows us to reduce its degree. Of course, once one has a good
control of the variety of a matrix in one variable, one will have a better
chance of describing the variety of a matrix in two variables, giving an
idea of 'algebraic sets'.

We have already observed that the variety of a matrix is unchanged by stable association, so we may take our matrix to be in linear form

$$C = A + Bx, \quad \text{where } A, B \in K_N. \tag{4}$$

At $x = \alpha$ we obtain $C(\alpha) = A + B\alpha$ and α lies in $\mathcal{V}(C)$ precisely when $C(\alpha)$ is singular. Let us denote the rank and nullity of $C(\alpha)$ by $r(\alpha)$, $n(\alpha)$ respectively; thus if C is $N \times N$, then $r(\alpha) + n(\alpha) = N$. Clearly both rank and nullity are unchanged on passing to an associated matrix, but this no longer holds for stable association. If C becomes $C \oplus I_s$, then $r(\alpha)$ increases by s while $n(\alpha)$ remains unchanged. At ∞, for C as in (4), we have $r(\infty) = rB$, $n(\infty) = N - rB$, so at ∞ the rank is unchanged by stable association.

We next examine how the variety changes when x is transformed. Let us fix $C = A + Bx$ and write n, r for the nullity and rank in terms of x, and n', r' for the same in terms of y, where y is obtained from x by a fractional linear transformation. It will be enough to consider three special cases, from which we know every fractional linear transformation can be built up (see e.g. A.3, 4.7).

(i) $y = x + \lambda$ ($\lambda \in K$), $C = A + Bx = A - B\lambda + By$.

$$r'(\infty) = r(\infty), \quad n'(\alpha) = n(\alpha - \lambda),$$

for $n'(\alpha)$ is the nullity of $A - B\lambda + B\alpha = A + B(\alpha - \lambda)$.

(ii) $y = \lambda x$ ($\lambda \in K^\times$), $C = A + B\lambda^{-1}y$, hence

$$r'(\infty) = r(\infty), \quad n'(\alpha) = n(\lambda^{-1}\alpha).$$

(iii) $y = x^{-1}$, $C' = Cy = Ay + B$. Here we have (for $N \times N$ matrices)

$$r'(\alpha) = r(\alpha^{-1}), \quad n'(\alpha) = n(\alpha^{-1}) \text{ if } \alpha \neq 0,$$

$$r'(\infty) = N - n(0), \quad n'(0) = N - r(\infty).$$

So in this case we have $n'(0) - r'(\infty) = n(0) - r(\infty)$.

Let us define the *defect* of C as

$$d(C) = \sum n(\alpha) - r(\infty), \tag{5}$$

where the sum is taken over all $\alpha \in E$. Of course this only makes sense if C has no ray singularities and only a finite number of point singularities, but we can modify the definition either (i) by summing only over point singularities, or (ii) by allowing only one point from each ray singularity. What we have shown may be summed up as

PROPOSITION 8.7.1. *For any linear matrix C over $K_k\langle x \rangle$ with a pure point singularity consisting of a finite number of points, the defect is unchanged by fractional linear transformations.* ∎

We shall throughout assume that the conditions of the specialization lemma hold. Thus k is infinite, and by a suitable transformation of x we may assume C to be non-singular at infinity, so it can be taken in the form

$$C = Ix - A,$$

and here A is unique up to conjugacy over k, by Prop. 1.4. In this case (with a pure point singularity), if C is $N \times N$, then its defect is given by

$$d(C) = \sum n(\alpha) - N. \tag{6}$$

In the commutative case this defect for a diagonalizable matrix is zero, and in any case it is non-positive. Whether this holds more generally, say over an EC-field, is bound up with the question whether every EC-field is CAC (or FAC). Below we examine some special cases.

1. $r(\infty) = 1$. Here $C = A + Bx$, where B has rank 1. By elementary row transformations we can reduce all rows of B after the first to zero. We now apply the reduction to echelon form to A, by elementary row transformations and column permutations, but not acting on the first row. Omitting the first row, we obtain $(I \quad a)$, with no zero rows, because C was full. Transferring the final column a to the first column and reducing the first row of A after the $(1, 1)$-entry to 0 by row transformations, we finally reach the form

$$C = \begin{pmatrix} a_1 - b_1 x & b_2 x & b_3 x & \dots & b_N x \\ a_2 & 1 & 0 & \dots & 0 \\ a_3 & 0 & 1 & \dots & 0 \\ \dots & \dots & & \dots & \\ a_N & 0 & 0 & & 1 \end{pmatrix}.$$

This is easily seen to be skew cyclic, being stably associated to

$$a_1 - b_1 x - b_2 x a_2 - \dots - b_N x a_N.$$

We thus have the equation

$$b_1 x + b_2 x a_2 + \dots + b_N x a_N = a_1. \tag{7}$$

Its solutions are ray singularities of C (for $N > 1$), since with any solution x_0 there are solutions $x_0 + \lambda x_1$ for any $\lambda \in \mathscr{C}_E(b_1, \dots, b_N)$, where x_1 satisfies the homogeneous equation

$$b_1 x + b_2 x a_2 + \dots + b_N x a_N = 0.$$

For $N = 1$, (7) reduces to a point singularity, so we may assume that $N > 1$, and of course take N minimal. This will be ensured by taking b_1, $\dots, b_N$ and $1, a_2, \dots, a_N$ linearly independent over k.

2. For a 2×2 matrix we can determine the variety with the help of Th. 5.5 and also answer Problem 2 affirmatively. Let $C = A + Bx$ be 2×2. If $rB = 2$, the singularities are just the singular eigenvalues of $-B^{-1}A$, and by Th. 5.5 there is at least one. Taking it at ∞, we may assume that C has the form $A + Bx$, where $rB = 1$. By the last example, we thus obtain a matrix

$$\begin{pmatrix} a_1 - b_1 x & b_2 x \\ a_2 & 1 \end{pmatrix}$$

and this matrix is singular precisely when

$$b_1 x + b_2 x a_2 = a_1.$$

Either $b_2 = 0$. Then we have another point singularity $x = b_1^{-1} a_1$. We see that the defect is zero in this case.

Or $b_2 \neq 0$. Then by a scale change we can take $b_2 = -1$ and obtain

$$b_1 x - x a_2 = a_1,$$

which gives a ray singularity.

Exercises

1. Show that if a variety in E^n, not passing through the origin, is defined by a single matrix, the latter can always be taken in the form $I - \sum A_i x_i$.

2. Let K be a field and $A_1, A_2 \in \mathfrak{M}_2(K)$, where A_1 is non-singular, and let E be an EC-field containing K. Show that the variety in E^2 defined by $I - A_1 x_1 - A_2 x_2$ meets every line $x_2 = c_2$. Give an example where the variety does not meet all lines $x_1 = c_1$.

3°. Does every matrix over an EC-field with a pure point spectrum have defect zero?

8.8 Nullstellensatz and elimination

In commutative algebraic geometry the Hilbert Nullstellensatz is a basic result; it is usually stated in two parts. The first part, in Zariski's formulation (see Zariski and Samuel [60], p. 165 or A.2, 9.10) is

NULL 1. *If k is a commutative field, $L \supseteq k$ a field extension and*

$$f\colon k[x_1, \ldots, x_n] \to L$$

a surjective homomorphism, then L is algebraic over k.

The proof depends essentially on the Noether normalization lemma. We note that when k is algebraically closed, the conclusion is that $L = k$ (the usual form of the theorem). From Null 1 one can deduce (by the Rabinowitsch trick):

NULL 2. *If k is algebraically closed and $\mathfrak{a}$ is an ideal in $k[x_1, \ldots, x_n]$, then $g \in k[x_1, \ldots, x_n]$ vanishes at all common zeros of $\mathfrak{a}$ if and only if $g^N \in \mathfrak{a}$ for some $N \geqslant 1$.*

In the non-commutative case, Null 1 gives rise to a problem first raised by Amitsur (in the case where K is central in L): Let L be a skew field, finitely generated as a ring over a subfield K; is L necessarily algebraic over K?

This problem is still open, but we can sidestep it by replacing 'surjection' by 'epimorphism'. Then we obtain the following version of the Nullstellensatz, where $\mathcal{V}_K(\mathcal{A})$ again denotes the variety defined by $\mathcal{A}$.

THEOREM 8.8.1. *Let K be an EC-field over k and write $F = K_k\langle X \rangle$. If $\mathcal{A}$ is a finite set of matrices over F which all become singular under some homomorphism of F into a K-field L, then $\mathcal{V}_K(\mathcal{A}) \neq \varnothing$. For an infinite set $\mathcal{A}$ the conclusion need not hold.*

Proof. The first part is merely a restatement of the definitions: Since K is an EC-field, a finite set of matrices which becomes singular under some homomorphism $F \to L$ already becomes singular in K. To establish the last part, let K be a countable EC-field; then K has an outer automorphism σ, say, by Th. 6.5.11. Now the elements $ax - xa^\sigma$ ($a \in K$), $1 - xy$ all become zero (i.e. singular) in the skew rational function field $K(x; \sigma)$, but not in K itself, because σ is not inner. ∎

We remark that in the commutative case any variety defined by a subset S of $k[x_1, \ldots, x_n]$ can also be defined by a finite set; for S may be replaced by the ideal $\mathfrak{a}$ it generates and $\mathfrak{a}$ has a finite generating set, by the Hilbert basis theorem. By contrast, in the general case not every set $\mathcal{A}$ of matrices can be replaced by a finite set, as the example given in the proof shows.

To find an analogue of Null 2 we shall need to use prime matrix ideals.

THEOREM 8.8.2. *Let K be an EC-field over k, B a matrix and $\mathcal{A}$ any matrix ideal over the tensor ring $F = K_k\langle X \rangle$. Then $B \in \sqrt{\mathcal{A}}$ if and only if B becomes singular at all points of $\mathcal{V}_L(\mathcal{A})$, for a suitable extension L of K. If $\mathcal{A}$ is finitely generated, it is enough to take $L = K$.*

Proof. Suppose that $B \in \sqrt{\mathcal{A}}$ and let $f: F \to L$ be any homomorphism under which all matrices in $\mathcal{A}$ become singular. Then the singular kernel $\mathcal{P}$ of f is a prime matrix ideal containing $\mathcal{A}$, hence $\mathcal{P} \supseteq \sqrt{\mathcal{A}}$, and so $B \in \mathcal{P}$. Therefore B becomes singular, as asserted.

Conversely, assume that $B \notin \sqrt{\mathcal{A}}$; since $\sqrt{\mathcal{A}}$ is the intersection of all prime matrix ideals containing $\mathcal{A}$ (see 4.4), there is a prime matrix ideal $\mathcal{P}$ such that $\mathcal{P} \supseteq \sqrt{\mathcal{A}}$ but $B \notin \mathcal{P}$. The field $L = F/\mathcal{P}$ is an extension of K in which each matrix of $\mathcal{A}$ becomes singular, but B maps to an invertible matrix, so B is not singular at all points of $\mathcal{V}_L(\mathcal{A})$.

If $\mathcal{A}$ is finitely generated, by $A_1, \ldots, A_r$ say, then the sentence

$$\mathrm{sing}\,(A_1) \wedge \ldots \wedge \mathrm{sing}\,(A_r) \wedge \mathrm{nonsing}\,(B)$$

holds in L, hence it also holds in K, and the conclusion follows. ∎

The situation may be described by saying that we have a correspondence $\mathcal{A} \mapsto \mathcal{V}(\mathcal{A})$ between finitely generated matrix ideals and certain subsets of K^n (the closed subsets in the rational topology) and this is an order-reversing bijection (a Galois connexion) if we confine ourselves to semiprime matrix ideals that can be obtained as radicals of finitely generated matrix ideals. However, the points of K^n are not enough to distinguish between general semiprime matrix ideals and here we need to take an extension of K. In fact a fixed extension will do: In the proof of Th. 8.2 the only extensions of K needed (for a finite X) were finitely generated over K, so we need only take a universal EC-field.

Let us consider more closely the finitely generated maximal matrix ideals of $F = K_k\langle X \rangle$, where $X = \{x_1, \ldots, x_n\}$. If $\mathcal{A}$ is such a matrix ideal, then by Th. 8.2, $\mathcal{V}_K(\mathcal{A})$ is not empty, say $a = (a_1, \ldots, a_n) \in \mathcal{V}_K(\mathcal{A})$. Then $x_i - a_i \in \mathcal{A}$ and hence the matrix ideal $\mathcal{M}_a$ generated by the $x_i - a_i$ is contained in $\mathcal{A}$. But $\mathcal{M}_a$ is clearly maximal, hence when K is an EC-field, any finitely generated maximal matrix ideal over F has the form

$$\mathcal{M}_a = (x_1 - a_1, \ldots, x_n - a_n) \quad (a_i \in K).$$

Conversely, every matrix ideal of this form is finitely generated and maximal. Thus the finitely generated maximal matrix ideals correspond to the points of K^n. Now we can show as in the commutative case that every algebraically closed set is the union of its points:

THEOREM 8.8.3. *Let K be an EC-field over k and $\mathcal{A}$ a finitely generated matrix ideal in $F = K_k\langle x_1, \ldots, x_n \rangle$. Then $\sqrt{\mathcal{A}} = \bigcap \mathcal{P}_\lambda$ is the intersection of all the finitely generated maximal matrix ideals containing $\mathcal{A}$; hence $\mathcal{V}(\mathcal{A}) = \mathcal{V}(\sqrt{\mathcal{A}})$ is the union of the $\mathcal{V}(\mathcal{P}_\lambda)$.*

Proof. Let $\{\mathcal{P}_\lambda\}$ be the family of all finitely generated maximal matrix ideals containing $\mathcal{A}$. We clearly have

$$\sqrt{\mathcal{A}} \subseteq \bigcap \mathcal{P}_\lambda. \tag{1}$$

To establish equality, take $B \notin \sqrt{\mathcal{A}}$; by Th. 8.2 there is a point $a \in K^n$ at which all matrices of $\mathcal{A}$ are singular but B is not. Hence the corresponding matrix ideal $\mathcal{M}_a$ contains $\mathcal{A}$, and so occurs among the $\mathcal{P}_\lambda$ on the right of (1), but $B \notin \mathcal{M}_a$. Hence B is not contained in the right-hand side of (1) and the equality follows. Taking varieties, we now obtain $\mathcal{V}(\mathcal{A}) = \bigcup \mathcal{V}(\mathcal{P}_\lambda)$. ∎

Sometimes a homogeneous form of Th. 8.1 is needed. Let us write $X_h = \{x_0, x_1, \ldots, x_n\}$ and consider $K_k\langle X_h \rangle$ as a graded algebra, with each x_i of degree 1. The only matrices allowed are homogeneous in the xs; such a matrix is always stably associated to one in linear form: $\sum_0^n A_i x_i$, where $A_i \in K_N$ for some $N \geqslant 1$.

THEOREM 8.8.4. *Let K be an EC-field over k, $F = K_k\langle X_h \rangle$ and denote by $\mathcal{J}$ the singular kernel of the homomorphism $\varphi \colon F \to K$ given by $x_i \mapsto 0$ $(i = 0, 1, \ldots, n)$. If $\mathcal{A}$ is a homogeneous finitely generated proper matrix ideal, then $\mathcal{V}_K(\mathcal{A}) = \varnothing$ if and only if $\sqrt{\mathcal{A}} = \mathcal{J}$.*

Proof. Since φ maps $\mathcal{A}$ to singular matrices, we have in any case $\sqrt{\mathcal{A}} \subseteq \mathcal{J}$. If $\mathcal{V}_K(\mathcal{A})$ contains a point $\alpha \neq 0$, then $\sqrt{\mathcal{A}} \subseteq \mathcal{M}_\alpha$ and so $\sqrt{\mathcal{A}} \neq \mathcal{J}$. Now we have

$$\sqrt{\mathcal{A}} = \bigcap \{\mathcal{M} | \mathcal{M} \text{ fin. gen. max.} \supseteq \mathcal{A}\}$$

by Th. 8.3, so if this is $\neq \mathcal{J}$, then $\mathcal{V}_K(\mathcal{A})$ contains some point $\neq 0$. ∎

For a slightly different point of view, leading again to a form of the Nullstellensatz, we require the notion of a generic point. Let $K \subseteq L$ be fields that are k-algebras. Given a prime matrix ideal $\mathcal{P}$ in $F = K_k\langle x_1, \ldots, x_n \rangle$, by a *generic point* for $\mathcal{P}$ over L we understand a point $c \in \mathcal{V}_L(\mathcal{P})$ such that every matrix over F which becomes singular at c is in $\mathcal{P}$, and hence is singular on all of $\mathcal{V}_L(\mathcal{P})$. When the role of the coefficient field is to be stressed, we may speak of a K-*generic point*. Every point of L^n is generic for some prime: given $c \in L^n$, we take $\mathcal{P}$ to be the singular kernel of the mapping from F to L given by $x_i \mapsto c_i$. Conversely, every prime matrix ideal has a generic point; for the proof we recall from 6.5 that an EC-field L containing K is called *semiuniversal over K* if every finitely presented field over K is embeddable in L. This means that L contains a copy of every quotient of F by a finitely generated prime matrix ideal (for varying n). Such semiuniversal fields always exist and may be taken countable when K is countable over k. For clearly the set of all finitely presented fields over K is countable; we now form their field coproduct over K and take an EC-field containing this field coproduct. In a similar way one can form a *universal* EC-field over K, i.e. an EC-field containing all finitely generated fields over K, but this field need not be countable, even if K is.

PROPOSITION 8.8.5. *Let K be a field which is a k-algebra, countably generated over k. Then every prime matrix ideal $\mathscr{P}$ in $F = K_k\langle x_1, \ldots, x_n \rangle$ has a generic point in any universal EC-field over K. If $\mathscr{P}$ is finitely generated, the point may be taken in any semiuniversal EC-field over K.*

Proof. The quotient $L_0 = F/\mathscr{P}$ is finitely generated, and when $\mathscr{P}$ is finitely generated, L_0 is even finitely presented over K, so we have an embedding $L_0 \to L$ in a universal or semiuniversal EC-field over K, respectively. The natural map $F \to L_0 \to L$ takes the members of $\mathscr{P}$ and no others to singular matrices. If $c \in L^n$ is the image of $x = (x_1, \ldots, x_n)$, then any matrix B over F is in $\mathscr{P}$ if and only if it is singular at c; this shows c to be a generic point for $\mathscr{P}$. ∎

This result allows us to deduce another form of the Nullstellensatz:

THEOREM 8.8.6. *Let $K \subseteq L$ be fields which are k-algebras, where K is countably generated over k and L is a universal EC-field over K. Given a prime matrix ideal $\mathscr{P}$ in $F = K_k\langle x_1, \ldots, x_n \rangle$, $\mathscr{P}$ is the precise set of matrices that are singular on all of $\mathscr{V}_L(\mathscr{P})$. More generally, if $\mathscr{A}$ is any matrix ideal in F, then $\sqrt{\mathscr{A}}$ is the set of matrices that become singular on $\mathscr{V}_L(\mathscr{A})$. If L is merely assumed semi-universal, the assertions hold for finitely generated matrix ideals.*

Proof. Given $\mathscr{P}$, we can by Prop. 8.5 find a generic point c over L. The members of $\mathscr{P}$ clearly become singular on $\mathscr{V}_L(\mathscr{P})$; on the other hand, if $B \notin \mathscr{P}$, then B is non-singular at c, hence B is not singular on all of $\mathscr{V}_L(\mathscr{P})$. Similarly, for general $\mathscr{A}$ we replace c by the set of all generic points for the primes containing $\mathscr{A}$ and use the formula $\sqrt{\mathscr{A}} = \bigcap\{\mathscr{P}|\mathscr{P}$ prime $\supseteq \mathscr{A}\}$. ∎

We observe that if $\mathscr{P}_1, \ldots, \mathscr{P}_r$ are distinct prime matrix ideals that are finitely generated, then their product $\mathscr{A} = \mathscr{P}_1 \ldots \mathscr{P}_r$ is again finitely generated and $\sqrt{\mathscr{A}} = \mathscr{P}_1 \cap \ldots \cap \mathscr{P}_r$. However, $\sqrt{\mathscr{A}}$ itself may not be finitely generated; thus we cannot limit ourselves to semiprime matrix ideals.

We conclude this section by discussing a fundamental result of algebraic geometry and its possible generalizations. The main theorem of elimination theory, in its simplest form, gives conditions for two equations to have a common solution (in some field extension) – namely the vanishing of the resultant. In its most general form it states that projective space P^n is complete (see Mumford [76], p. 33). We recall that a variety X is said to be *complete* if for every variety Y the projection $p_2: X \times Y \to Y$ carries closed sets into closed sets. Thus the theorem in question asserts that the projection

$$p_2: P^n \times Y \to Y \qquad (2)$$

maps closed sets to closed sets, for any Y. Being closed is a local property, so we can cover Y by affine open sets and it will be enough to prove the result for affine Y. If Y is closed in A^m, then $P^n \times Y$ is closed in $P^n \times A^m$, so we can without loss of generality take $Y = A^m$.

Every closed subset Z of $P^n \times A^m$ is given by a finite set of equations

$$f_i(u; y) = 0 \quad (i = 1, \ldots, t),$$

where the us are homogeneous coordinates in P^n and the ys coordinates in A^m. For any point y^0 of A^m, $p_2^{-1}(y^0)$ consists of all non-zero solutions $u = u^0$ of $f_i(u; y^0) = 0$. Hence we have

$$y^0 \in p_2(Z) \Leftrightarrow \text{the equations } f_i(u; y^0) = 0 \text{ have a solution } u^0 \neq 0. \quad (3)$$

It has to be shown that the set T of points y^0 of A^m satisfying the two sides of (3) is closed, i.e. we have to find a set of polynomials $g_j(y)$ such that

$$y^0 \in p_2(Z) \Leftrightarrow g_j(y^0) = 0 \text{ for } j = 1, \ldots, r.$$

These polynomials are precisely the ones obtained by the usual elimination process (see Mumford loc. cit.)

In the non-commutative case the closed subset Z of $P^n \times A^m$ is given as a set of singularities

$$\text{sing}\,(A_\lambda(u; y)), \; \lambda \in \Lambda, \; u \in P^n, \; y \in A^m. \qquad (4)$$

Here we may without loss of generality take the A_λ to be linear homogeneous in the us:

$$A_\lambda = \sum A_{\lambda i}(y)u_i.$$

If we denote by $\mathscr{A}$ the matrix ideal homogeneous in u generated by all the $A_\lambda(u; y)$ over $K_k\langle u; y \rangle$, the tensor ring in variables y and homogeneous variables u, then in analogy to (3) we have, for any $y^0 \in K^m$,

$$y^0 \in p_2(Z) \Leftrightarrow \text{sing}(A_\lambda(u'; y^0)) \quad \text{for some } u' \in \mathbf{P}^n(E), \text{for suitable } E \supseteq K.$$

(5)

Let us define, for $R = K_k\langle y \rangle$, and the above matrix ideal $\mathscr{A}$ in $K_k\langle u; y \rangle$,

$$\mathscr{A}^* = \{ B \in \mathfrak{M}(R) | (B \oplus I_r)u_i \in \mathscr{A} \text{ for } i = 0, 1, \ldots, n \text{ and some } r \}. \quad (6)$$

It is easily verified that $\mathscr{A}^*$ is a matrix ideal in R. Let $\{B_\mu(y)\}$ be a generating set of $\mathscr{A}^*$. Then

$$y^0 \in p_2(Z) \Rightarrow B_\mu(y^0) \text{ is singular for all } \mu. \quad (7)$$

For if $y^0 \in p_2(Z)$, then all the $A_\lambda(u; y^0)$ have a common singularity $u' \in \mathbf{P}^n(E)$. Suppose that some B_μ is non-singular at y^0. By definition of $\mathscr{A}^*$ we have

$$(B_\mu(y) \oplus I)u_i = \nabla\{A_\lambda(u; y)\}, \quad (8)$$

where the right-hand side is a determinantal sum of terms. If we put $y = y^0$, $u = u'$, then the right-hand side of (8) becomes singular, while $B_\mu(y^0)$ is non-singular; hence $u_i' = 0$ for all i, which is a contradiction, because $u' \in \mathbf{P}^n(E)$. So all the B_μ are singular at y^0, as claimed.

Thus (7) gives a necessary condition for a common singularity. However, there is no evidence so far to show that it is an equivalence. It is possible that $\mathscr{A}^*$, as defined in (6), is too small to be relevant.

Exercises

1. Given an extension E/K of fields that are k-algebras, a point c of E^n is said to be *finitely presented* over K if the prime matrix ideal in $K_k\langle x_1, \ldots, x_n \rangle$ with c as generic point is finitely generated. Show that c is finitely presented if and only if $K(c_1, \ldots, c_n)$ has a finite presentation over K. Deduce that every point in K^n is finitely presented over K.

2°. Given a field E and a point α of E^n, determine the subfields of E over which α is finitely presented. Is there always a least such subfield?

3°. If $\mathscr{A}$ is a finitely generated matrix ideal in $K_k\langle x_1, \ldots, x_n \rangle$, is $\sqrt{\mathscr{A}}$ always an intersection of finitely many prime matrix ideals?

4. Show that for $a \neq 0$, the matrix $A = \begin{pmatrix} 1 & a^{-1} \\ a & 1 \end{pmatrix}$ has the eigenvalues $0, -2$. Thus if $B = \begin{pmatrix} 1 & b^{-1} \\ b & 1 \end{pmatrix}$, then A, B have the same eigenvalues, but verify that $A \otimes I - I \otimes B$ is non-singular unless $ab = ba$.

5. Put $A = \begin{pmatrix} 0 & 1 \\ ab & 0 \end{pmatrix}$, $B = \begin{pmatrix} 0 & b \\ a & 0 \end{pmatrix}$; show that $A \otimes I - I \otimes B$ is singular, but A, B have no common singular eigenvalue, unless $ab = ba$.

6. Let R be a k-algebra and K an R-field with singular kernel $\mathcal{P}$. If S is a set of square matrices over $R_k\langle X \rangle$ and $(\mathcal{P}, S)$ denotes the matrix ideal of $R_k\langle X \rangle$ generated by $\mathcal{P}$ and S, show that the matrices of S have a common singularity in some extension field E of K if and only if $(\mathcal{P}, S) \cap \mathfrak{M}(R) = \mathcal{P}$. Verify that $(S) \cap \mathfrak{M}(R) \subseteq \mathcal{P}$ is necessary for a common singularity but not sufficient. (Hint. Consider $R = k[a]$, $S = (1 - ax)$.)

7°. Show that two matrices A, B over a commutative field k have a common eigenvalue in some extension field of k if and only if $A \otimes I - I \otimes B$ is singular. Given a skew field K and matrices A, B over K, find a matrix C whose entries are functions of the entries of A and B such that A and B have a common singular eigenvalue in some extension of K if and only if C is singular (it may be necessary to allow for a family $\{C_\lambda\}$ of matrices, possibly infinite, such that A, B have a common singular eigenvalue if and only if all the C_λ are singular).

Notes and comments

The development of general skew fields made it natural to replace equations by singularities, though this does not always make the solution easier. The linearization process has long been used informally, e.g. in the theory of differential equations (for an early explicit use in the study of group rings, see Higman [40]). It was first used in the present sense in Cohn [72', 73, 76] and the first edition of FR.

The special case of Th. 1.6 of extensions in which every element is right algebraic was proved by R. Baer [27]; the rest of 8.1 was in *Skew Field Constructions* or is new. The problem of finding right roots of equations was solved for the real quaternions $\mathbf{H}$ by Niven [41] and Eilenberg and Niven [44] gave a topological proof that every f in $\mathbf{H_R}\langle x \rangle$ with a unique term of highest degree has a zero. This also follows from the more general theorem of R. M. W. Wood [85] that $\mathbf{H}$ is a CAC-field. L. G. Makar-Limanov [75, 85] has constructed a PAC-field of infinite degree over its centre. The metro-equation (2) of 8.1 (which arose in conversation with Amitsur on the Paris Metro on 28 June, 1972) has often been considered e.g. in connexion with the Weyl algebra. For an application to the geometry of the tetrahedron see Bilo [80].

The reduction to diagonal form in 8.3 is taken from Cohn [73], where left and right eigenvalues are introduced, the results of 8.2 are proved and the normal form of 8.3 is derived, except for the final step depending on Prop. 3.5; this result and its application are due to W. S. Sizer [75]. The

Cayley–Hamilton theorem for skew cyclic matrices was proved in Cohn [76]. The following characterization of skew cyclic matrices was obtained by M. L. Roberts [82]: A matrix A is skew cyclic if and only if A is conjugate over the centre to a matrix C with non-zero entries on the subdiagonal (i.e. $c_{i+1\,i} \neq 0$) and zeros below the subdiagonal.

Th. 4.2 is taken from Cohn [85], but parts of it were well known, cf. e.g. Jacobson [56], p. 241. V-rings (named after O. E. Villamayor) have been studied by Faith [81], see also Cozzens and Faith [75], where Prop. 4.3 is proved. The rather curious example of a V-ring in Th. 4.5 was constructed by Resco [87] to answer a question of Cozzens and Faith [75], who asked whether a tensor product of two simple V-rings is always a V-ring. In Johns [77] it was claimed that a right Noetherian right annihilator ring is right Artinian; the claim was based on a result of R. P. Kurshan [70] which turned out to be false, and the counter-example presented in the text is due to Faith and Menal [92]. Amitsur [56] has shown that a field which is algebraic over an uncountable centre is also matrix-algebraic (Jacobson [56], p. 247).

Most of 8.5 first appeared in the Skew Field Constructions; Th. 5.4 is taken from Cohn [73'''], though special cases had often been noted before, e.g. Johnson [44]. The existence of singular eigenvalues for a matrix with indeterminate entries (Th. 5.6) was proved in Cohn [75'].

The work of 8.6 was essentially new in Skew Field Constructions, based on Cohn [77'] and developed further in Cohn [85'], where the examples of 8.7 also appeared. The non-commutative Nullstellensatz was first obtained by W. H. Wheeler [72], cf. Hirschfeld and Wheeler [75], where the result is stated in terms of so-called d-prime ideals (kernels of homomorphisms into fields) taking the place of prime matrix ideals. The present version is taken from Cohn [75]. The remarks on elimination have not been published before (but see Cohn [88]).

9

Valuations and orderings on skew fields

Normal subgroups can be used to decompose groups, and this is an important tool in the analysis of groups. Ideals play a similar role in ring theory, but there is no direct analogue in fields. The nearest equivalent is a general valuation, which allows a field to be analysed into a group, the 'value group', and a residue-class field. Thus valuations form a useful tool in commutative field theory, but there is no method of construction in general use, mainly because in most cases all the valuations are explicitly known, e.g. for algebraic number fields, function fields of one variable or even two variables (see e.g. Cohn [91], Ch. 5). Our aim in this chapter is to describe a general method of construction, using subvaluations, which can be used even in the non-commutative case.

We begin by recalling the basic notions in 9.1, which still apply to skew fields, and then in 9.2 explain the special case of an abelian value group, which presents a close analogy to the commutative case while being sufficiently general to include some interesting applications. In the commutative case a ring R with a valuation v is an integral domain and v extends in a unique way to the field of fractions of R. In the general case neither existence nor uniqueness is ensured; what is needed here is a valuation on all the square matrices over R and 9.3 introduces the study of such matrix valuations and explains the way they determine valuations on epic R-fields. In 9.4 a method of constructing such matrix valuations is described, based on the principle of the Hahn–Banach theorem (that a closed subspace disjoint from a convex body can be separated from this body by a hyperplane), and this is used in 9.5 to construct matrix valuations on firs.

The last two sections are devoted to ordered fields. In 9.6 the analogue of the Artin–Schreier theory is developed for skew fields, giving conditions under which a skew field (or even a ring) can be ordered,

420

while 9.7 describes matrix cones on R, which are then used to order epic R-fields. As an application it is shown how free fields may be ordered.

9.1 The basic definitions

To discuss valuations we shall need to deal with ordered groups. It will be convenient to use additive notation, though our groups need not be abelian. By an *ordered group* Γ we shall understand a group with a total ordering $x \geq y$, which is preserved by the group operation:

$$x \geq y, x' \geq y' \Rightarrow x + x' \geq y + y' \text{ for all } x, x', y, y' \in \Gamma.$$

Sometimes Γ will be augmented by a symbol ∞ to form a monoid with the operation

$$x + \infty = \infty + x = \infty \text{ for all } x \in \Gamma \cup \{\infty\},$$

and the ordering $\infty > x$ for all $x \in \Gamma$.

Let R be a ring. By a *valuation* on R with values in an ordered group Γ, the *value group*, we shall understand a function v on R with values in $\Gamma \cup \{\infty\}$ subject to the conditions:

(V.1) $v(x) \in \Gamma \cup \{\infty\}$ *and* v *assumes at least two values,*

(V.2) $v(xy) = v(x) + v(y),$

(V.3) $v(x + y) \geq \min \{v(x), v(y)\}$ *(ultrametric inequality).*

The set

$$\ker v = \{x \in R | v(x) = \infty\}$$

is easily verified to be an ideal of R, which is proper by (V.1). If $\ker v = 0$, v is said to be *proper*; e.g. on a field every valuation is proper, because 0 is the only proper ideal. In general $R/\ker v$ is an integral domain, by (V.2). Further it is clear that $v(0) = \infty$ and if $u \notin \ker v$, then $v(u) = v(u \cdot 1) = v(u) + v(1)$, hence $v(1) = 0$. It also follows that $2v(-1) = v((-1)^2) = 0$, so $v(-1) = 0$, and by another application of (V.2) we obtain

(V.4) $v(-x) = v(x).$

If Γ can be embedded in $\mathbf{R}$ (which happens precisely when Γ is Archimedean ordered, see e.g. Cohn [91], Prop. 1.6.3, p. 33), we can define a metric on R by choosing a real constant c between 0 and 1 and defining

$$d(x, y) = c^{v(x-y)}.$$

It is easily verified, using (V.1–3), that (M.1–4) below hold, so we have a metric on $R/\ker v$, turning R into a topological ring, with a Hausdorff topology if and only if $\ker v = 0$:

(M.1) $d(x, y) \geqslant 0$ *with = if and only if $x \equiv y$* (mod ker v),

(M.2) $d(x, y) = d(y, x)$,

(M.3) $d(x, y) + d(y, z) \geqslant d(x, z)$ *(triangle inequality)*,

(M.4) $d(x + a, y + a) = d(x, y)$ *(translation invariance)*.

When v is proper, the function $|x| = d(x, 0)$ is also called a *norm* or *absolute value* on R. As with every metric space, one can form the completion of R, which plays an important role in commutative field theory, but which will not concern us here.

We remark that the triangle inequality (M.3) actually holds in the stronger form:

(M.3') $d(x, z) \leqslant \max \{d(x, y), d(y, z)\}$.

Whereas (M.3) tells us that in any triangle the sum of any two sides is at least equal to the third, (M.3') shows that every triangle is isosceles; for if $d(x, y) < d(y, z)$ say, then by (M.3'), $d(x, z) \leqslant d(y, z)$ and $d(y, z) \leqslant \max \{d(y, x), d(x, z)\} = d(x, z)$ (using (M.2)), hence $d(x, z) = d(y, z)$. In terms of the original valuation this states that

$$\text{If } v(x + y) > \min \{v(x), v(y)\}, \text{ then } v(x) = v(y). \tag{1}$$

Of course this is easily proved directly, for any value group Γ.

Every integral domain has the *trivial* valuation

$$v(x) = \begin{cases} 0 & \text{if } x \neq 0, \\ \infty & \text{if } x = 0, \end{cases} \tag{2}$$

corresponding to the discrete topology. Any other valuation is called *non-trivial*.

We begin by showing that any proper valuation on an Ore domain can be extended to its field of fractions:

PROPOSITION 9.1.1. *Let R be a right Ore domain with field of fractions K. Then any proper valuation v on R has a unique extension to K.*

Proof. Every element of K has the form $u = ab^{-1}$, $a, b \in R$, so if an extension w of v to K exists, then $v(a) = w(ub) = w(u) + v(b)$, hence $w(u) = v(a) - v(b)$. If we also have $u = a_1 b_1^{-1}$, then $ap = a_1 p_1$, $bp = b_1 p_1$ for some $p, p_1 \in R^{\times}$. Hence $v(a) - v(b) = v(ap) - v(bp) = v(a_1 p_1) - v(b_1 p_1) = v(a_1) - v(b_1)$, and this shows w to be independent of the choice of a and b. It remains to show that w is a valuation on K. (V.1) is clear; to prove (V.2), given ab^{-1}, cd^{-1}, let $bp = cq = m$, say; then we can replace ab^{-1} by $ap \cdot (bp)^{-1} = apm^{-1}$ and cd^{-1} by $cq(dq)^{-1} = m(dq)^{-1}$, hence $ab^{-1} \cdot cd^{-1} = apm^{-1} \cdot m(dq)^{-1}$ and $w(ab^{-1} \cdot cd^{-1}) =$

$w(ap(dq)^{-1}) = v(ap) - v(dq) = v(ap) - v(m) + v(m) - v(dq) = w(ab^{-1})$
$+ w(cd^{-1})$. (V.3) follows similarly, by bringing x and y to a common denominator. ∎

Let us now consider a valuation v on a field K. With v we associate the subset of K defined by

$$V = \{x \in K \,|\, v(x) \geq 0\}. \tag{3}$$

It is clear from (V.1–3) that V is a subring of K; in fact it has two further properties: It is *total*, i.e. for every $x \in K$, either $x \in V$, or $x \neq 0$ and $x^{-1} \in V$. Secondly it is *invariant*, i.e. for every $a \in V$ and $c \in K^{\times}$, $c^{-1}ac \in V$. A total invariant subring of a field is called a *valuation ring* and what we have said can be summed up as follows:

PROPOSITION 9.1.2. *For any valuation v on a field K, the set V defined by (3) is a valuation ring.* ∎

When K is commutative, the condition of invariance becomes vacuous and the above term agrees with the usage in the commutative case.

Conversely, given any skew field K and a valuation ring V in K, we can form a valuation giving rise to it as follows. In V we define the group of units as

$$U = \{x \in K \,|\, x \in V \text{ and } x^{-1} \in V\}. \tag{4}$$

Briefly we may write (with a slight abuse of notation) $U = V \cap V^{-1}$. Because V is invariant in K, it follows that U is a normal subgroup of $K^{\times}$. The quotient $K^{\times}/U$ will be denoted by Γ and written additively. On $K^{\times}$ we have a preordering by divisibility: $a|b$ if and only if $ba^{-1} \in V$. Two elements a, b are equivalent in this preorder, i.e. $a|b$ and $b|a$, precisely when they lie in the same coset of U. Thus the divisibility preordering on $K^{\times}$ defines an ordering of $\Gamma = K^{\times}/U$, which is a total ordering, because V was a total subring of K. Now it is easily checked that the natural map $v: K^{\times} \to \Gamma$ together with the rule $v(0) = \infty$ defines a valuation on K whose valuation ring is just the ring V with which we started.

There is a third way of describing valuations which is sometimes useful. By a *place* on a field K with values in a field L we mean a homomorphism φ from a subring V of K to L, extended to all of K by writing $\varphi(x) = \infty$ for $x \in K\backslash V$, such that
(P.1) *if* $\varphi(x) \neq \infty$, *then* $\varphi(c^{-1}xc) \neq \infty$ *for all* $c \in K^{\times}$,
(P.2) *if* $\varphi(x) = \infty$, *then* $x \neq 0$ *and* $\varphi(x^{-1}) = 0$.
It is clear from (P.1–2) that V is necessarily a valuation ring in K. Conversely, every valuation ring V in K defines a place on K as follows.

The non-units in V form an ideal, for in terms of the corresponding valuation v the set of non-units is given by

$$\mathfrak{m} = \{x \in K | v(x) > 0\},$$

and this is an ideal by (V.1–3). We have the residue-class field $V/\mathfrak{m}$ and the natural homomorphism $V \to V/\mathfrak{m}$ satisfies (P.1–2) because V is a valuation ring. Thus we have the following triple correspondence on any skew field:

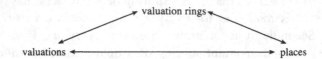

valuation rings

valuations ←————————————————————→ places

Two valuations v, v' on K with value groups Γ, Γ' are said to be *equivalent* if there is an isomorphism $\theta \colon \Gamma \to \Gamma'$ such that $v'(x) = v(x)\theta$ for all $x \in K$. Two places φ, φ' on K with values in L, L' are called *equivalent* if there is an isomorphism $f \colon L \to L'$ such that $\varphi'(x) = \varphi(x)f$ for all $x \in K$. It is easily verified that two valuations are equivalent if and only if they define equivalent places, or also if and only if they have the same valuation ring. Thus for any field K we have a triple correspondence (a 'trijection') between (i) equivalence classes of valuations, (ii) equivalence classes of places and (iii) valuation rings on K. This shows the advantage of dealing with the valuation ring, rather than the valuation or the place, either of which is only determined up to equivalence.

The simplest non-trivial case is that where the value group is $\mathbf{Z}$, the additive group of integers. A valuation with value group $\mathbf{Z}$ will be called *principal*. For example, the p-adic valuation on $\mathbf{Q}$, for any prime number p, is a principal valuation (see e.g. A.2, 8.1). In the case of a principal valuation the valuation ring is a principal ideal domain; conversely, a valuation ring which is a principal ideal domain gives rise to a principal valuation (see e.g. Cohn [91], Th. 1.4.2 or Ex. 1).

Let K be any field with a valuation v; it allows us to 'decompose' K as follows. The valuation may be regarded as a homomorphism from $K^\times$ to the value group Γ; its kernel is the group U of units in the valuation ring. Thus we have

$$\Gamma \cong K^\times / U. \tag{5}$$

Let V be the valuation ring and $\mathfrak{m}$ its maximal ideal; clearly every element of the form $1 + x$, where $x \in \mathfrak{m}$, is a unit. It is called a 1-unit (Einseinheit) and the group of all 1-units is written $1 + \mathfrak{m}$ or U_1. Let us write k for the residue-class field; thus $k = V/\mathfrak{m}$, and in the natural mapping from V to k an element maps to 1 precisely when it is a 1-unit. The residue-classes mod U_1 correspond to the different elements of k, so that we have an isomorphism

$$k^\times \cong U/U_1. \tag{6}$$

All these facts are conveniently summed up in the following commutative diagram with exact rows and columns:

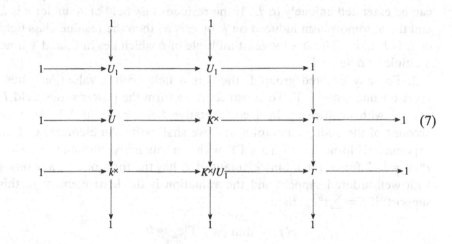

$$(7)$$

In the special case of a principal valuation Γ is infinite cyclic and so the horizontal sequences in (7) split. This is the case that occurs most frequently in the commutative theory and we then have the isomorphism

$$K^\times/U_1 \cong k^\times \times \Gamma. \tag{8}$$

Here the diagram (7) does not add much to our knowledge that is not already clear from (8). By contrast, in the general case, especially when Γ is non-abelian, (7) provides more information about $K^\times$. In particular, the third row will be of interest in what follows.

To end this section we give some examples of valuations on skew fields.

1. Let K be any field and consider the rational function field $K(t)$ formed with a central indeterminate t. Every element of $K(t)$ is either 0 or of the form $\varphi = t^\nu f g^{-1}$, where f, g are polynomials in t with non-zero constant term. The function $v(\varphi) = v$ is easily verified to be a valuation, called the *t-adic valuation* on $K(t)$. Its value group is $\mathbf{Z}$ and its residue-class field is K.

The same construction can be used in the more general situation of a skew function field $K(t; \alpha)$, where α is an automorphism of K. More generally, let K be a field with a valuation v and an automorphism α such that $v(a^\alpha) = v(a)$ for all $a \in K$, and form the skew function field $L = K(t; \alpha)$ as before. Now select an element δ in the value group Γ (or in an ordered extension of Γ) and define a valuation on the skew polynomial ring $K[t; \alpha]$ by the rule

$$w\left(\sum t^i a_i\right) = \min_r \{r\delta + v(a_r)\}.$$

It is easily checked that this is a valuation on $K[t; \alpha]$ and by Prop. 1.1 it can be extended uniquely to L. If the residue-class field of K under v is k and the automorphism induced on k by α is $\bar{\alpha}$, then the residue-class field of L is $k(u; \alpha^{-r})$ if $r\delta$ is the least multiple of δ which lies in Γ, and k if no multiple of δ lies in Γ.

2. For any ordered group Γ there is a field with a valuation whose precise value group is Γ. To construct it we form the power series field L over Γ with coefficients in a commutative field k, as in 2.4. To take account of the additive notation in Γ we shall write the elements of Γ in exponential form as t^α ($\alpha \in \Gamma$) with an auxiliary variable t. Thus $t^\alpha t^\beta = t^{\alpha+\beta}$ for $\alpha, \beta \in \Gamma$, any element of L has the form of a power series with well-ordered support and the valuation is the least element in this support: if $f = \sum t^\alpha c_\alpha$, then

$$v(f) = \min \{\alpha \in \Gamma | c_\alpha \neq 0\}.$$

3. Consider the free field $F = k \langle\!\langle x, y \rangle\!\rangle$; we may ask whether there is anything like an x-adic valuation on it. A straightforward approach yields nothing, because we cannot write every element in the form $x^v fg^{-1}$ as under 1. However, we can proceed as follows: We shall want to have $v(x) > 0 = v(y)$ and this will entail $v(c^{-1}xc) > 0 = v(c^{-1}yc)$ for any $c \neq 0$. Let us define

$$y_r = x^{-r} y x^r \quad (r \in \mathbf{Z}). \tag{9}$$

By Lemma 5.5.6, the subfield E generated by the y_r over k is free on these generators; it has the shift automorphism $\sigma: y_i \mapsto y_{i+1}$ and $F = k \langle\!\langle x, y \rangle\!\rangle \cong E(x; \sigma)$, so we can take the x-adic valuation on E as valuation on F. Here the value group is still $\mathbf{Z}$, but the residue-class field is now $E = k \langle\!\langle y_r | r \in \mathbf{Z} \rangle\!\rangle$, and in this sense it is more complicated than our initial field F. Later we shall find valuations of F with simpler residue-class fields, but this is achieved at the cost of complicating the value group.

4. If we have a commutative field extension $F \subseteq E$ and v is a valuation on F, then there always exists a valuation on E extending v, by Chevalley's extension lemma (see A.2, Lemma 8.4.3, p. 290). There is a generalization to the skew case, which provides an extension under suitable conditions (see 9.2 below), but there may be no extension in this

case. To give an example, let D be a field with a non-trivial valuation v, which is trivial on a subfield K of D. Then $v(a) > 0$ for some $a \in D^{\times}$; it follows that a is not algebraic over K, hence D can be enlarged to a field E containing an element b such that $ab = ba^{-1}$, by Lemma 5.5.4. If v could be extended to E, we would have $v(a^{-1}) < 0$, but $a^{-1} = b^{-1}ab$, so $v(a^{-1}) = v(b^{-1}ab) = -v(b) + v(a) + v(b) > 0$. This contradiction shows that v cannot be extended to E.

Exercises

1. Show that a valuation is $\mathbf{Z}$-valued if and only if its valuation ring is a principal ideal domain (Hint. Show that every element of the valuation ring V can be written as $p^r u$ (or 0), where pV is the maximal ideal and u is a unit.)

2. Let K be a skew field of characteristic 0 with a $\mathbf{Z}$-valued valuation v. If the residue-class field is a commutative field of characteristic p and $v(p) = 1$, show that K is commutative. Does this remain true without the condition $v(p) = 1$?

3. Show that an EC-field over a ground field k has no non-trivial valuations.

4. (A. H. Schofield) Let K be a field with an automorphism α and an α-derivation δ, put $U = K(t; \alpha, \delta)$ and $C = \mathscr{C}(U) \cap K$. Using the valuation $-\deg$, where deg is the degree in t, show that any finite-dimensional C-algebra in U is contained in K.

9.2 Abelian and quasi-commutative valuations

A central problem in valuation theory concerns the extendibility of valuations to extension fields. For a commutative field extension $F \subseteq E$, every valuation on F has at least one extension to E, by Chevalley's extension lemma, but as the example in 9.1 showed, in the general case there may be no extension. The general problem is difficult, since it may involve an extension of the ordered value group; one way of simplifying the problem is to assume the value group to be abelian. In that case the problem turns out to be quite tractable; an extension does not always exist, but there is a simple criterion.

Thus we specialize by considering valuations with abelian value group, more briefly *abelian* valuations. Such a valuation on K is zero on all commutators and so must be trivial on the derived group of $K^{\times}$; we shall denote this derived group, i.e. the subgroup generated by all commutators in K, by K^c. Then we have the following simple condition for a valuation to be abelian:

PROPOSITION 9.2.1. *Let K be a field with a valuation v and denote its valuation ring by V. Then v is abelian if and only if $v(a) = 0$ for all $a \in K^c$ or equivalently, $V \supseteq K^c$.*

Proof. By definition $v: K^\times \to \Gamma$ is a group homomorphism whose kernel is the group U of units in V. Its image is abelian precisely when $U \supseteq K^c$; when this is so, then $V \supseteq K^c$, and conversely, when $V \supseteq K^c$, then $V^{-1} \supseteq K^c$, so $U = V \cap V^{-1} \supseteq K^c$. ∎

In order to describe extensions we shall need the notion of domination. Let K be a field and consider pairs $P = (R, \mathfrak{a})$ consisting of a subring R of K and a proper ideal $\mathfrak{a}$ of R. Given two such pairs P and $P' = (R', \mathfrak{a}')$, we shall say that P' *dominates* P, in symbols $P' \geq P$ or $P \leq P'$, if $R' \supseteq R$ and $\mathfrak{a}' \supseteq \mathfrak{a}$. If a pair $(R, \mathfrak{a})$ is such that $R \supseteq K^c$, then since K^c is a group, every element of K^c is a unit in R and so $K^c \cap \mathfrak{a} = \varnothing$ (because $\mathfrak{a}$ is proper). It will be convenient to single out the essential step in the construction as a separate lemma.

LEMMA 9.2.2. *Let K be a field, R a subring containing K^c and $\mathfrak{a}$ a proper ideal of R. Then R and $\mathfrak{a}$ are invariant in K. Further, there is a subring V with a proper ideal $\mathfrak{m}$ such that $(V, \mathfrak{m})$ is maximal among pairs dominating $(R, \mathfrak{a})$, and any such maximal pair $(V, \mathfrak{m})$ consists of a valuation ring in K and its maximal ideal.*

Proof. If $a \in R^\times$, $b \in K^\times$, then $b^{-1}ab = a \cdot a^{-1}b^{-1}ab \in R$, because all commutators lie in R. Thus R is invariant and the same argument applies to $\mathfrak{a}$.

Now consider the family of all pairs dominating $(R, \mathfrak{a})$. It is clear that this family is inductive, hence by Zorn's lemma there is a maximal pair, $(V, \mathfrak{m})$ say. By maximality, $\mathfrak{m}$ is a maximal ideal in V and since $V \supseteq R \supseteq K^c$, it follows that V and $\mathfrak{m}$ are invariant, so it only remains to show that V is a valuation ring. Thus we must show that V is a total subring of K. Take $c \in K^\times$; if $c \notin V$, then $V[c] \supset V$, so if the ideal $\mathfrak{m}'$ generated by $\mathfrak{m}$ in $V[c]$ were proper, we would have $(V[c], \mathfrak{m}') > (V, \mathfrak{m})$, in contradiction with the maximality of the latter. Hence $\mathfrak{m}' = V[c]$ and we have an equation

$$a_0 + a_1 c + \ldots + a_m c^m = 1, \quad a_i \in \mathfrak{m}. \tag{1}$$

Here we were able to collect powers of c on the right because of the invariance of $\mathfrak{m}$, using the equation $cr = crc^{-1} \cdot c$.

Similarly, if $c^{-1} \notin V$, we have an equation

$$b_0 + b_1 c^{-1} + \ldots + b_n c^{-n} = 1, \quad b_j \in \mathfrak{m}. \tag{2}$$

Let us choose m and n as small as possible and suppose that $m \geq n$, say. Multiplying (2) on the right by c^m, we obtain

$$(1 - b_0)c^m = b_1 c^{m-1} + \ldots + b_n c^{m-n}. \tag{3}$$

By the invariance of V we have $xc = c \cdot x\gamma$ for all $x \in V$, where $\gamma = \gamma(c)$ is an automorphism of V which maps $\mathfrak{m}$ onto itself. If we multiply (3) by $a_m \gamma^m$ on the right, we obtain

$$(1 - b_0)a_m c^m = b_1 (a_m \gamma)c^{m-1} + \ldots + b_n (a_m \gamma^n)c^{m-n}. \tag{4}$$

Now multiply (1) by $(1 - b_0)$ on the left and substitute from (4); we obtain an equation of the same form as (1), with m replaced by $m - 1$. This contradicts the minimality of m, and so either c or c^{-1} lies in V and V is total; hence V is a valuation ring, with maximal ideal $\mathfrak{m}$. ∎

This result allows us to prove the following extension theorem for abelian valuations:

THEOREM 9.2.3. *Let* $K \subset L$ *be an extension of fields. Given an abelian valuation* v *on* K, *there is an extension of* v *to an abelian valuation on* L *if and only if there is no equation*

$$\sum a_i c_i = 1, \text{ where } a_i \in K, \ v(a_i) > 0, \text{ and } c_i \in L^c. \tag{5}$$

Proof. For any $a_i \in K$ such that $v(a_i) > 0$ and any $c_i \in L^c$, an abelian extension w of v to L must satisfy

$$w(a_i c_i) = w(a_i) = v(a_i) > 0,$$

hence when (5) holds, we have

$$w(1) \geq \min_i \{w(a_i c_i)\} > 0,$$

a contradiction. Conversely, if no equation (5) holds, this means that if V is the valuation ring of v, with maximal ideal $\mathfrak{m}$, then $\mathfrak{m}L^c = \{\sum a_i c_i | a_i \in K, \ v(a_i) > 0, \ c_i \in L^c\}$ is a proper ideal in VL^c; thus $(VL^c, \mathfrak{m}L^c)$ is a pair in L and by Lemma 2.2 there is a maximal pair $(W, \mathfrak{n})$ dominating it. Further, W is a valuation ring satisfying $W \cap K \supseteq V$, $\mathfrak{n} \cap K \supseteq \mathfrak{m}$, hence $W \cap K = V$ and so W defines the desired extension. ∎

Sometimes we shall want to restrict the class of valuations even further; thus one might consider valuations with abelian value group and commutative residue-class field. From the bottom row of the diagram (7) of 9.1 we see that this means that $K^\times/U_1$ is an extension of one abelian group by another. It is convenient to take the special case where $K^{\times}/U_1$ is itself abelian and we shall define a valuation v on a field K to be *quasi-commutative* if the quotient $K^{\times}/U_1$ is abelian. Below are some equivalent ways of expressing this condition:

THEOREM 9.2.4. *Let K be a field with a valuation v, having valuation ring V, maximal ideal $\mathfrak{m}$ and group of 1-units U_1. Then the following conditions are equivalent:*
   (a) *v is quasi-commutative,*
   (b) *$K^c \subseteq 1 + \mathfrak{m} = U_1$,*
   (c) *$v(1 - a) > 0$ for all $a \in K^c$.*
Moreover, when (a)–(c) hold, then the value group and residue-class field are commutative.

This is an easy consequence of the definitions, with the last part following from the diagram (7) of 9.1. ∎

We remark that a quasi-commutative valuation on a field K is trivial precisely when K is commutative. For quasi-commutative valuations we have an extendibility criterion similar to that of Th. 2.3:

THEOREM 9.2.5. *Let $K \subset L$ be an extension of fields. Given a quasi-commutative valuation v on K, v has a quasi-commutative extension to L if and only if there is no equation in L of the form*

$$\sum a_i p_i + \sum b_j(q_j - 1) = 1, \tag{6}$$

where $a_i, b_j \in K$, $v(a_i) > 0$, $v(b_j) \geq 0$, $p_i, q_j \in L^c$.

Proof. If there is a quasi-commutative extension w of v to L, then for a_i, b_j, p_i, q_j as above we have

$$w\left(\sum a_i p_i + \sum b_j(q_j - 1)\right) \geq \min \{v(a_i) + w(p_i), v(b_j) + w(q_j - 1)\} > 0,$$

because $v(a_i) > 0$, $w(q_j - 1) > 0$. Hence there can be no equation of the form (6). Conversely, assume that no equation (6) holds and consider the set $\mathfrak{Q}$ of all expressions $\sum a_i p_i + \sum b_j(q_j - 1)$, where a_i, b_j, p_i, q_j are as before. It is easily checked that $\mathfrak{Q}$ is closed under addition and contains the maximal ideal of the valuation ring V of v in K. Moreover, $\mathfrak{Q}$ is invariant in L: $u^{-1}\mathfrak{Q}u = \mathfrak{Q}$ for $u \in L^{\times}$, because $u^{-1}a_i p_i u = a_i \cdot a_i^{-1} u^{-1} a_i u \cdot u^{-1} p_i u \in VL^c$, and similarly for the other terms. In the same way we can verify that $\mathfrak{Q}$ admits multiplication. We now define

$$W = \{a \in L \,|\, a\mathfrak{Q} \subseteq \mathfrak{Q}\}. \tag{7}$$

It is clear that W is a subring of L containing L^c and V, for if $c \in V$ and we multiply the expression on the left of (6) by c, we obtain $\sum ca_ip_i + \sum cb_j(q_j - 1)$; this is of the same form, because $v(ca_i) = v(c) + v(a_i) > 0$ and $v(cb_j) \geqslant 0$. Further, $\mathfrak{Q}$ is an ideal in W, because $c\mathfrak{Q} \subseteq \mathfrak{Q}$ for all $c \in W$, by the definition of W, and $\mathfrak{Q}c = c \cdot c^{-1}\mathfrak{Q}c = c\mathfrak{Q} \subseteq \mathfrak{Q}$. By hypothesis, $1 \notin \mathfrak{Q}$, hence $\mathfrak{Q}$ is a proper ideal of W. Thus W is a subring of L containing L^c and the valuation ring of v. By Lemma 2.2, there exists a maximal pair $(T, \mathfrak{p})$ dominating $(W, \mathfrak{Q})$, and T is a valuation ring such that $T \supseteq W \supseteq L^c$, while $1 + \mathfrak{p} \supseteq 1 + \mathfrak{Q} \supseteq L^c$. Hence the valuation w defined by T extends v and is quasi-commutative, by Th. 2.4. ∎

To apply this result we shall need fields of infinite degree over their centres, with a quasi-commutative valuation. Such fields may be constructed as follows. Let k be any commutative field of characteristic zero and form the Weyl field $K = W_1(k)$, generated by s, t over k with the defining relation $st - ts = 1$. Its centre is k, as we have seen in 6.1. Putting $x = s^{-1}$, we can write the above relation as

$$tx = x(t + x). \tag{8}$$

On K we have the x-adic valuation v, obtained by writing any element as a formal Laurent series in x with coefficients in $k(t)$ and taking the exponent of the lowest power of x. Thus any $f \in K$ has the form

$$f = x^{v(f)}f_0,$$

where f_0 is a power series in x with coefficients in $k(t)$ and non-zero term independent of x. We can pull the coefficients through to the right, using the shift map $\sigma\colon \varphi(t) \mapsto \varphi(t + x)$ of $k(t)$ into K based on (8). Although the image is not in $k(t)$, the terms of lowest degree remain unchanged, thus if $f_0 = a_0 + xa_1 + \ldots$, then $f_0^\sigma = a_0 + xa_1' + \ldots$ with the same a_0. We claim that v is quasi-commutative. Given $f, g \in K^\times$, let $f = x^r f_0$, $g = x^s g_0$ say; then their commutator is given by

$$(f, g) = f_0^{-1}x^{-r}g_0^{-1}x^{-s}x^r f_0 x^s g_0;$$

let $f_0 = a_0 + xf_1$, $g_0 = b_0 + xg_1$, so that $f_0^{\sigma^s} = a_0 + xf_1^*$, $g_0^{\sigma^r} = b_0 + xg_1^*$ for some f_1^*, g_1^*; then

$$(f, g) = (a_0 + xf_1)^{-1}(b_0 + xg_1^*)^{-1}(a_0 + xf_1^*)(b_0 + xg_1).$$

When we apply the residue-class map, the right-hand side reduces to 1, because the residue-class field $k(t)$ is commutative. Hence $v((f, g) - 1) > 0$ and this shows v to be quasi-commutative.

To illustrate Th. 2.5 we now show that the free field has a quasi-commutative valuation. We begin by proving an extension theorem:

THEOREM 9.2.6. *Let K be a field with an infinite centre C such that $[K:C] = \infty$ and let X be any set. Then any quasi-commutative valuation v on K can be extended to a quasi-commutative valuation on the free field $K_C \langle\!\langle X \rangle\!\rangle$.*

Proof. Writing $L = K_C \langle\!\langle X \rangle\!\rangle$, we have to show that there is no equation

$$\sum a_i p_i + \sum b_j (q_j - 1) = 1, \tag{9}$$

where $a_i, b_j \in K$, $v(a_i) > 0$, $v(b_j) \geqslant 0$, $p_i, q_j \in L^c$. Suppose we have such an equation. Since each p_i and q_j is a product of commutators, there is a finite set of elements of $L^\times$, $c_1, \ldots, c_N$ such that each p_i and q_j is a product of commutators formed from the cs. By the specialization lemma (Prop. 6.2.7) there is a specialization from L to K which is defined on all the cs and their inverses. By applying this specialization we obtain an equation (9) with $p_i, q_j \in K^c$, but this contradicts the fact that there is a quasi-commutative valuation on K. Hence no equation (9) can exist, so we can apply Th. 2.5 to obtain a quasi-commutative extension of v to L. ∎

In particular, the example just constructed allows us to form $W_1(k)_k \langle\!\langle X \rangle\!\rangle$ with a quasi-commutative valuation, and it has $k \langle\!\langle X \rangle\!\rangle$ as a subfield, by Prop. 5.4.3. We thus obtain

COROLLARY 9.2.7. *For any commutative field k of characteristic zero and any set X the free field $k \langle\!\langle X \rangle\!\rangle$ has a quasi-commutative valuation over k.* ∎

The restriction on the characteristic can be lifted by a slight modification of the construction (see Ex. 2).

In any ordered group a subset is called *convex* if with any $a \leqslant b$ it contains all x satisfying $a \leqslant x \leqslant b$. In the elementary theory most valuations are real-valued; since $\mathbf{R}$ as ordered group has no convex subgroups other than 0 and itself, it follows that no two real-valued valuations on a given field can be comparable. By contrast, in the general case it can happen that valuations are comparable, and this leads to an interesting decomposition theorem, which persists even in the non-commutative case, although there is then an extra condition.

Let K be a field with a valuation v, with valuation ring V, maximal ideal $\mathfrak{m}$ and residue-class field $K_v = V/\mathfrak{m}$. A second valuation v' on K with valuation ring V' is said to be *subordinate* to v, $v' \leqslant v$, if $V' \supseteq V$. Writing $\mathfrak{m}'$ for the maximal ideal of V' and U, U' for the unit groups in V, V', we see that $U' \supseteq U$, $\mathfrak{m}' \subseteq \mathfrak{m}$, and if the value groups are Γ, Γ'

respectively, then $\Gamma = K^\times/U$,

$$\Gamma' = K^\times/U' \cong (K^\times/U)/(U'/U) = \Gamma/\Delta,$$

where $\Delta = U'/U$ is a convex subgroup of Γ. Thus if $\varphi\colon \Gamma \to \Gamma'$ is the natural homomorphism, then $v'(x) = \varphi(v(x))$ for all $x \in K$. In this case we can decompose the valuation v' as φv and we see that its residue-class field $K_{v'}$ has a valuation with value group $\Delta = \ker \varphi$ and residue-class field K_v. The situation may be illustrated by the accompanying diagram.

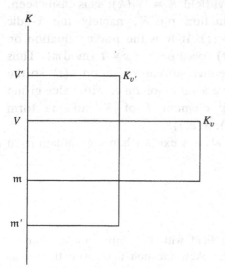

In the natural homomorphism $V' \to V'/\mathfrak{m}' = K_{v'}$ the ring V corresponds to a valuation ring in $K_{v'}$ with residue-class field K_v. We may think of the place $K \to K_v$ as being composed of the places $K \to K_{v'}$ and $K_{v'} \to K_v$. Conversely, if we have a valuation v' on K and a valuation on the residue-class field $K_{v'}$, we may ask when these valuations can be composed to yield a valuation on K to which v' is subordinate.

Consider first the commutative case. Thus K is a commutative field with a valuation v', valuation ring V' and residue-class field $K_{v'}$ and w is a valuation on $K_{v'}$ with valuation ring W. If V denotes the inverse image of W in K, then $V' \supseteq V \supset \mathfrak{m} \supseteq \mathfrak{m}'$, where $\mathfrak{m}$ is the set of non-units in V. Any element x of K not in V either lies in V', in which case $x^{-1} \in \mathfrak{m}$, or $x \notin V'$, in which case $x^{-1} \in \mathfrak{m}' \subseteq \mathfrak{m}$. So in any case V is a valuation ring with maximal ideal $\mathfrak{m}$. The valuation v of K corresponding to V is called the *concatenation* of v' and w; by definition we have $v' \leqslant v$.

If we consider the same situation for a skew field K, we find again that V is a total subring contained in V', but it may not be invariant. Let $\xi \in K_{v'}$, say $\xi = \bar{x}$, where $x \in K$ and the bar indicates the residue-class map. Then $v'(x) \geqslant 0$, hence for any $a \in K^\times$, $v(a^{-1}xa) = -v(a) + v(x) + v(a) \geqslant 0$, so $a^{-1}xa$ maps to an element of $K_{v'}$ which we shall denote by $\xi\alpha_a$. It is clear that for any $a \in K^\times$ the map $\alpha_a\colon \xi \mapsto \xi\alpha_a$ is an automorphism of $K_{v'}$ and $a \mapsto \alpha_a$ is a homomorphism $K^\times \to \operatorname{Aut}(K_{v'})$. We claim that the inverse image V of W in K is invariant precisely when

$$w(\xi) \geqslant 0 \Rightarrow w(\xi\alpha_a) \geqslant 0, \quad \text{for all } \xi \in K_{v'} \text{ and } a \in K^\times. \tag{10}$$

For in terms of K this condition just states that $x \in V \Rightarrow a^{-1}xa \in V$. Thus

(10) is necessary and sufficient for V to be a valuation ring, and hence to define a concatenation of v' and v. So we have

THEOREM 9.2.8. *Let K be a field with a valuation v, with residue-class field E. If w is a valuation on E, then the concatenation of v and w exists if and only if the induced automorphisms α_a $(a \in K^\times)$ of E preserve the sign of the valuation w, i.e. (10) holds.* ∎

As an illustration consider the Weyl field $K = W_1(k)$; as is easily seen, we have a quasi-commutative valuation on K, namely the x-adic valuation v with residue-class field $k(t)$. If w is the t-adic valuation on $k(t)$, then α_x is the identity on $k(t)$, because $t + x \equiv t$ (mod m). Thus every element of $K^\times$ induces the identity automorphism on $k(t)$, so the condition of Th. 2.8 holds and we have a valuation on K with value group $\mathbf{Z}^2$, ordered lexicographically. Any element f of $K^\times$ of the form $f = x^{v(f)}(a_0 + xf_1)$ has the value $(v(f), w(a_0))$.

More generally the concatenation always exists when the valuation on K is quasi-commutative.

Exercises

1. Let $F = k(t)$ be a rational function field with the shift automorphism σ: $f(t) \mapsto f(t + 1)$. Define $K = F(x; \sigma)$, the skew function field, with the x-adic valuation v. Show that v has an abelian value group and a commutative residue-class field but is not quasi-commutative.

2. Let k be a commutative field and let E be the Hilbert field on k, with relation $uv = vu^2$. Show that the x-adic valuation on E, where $x = u - 1$, is quasi-commutative.

3. Show that for a quasi-commutative valuation the concatenation with any valuation exists.

4. Let F be a commutative field of characteristic $\neq 2$ with an automorphism $\alpha \neq 1$ and put $K = F((t; \alpha))$. Verify that the t-adic valuation on K has residue-class field F, and use it to show that no equation $x^2 + xa_1 + a_2 = 0$ over K has more than two roots.

5. Let k be a commutative algebraically closed field and E a proper commutative extension of k. Show that there is a non-trivial valuation on E trivial on k and with residue-class field k.

6. Let K be a field with an abelian valuation v with residue-class field k which is

commutative algebraically closed. Given an extension L/K such that v extends to an abelian valuation w on L, show that w can be chosen so as to have the same residue-class field.

9.3 Matrix valuations on rings

We have seen in 9.1 that any valuation v on a commutative ring R leads to a residue-class ring $R/\mathfrak{a}$ which is an integral domain with a field of fractions to which v can be extended. Such a result cannot be expected for general rings; here the theory of Ch. 4 suggests that we define valuations on matrices over R, and in this way a generalization can indeed be found. However, we shall need to restrict ourselves to the case of an abelian value group, mainly because for matrices over a general ring the determinant function (Dieudonné determinant) has values in an abelian group.

DEFINITION. A *matrix valuation* on a ring R is a function V on $\mathfrak{M}(R)$ with values in $\Gamma \cup \{\infty\}$, where Γ is an abelian ordered group, satisfying the following conditions:

(MV.1) $VA = \infty$ *for every non-full matrix A,*
(MV.2) VA *is unchanged if a row or column of A is multiplied by* -1,
(MV.3) $V(A \oplus B) = VA + VB$,
(MV.4) $V(A \nabla B) \geqslant \min\{VA, VB\}$ *whenever the determinantal sum $A \nabla B$ is defined,*
(MV.5) $V1 = 0$.

It is clear that the restriction of a matrix valuation to R, *qua* set of 1×1 matrices, is a valuation on R, hence the set $\ker V = \{x \in R | Vx = \infty\}$ is an ideal of R and V induces a proper valuation on $R/\ker V$. More generally, the set $\operatorname{Ker} V = \{A \in \mathfrak{M}(R) | VA = \infty\}$ is a prime matrix ideal of R (see Ex. 1).

We note some immediate consequences of the definition.
1. *If $A \nabla B$ is defined and $VA \neq VB$, then $V(A \nabla B) = \min\{VA, VB\}$.*

For if $C = A \nabla B$, then $VC \geqslant \min\{VA, VB\}$ by (MV.4). By hypothesis, $VA \neq VB$, say $VA < VB$, and so $VC \geqslant VA$. Suppose that $VC > VA$ and write B^- for the matrix obtained from B by changing the sign of the column operated on, so that $A = C \nabla B^-$; using (MV.4) once more, we find $VA \geqslant \min\{VC, VB\}$, where we have used (MV.2). Since $VC > VA$, it follows that $VA \geqslant VB$, which is a contradiction; hence $VC = VA$, as claimed.

2. $V(AE) = V(EA) = VA$ *for any elementary matrix E.*
This amounts to saying that VA is unchanged if we add a left multiple of one row or a right multiple of one column to another. Suppose we add a multiple of the second column to the first:

$$(A_1 + A_2\lambda \ A_2 \ A_3 \ \ldots \ A_n) = A \ \nabla \ (A_2 \ \ldots \ A_n)\begin{pmatrix} \lambda & \\ & I \\ 0 & \end{pmatrix}.$$

Here the second matrix on the right is non-full, so its value is ∞ and the other two must have equal value, by 1.

Let us call two matrices *E-associated* if we can pass from one to the other by elementary operations on rows and columns, i.e. multiplying on the left or right by elementary matrices, and *stably E-associated* if they become E-associated after forming diagonal sums with I. Then we have as an immediate consequence of 2 (and (MV.3, 5)):

3. $VA = VB$ *whenever B is stably E-associated to A.*

4.
$$V\begin{pmatrix} A & C \\ 0 & B \end{pmatrix} = VA + VB = V\begin{pmatrix} A & 0 \\ D & B \end{pmatrix}.$$

Let $B = (b \ \ B')$, $C = (c \ \ C')$; then

$$\begin{pmatrix} A & C \\ 0 & B \end{pmatrix} = \begin{pmatrix} A & 0 & C' \\ 0 & b & B' \end{pmatrix} \nabla \begin{pmatrix} A & c & C' \\ 0 & 0 & B' \end{pmatrix};$$

if A, B have orders m, n respectively, then the second matrix on the right has an $n \times (m + 1)$ block of zeros and so is not full, hence V has the same value on the other two matrices. In this way we can replace all the columns of C by 0 and finally obtain $V(A \oplus B)$, which has the value $VA + VB$ by (MV.3). The other equation follows in the same way.

5. $V(AB) = VA + VB$.
This follows because in the following chain all matrices have the same value:

$$AB, \begin{pmatrix} AB & 0 \\ 0 & I \end{pmatrix}, \begin{pmatrix} AB & A \\ 0 & I \end{pmatrix}, \begin{pmatrix} A & -AB \\ I & 0 \end{pmatrix}, \begin{pmatrix} A & 0 \\ I & B \end{pmatrix}, \begin{pmatrix} A & 0 \\ 0 & B \end{pmatrix}.$$

A ring homomorphism can be used to transfer matrix valuations:

PROPOSITION 9.3.1. *Let $f: R \to R'$ be a ring homomorphism. If V is a matrix valuation on R', then V^f, defined by*

$$V^f A = V(Af) \quad \text{for } A \in \mathfrak{M}(R),$$

is a matrix valuation on R.

V^f will be called the matrix valuation defined by *pullback* along f. The proof is a routine verification, which may be left to the reader. ∎

If R is a commutative ring with a valuation v, then the function V defined on $\mathfrak{M}(R)$ by the equation

$$VA = v(\det A) \tag{1}$$

is a matrix valuation, as is easily checked. The correspondence $v \leftrightarrow V$ is a bijection between matrix valuations and valuations, taking v to be defined in terms of V as the restriction $V|R$. For this reason matrix valuations are not needed in the commutative case. In the general case this argument cannot be used, because we do not then have a determinant. But for skew fields we have at least the Dieudonné determinant. We briefly recall the definition (see A.3, 11.2 for details).

Let K be any field; the quotient of its multiplicative group $K^\times$ by its derived group K^c is called the *abelianization* and is written K^{ab}. Thus we have

$$K^{ab} = K^\times/K^c. \tag{2}$$

By definition, the natural map $\delta: K^\times \to K^{ab}$ is universal for homomorphism to abelian groups. The basic theorem on Dieudonné determinants states that for any field K and any $n \geqslant 1$ there is a homomorphism

$$\mathrm{Det}: \mathbf{GL}_n(K) \to K^{ab}, \tag{3}$$

which is universal for homomorphisms from $\mathbf{GL}_n(K)$ to abelian groups. The image of a matrix A is written $\mathrm{Det}(A)$ or $D(A)$ and is called the *Dieudonné determinant* of A. To find its value we note that for any $A \in \mathbf{GL}_n(K)$ there exist products of elementary matrices U, V, briefly *E-matrices*, such that

$$UAV = I \oplus \alpha, \quad \text{where } \alpha \in K^\times. \tag{4}$$

Here α will in general depend on the mode of reduction, but its residue-class mod K^c is completely determined by A, and it is just the Dieudonné determinant of A (see A.3, Th. 11.2.6, p. 442).

Our aim will be to show that any abelian valuation on a field gives rise to a matrix valuation via the Dieudonné determinant; here we shall need the following lemma to replace the additivity of the ordinary determinant:

LEMMA 9.3.2. *Let $C = A \nabla B$ be a matrix equation over a field, where ∇ is taken over the first column, say, and A is non-singular. Then there are invertible matrices P, Q, where $Q = 1 \oplus Q'$ and P is an E-matrix, such that*

$$A = P\begin{pmatrix} a_{11} & 0 \\ a & I \end{pmatrix}Q, \quad B = P\begin{pmatrix} b_{11} & 0 \\ b & I \end{pmatrix}Q, \quad C = P\begin{pmatrix} c_{11} & 0 \\ c & I \end{pmatrix}Q, \quad (5)$$

where a, b, c are columns satisfying

$$c = a + b, \text{ and } c_{11} = a_{11} + b_{11}. \quad (6)$$

Proof. We note that the relation $C = A \nabla B$ is unaffected by elementary row operations (multiplying on the left by an E-matrix P).

Since A is non-singular, we can permute the rows so that the cofactor of the $(1, 1)$-entry in A is non-singular and now by row operations we can reduce the first row of A to $(a_{11} \ 0 \ \ldots \ 0)$. Thus we have

$$A = P\begin{pmatrix} a_{11} & 0 \\ a & A_{11} \end{pmatrix} = P\begin{pmatrix} a_{11} & 0 \\ a & I \end{pmatrix}\begin{pmatrix} 1 & 0 \\ 0 & A_{11} \end{pmatrix}.$$

where A_{11} is non-singular. Writing $Q' = A_{11}$, we see that A has the form (5), where $Q = 1 \oplus Q'$, and similarly for B, C and now (6) follows. ∎

To deal with the case where A is singular, we prove a more general result, sometimes known as the magic lemma:

LEMMA 9.3.3. *Let R be a ring, Σ a set of square matrices over R such that $AB \in \Sigma$ if and only if $A, B \in \Sigma$ and $\varphi: \Sigma \to \Gamma$ a map to an abelian group Γ such that $\varphi(AB) = \varphi(A) + \varphi(B)$ and $\varphi(E) = 0$ for any elementary matrix E. Then in any matrix equation*

$$C = A \nabla B,$$

where $B, C \in \Sigma$ and A is non-full, we have $\varphi(B) = \varphi(C)$.

Proof. Assume that the determinantal sum is relative to the first column and let A_1, B_1 be the first columns of A, B. Since A is non-full, we have $A = PQ$, where P has $n - 1$ columns and Q has $n - 1$ rows if A is $n \times n$. Writing $Q = (Q_1 \ Q')$, where $Q' \in R_{n-1}$, we have

$$B = (B_1 \ PQ') = (B_1 \ P)(1 \oplus Q'). \quad (7)$$

Since $A_1 = PQ_1$, we can write C as

$$C = (A_1 + B_1 \ PQ') = (PQ_1 + B_1 \ P)(1 \oplus Q')$$

$$= (B_1 \ P)\begin{pmatrix} 1 & 0 \\ Q_1 & I \end{pmatrix}\begin{pmatrix} 1 & 0 \\ 0 & Q' \end{pmatrix};$$

now a comparison with (7) shows that $\varphi(C) = \varphi(B)$. ∎

We can now establish the connexion between valuations and matrix valuations for fields:

THEOREM 9.3.4. *Let K be a field with an abelian valuation v. Then v can be extended to a matrix valuation V on K by defining*

$$VA = \begin{cases} v(\operatorname{Det} A) & \text{if A is non-singular,} \\ \infty & \text{if A is singular.} \end{cases} \tag{8}$$

This correspondence $v \leftrightarrow V$ is a bijection.

Proof. That we have a bijection is clear since v is determined by V as its restriction to K; it remains to verify (MV.1–5). (MV.1) and (MV.5) are clear from the definition and (MV.2, 3) follow because Det is a homomorphism. To prove (MV.4) suppose first that A is non-singular. Then we have expressions (5) for A, B, C and P is an E-matrix, so $VP = 0$, while $\operatorname{Det}(A) = a_{11}^{\delta}q$, where $q = \operatorname{Det}(Q)$. Hence $VA = v(\operatorname{Det} A) = v(a_{11}) + v(q)$, and similarly, $VB = v(b_{11}) + v(q)$, $VC = v(c_{11}) + v(q)$. By (6) we have $c_{11} = a_{11} + b_{11}$, hence

$$v(c_{11}) \geqslant \min \{v(a_{11}), v(b_{11})\},$$

and so $VC \geqslant \min \{VA, VB\}$, which proves (MV.4) when A is non-singular. When A is singular, (MV.4) follows by Lemma 3.3. Thus V is indeed a matrix valuation on K. ■

Let R be any ring and K an epic R-field. Any valuation v on K defines a matrix valuation V on K, by Th. 3.4, and by pullback we get a matrix valuation on R. We now turn to the inverse problem: Given a matrix valuation V on R, how can we find an epic R-field with a valuation giving rise to V?

Let R be a ring with a matrix valuation V; we define its *singular kernel* as

$$\mathscr{P} = \operatorname{Ker} V = \{A \in \mathfrak{M}(R) | VA = \infty\}. \tag{9}$$

It is easily verified that $\mathscr{P}$ is a prime matrix ideal. We form its localization $R_{\mathscr{P}}$ with residue-class field $K = R/\mathscr{P}$. In order to define a valuation on K we first define a function W on the set S of all admissible matrices $A = (A_0 \quad A_* \quad A_\infty)$ by putting

$$WA = V(A_0 \quad A_*) - V(A_* \quad A_\infty). \tag{10}$$

Clearly this is a well-defined element of $\Gamma \cup \{\infty\}$ which by (9) is equal to ∞ precisely when $(A_0 \quad A_*) \in \mathscr{P}$. We recall from 4.2 that if A, B are admissible for p, q respectively, in K, then the matrix

$$A \cdot B = \begin{pmatrix} 0 & 0 & A_0 & A_* & A_\infty \\ B_0 & B_* & B_\infty & 0 & 0 \end{pmatrix} \qquad (11)$$

is admissible for pq. This defines an associative multiplication on S, so S forms a semigroup under this multiplication. The subset S_1 of all admissible matrices whose numerator is not in $\mathscr{P}$ is again closed under the multiplication (11), as well as the operation

$$\bar{A} = (A_\infty \quad A_* \quad A_0), \qquad (12)$$

which defines p^{-1} if A defines p.

Let us write $f \colon S_1 \to K^\times$ for the map which associates with any $A \in S_1$ the element $p \in K^\times$ which is determined by A. The above remarks show that $(A \cdot B)^f = A^f B^f$ and $\bar{A}^f = (A^f)^{-1}$. We also note that

$$W\bar{A} = V(A_\infty \quad A_*) - V(A_* \quad A_0) = -WA$$

and

$$W(A \cdot B) = V\begin{pmatrix} 0 & 0 & A_0 & A_* \\ B_0 & B_* & B_\infty & 0 \end{pmatrix} - V\begin{pmatrix} 0 & A_0 & A_* & A_\infty \\ B_* & B_\infty & 0 & 0 \end{pmatrix}$$

$$= V(B_0 \quad B_*) + V(A_0 \quad A_*) - V(B \quad B_\infty) - V(A_* \quad A_\infty)$$

$$= WA + WB.$$

We thus have two homomorphisms from S_1 as a semigroup, namely $W \colon S_1 \to \Gamma$ and $f \colon S_1 \to K^\times$, and our aim is to define a homomorphism from $K^\times$ to Γ by factoring W by f. Firstly we note that f is surjective; we claim further that $\ker f \subseteq \ker W$. For if $A^f = 1$, then A is admissible for 1 and the equation $Au = 0$ with $u_\infty = u_0 = 1$ shows that $(A_0 + A_\infty \quad A_*) \in \mathscr{P}$. Hence $V(A_0 + A_\infty \quad A_*) = \infty$ and so

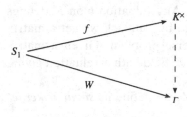

$$V(A_0 \quad A_*) = V(A_\infty \quad A_*) = V(A_* \quad A_\infty),$$

It follows that $WA = 0$, as claimed.

Therefore, if $A^f = B^f$, then $(A \cdot \bar{B})^f = A^f (B^f)^{-1} = 1$, so $W(A \cdot \bar{B}) = 0$ and $WA = -W\bar{B} = WB$. Thus we have indeed a homomorphism v from $K^\times$ to Γ such that $v(A^f) = WA$. Explicitly, if $p \in K^\times$ is determined by the admissible matrix A, we put

$$v(p) = WA = V(A_0 \quad A_*) - V(A_* \quad A_\infty).$$

Further we put $v(0) = \infty$ and claim that v is a valuation on K. (V.1) is clear and (V.2) follows since v is a homomorphism. To verify (V.3), suppose first that $v(p) \geq 0$. Then any admissible matrix A for p satisfies $V(A_0 \quad A_*) \geq V(A_* \quad A_\infty)$. By (MV.4) it follows that $V(A_0 + A_\infty \quad A_*) \geq V(A_* \quad A_\infty)$, and since $(A_0 + A_\infty \quad A_* \quad A_\infty)$ is an admissible matrix for $p - 1$, we have shown that $v(p) \geq 0$ implies $v(p - 1) \geq 0$.

Now to prove (V.3), take $a, b \in K$, where $v(a) \leq v(b)$ say. If $v(a) = \infty$, then $a = 0$ and (V.3) is clear; otherwise we have $v(ba^{-1}) \geq 0$, and so

$$v(b - a) = v(ba^{-1} - 1) + v(a) \geq v(a) = \min\{v(a), v(b)\},$$

so (V.3) is established.

Thus we have indeed a valuation v on K. By Th. 3.4 it gives rise to a matrix valuation on K which will again be denoted by v, and if $\lambda: R \to K$ is the natural homomorphism, then it is clear that v^λ is just V. Thus we have proved the following existence theorem for matrix valuations:

THEOREM 9.3.5. *Let R be any ring. Then each matrix valuation V on R determines an associated epic R-field K with an abelian valuation v, and conversely, every epic R-field with an abelian valuation v determines a matrix valuation on R in this way. This correspondence between abelian valuations on epic R-fields and matrix valuations on R is bijective, and the singular kernel of the R-field agrees with the singular kernel of the corresponding matrix valuation.* ∎

Exercises

1. Verify that for any matrix valuation V, the set $\operatorname{Ker} V = \{A \in \mathfrak{M}(R) | VA = \infty\}$ is a prime matrix ideal. Show that

$$\mathcal{P}_0 = \{A \in \mathfrak{M}(R) | VA > 0\}$$

is also a prime matrix ideal and that $\bigcap \mathcal{P}_0^n \subseteq \operatorname{Ker} V$, where $\mathcal{P}_0^n$ is defined as the matrix ideal generated by $A_1 \oplus \ldots \oplus A_n$ ($A_i \in \mathcal{P}_0$).

2. Let R be a weakly finite ring. Use the magic lemma to show that the determinantal sum of an invertible and a non-full matrix is invertible.

3. Show that in the free field $k \langle\!\langle x, y \rangle\!\rangle$, x cannot be expressed as a sum of products of multiplicative commutators. (Hint. Find an abelian valuation such that $v(x) \neq 0$.)

4. If the value group Γ is not assumed abelian, show that a matrix valuation defined by (MV.1–5) has nevertheless commuting values. (Hint. Use the fact that $A \oplus B$ is E-associated to $B \oplus A$.)

9.4 Subvaluations and matrix subvaluations

There is a basic method of constructing valuations via an auxiliary function called a subvaluation, rather in the way linear functionals are constructed from sublinear functionals by the Hahn–Banach theorem. In the commutative case this route is hardly ever used, probably because there the valuations are usually known explicitly, but that is not so for skew fields. It will illustrate the construction if we first describe it for commutative fields. As a matter of fact, subvaluations were first introduced in the 1930s by Mahler (under the name 'pseudo-valuations'), who classified all pseudo-valuations on algebraic number fields.

Let K be a commutative field and A a subring generating it. As we saw in Prop. 1.1, any valuation v on K is completely determined by its values on A; more generally, given any valuation v on a commutative ring A, we can use v to define a valuation on the epic A-field generated by $A/\ker v$. But if the values of v on a generating set of A are assigned, it will not in general be possible to reconstruct v from them. We therefore proceed as follows. ·

Let X be a generating set of A as a commutative ring, take any family of real numbers indexed by X, $\lambda(x)$ $(x \in X)$, and define

$$v(a) = \sup\left\{\sum_1^n \lambda(x_i) | a = x_1 \ldots x_n, x_i \in X\right\}. \tag{1}$$

The function v is defined for all $a \in A$ that can be written as products of elements of X. It takes values in $\mathbf{R} \cup \{\infty\}$ and the definition (1) is so designed that for any a, b on which v is defined,

$$v(ab) \geq v(a) + v(b). \tag{2}$$

This rule, which is easily checked, will replace the law (V.2). We express (2) by saying that v is *submultiplicative*. In order to have a function defined on all of A, satisfying (V.3), we now define μ by

$$\mu(a) = \sup\{\min_i (v(b_i) | a = b_1 + \ldots + b_r)\}. \tag{3}$$

Then μ still satisfies (2) and in addition,

$$\mu(a + b) \geq \min\{\mu(a), \mu(b)\}.$$

So μ has the following properties:
(SV.1) $\mu(a)$ *is defined for all* $a \in A$ *and* $\mu(a) \in \mathbf{R}$ *or* $\mu(a) = \infty$,
(SV.2) $\mu(ab) \geq \mu(a) + \mu(b)$,
(SV.3) $\mu(a + b) \geq \min\{\mu(a), \mu(b)\}$.
A function μ on a ring A satisfying (SV.1–3) is called a *subvaluation* on

A. If we had equality in (SV.2) and $\mu(a)$ is not identically ∞, we would have a valuation. The subvaluation μ will be called *radical* if

$$\mu(a^n) = n\mu(a) \text{ for all } a \in A \text{ and all } n \geq 1. \tag{4}$$

With every subvaluation μ we associate the least radical subvaluation majorizing it, called its *root*, which may be defined by

$$\mu^*(a) = \lim_{n \to \infty} \frac{1}{n}\mu(a^n). \tag{5}$$

This limit always exists, for by (SV.2) we have $\mu(a^{n+1}) \geq \mu(a^n) + \mu(a)$, hence

$$\frac{\mu(a^{n+1})}{n+1} \geq \frac{n}{n+1}\frac{\mu(a^n)}{n} + \frac{1}{n+1}\mu(a).$$

This shows that $\mu(a^n)/n$ is ultimately monotone increasing and so has a limit, possibly $+\infty$, as $n \to \infty$. Moreover, μ^* is easily seen to be a subvaluation and $\mu^*(a^r) = \lim(\mu(a^{rm})/n) = r \cdot \lim(\mu(a^n)/n) = r \cdot \mu^*(a)$. If f is any radical subvaluation majorizing μ, then $f(a) \geq \mu(a)$, hence $f(a) = f(a^n)/n \geq \mu(a^n)/n$ and on passing to the limit we find that $f(a) \geq \mu^*(a)$. Thus μ^* is indeed the least radical subvaluation majorizing μ.

As we remarked, it is possible that $\mu^*(a) = \infty$; but then from the way μ was defined, we might have $\mu(a) = \infty$ for all a, even when all the $\lambda(x)$ are bounded. However, for a subvaluation with a finite root we obtain a whole family of valuations by the following existence theorem:

THEOREM 9.4.1. *Let μ be any subvaluation on a commutative ring R and denote by μ^* its root defined by (5). Then*

$$\mu^*(a) = \inf\{v(a)|v(a) \geq \mu(a)\}, \tag{6}$$

where v ranges over all valuations on R majorizing μ.

The proof does not use the commutativity of R, but as we saw earlier, in the general case it does not lead to a valuation on an epic R-field. We shall therefore reformulate the result in terms of matrix valuations, with a proof which will include the above result.

Our first task is to define an analogue of subvaluations. By a *matrix subvaluation* on a ring R we understand a real-valued function μ on $\mathfrak{M}(R)$ satisfying (MV.1, 2, 4) and in place of (MV.3) and (MV.5),

(MV.3') $\mu(A \oplus B) \geq \mu A + \mu B$,

(MV.5') $\mu(1_0) = 0$,

where 1_0 is the unique 0×0 matrix over R. A function satisfying (MV.3') is again called *submultiplicative*.

A diagonal sum $A_1 \oplus \ldots \oplus A_n$ with $A_i = A$ will be abbreviated as $\oplus^n A$. Now a matrix subvaluation μ is said to be *radical* if it satisfies

$$\mu\left(\overset{n}{\underset{}{\oplus}} A\right) = n \cdot \mu A \quad \text{for all } A \in \mathfrak{M}(R) \text{ and } n = 1, 2, \ldots . \tag{7}$$

As before we can from any matrix subvaluation μ construct its *root*,

$$\mu^*(A) = \lim_{n \to \infty} \frac{\mu\left(\overset{n}{\underset{}{\oplus}} A\right)}{n}, \tag{8}$$

which is the least radical matrix subvaluation majorizing μ.

A matrix subvaluation μ is said to be *regular* at a matrix A if

$$\mu(A \oplus X) = \mu A + \mu X \quad \text{for all } X \in \mathfrak{M}(R). \tag{9}$$

If μ is regular at every matrix of a subset S of $\mathfrak{M}(R)$, it is said to be *regular* on S. Such a function is *multiplicative* on S, i.e. (9) holds for all $A, X \in S$.

Our aim will be to construct a matrix subvaluation majorizing μ that is regular on a prescribed set. Clearly a matrix valuation is just a matrix subvaluation that is ∞ on some prime matrix ideal and regular on its complement in $\mathfrak{M}(R)$. So our aim will be gradually to enlarge the set on which our given matrix subvaluation is regular.

Let μ be any matrix subvaluation on a ring R and define the *domain* of μ as

$$\Sigma = \{A \in \mathfrak{M}(R) | \mu A < \infty\}.$$

We fix a subset S of Σ and define a function μ' by

$$\mu'(X) = \sup\left\{\mu(X \oplus A_1 \oplus \ldots \oplus A_r) - \sum_1^r \mu(A_i)\right\}, \quad \text{for any } X \in \mathfrak{M}(R), \tag{10}$$

where $\{A_1, \ldots, A_r\}$ ranges over all finite subsets of S. Since $S \subseteq \Sigma$, μ' is certainly defined, if we allow the value ∞; it is called the *regularization* of μ over S. By (MV.3$'$) we have

$$\mu'(X) \geqslant \mu(X). \tag{11}$$

We claim that μ' satisfies (MV.1, 2, 4, 3$'$).

(MV.1 and 2) are clear because they hold for μ. To prove (MV.4), we note that

$$(X \nabla Y) \oplus A = (X \oplus A) \nabla (Y \oplus A),$$

as is easily checked, where the determinantal sums are both over the
same column. Let us write $A = \bigoplus_1^r A_i$, $B = \bigoplus_1^s B_j$; then we have

$$\mu((X \nabla Y) \oplus A \oplus B) \geqslant \min \{\mu(X \oplus A \oplus B), \mu(Y \oplus A \oplus B)\}$$

$$\geqslant \min \left\{\mu(X \oplus A) + \sum_1^s \mu(B_j), \mu(Y \oplus B) + \sum_1^r \mu(A_i)\right\}.$$

Hence

$$\mu((X \nabla Y) \oplus A \oplus B) - \sum \mu(A_i) - \sum \mu(B_j)$$

$$\geqslant \min \left\{\mu(X \oplus A) - \sum \mu(A_i), \mu(Y \oplus B) - \sum \mu(B_j)\right\}.$$

If we now take the sup as the A_i, B_j range over S for varying r, s we find
that the right-hand side becomes $\min \{\mu'(X), \mu'(Y)\}$, while the left-hand
side is at most $\sup \{\mu((X \nabla Y) \oplus C) - \sum \mu(C_k)\}$, where $C = \bigoplus_1^t C_k$, and
this is just $\mu'(X \nabla Y)$. Thus we reach the conclusion

$$\mu'(X \nabla Y) \geqslant \min \{\mu'(X), \mu'(Y)\},$$

which proves (MV.4). Finally, to establish (MV.3') we have, with A, B
as before,

$$\mu(X \oplus Y \oplus A \oplus B) \geqslant \mu(X \oplus A) + \mu(Y \oplus B).$$

Hence

$$\mu(X \oplus Y \oplus A \oplus B) - \sum \mu(A_i) - \sum \mu(B_j)$$

$$\geqslant \mu(X \oplus A) - \sum \mu(A_i) + \mu(Y \oplus B) - \sum \mu(B_j),$$

and taking suprema as before, we obtain

$$\mu'(X \oplus Y) \geqslant \mu'(X) + \mu'(Y),$$

i.e. (MV.3'). By contrast, (MV.5') need not hold for μ'; in fact, as the
definition of μ' in (10) shows, (MV.5') holds precisely when μ is
multiplicative on S. Our aim will be to use μ' to enlarge the domain of
regularity. As a first step we shall show how to manufacture a matrix
subvaluation that is regular on S from one that is merely multiplicative on
S. Let us call a set of matrices *diagonally closed* if it contains 1_0 and is
closed under diagonal sums.

LEMMA 9.4.2. *Let μ be a matrix subvaluation on a ring R, with domain
Σ. Then for any subset S of Σ,*

$$\mu'(X \oplus C) = \mu'(X) + \mu(C) \quad \text{for all } X \in \mathfrak{M}(R), C \in S, \quad (12)$$

where μ' is the regularization of μ on S. If, further, S is diagonally closed, then the following conditions are equivalent:

  (a) *there exists a matrix subvaluation μ_S regular on S and such that*

$$\mu_S|S = \mu|S \text{ and } \mu X \leq \mu_S X$$

$$\leq \sup_A \{\mu(X \oplus A) - \mu A\} \text{ for all } X \in \mathfrak{M}(R), A \in S,$$

$$(13)$$

  (b) *μ is multiplicative on S,*
  (c) *the regularization μ' satisfies $\mu'(1_0) < \infty$,*
When (a)–(c) hold, μ_S is the regularization of μ on S. If, moreover, μ is radical, then so is μ_S.

Proof. For any $C \in S$ we have, by the definition of μ',

$$\mu'(X \oplus C) = \sup \left\{ \mu(X \oplus C \oplus A_1 \oplus \ldots \oplus A_r) - \sum_1^r \mu(A_i) \right\}. \quad (14)$$

In (10) the expression in brackets increases if we enlarge the family $A_1, \ldots, A_r$. Since we are taking the supremum, this will be unchanged if we only consider families including C. Thus we can write (10) as

$$\mu'(X) = \sup \left\{ \mu(X \oplus C \oplus A_1 \oplus \ldots \oplus A_r) - \mu C - \sum_1^r \mu(A_i) \right\}, \quad (15)$$

and a comparison with (14) shows that

$$\mu'(X \oplus C) = \mu'(X) + \mu C,$$

which is the equation (12).

  Assume now that S is diagonally closed and that (a) holds. Then μ_S is multiplicative on S, hence so is μ, by (13), and (b) follows. If (b) holds, then from the definition (10) we see that $\mu'(1_0) = 0$ and so (c) follows. Finally, when (c) holds and we define

$$\mu_S(X) = \mu'(X) - \mu'(1_0) \quad \text{for all } X \in \mathfrak{M}(R), \quad (16)$$

then by putting $X = 1_0$ in (12), we obtain

$$\mu_S(C) = \mu(C) \quad \text{for all } C \in S, \quad (17)$$

which proves the first part of (13). Inserting this value in (12) and remembering (16), we find that

$$\mu_S(X \oplus C) = \mu_S(X) + \mu_S(C) \quad \text{for } X \in \mathfrak{M}(R), C \in S.$$

Thus μ_S is regular on S, and hence is multiplicative on S; by (17), so is μ, and so $\mu'(1_0) = 0$, whence $\mu_S = \mu'$ by (16). Now for $A = \bigoplus_1^r A_i$, where $A_i \in S$, we have

$$\mu(X \oplus A) - \mu A = \mu(X \oplus A_1 \oplus \ldots \oplus A_r) - \sum \mu(A_i)$$

$$\geqslant \mu X + \mu(A_1 \oplus \ldots \oplus A_r) - \sum \mu(A_i).$$

Taking the sup on both sides, we obtain $\mu'(X) = \sup \{\mu(X \oplus A) - \mu A\} \geqslant \mu X$, i.e.

$$\mu X \leqslant \mu_S X \leqslant \sup \{\mu(X \oplus A) - \mu A\},$$

and this proves the rest of (13).

It remains to show that μ_S is a matrix subvaluation. Now μ_S equals μ', which satisfies (MV.1, 2, 4) and also (MV.3' and 5'), because $\mu'(1_0) = 0$; so μ_S is a matrix subvaluation, as claimed.

We saw that when (c) holds, then $\mu'(1_0) = 0$, so $\mu_S = \mu'$ by (16). Suppose now that μ is radical. In (10) replace X by $\bigoplus^n X$ and bear in mind that the quantity in brackets can only increase when $\{A_1, \ldots, A_r\}$ is enlarged. We thus have

$$\mu'\left(\bigoplus^n X\right) = \sup \left\{\mu\left(\bigoplus^n (X \oplus A_1 \oplus \ldots \oplus A_r)\right) - n \cdot \sum \mu(A_i)\right\}.$$

Since μ is radical, we obtain

$$\mu'\left(\bigoplus^n X\right) = n \cdot \sup \left\{\mu(X \oplus A_1 \oplus \ldots \oplus A_r) - \sum \mu(A_i)\right\} = n \cdot \mu'(X),$$

and this shows $\mu_S = \mu'$ to be radical as asserted. ∎

We can use this result as follows to enlarge the domain of regularity. Suppose that μ is a radical matrix subvaluation and S is a diagonally closed subset of its domain Σ, on which μ is multiplicative. By Lemma 4.2 we can form its regularization μ_S on S and this will be regular on S and again radical. Now choose any matrix $C \in \Sigma \setminus S$ and let S' be the least diagonally closed subset containing C and S. Any member of S' is a diagonal sum of terms in S or equal to C, and so is E-associated to $\bigoplus^n C \oplus A$ for some $A \in S$ and $n \geqslant 0$. By regularity we have

$$\mu_S\left(\bigoplus^n C \oplus A\right) = \mu_S\left(\bigoplus^n C\right) + \mu_S(A)$$

$$= n \cdot \mu_S(C) + \mu_S(A).$$

Thus μ_S is multiplicative on S'. If we adjoin all the matrices in Σ in turn we obtain in this way a matrix valuation majorizing μ; of course this valuation may well depend on the order in which the matrices are adjoined.

We well-order Σ as $\{A_\alpha | \alpha < \tau\}$ for some ordinal τ and write S_α for the least diagonally closed set containing all A_β for $\beta < \alpha$; further we write μ_α for μ_{S_α}. Then μ_τ is multiplicative on Σ and so is a matrix valuation; moreover, we have $\mu_\tau \geqslant \mu$ and if μ is multiplicative on some diagonally closed subset S of Σ, then $\mu_\tau = \mu$ holds on S. We thus have

THEOREM 9.4.3. *Let μ be a radical matrix subvaluation on a ring R, with domain Σ and multiplicative on some diagonally closed subset S of Σ. Then there is a matrix valuation v on R with domain contained in Σ, such that $v|S = \mu|S$ and*

$$\mu(X) \leqslant v(X) \leqslant \sup_A \{\mu(X \oplus A) - \mu(A)\}, \quad \text{where } A \text{ ranges over } S.$$

For the proof we need only choose a well-ordering of Σ in which the matrices in S come first, and then apply Lemma 4.2. ∎

This theorem holds in particular if we take S to consist of all diagonal sums of a fixed matrix C. In this way we see that there is for any matrix C and any matrix subvaluation μ satisfying $\mu(\bigoplus^n C) = n \cdot \mu C$, a matrix valuation majorizing μ and agreeing with μ on C. For when we form the root μ^* to obtain a radical matrix subvaluation, we clearly have $\mu^*(C) = \mu(C)$ and so we can apply Th. 4.3.

We can now achieve our objective, of finding an analogue of Th. 4.1.

THEOREM 9.4.4. *Let μ be any matrix subvaluation on a ring R. Then its root μ^* satisfies*

$$\mu^* = \inf\{V | V \geqslant \mu\}, \tag{18}$$

where V ranges over all real-valued matrix valuations majorizing μ. Further, if μ is multiplicative on a diagonally closed subset S of $\mathfrak{M}(R)$, then there is a matrix valuation $V \geqslant \mu$ such that $V|S = \mu|S$.

Proof. It is easily checked that both sides of (18) are radical matrix subvaluations, and for any matrix valuation V such that $V \geqslant \mu$, we have $V \geqslant \mu^*$, hence $\inf\{V | V \geqslant \mu\} \geqslant \mu^*$. If this inequality were strict at C, say, then $\mu^*(C) < VC$ for all V considered, but $\mu(C) \leqslant \mu^*(C) \leqslant \infty$, so by what has been said, we could find a matrix valuation $V \geqslant \mu$ such that $\mu^*(C) = V(C)$, and this contradicts the hypothesis. Thus we have equality in (18). ∎

We remark that in all the results of this section we have had to limit ourselves to *real*-valued valuations because of the need to form suprema, which presupposes that we are operating in **R** (see A.2, Th. 6.6.5, p. 218).

The results of this section can be used to extend valuations on field coproducts; we shall not go into the proofs (which can be found in Cohn [89']), but merely quote the result. Let (K, v) be a field K with a valuation v; an extension (L, w) is said to be *unramified* if the value group has not been enlarged. Further, K is said to be *v-closed* in L if for each $c \in L$ there exists $c_0 \in K$ such that

$$w(c - c_0) \geqslant w(c - a) \quad \text{for all } a \in K.$$

Such a c_0 may be regarded as a 'closest neighbour' of c in K.

With these definitions we have the following extension theorem for valuations on field coproducts:

THEOREM 9.A. *Let (K_λ, v_λ) be a family of valuated fields with a common subfield E such that $v = v_\lambda|E$ is independent of λ and assume that K_λ/E is an unramified extension and E is v-closed in each K_λ. Then the field coproduct $L = \circ_E K_\lambda$ has a unique valuation extending all the v_λ.* ∎

Exercises

1. Give a direct proof of Th. 4.1 (after the pattern of the proof of Th. 4.4).

2. Let V be a matrix subvaluation on a field K such that for any invertible matrix A over K, $V(A^{-1}) = -VA$. Show that V is a matrix valuation.

3. Show that for any matrix subvaluation μ on a ring R the sets $\text{Ker } \mu = \{A \in \mathfrak{M}(R)|VA = \infty\}$ and $\{A \in \mathfrak{M}(R)|VA > 0\}$ are matrix ideals.

9.5 Matrix valuations on firs

We shall now use the method of the last section to construct matrix valuations on a ring R. These matrix valuations will be positive real-valued on R and so define a valuation on a certain epic R-field with a valuation ring containing the image of R. Although the method of construction is quite general, for it to be useful we shall need to restrict the class of rings; interesting conclusions can be reached for tensor rings, but to begin with we shall impose no restrictions.

Let R be any ring. To construct a matrix subvaluation on R we can imitate the method used for subvaluations in 9.4. Thus let $\{A_i\}$ be a family of square matrices over R and for each A_i define a real number $\lambda(A_i)$ such that $\lambda(A_i) \geqslant 0$ with equality if and only if A_i is invertible. If A_i' is any matrix obtained from A_i by elementary transformations or changing the sign of a row or column, we put $\lambda(A_i') = \lambda(A_i)$ and denote the family of all possible A_i' by S. Next we define a function v on $\mathfrak{M}(R)$ by putting for any $C \in \mathfrak{M}(R)$,

$$v(C) = \begin{cases} \sup \left\{ \sum_1^r \lambda(C_i) \,\middle|\, C \oplus I = C_1 \oplus \ldots \oplus C_r, \, C_i \in S \right\} & \text{if } C \text{ is full,} \\ \infty & \text{if } C \text{ is non-full.} \end{cases}$$

(1)

We note that for $C = I$ the C_i in (1) must all be unit-matrices, hence we have $v(I) = 0$. More generally, the definition (1) shows that

$$v(I) = 0, \quad v(C \oplus I) = v(C) \quad \text{for any } C \in \mathfrak{M}(R). \tag{2}$$

This function v is the least submultiplicative function satisfying (2) and majorizing λ:

LEMMA 9.5.1. *Let R be a ring and λ a function defined on a certain set S of matrices of R. If v is defined in terms of λ by (1), then (i) v is submultiplicative and satisfies (2), (ii) $v \geqslant \lambda$, (iii) any submultiplicative function satisfying (2) and majorizing λ also majorizes v.*

Proof. Given matrices C, D and any fixed $\varepsilon > 0$, we can find decompositions $C \oplus I = A_1 \oplus \ldots \oplus A_r$, $D \oplus I = B_1 \oplus \ldots \oplus B_s$, $A_i, B_j \in S$, such that

$$v(C) < \sum \lambda(A_i) + \varepsilon, \quad v(D) < \sum \lambda(B_j) + \varepsilon;$$

now $C \oplus D \oplus I$ is E-associated to $A_1 \oplus \ldots \oplus A_r \oplus B_1 \oplus \ldots \oplus B_s$, therefore

$$v(C \oplus D) \geqslant \sum \lambda(A_i) + \sum \lambda(B_j) > v(C) + v(D) - 2\varepsilon.$$

Since ε was arbitrary, it follows that v is submultiplicative. Further, v satisfies (2), as we have seen. Now let $C \in S$, so that $\lambda(C)$ is defined; by taking the decomposition $C = C$ in (1), we see that $v(C) \geqslant \lambda(C)$, so (ii) holds. Finally, if f is a submultiplicative function satisfying (2) and majorizing λ, then for any C such that $C \oplus I = C_1 \oplus \ldots \oplus C_r$ we have $f(C) = f(C \oplus I) \geqslant \sum f(C_i) \geqslant \sum \lambda(C_i)$, hence $f(C) \geqslant \sup \{\sum \lambda(C_i)\} = v(C)$ and (iii) follows. ∎

Thus in v we have a function satisfying (MV.1, 2, 3', 5). In order to satisfy also (MV.4) and so obtain a matrix subvaluation, we form

$$\mu(C) = \sup\{\min_i\{v(C_i)|C = \nabla_i C_i\}\}, \tag{3}$$

where the sup is taken over all possible decompositions as a (repeated) determinantal sum.

LEMMA 9.5.2. *Let v be any function satisfying (MV.1, 2, 3', 5') and (2). Then the function μ defined by (3) is a matrix subvaluation and $\mu \geqslant v$, while any matrix subvaluation majorizing v also majorizes μ.*

Proof. Since we always have the decomposition $C = C$, it follows that $\mu(C) \geqslant v(C)$; in particular, μ satisfies (MV.1). (MV.2) holds because for any matrix C, C and C^-, obtained by changing the sign of a row or columm, have the same decompositions. (MV.5') is clear because $\mu(1_0) = v(1_0) = 0$.

To prove (MV.4), let $C = A \nabla B$, choose $\varepsilon > 0$ and take decompositions

$$A = \nabla_i A_i, \quad B = \nabla_j B_j, \tag{4}$$

such that

$$\mu(A) \leqslant \min_i\left\{v(A_i)|A = \nabla_i A_i\right\} + \varepsilon, \quad \mu(B) \leqslant \min_j\left\{v(B_j)|B = \nabla_j B_j\right\} + \varepsilon. \tag{5}$$

Then

$$\mu(C) \geqslant \min_{ij}\left\{v(A_i), v(B_j)|C = \nabla_i A_i \nabla_j B_j\right\} \geqslant \min\{\mu(A), \mu(B)\} - \varepsilon.$$

Since ε was arbitrary, it follows that $\mu(C) \geqslant \min\{\mu(A), \mu(B)\}$ and (MV.4) follows.

It remains to prove (MV.3'). We again take $\varepsilon > 0$ and find determinantal decompositions (4). Now we have $A \oplus B = \nabla_j(A \oplus B_j) = \nabla_i \nabla_j(A_i \oplus B_j)$, hence

$$\mu(A \oplus B) \geqslant \min_{ij}\{v(A_i \oplus B_j)\} \geqslant \min_{ij}\{v(A_i) + v(B_j)\}$$

$$\geqslant \mu(A) + \mu(B) - 2\varepsilon,$$

by (5). Since ε was arbitrary, we conclude that $\mu(A \oplus B) \geqslant \mu(A) + \mu(B)$, so (MV.3') holds. This shows that μ is a matrix subvaluation. Now let f

be any subvaluation such that $f \geqslant v$. Then for any decomposition $C = \nabla_i C_i$ we have

$$f(C) \geqslant \min_i \{f(C_i)\} \geqslant \min_i \{v(C_i)\},$$

and taking the sup over all determinantal decompositions of C, we find that $f(C) \geqslant \mu(C)$. Thus any matrix subvaluation majorizing v also majorizes μ. ∎

By combining these two lemmas we have a process of forming from any function λ on certain matrices over R the least matrix subvaluation μ associated with λ, with the following properties:

THEOREM 9.5.3. *Let R be any ring and λ a function defined on certain non-invertible square matrices over R, with real positive values. Define v by (1) in terms of λ and μ by (3) in terms of v. Then $\mu \geqslant v$, μ is a matrix subvaluation on R and any matrix subvaluation majorizing λ also majorizes μ.* ∎

Given any family of non-invertible square matrices $\{A_i\}$ over R and any function λ with positive real values defined on it, we can in this way construct the least matrix subvaluation majorizing λ. With each matrix subvaluation μ we can associate two matrix ideals, its *singular kernel* given by

$$\mathscr{P}_0 = \mathrm{Ker}\, \mu = \{A \in \mathfrak{M}(R) | \mu A = \infty\},$$

and its *associated matrix ideal* defined as

$$\mathscr{P}_1 = \{A \in \mathfrak{M}(R) | \mu A > 0\}.$$

The next result clarifies the relation between these concepts:

THEOREM 9.5.4. *Let R be a ring and $\mathscr{A}$ a proper matrix ideal generated by a family $\{A_i\}$ of matrices. For any real-valued function λ on $\{A_i\}$ let μ be the least matrix subvaluation majorizing λ. Then*

$$\mu(I) = 0, \quad \mu(C \oplus I) = \mu I \quad \text{for all } C \in \mathfrak{M}(R). \tag{6}$$

Moreover, $\mathscr{P}_1$, the matrix ideal associated with μ, is $\mathscr{A}$ and if the values of λ are positive and bounded, and bounded away from 0, then its singular kernel is given by

$$\mathscr{P}_0 = \bigcap \mathscr{A}^n. \tag{7}$$

Proof. Since $\mathscr{A}$ is proper, the terms C_i in any determinantal decomposition $I = \nabla_i C_i$ cannot all lie in $\mathcal{A}$, say $C_1 \notin \mathcal{A}$. Then from the definition (1), $v(C_1) = 0$, because the sup is taken over the empty family, hence $\mu(I) = 0$. A similar argument shows that $\mu(C \oplus I) = \mu C$.

Now any matrix C belongs to $\mathcal{A}$ precisely when there is an equation

$$C \oplus I = B_1 \nabla \ldots \nabla B_n,$$

where $B_i = A_j \oplus D$ for some D, or B_i is non-full.

Hence $\lambda(B_i) > 0$, so $v(B_i) > 0$ and $\mu C > 0$; thus $C \in \mathcal{P}_1$. Conversely, if $C \in \mathcal{P}_1$, then $\mu C > 0$, hence $C = C_1 \nabla \ldots \nabla C_n$ and $v(C_i) > 0$ for all i. Thus each C_i has the form $C_i \oplus I = A_1 \oplus \ldots \oplus A_s$, where A_i is in the original family and $s > 0$, hence $C_i \in \mathcal{A}$ and so $C \in \mathcal{A}$. This shows that $\mathcal{P}_1 = \mathcal{A}$.

It remains to prove (7). Assume that δ, M exist such that

$$0 < \delta < \lambda(A_i) < M \quad \text{for all } i.$$

If $A \in \mathcal{A}^n$, then $A \oplus I = \nabla_i B_i$, where each B_i is either non-full or a diagonal sum of n terms in $\mathcal{A}$. It follows that $\mu(B_i) > n\delta$ and so $\mu A > n\delta$. Suppose that $A \notin \mathcal{P}_0$; then μA is finite, so we can find n to satisfy $n\delta > \mu A$ and hence $A \notin \mathcal{A}^n$. This proves that $\bigcap \mathcal{A}^n \subseteq \mathcal{P}_0$. Conversely, if $A \in \mathcal{P}_0$, then $\mu A = \infty$, so for any n, $\mu A > nM$ and we can write $A = \nabla_i B_i$, where $v(B_i) > nM$. Hence $B_i \oplus I$ is a diagonal sum of more than n matrices in $\mathcal{A}$ and so $A \in \mathcal{A}^n$. This holds for all n, hence $A \in \bigcap \mathcal{A}^n$ and (7) is established. ∎

This result gives a clue to the kind of conditions on a ring needed to allow a matrix valuation to be defined on it. We shall need the following definitions:

DEFINITION 1. A matrix ideal $\mathcal{A}$ in a ring R is said to be *small* if $\bigcap \mathcal{A}^n = \mathcal{N}$, where $\mathcal{N}$ is the set of all non-full matrices.

This definition, although quite general in form, is only significant in the case where the set $\mathcal{N}$ is itself a matrix ideal. For example, a ring can have no small matrix ideals unless the determinantal sum of any two non-full matrices is again non-full. We shall want to use the above definition mainly for firs, where $\mathcal{N}$ is actually a prime matrix ideal, as we saw in 4.5. In a commutative principal ideal domain a prime matrix ideal corresponds to a prime ideal pR and we have $\bigcap p^n R = 0$; from this it easily follows that every proper ideal in R is small. It seems likely that this holds for every fir, but so far this has only been proved for tensor rings (see Th. 5.8 below).

DEFINITION 2. A matrix ideal $\mathcal{A}$ in a ring R is said to be *-*small* if $\mathcal{A}$ is proper and for any full matrix C there exists $r \in \mathbf{N}$ such that

$$\overset{n}{\bigoplus} C \notin \mathcal{A}^{rn} \quad \text{for all } n \geq 1.$$

We record two obvious properties:

PROPOSITION 9.5.5. *Let R be any ring. (i) Any *-small matrix ideal is small. (ii) Given matrix ideals $\mathcal{B} \subseteq \mathcal{A}$, if $\mathcal{A}$ is small or *-small, then so is $\mathcal{B}$.*

Proof. (i) is clear. To prove (ii), we have $\mathcal{B}^n \subseteq \mathcal{A}^n$, hence $\bigcap \mathcal{B}^n \subseteq \bigcap \mathcal{A}^n$, and so whenever $\mathcal{A}$ is small, $\mathcal{B}$ is so too. Now assume that $\mathcal{A}$ is *-small and take any full matrix C. There exists $r > 0$ such that $\bigoplus^n C \notin \mathcal{A}^{rn}$ for all n, hence $\bigoplus^n C \notin \mathcal{B}^{rn}$ for all n, and this shows $\mathcal{B}$ to be *-small. ∎

The role played by the notion of smallness in the construction of matrix valuations is clarified by our next result:

THEOREM 9.5.6. *Let R be a ring in which the set $\mathcal{N}$ of all non-full matrices forms a matrix ideal. Given a real-valued radical matrix sub-valuation μ on R with singular kernel $\mathcal{N}$ and associated prime matrix ideal $\mathcal{P}_1$, if the values μA for $A \in \mathcal{P}_1$ have a positive lower bound, then $\mathcal{P}_1$ is *-small. Conversely, given a *-small matrix ideal $\mathcal{A}$ on R and a real function λ on a generating set of $\mathcal{A}$, with a finite upper bound, let μ be its associated matrix subvaluation. Then its root μ^* is finite on the set of all full matrices.*

Proof. By definition $\mu A > 0$ for $A \in \mathcal{P}_1$ and by hypothesis there exists $\delta > 0$ such that $\mu A \geq \delta$. Let $\mu A = \alpha$ and suppose that $\bigoplus^n A \in \mathcal{P}_1^{rn}$; then $\mu(\bigoplus^n A) = n\alpha \geq rn\delta$. By choosing $r > \alpha/\delta$, we ensure that $\bigoplus^n A \notin \mathcal{P}_1^{rn}$ for all n, and the first assertion follows.

For the converse let $\{A_i\}$ be a generating set of $\mathcal{A}$ on which λ is defined, and suppose that $\mu(A_i) \leq M$. If C is a full matrix such that $\mu^*(C) = \infty$, this means that for any $r \in \mathbf{N}$ there exists n such that $\mu(\bigoplus^n C) > rnM$. As in the first part of the proof it follows that $\bigoplus^n C \in \mathcal{A}^{rn}$, but this contradicts the fact that $\mathcal{A}$ is *-small. ∎

By invoking Th. 4.4 we see that this result provides a means of obtaining a matrix valuation from a *-small matrix ideal. By Prop. 5.5 it is enough to show that every maximal matrix ideal is *-small. We shall now show this for the case of tensor rings.

Let K be any field and k a central subfield and consider the tensor K-ring on a set X: $R = K_k\langle X \rangle$. We know that this ring is a fir and so has a universal field of fractions $U = K_k \langle\!\langle X \rangle\!\rangle$. Let $a = (a_x) \in K^X$, thus a is a function on X with values in K; we obtain a K-ring homomorphism from R to K by mapping x to a_x. The singular kernel is a prime matrix ideal $\mathcal{M}_a$ which is generated by the family $(x - a_x)$. It is clear that $\mathcal{M}_a$ is a maximal matrix ideal in R and the next result shows it to be *-small; moreover, it yields a matrix valuation associated with $\mathcal{M}_a$.

PROPOSITION 9.5.7. *Let $R = K_k\langle X \rangle$ be a tensor K-ring on a set X and $a = (a_x) \in K^X$. Then the matrix ideal $\mathcal{M}_a$ generated by all $x - a_x$ ($x \in X$) is a *-small prime matrix ideal and there is a matrix valuation V_a on R with singular kernel $\mathcal{N}$, the set of all non-full matrices, and associated matrix ideal $\mathcal{M}_{\dot{a}}$.*

Proof. On replacing x by $x - a_x$ we may take $a = 0$. We write again $U = K_k \langle\!\langle X \rangle\!\rangle$ and form the rational function field $U(t)$ with a central indeterminate t. On $U(t)$ we have the t-adic valuation v, which induces a matrix valuation on $R[t]$, again denoted by v. Now for any $A \in \mathfrak{M}(R)$ we define

$$V(A) = v(A_t), \quad \text{where } A_t = A(tx). \tag{8}$$

Thus the value of A is defined as the t-adic value of the matrix $A(tx)$ obtained from A by replacing each $x \in X$ by tx. To verify that V is a matrix valuation on R with associated prime matrix ideal $\mathcal{M}_0$, generated by all $x \in X$, we note first that v is a matrix valuation on $R[t]$ and $x \mapsto tx$ defines a homomorphism $R \to R[t]$, hence (8) defines a matrix valuation on R by pullback. Further, $VA > 0$ precisely when $A(0)$ is non-invertible, i.e. when the matrix obtained from A_t by putting $t = 0$ is non-invertible. But this is just the matrix ideal associated with the t-adic valuation, namely $\mathcal{M}_0$. By Th. 5.6, $\mathcal{M}_0$ is *-small, and it follows that the singular kernel is $\mathcal{N}$. ∎

This result can be extended to any prime matrix ideal of the tensor ring:

THEOREM 9.5.8. *Let $R = K_k\langle X \rangle$ be the tensor K-ring on X as before and let $\mathcal{P}$ be any prime matrix ideal on R. Then there exist an extension field L of K and $a \in L^X$ such that $\mathcal{P} = \mathcal{M}_a \cap \mathfrak{M}(R)$, where $\mathcal{M}_a = (x - a_x)$ is a maximal matrix ideal in $L_k\langle X \rangle$. Hence $\mathcal{P}$ is *-small and there is a matrix valuation on R with associated matrix ideal $\mathcal{P}$.*

Proof. Let $L = R/\mathscr{P}$, the residue-class field of the localization $R_{\mathscr{P}}$, and denote the image of $x \in X$ in L by a_x. The ring $S = L_k\langle X \rangle$ is again a tensor ring, containing R as subring. We write $\mathcal{M}_a$ for the matrix ideal of S generated by all $x - a_x$ ($x \in X$), clearly a maximal matrix ideal of S. If the matrix ideal of S generated by $\mathscr{P}$ is denoted by $\mathscr{P}'$, then by the definition of a we have $\mathcal{M}_a \subseteq \mathscr{P}'$, and $\mathscr{P}'$ is proper, because the natural homomorphism $R \to L$ can be extended to S. Hence $\mathscr{P}' = \mathcal{M}_a$ by the maximality of $\mathcal{M}_a$ and it follows that $\mathcal{M}_a \cap \mathfrak{M}(R) \supseteq \mathscr{P}$. But the natural map $R \to S$ induces a homomorphism of local rings $R_{\mathscr{P}} \to S_{\mathcal{M}_a}$ which yields an isomorphism of residue-class fields; therefore $\mathcal{M}_a \cap \mathfrak{M}(R) = \mathscr{P}$, as claimed. Let V_a be the matrix valuation on S corresponding to $\mathcal{M}_a$ which exists by Prop. 5.7. Its associated matrix ideal is $\mathcal{M}_a$, hence the restriction $V_a|R$ is a matrix valuation on R, with associated matrix ideal $\mathscr{P}$, and by Th. 5.6, $\mathscr{P}$ is *-small. ∎

This result shows that the matrix subvaluation μ obtained from a proper matrix ideal $\mathscr{A}$ of $K_k\langle X \rangle$ has a finite root μ^* and so, by Th. 4.4,

$$\mu^*(A) = \inf\{VA \,|\, VA \geq \mu A\}. \tag{9}$$

We recall from (3) of 4.4 that for any matrix ideal $\mathscr{A}$ we have

$$\sqrt{\mathscr{A}} = \bigcap\{\mathscr{P} \text{ prime} \,|\, \mathscr{P} \supseteq \mathscr{A}\}. \tag{10}$$

By Th. 5.8, each $\mathscr{P}$ in (10) arises by restriction from some $\mathcal{M}_a$; we thus have a family of matrix valuations associated with $\mathscr{A}$. It is clear that $\inf(V_a)$ defines another matrix subvaluation, and one would expect this to be in some sense equivalent to μ^*; whether it actually equals μ^* is not known.

Th. 5.8 leaves open the question whether every maximal matrix ideal of $K_k\langle X \rangle$ is necessarily of the form $\mathcal{M}_a$ for $a \in K^X$. If we look to the commutative case for guidance, we find that when we take (i) X finite and (ii) k algebraically closed, then every maximal ideal of $k[X]$ is finitely generated (by the Hilbert basis theorem) and is of the form $(x - a_x|x \in X)$, by the Hilbert Nullstellensatz (see A.2, Th. 9.10.3, p. 351). In the non-commutative case we need to assume finite generation explicitly and replace algebraically closed fields by existentially closed skew fields. As we saw in 8.8, for any EC-field K over k, a finitely generated maximal matrix ideal of $K_k\langle X \rangle$ must be of the form $\mathcal{M}_a = (x - a_x|x \in X)$.

Exercises

1. Let $k\mathbf{Q}$ be the group algebra over k of the additive group $\mathbf{Q}$ of rational numbers. Show that the augmentation ideal of $k\mathbf{Q}$ is idempotent, and deduce that the singular kernel of the augmentation map $k\mathbf{Q} \to k$ is not small.

2. Verify that in a commutative principal ideal domain every proper matrix ideal is *-small.

3. Let R be an Hermite ring in which the determinantal sum of any two non-full matrices (where defined) is again non-full. Show that the set $\mathcal{N}$ of all non-full matrices over R is a matrix ideal, and give a condition for $\mathcal{N}$ to be prime. (R is *Hermite* if for every matrix A with a left inverse, (A, A') is square invertible for some A'; see FR, 0.4.)

4°. Is it true that every proper matrix ideal in a fir is small, or *-small?

5. A matrix ideal $\mathcal{P}$ in a ring R is called *superficial* if it is small and for any $m, n \geq 0$ and any matrices A, B, $A \notin \mathcal{P}^{m+1}$, $B \notin \mathcal{P}^{n+1} \Rightarrow A \oplus B \notin \mathcal{P}^{m+n+1}$. Show that every superficial matrix ideal is prime and *-small.

6. Show that the matrix subvaluation associated with a prime matrix ideal $\mathcal{P}$ is a matrix valuation if and only if $\mathcal{P}$ is superficial.

7. Let $K_k \langle X \rangle$ be a tensor K-ring on a set X, where $|X| > 1$, and let $\mathcal{M}_a$ be a maximal matrix ideal, with the corresponding matrix valuation V_a. Show that the residue-class field for V_a is not an epic R-field.

9.6 Ordered rings and fields

The theory of orderings on a field presents some analogies to the theory of valuations; thus Th. 2.3 and 2.4 have an analogue in Th. 6.4. These results are of intrinsic interest and also serve as a preparation for the study of matrix cones in 9.7. We begin by recalling the basic results, but assume that the reader has some familiarity with the commutative case.

An *ordered ring* is a non-trivial ring R with a total ordering '>' which is preserved by the ring operations:
(O.1) *If* $x_1 > x_2$, $y_1 > y_2$, *then* $x_1 + y_1 > x_2 + y_2$,
(O.2) *If* $x > 0$, $y > 0$, *then* $xy > 0$.
As usual we write $x \geq y$ to mean '$x > y$ or $x = y$' and use $\leq$, $<$ for the opposite ordering. An element x is called *positive* if $x > 0$, *negative* if $x < 0$. From (O.2) we see that an ordered ring is necessarily an integral domain. Further, the square of any non-zero element is positive, for if $a > 0$ then $a^2 > 0$ and if $a < 0$ then $-a > 0$ and so $a^2 = (-a)^2 > 0$. Since any positive integer n is a sum of squares 1^2, it follows that any ordered ring has characteristic zero. We shall also need the following general property:

LEMMA 9.6.1. *In any ordered ring, if* $x_1 \ldots x_r > 0$, *then* $x_{1'} \ldots x_{r'} > 0$ *for any permutation* $(1', \ldots, r')$ *of* $(1, \ldots, r)$.

Proof. $x_1 \ldots x_r > 0$ holds if and only if $x_i \neq 0$ for $i = 1, \ldots, r$ and the number of negative xs is even. ∎

In any ordered ring R the set

$$P = \{x \in R | x > 0\} \tag{1}$$

is called the *positive cone*. By (O.1, 2) it is closed under addition and multiplication and, writing $-P = \{-x | x \in P\}$, we have

$$P \cap -P = \varnothing, \quad P \cup -P = R^\times. \tag{2}$$

Conversely, every additively and multiplicatively closed subset P of $R^\times$ satisfying (2) is the positive cone of an ordering on R, as is easily checked.

A commutative ordered ring, being an integral domain, naturally has a field of fractions and the ordering can be extended to it. This holds more generally for Ore domains:

THEOREM 9.6.2. *Let R be a right Ore domain and K its field of fractions. If R is an ordered ring, then the ordering can be extended to K in just one way.*

Proof. Every element of K has the form as^{-1}, where $a, s \in R$, $s \neq 0$. If there is an ordering on K extending that of R, then as^{-1} must have the same sign as $as = as^{-1} \cdot s^2$, so the ordering is unique if it exists at all. Now the rule just given defines an ordering. For if $as^{-1} = a's'^{-1}$, let $su = s'u'$; then $au = a'u'$. Suppose that $as > 0$; then $as'u'u = asu^2 > 0$, hence $au \cdot s'u' > 0$, so $a'u' \cdot s'u' > 0$ and hence $a's' > 0$.

Now let $x, y > 0$ in K, say $x = as^{-1}$, $y = bs^{-1}$ with a common denominator s; then $(x + y)s^2 = (a + b)s = as + bs > 0$. Similarly, if $x = as^{-1}$, $y = bt^{-1}$ and $x, y > 0$, then $s^{-1}b = b_1s_1^{-1}$, and $xy = as^{-1}bt^{-1} = ab_1(ts_1)^{-1}$. Now ab_1ts_1 has the same sign as atb_1s_1, or also as $atbs$, or also $as \cdot bt$, which is positive, so $xy > 0$. ∎

To study the existence of orderings we shall need the general notion of a cone. Given any elements $a_1, \ldots, a_n$ of a ring, we shall write $[a_1a_2 \ldots a_n]$ for the set of all products of $a_1, a_2, \ldots, a_n$ in all possible orders; here the a_i need not all be distinct; for example, in an ordered ring, any member of $[a_1^2a_2^2 \ldots a_n^2]$ is positive, if $a_i \neq 0$.

We now define a *cone* on a ring R as a subset P of R which is closed under addition and, for any $a_1, \ldots, a_n \in R^\times$, $u_1, \ldots, u_m \in P$, contains the whole of $[a_1^2 \ldots a_n^2u_1 \ldots u_m]$; P is said to be *proper* if $0 \notin P$, *total* if

$P \cup -P \supseteq R^\times$. The definition shows that P always contains all squares of non-zero elements.

It is clear that the positive cone of an ordering on R is a proper total cone. Moreover, the intersection of any family of cones on R is again a cone, so we can speak of the cone *generated* by a given subset of R. For example, the least cone on R is the cone P_0 generated by the empty set; it consists of all sums of members of $[a_1^2 \ldots a_n^2]$, for any $a_i \in R^\times$. Of course, P_0 may be improper, e.g. if R is not an integral domain, or a finite field. If the least cone in R is proper, R is said to be *formally real*; this then means that no sum of terms from any sets $[a_1^2 \ldots a_n^2]$ is zero.

Given a ring R and any proper cone P on R, not necessarily total, if we define $x > y$ to mean $x - y \in P$, we obtain a partial ordering of R in which all squares are positive; it is a total ordering if and only if P is total. Thus when we speak of a partially ordered ring (or field), it is always understood that all non-zero squares are positive.

We remark that **Z** (and hence **Q**) has a unique ordering, because there is only one proper cone. The real field **R** also has a unique ordering, because every element is either a square or the negative of a square.

Our aim will be to show that any formally real ring can be ordered. The essential properties are contained in the following lemma, where we write $P + aP = \{p + aq \,|\, p, q \in P\}$.

LEMMA 9.6.3. *Let R be a ring and P a proper cone in R. Then (i) the cone generated by P and an element a of R is proper if and only if $0 \notin P + aP$, (ii) every proper cone on R is contained in a maximal proper cone and (iii) a proper cone is maximal if and only if it is total.*

Proof. (i) Let (P, a) be the cone generated by P and a; it consists of sums of terms which are products of elements of P and a in some order. Since $a^2 \in P$, we can write any element of P as $u_0 + u_1$ where u_0 is the sum of products involving an even number of factors a, and so belonging to P, while u_1 is the sum of products with an odd number of factors a, hence $au_1 \in P$. If $0 \in (P, a)$, we have $u_0 + u_1 = 0$, hence $au_0 + au_1 = 0$ and $u_0, au_1 \in P$, so if (P, a) is improper, then $0 \in P + aP$. The converse is clear.

That a proper cone is contained in a maximal proper cone is clear by Zorn's lemma, so (ii) holds. To prove (iii), let P be a maximal proper cone. If P is not total, say $a \in R^\times \backslash P \cup -P$, consider (P, a); since P was maximal, (P, a) is improper, so $u + av = 0$ for some $u, v \in P$, and similarly, $x - ay = 0$ for some $x, y \in P$, hence

$$0 = xu + xav = xu + (ay)(av) \in P,$$

which is a contradiction; thus P is total. Conversely, if P is total, then $P \cup -P = R^\times$ and so P is clearly maximal. ∎

This lemma allows us to state conditions for an ordering to exist; we shall slightly more generally give conditions for an ordering on a subring or a partial ordering to be extended. It is clear that any ordering on a ring R defines an ordering on any subring by restriction.

THEOREM 9.6.4. *Let R be a ring. Then (i) any partial ordering $>$ on R can always be extended to a total ordering $>''$ such that $x > y \Rightarrow x >'' y$; (ii) if R has an ordered subring A, then the ordering of A can be extended to one of R if and only if no sum of terms from the sets*

$$[c_1^2 \ldots c_m^2 a_1 \ldots a_n], \quad \text{where } c_i \in R^\times, \, a_j > 0 \text{ in } A, \tag{3}$$

vanishes. In particular, R can be ordered if and only if it is formally real.

Proof. (i) Given a partial ordering on R, let P be its cone; by Lemma 6.3 P is contained in a maximal proper cone P', which is total and so defines an ordering of R extending the partial ordering. If for P we take the least cone, P_0, we see that there is an ordering on R precisely when P_0 is proper, i.e. when R is formally real.

To prove the rest of (ii) we note that the positive cone P of A consists of all sums of terms from the sets (3), so the hypothesis just states that P is proper. In any ordering of R extending that of A the positive cone contains P, so if an extension exists, P must be proper. Conversely, when P is proper, then it is contained in a maximal proper cone P' which is total, again by Lemma 6.3, so P' defines an ordering which by construction extends that of A. ∎

In a field the definition of a cone can still be simplified. It is clear that any cone in a field K is closed under sums and products and contains all non-zero squares. Conversely, a subset P of a field with these properties is a cone, for in the first place P contains all commutators

$$(a, b) = a^{-1}b^{-1}ab = a^{-2}(ab^{-1})^2 b^2 \in P.$$

Hence if $ba \in P$, then $ab = ba(a, b) \in P$ and it follows that P satisfies the conditions for a cone. In particular, the least cone in a field is generated by all sums of products of squares. This leads to a simple criterion for a field to be formally real; we state the result again in slightly more general form in terms of extensions:

THEOREM 9.6.5. *Given a field extension L/K, if K is ordered, then the ordering can be extended to L if and only if there is no equation*

$-1 = \sum p_i u_i$, where $p_i \in K$, $p_i > 0$ and u_i is a product of squares in L.

$$(4)$$

In particular, a field is formally real if and only if -1 cannot be written as a sum of products of squares.

Proof. It is clear that no equation (4) can exist if L has an ordering extending that of K. Conversely, when no such equation exists, let P be the set of all sums $\sum p_i u_i$ as in (4) where $p_i u_i \neq 0$. Then P is a cone, which we claim is proper. For if not, then we have an equation

$$p_0 u_0 + p_1 u_1 + \ldots + p_r u_r = 0,$$

where $p_i > 0$ in K and u_i is a product of squares, for $i = 0, 1, \ldots, r$. Hence

$$-1 = p_0^{-1} p_1 u_1 u_0^{-1} + \ldots + p_0^{-1} p_r u_r u_0^{-1},$$

which is a forbidden equation, by hypothesis. Hence P is proper and we can again use Lemma 6.3 to find a maximal cone, leading to an ordering of L. ∎

For commutative fields this reduces to the well known condition of Artin and Schreier: A commutative field is formally real if and only if -1 cannot be written as a sum of squares.

We conclude with a theorem of Albert on algebraic elements in ordered fields.

THEOREM 9.6.6. *Let K be an ordered field. Then the centre of K is relatively algebraically closed in K.*

Proof. Let C be the centre and suppose that $a \in K$ and a is algebraic over C but not in C. Then its minimal polynomial over C,

$$f = x^n + \lambda_1 x^{n-1} + \ldots + \lambda_n \quad (\lambda_i \in C),$$

is of degree $n > 1$. Since char $K = 0$, we can replace a by $a' = a + (1/n)\lambda_1$; its minimal polynomial is obtained by replacing x by $x - (1/n)\lambda_1$ and expanding in powers of x. In the resulting polynomial the coefficient of x^{n-1} is zero. Now by Cor. 3.4.5 we can factorize this polynomial as

$$(x - a_1) \ldots (x - a_n),$$

where the a_i are conjugates of a'. Comparing coefficients of x^{n-1} we have

$$a_1 + \ldots + a_n = 0.$$

If $a' > 0$, then each conjugate a_i is positive and $\sum a_i > 0$; similarly, $\sum a_i < 0$ when $a' < 0$, so $a' = 0$, which contradicts the fact that $a \notin C$, and so the conclusion follows. ∎

Exercises

1. Show without using Th. 6.4 that every formally real ring is an integral domain.

2. Show that a proper cone P on a ring is an intersection of maximal cones if and only if $p, ap \in P$ implies $a \in P$.

3. Let R be an integral domain with an ordered subring A such that for any $r \in R$ there exist $a, b \in R^\times$ such that $ar, rb \in A$. If the positive cone of A is P, and $P' = \{x \in R | ax \in P \text{ for some } a \in P\}$, show that $x \in P'$ if and only if $axb \in P$ for some $a, b \in P$. Verify that P' is a proper total cone on R and deduce that the ordering on A can be extended to one of R in just one way.

4. Show that a field K can be ordered if and only if $K^\times$ contains a subgroup of index 2, closed under addition.

5. Show that an element c of a field K is positive under every ordering of K if and only if c is a sum of products of squares.

6. (A. A. Albert [57]) Show that an ordered PI-ring R is commutative. (Hint. Use Ex. 1 of 7.2 to show that R is an Ore domain; now apply Kaplansky's PI-theorem and Th. 6.6.)

7. Let P be a cone on a field K. If P is total and $P \neq K^\times$, show that $P \cap -P = \varnothing$.

8. In an ordered field show that if a commutes with b^n, where $n > 0$, then a commutes with b.

9. Let P be a cone on a field K. If $-1 \in P$, show that $P = K^\times$.

9.7 Matrix cones and orderings on skew fields

In Th. 6.2 we have seen that for a commutative integral domain R (or even an Ore domain) with field of fractions K every ordering of R extends to a unique ordering of K. For general skew fields this is no longer so; if R is an ordered ring and K is an epic R-field, then the ordering may or may not be extendible to K, and if it is, the extension may not be unique. It turns out that we need to order not just the elements of R but the square matrices over R. This is most easily expressed in terms of matrix cones.

For any subset Π of $\mathfrak{M}(R)$ define

$$-\Pi = \{A \in \mathfrak{M}(R)| -1 \oplus A \in \Pi\}, \quad \Pi^+ = \Pi\backslash(\Pi \cap -\Pi).$$

A subset Π of $\mathfrak{M}(R)$ is called a *matrix cone* if

(M.1) Π *contains all non-full matrices*,

(M.2) Π *contains* $A \oplus A$, *for all* $A \in \mathfrak{M}(R)$,

(M.3) *If* $A, B \in \Pi$, *then* $A \oplus B \in \Pi$,

(M.4) *If* $A, B \in \Pi$ *and* $C = A \nabla B$ *with respect to some column (or row) is defined, then* $C \in \Pi$,

(M.5) *If* $A \oplus 1 \in \Pi$, *then* $A \in \Pi$,

(M.6) *If* $A \in \Pi \cap -\Pi$ *and* $B \in \mathfrak{M}(R)$, *then* $A \oplus B \in \Pi$.

As a first consequence we note that $A \in \Pi$ implies $EA, AE \in \Pi$ for any elementary matrix E of the same order as A. For if $E = I + \lambda e_{21}$ say, and $A = (a_1 \quad a_2 \quad \ldots \quad a_n)$, then

$$AE = (a_1 + a_2\lambda \quad a_2 \quad \ldots \quad a_n) = A \nabla (a_2\lambda \quad a_2 \quad a_3 \quad \ldots \quad a_n).$$

By (M.1) the second term on the right is in Π as well as the first, so $AE \in \Pi$ by (M.4); similarly for other columns or for rows.

It follows that if $A \in \Pi$, then the result of interchanging two columns and changing the sign of one of them still lies in Π, and likewise (by a double application) for the result of multiplying two columns of A by -1.

With any matrix cone Π we associate the set

$$\Pi^0 = \Pi \cap -\Pi,$$

which is easily seen to be a matrix ideal. If, further,

(M.7) $A \oplus B \in \Pi^0$ *implies* $A \in \Pi$ *or* $B \in \Pi$,

(M.8) $-1 \notin \Pi$,

then Π is said to be *proper*. Clearly (M.7 and 8) hold when the associated matrix ideal Π^0 is prime. Conversely, if Π is proper, so that (M.7, 8) hold, then Π^0 is prime. To establish this fact it will be enough to show that (M.7) is equivalent to

(M.7') $A \oplus B \in \Pi^0$ *implies* $A \in \Pi^0$ *or* $B \in \Pi^0$.

Assume (M.7) and suppose that $A \oplus B \in \Pi^0$. Then $A \in \Pi$ or $B \in \Pi$; further, $-1 \oplus A \oplus B \in \Pi^0$, hence $-1 \oplus A \in \Pi$ or $B \in \Pi$, i.e. $A \in -\Pi$ or $B \in \Pi$. Similarly $A \in \Pi$ or $B \in -\Pi$, and since $-1 \oplus -1 \oplus A \oplus B \in \Pi^0$, we conclude that $A \in -\Pi$ or $B \in -\Pi$. Thus we have

$$(A \in \Pi \vee B \in \Pi) \wedge (A \in \Pi \vee B \in -\Pi) \wedge (A \in -\Pi \vee B \in \Pi)$$

$$\wedge (A \in -\Pi \vee B \in -\Pi),$$

which by the distributive law reduces to

$$(A \in \Pi \wedge A \in -\Pi) \vee (B \in \Pi \wedge B \in -\Pi),$$

from which (M.7′) follows. The converse is clear.

From the definition it is clear that any intersection of matrix cones is again a matrix cone, so we can speak of the matrix cone *generated* by a given matrix ideal (or indeed, by any set of square matrices). An intersection of proper matrix cones need not be proper, since (M.7) may fail; however, the intersection of a family of matrix cones associated with a fixed prime matrix ideal $\mathscr{P}$ is a matrix cone which is again associated with $\mathscr{P}$ and so is proper.

We note some elementary properties of matrix cones.

LEMMA 9.7.1. *Let Π be a matrix cone over a ring R and $A, B \in \mathfrak{M}(R)$. Then*

(i) *If $A \in \Pi^+$ and $B \in \Pi$ and $C = A \nabla B$ is defined, then $C \in \Pi^+$.*

(ii) *If $A \in \Pi^+$ and E is an elementary matrix of the same order as A, then $AE, EA \in \Pi^+$. In particular, if any two columns (or rows) of A are interchanged and the sign of one is changed, the result is still in Π^+.*

(iii) *If $A \oplus B \in \Pi$, then $\begin{pmatrix} A & C \\ 0 & B \end{pmatrix}, \begin{pmatrix} A & 0 \\ D & B \end{pmatrix} \in \Pi$ for all C, D of appropriate size.*

(iv) *If $A \oplus B \in \Pi^+$, then $\begin{pmatrix} A & C \\ 0 & B \end{pmatrix}, \begin{pmatrix} A & 0 \\ D & B \end{pmatrix} \in \Pi^+$ for all C, D of appropriate size.*

(v) *If Π is proper and $A, B \in \Pi^+$, then $A \oplus B \in \Pi^+$.*

Proof. (i) Let $A \in \Pi^+$, $B \in \Pi$ and suppose that $A \nabla B$ is defined with respect to the first column say: $A = (a, A')$, $B = (b, A')$. By (M.4), $A \nabla B \in \Pi$; we have to show that $A \nabla B \notin -\Pi$, so assume that $A \nabla B = (a + b, A') \in -\Pi$. By (M.2) and (M.5), $-1 \oplus (-b, A') \in \Pi$. Thus by (M.4),

$$-1 \oplus A = (-1 \oplus (a + b, A')) \nabla (-1 \oplus (-b, A')) \in \Pi,$$

but then $A \in -\Pi$, which contradicts the hypothesis, hence $A \nabla B \notin -\Pi$. Now (ii) follows as before, by the remarks made earlier.

To prove (iii) we build up C in $\begin{pmatrix} A & C \\ 0 & B \end{pmatrix}$ one column at a time, using (M.4) and (M.1), and similarly for $\begin{pmatrix} A & 0 \\ D & B \end{pmatrix}$, while (iv) follows by using (ii).

To prove (v), suppose that Π is proper and let $A, B \in \Pi^+$; then $A \oplus B \in \Pi$ by (M.3), and we have to show that $A \oplus B \notin -\Pi$. If this is not so, then $-1 \oplus A \oplus B \in \Pi$, hence $-1 \oplus A \oplus B \in \Pi \cap -\Pi = \Pi^0$, so by (M.8) and (M.7'), $A \in \Pi^0$ or $B \in \Pi^0$, which is a contradiction, hence (v) follows. ∎

The next result shows how matrix cones can be used to construct orderings. We shall call a matrix cone *total* if $\Pi \cup -\Pi = \mathfrak{M}(R)$.

THEOREM 9.7.2. *Let R be any ring and Π a proper matrix cone on R, with associated prime matrix ideal Π^0. If $K = R/\Pi^0$ is the corresponding epic R-field, then the set*

$$P(\Pi) = \{a \in K \mid \text{for some admissible } A = (A_0 \quad A_* \quad A_\infty) \text{ for } a,$$

$$(A_0 \quad A_*) \oplus (-A_\infty \quad A_*) \in \Pi^+\} \quad (1)$$

is the positive cone of a partial ordering of K; it is a total ordering if and only if Π is total.

Proof. Suppose that $a, b \in K$ and denote admissible matrices for them by $A = (A_0 \quad A_* \quad A_\infty)$, $B = (B_0 \quad B_* \quad B_\infty)$. Then an admissible matrix for $a + b$ is

$$\begin{pmatrix} A_0 & A_* & A_\infty & 0 & 0 \\ B_0 & 0 & -B_\infty & B_* & B_\infty \end{pmatrix},$$

and we have to show that $a, b \in P(\Pi)$ implies $a + b \in P(\Pi)$. Thus we need to show that

$$\begin{pmatrix} A_0 & A_* & A_\infty & 0 \\ B_0 & 0 & -B_\infty & B_* \end{pmatrix} \oplus \begin{pmatrix} 0 & A_* & A_\infty & 0 \\ -B_\infty & 0 & -B_\infty & B_* \end{pmatrix} \in \Pi^+. \quad (2)$$

The first term can be written as a determinantal sum,

$$\begin{pmatrix} A_0 & A_* & A_\infty & 0 \\ B_0 & 0 & -B_\infty & B_* \end{pmatrix} = \begin{pmatrix} A_0 & A_* & A_\infty & 0 \\ 0 & 0 & -B_\infty & B_* \end{pmatrix}$$

$$\nabla \begin{pmatrix} 0 & A_* & A_\infty & 0 \\ B_0 & 0 & -B_\infty & B_* \end{pmatrix}, \quad (3)$$

while the second can by an interchange of columns be brought to the form

$$\begin{pmatrix} -A_\infty & A_* & 0 & 0 \\ B_\infty & 0 & -B_\infty & B_* \end{pmatrix}. \quad (4)$$

The diagonal sum of (4) with the first term on the right of (3) gives

$$\begin{pmatrix} A_0 & A_* & A_\infty & 0 \\ 0 & 0 & -B_\infty & B_* \end{pmatrix} \oplus \begin{pmatrix} -A_\infty & A_* & 0 & 0 \\ B_\infty & 0 & -B_\infty & B_* \end{pmatrix}.$$

Since $(A_0 \ \ A_*) \oplus (-A_\infty \ \ A_*)$, $(-B_\infty \ \ B_*) \oplus (-B_\infty \ \ B_*) \in \Pi^+$, this term lies in Π^+. The second term on the right of (3) together with (4) gives

$$\begin{pmatrix} -A_\infty & A_* & 0 & 0 \\ B_\infty & 0 & B_0 & B_* \end{pmatrix} \oplus \begin{pmatrix} -A_\infty & A_* & 0 & 0 \\ B_\infty & 0 & -B_\infty & B_* \end{pmatrix},$$

and this lies in Π^+ because $(-A_\infty \ \ A_*) \oplus (-A_\infty \ \ A_*)$, $(B_0 \ \ B_*) \oplus (-B_\infty \ \ B_*) \in \Pi^+$. It follows that (2) holds, and this shows $P(\Pi)$ to be closed under sums. Next consider a product ab; an admissible matrix is

$$\begin{pmatrix} B_0 & B_* & B_\infty & 0 & 0 \\ 0 & 0 & A_0 & A_* & A_\infty \end{pmatrix},$$

and to show that $ab \in P(\Pi)$ we need to verify that

$$\begin{pmatrix} B_0 & B_* & B_\infty & 0 \\ 0 & 0 & A_0 & A_* \end{pmatrix} \oplus \begin{pmatrix} 0 & B_* & B_\infty & 0 \\ -A_\infty & 0 & A_0 & A_* \end{pmatrix} \in \Pi^+.$$

By appropriate elementary transformations, and transforming away non-diagonal terms by Lemma 7.1 (iv), we can reduce this to

$$\begin{pmatrix} B_0 & B_* & 0 & 0 \\ 0 & 0 & A_0 & A_* \end{pmatrix} \oplus \begin{pmatrix} -B_\infty & B_* & 0 & 0 \\ 0 & 0 & -A_\infty & A_* \end{pmatrix} \in \Pi^+.$$

If $a, b \in P(\Pi)$, it is clear by (M.3) that $ab \in P(\Pi)$; when $b = a$, then $a^2 \in P(\Pi)$ by (M.2). Thus $P(\Pi)$ is a matrix cone on K. It is proper because 0 has the admissible matrix $(0 \ \ 1)$ and $0 \oplus -1 \in \Pi \cap -\Pi$, so $0 \notin P(\Pi)$. Hence $P(\Pi)$ defines a partial ordering on K.

Finally, if Π is total, then for any $a \in K^\times$ with admissible matrix $A = (A_0 \ \ A_* \ \ A_\infty)$, either $(A_0 \ \ A_*) \oplus (-A_\infty \ \ A_*) \in \Pi^+$ or $-1 \oplus (A_0 \ \ A_*) \oplus (-A_\infty \ \ A_*) \in \Pi^+$. In the second case, we have by elementary transformations, $1 \oplus (A_0 \ \ A_*) \oplus (A_\infty \ \ A_*) \in \Pi^+$, hence $(A_0 \ \ A_*) \oplus (A_\infty \ \ A_*) \in \Pi^+$ and so $-a \in P(\Pi)$; thus $P(\Pi)$ is total and so defines an ordering on K. ∎

Over a commutative ordered field we can define an ordering of matrices by the rule: $A > 0$ if and only if $\det A > 0$. With this definition it is easily checked that for any partially ordered field K the set of all matrices A such that $A > 0$ is a proper matrix cone, which is total precisely when we have a total ordering.

Assume now that K is a skew field. On K we have the Dieudonné determinant $\mathrm{Det}: \mathbf{GL}(K) \to K^{\mathrm{ab}}$. If K is an ordered field, or even just partially ordered, then its positive cone contains all squares and hence all commutators. Thus each residue-class in K^{ab} is either positive or negative and so we can again define a matrix A to be positive if $\text{Det}\, A$ is positive. In this way each partial ordering of an epic R-field leads to a proper matrix cone on R. As expected, this correspondence between proper matrix cones and partial orderings is a bijection:

THEOREM 9.7.3. *Let R be a ring and K an epic R-field with natural homomorphism $\lambda: R \to K$. Then for any partial ordering of K with positive cone P, the set*

$$\mathcal{M}(P) = \{A \in \mathfrak{M}(R) | \text{Det}\, A^\lambda \geq 0\}$$

is a proper matrix cone with associated prime matrix ideal $\text{Ker}\,\lambda$; moreover, $P(\mathcal{M}(P)) = P$.

Proof. (M.1) is clear and (M.2, 3) follow by the multiplication of determinants in diagonal sums; (M.5, 6) are also clear, so it remains to prove (M.4). Let $C = A \nabla B$, where $A, B \in \mathfrak{M}(R)$ and the determinantal sum is with respect to the first column. By Lemma 3.2 we can apply elementary transformations so as to reduce A, B, C to the form $a \oplus D$, $b \oplus D$, $c \oplus D$, where $c = a + b$. Hence $\text{Det}\, A^\lambda = a^\lambda \cdot \delta$, $\text{Det}\, B^\lambda = b^\lambda \cdot \delta$, $\text{Det}\, C^\lambda = c^\lambda \cdot \delta$, where $\delta = \text{Det}\, D^\lambda$. Now $a^\lambda\delta, b^\lambda\delta \geq 0$, hence $c^\lambda\delta = a^\lambda\delta + b^\lambda\delta \geq 0$, and so (M.4) holds. This shows $\mathcal{M}(P)$ to be a matrix cone.

For any $A \in \mathfrak{M}(R)$, A^λ is singular if and only if $\text{Det}\, A^\lambda = 0$ and this just means that $A \in \mathcal{M}(P) \cap -\mathcal{M}(P)$. Moreover, $a \in P(\mathcal{M}(P)) \Leftrightarrow (A_0 \quad A_*) \oplus (-A_\infty \quad A_*) \in \mathcal{M}(P)$ for an admissible matrix A for a, while Cramer's rule in the form

$$(A_0 \quad A_*) = (-A_\infty \quad A_*)\begin{pmatrix} a & 0 \\ -u_* & I \end{pmatrix}$$

shows that a is stably associated to $(A_0 \quad A_*) \oplus (-A_\infty \quad A_*)^{-1} = (A_0 \quad A_*) \oplus (-A_\infty \quad A_*) \cdot (-A_\infty \quad A_*)^{-2}$. Thus $a \in P(\mathcal{M}(P)) \Leftrightarrow a \in P$. ∎

This result leads to a criterion for an epic R-field to be orderable.

THEOREM 9.7.4. *Let R be a ring and K an epic R-field with natural homomorphism $\lambda: R \to K$. Then K can be ordered if and only if the matrix cone on R generated by $\text{Ker}\,\lambda$ is proper, with associated prime matrix ideal $\text{Ker}\,\lambda$.*

Proof. Put $\text{Ker}\,\lambda = \mathcal{P}$ and let Π be the set of all matrices C such that $C \oplus I$ is a determinantal sum of matrices which are either in $\mathcal{P}$ or of the form $A \oplus A$ or a diagonal sum of such matrices. Then it is clear that Π satisfies (M.1–5) and we have a matrix cone if (M.6) also holds. This will certainly be true if

$$\Pi \cap -\Pi = \mathscr{P}. \tag{5}$$

We note that in any case $\Pi \cap -\Pi \supseteq \mathscr{P}$. Suppose first that K can be totally ordered and let Π' be the corresponding matrix cone. Then Π' is proper, $\Pi' \supseteq \Pi$ and $\Pi' \cap -\Pi' = \mathscr{P}$, hence (5) holds and Π is a proper matrix cone with associated matrix ideal $\mathscr{P}$. Conversely, if Π is a proper matrix cone satisfying (5), then it defines a partial ordering of K, and by Th. 6.4 this can be extended to a total ordering. ■

As an application we shall prove that the free field on any ordered field can be ordered.

THEOREM 9.7.5. *Let K be an ordered field with centre k and X any ordered set. Then the free field $K_k\langle\!\langle X \rangle\!\rangle$ can be ordered so as to extend the orderings on K and X.*

Proof. It will be enough to find an ordering of $K_k\langle\!\langle X \rangle\!\rangle$ to extend the ordering of K, because every permutation of X induces an automorphism of $K_k\langle\!\langle X \rangle\!\rangle$, so we can apply a permutation to X to ensure that the ordering on X agrees with the given one.

By Th. 6.6 k is relatively algebraically closed in K, so $[K:k] = \infty$ unless $K = k$. Let us for the moment assume that $K \neq k$ and write $R = K_k\langle X \rangle$. We note that k is infinite, since it is of characteristic zero, so all the hypotheses of the specialization lemma are satisfied. We denote by $\mathcal{N}$ the set of all non-full matrices over R; we have to find a proper matrix cone Π of R with associated matrix ideal $\mathcal{N}$. For any K-ring homomorphism $\lambda\colon R \to K$ consider the set

$$\Pi_\lambda = \{A \in \mathfrak{M}(R) | \mathrm{Det}\,(A^\lambda) \geqslant 0 \text{ in } K\}.$$

It is clear that Π_λ is a proper matrix cone containing the positive cone P of K, considered as set of 1×1 matrices. Let $\Pi = \bigcap \Pi_\lambda$, where the intersection is taken over all Π_λ corresponding to all K-ring homomorphisms from R to K. Then Π is again a proper matrix cone; we claim that

$$\Pi \cap -\Pi = \mathcal{N}. \tag{6}$$

Clearly any non-full matrix belongs to $\Pi \cap -\Pi$. Conversely, if A is full, then by the specialization lemma, A^λ is non-singular for some homomorphism λ, hence $\mathrm{Det}\,(A^\lambda) \neq 0$ and it follows that $A \notin \Pi_\lambda \cap -\Pi_\lambda$;

therefore $A \notin \Pi \cap -\Pi$, and so (6) holds. Now by Th. 7.4, $R/\mathcal{N}$ can be ordered, and since $\Pi \supseteq P$, the ordering extends that of K.

When $K = k$, we take K to be the Hilbert field on k (see 6.1). The polynomial ring $k[u]$ can be ordered by assigning to each polynomial the sign of its lowest term, and this ordering can be extended to $F = k(u)$ by Th. 6.2. Now F has the order-preserving endomorphism $\alpha: f(u) \mapsto f(u^2)$ and on the skew polynomial ring $F[v; \alpha]$ we can extend the ordering of F by taking again the sign of the lowest term; this ordering again extends to $K = F(v; \alpha)$ and clearly K is an ordered field with k as centre, by 6.1. By Th. 6.4.6, $k \langle\!\langle X \rangle\!\rangle$ as a subfield of $K_k \langle\!\langle X \rangle\!\rangle$ can be ordered. ∎

Exercises

1. Show that a matrix cone satisfying (M.7) but not (M.8) must equal $\mathfrak{M}(R)$.

2. Let Π_λ be a family of matrix cones on a ring R, where Π_λ is associated with the matrix ideal $\mathscr{P}_\lambda$. Show that $\bigcap \Pi_\lambda$ is proper if and only if $\bigcap \mathscr{P}_\lambda$ is prime.

3. (G. Révész [83]) Let R be a partially ordered ring with positive cone P and let K be a field of fractions of R with singular kernel $\mathscr{P}$. Show that this partial order can be extended to an ordering of K if and only if there is a proper total matrix cone Π associated with $\mathscr{P}$ and satisfying $\Pi \supseteq P$. Show also that this ordering is unique if every matrix cone Π associated with $\mathscr{P}$ and satisfying $\Pi \supseteq P$ is total.

4. (G. Révész [83]) Let $F = k\langle x, y \rangle$ be the free algebra over an ordered field k on x and y and let G be the subalgebra generated by x, xy, xy^2. Verify that G has a total ordering extending that of k for which $xy < x < xy^2$, but that this ordering cannot be extended to the universal field of fractions of G.

5. Let K be an ordered field with centre k. Show that the field coproduct $K \underset{k}{\circ} K$ can be totally ordered. (Hint. Use Lemma 5.5.6.)

6. Let G be an ordered group with a subgroup H. Show that the free product $G \underset{H}{*} G$ amalgamating H can be ordered to extend the order on G. (Hint. Use Ex. 5, cf. Cohn [85″].)

Notes and comments

Kurt Hensel introduced p-adic valuations in number fields (see Hensel [08]), probably by analogy with places in function fields, but the study of valuations with a general value group (not necessarily **R**) was begun by W. Krull [31]. Valuations on skew fields were first defined by Schilling

[45], although in his book (Schilling [50]) he confines his attention to the commutative case. A further generalization has been studied by Mathiak [77, 81], in which the values form an ordered group with a group of order automorphisms; this corresponds to the case where the ring of the valuation is a total subring of the field, not necessarily invariant.

The notion of domination and Chevalley's lemma first appeared in Chevalley [51] and it received a coherent treatment in Bourbaki [64]. The extension of Chevalley's lemma to abelian valuations on skew fields first occurs in Cohn and Mahdavi-Hezavehi [80], and was also obtained independently by M. Krasner [a], while quasi-commutative valuations were defined in Cohn [88']. Sections 9.1–2 are based on the exposition in A.3, Ch. 9.4.

The notion of pseudo-valuation (here called subvaluation) was developed by Mahler in a series of papers in the mid-1930s in *Acta Mathematica*. In particular he showed that in any algebraic number field or algebraic function field of one variable any subvaluation (in the present terminology) is the infimum of a *finite* family of valuations. In Cohn [54] it was shown that any radical subvaluation on a field is an infimum of a family of valuations; this result was proved in the general context of commutative rings (as in Th. 4.1) by Bergman [71], who also points out the parallel with the well known formula for the radical of an ideal in a commutative ring:

$$\sqrt{\mathfrak{a}} = \bigcap \{\mathfrak{p} \text{ prime } | \mathfrak{p} \supseteq \mathfrak{a}\}.$$

Matrix valuations were defined and studied by M. Mahdavi-Hezavehi [79, 82], where Th. 3.5 is proved. The first complete proof of Th. 4.4 appeared in Cohn [89'], where the result is used to construct valuations on field coproducts. The results in 9.5 are taken from Cohn [86].

The theory of formally real commutative fields was developed by Artin and Schreier [26]. The extendibility criterion of Th. 6.5 in the commutative case (there is no equation $\sum p_i a_i^2 = -1$, $p_i > 0$) was obtained by Serre [49]; in the general case the orderability condition was proved independently by Pickert [51] and Szele [52], and the criterion for extendibility by Fuchs [58], where forms of Th. 6.4 are proved, as well as Lemma 6.3 on which it is based. The case of formally real rings is studied by R. E. Johnson [52] and V. D. Podderyugin [54]. Theorem 6.6 was proved by A. A. Albert [40]. Section 9.7 follows essentially Révész [83]; for Th. 7.5 see also Cohn [85''].

For a commutative field that is not formally real, the *level* has been defined as the least integer n such that -1 can be written as a sum of n squares. A. Pfister [65] has shown that the level is always a power of 2, and all powers can occur. For skew fields Scharlau and Tschimmel [83] define the level as the least integer n such that -1 can be written as a sum of n products of squares, and they show that for every $n \geqslant 1$ there is a field of level n, constructed as an iterated Laurent series ring in two variables over a commutative field.

STANDARD NOTATIONS

Besides the usual signs of logic, $\wedge$, $\vee$, $\neg$, $\Rightarrow$, $\Leftrightarrow$, $\forall$, $\exists$ (and, or, not, implies, is equivalent to, for all, there exists) and the signs $\mathbf{N}$, $\mathbf{Z}$, $\mathbf{Z}/p$, $\mathbf{Q}$, $\mathbf{R}$, $\mathbf{C}$, $\mathbf{H}$, $\mathbf{F}_q$ (natural numbers, integers, integers mod p, rational, real, complex numbers, quaternions and the Galois field of q elements), the following signs are often used without further explanation, where S, T, X are sets, R denotes a ring, and A a commutative ring.

$|S|$ cardinal of the set S,

$\mathcal{P}(S)$ set of all subsets of S,

$S\backslash T$ complement of T in S,

$\prod R_\lambda$ Cartesian product of the family of rings (R_λ),

R^X set of all mappings from X to R,

$R^{(X)}$ restricted direct product (direct sum) of copies of R,

$A[X;\Phi]$ commutative A-algebra generated by X with defining relations Φ,

$R\langle X;\Phi\rangle$ R-ring generated by X with defining relations Φ,

$R/\mathfrak{a}$ residue-class ring of R (mod $\mathfrak{a}$),

E/D field extension $D \subseteq E$,

$\mathscr{C}(R)$ centre of R,

$\mathscr{C}_R(X)$ centralizer of X in R,

R° opposite ring,

R^+ additive group of the ring R,

$_R M$, M_R left, right R-module,

$R^{\times} = R\backslash\{0\}$ set of all non-zero elements of R,

U(R) group of units of R (in 2.6 $U(L)$ is used to denote the universal associative envelope of the Lie algebra L),

J(R) Jacobson radical of R,

$(x, y) = x^{-1}y^{-1}xy$ multiplicative commutator,

$[x, y] = xy - yx$ additive commutator,

$^m R^n$ set of all $m \times n$ matrices over R,

$^m R = {}^m R^1$, $R^n = {}^1 R^n$ set of column, resp. row, vectors over R,

$\mathfrak{M}_n(R) = R_n$ set of all $n \times n$ matrices over R,

$\mathfrak{M}(R)$ set of all square matrices over R,

GL$_n(R)$ group of all invertible $n \times n$ matrices over R,

SL$_n(A)$ group of all $n \times n$ matrices of determinant 1 over A,

I_m $m \times m$ unit matrix,

C^T transpose of the matrix C,

diag$(a_1, \ldots, a_n)$ diagonal matrix,

e_{ij} matrix with (i, j)-entry 1, the rest 0,

e_i ith column of unit matrix,

δ_{rs} Kronecker delta (1 if $r = s$, 0 otherwise),

$A \oplus B$ diagonal sum of matrices A, B,

$\oplus^r C = C_1 \oplus \ldots \oplus C_r$,   where $C_i = C$,

hd$_R(M)$ homological dimension of the R-module M,

l.gl.dim.(R), r.gl.dim.(R) left, right global dimension of R.

$H^2(G, A)$ second cohomology group of G with coefficients in A.

LIST OF SPECIAL NOTATIONS USED
THROUGHOUT THE TEXT

$V_{n,m}$ universal non-IBN ring, 21

U_m universal non-UGN ring, 21

W_n 'universal' non-WF ring, 21

$K_E\langle X \rangle$ free K-ring on X, centralizing E, 38

$K_E\langle\langle X \rangle\rangle$ free power series ring, 38

Rg category of rings and homomorphisms, 41

Rg$_n$ category of matrix rings and homomorphisms compatible with the matrix structure, 42

$R \underset{K}{*} S$ coproduct of rings over K, 42

$\mathfrak{W}_n(R)$, $\mathfrak{W}_n(R; A)$ matrix reduction functor, 43f.

$\deg f$ degree of a polynomial, 48

$A[t]$, $A[t; \alpha, \delta]$ (skew) polynomial ring, 49

$\mathbf{T}_n(R)$ upper triangular matrix ring over R, 50

$K(t; \alpha, \delta)$ skew function field, 50

$A[t, t^{-1}; \alpha]$ skew Laurent polynomial ring, 55

$\Delta(f)$ divergence of a skew polynomial, 59

$I(e): a \mapsto eae^{-1}$ inner automorphism, 61

$K[\![t; \alpha]\!]$, $K((t; \alpha))$ skew power, resp. Laurent series ring, 66

$o(f)$ order of a power series, 67

KM, KG monoid ring, group ring, 71

$\mathscr{D}(f)$ support of a function, 74

$k((M))$ power series ring over an ordered monoid, 74

$G(R)$ graded ring associated with a filtered ring, 84

$[E:K]_L$, $[E:K]_R$ left, right degree, 94

λ_a, ρ_a left, right multiplication, 96

Gal(E/K) Galois group of a Galois extension E/K, 100

$x^{\rho} = x^p - x$, $V(x) = (t + x)^{\rho} - t^{\rho}$, 137f.

$\mathscr{F}_R$ category of epic R-fields and specializations, 155

Ker   singular kernel of a homomorphism, 156, of a matrix valuation, 439

R_Σ localization at a set Σ of matrices, 157

$(A_0 \quad A_* \quad A_\infty)$ admissible matrix, 159

$\begin{pmatrix} a' & a^* \\ a^0 & {}'a \end{pmatrix}$ admissible block, 159

$A \nabla B$ determinantal sum of matrices, 163

$R/\mathscr{P}$ epic R-field defined by a prime matrix ideal $\mathscr{P}$, 172

Spec(R) field spectrum of the ring R, 173

(X) matrix ideal generated by a set X of matrices, 174

$\sqrt{\mathscr{A}}$ radical of matrix ideal, 174

rA (inner) rank of A, 179

$\mathscr{U}(R)$ universal field of fractions (when it exists), 182

$\mathscr{P}(R)$ monoid of projectives, 186

$\tau(M)$ trace ideal of a module, 187

G_+ positive cone of an ordered group, 189

$\coprod A_\lambda$ coproduct in a category, 203

$\underset{K}{*}\, R_\lambda$ coproduct of a family of K-rings, 205

$^\otimes M^\otimes$ induced bimodule, 211

$\underset{K}{\circ}\, K_\lambda$ field coproduct of fields, 223

$D_K \langle\!\langle X \rangle\!\rangle$ free field, 224

$\mathbf{T}(M)$ or $K_E \langle M \rangle$ tensor ring on a K-bimodule M, 224, 226

$\Omega_K(R)$ universal derivation bimodule, 251

$D_k \langle\!\langle X; \Phi \rangle\!\rangle$ skew field presentation, 279

$\mathscr{D}(A)$ singularity support of a matrix, 301, 405

$\text{sing}\,(A)$, $\text{nonsing}\,(A)$ A is (non-) singular, 309

$\mathscr{R}(X; D)$ set of rational expressions, 336

$\text{dom}\, f$ domain of f, 336

$D_k(X; E)$ rational function field, 337

$k \langle X \rangle_d$ generic matrix ring, 343

$\mathscr{C}(D)$ domain of definition of an X-field, 344

$\mathfrak{A}(D)$ subset of vanishing expressions, 344

$k \langle\!\langle X \rangle\!\rangle_d$ generic division algebra, 344

$\bigwedge_S R_s$ rational meet of a family of X-rings, 348

$\text{Ess}\,(S)$ set of essential indices, 358

$\text{Supp}_S(t)$ subset supported by t, 359

$t \propto U$ support relation, 362

$a \| b$ a is total divisor of b, 380

$\xi \xrightarrow{\,\,} \eta$ specialization over K, 401f.

$\mathscr{V}_E(A)$ variety defined by A, 408

K^c derived group of $K^\times$, 427

$K^{ab} = K^\times / K^c$ abelianization of K, 437

$\text{Det}\, A$ Dieudonné determinant of A, 437

$P(\Pi)$ cone associated with Π, 465

$\mathcal{M}(P)$ matrix cone associated with P, 467

BIBLIOGRAPHY AND AUTHOR INDEX

This list contains, in addition to the papers quoted in the text, some relevant work not explicitly referred to, but no attempt at complete coverage has been made. In particular, the numerous papers on finite-dimensional division algebras, a branch much better documented, has for the most part been left aside.

Page references at the end of entries indicate the places in the text where the entry is quoted or implicitly referred to. Other references are listed after the author's name. Titles in Russian are marked (R).

A.1, 2, 3 refer to the author's *Algebra*, Second Edition, Volumes 1–3, J. Wiley and Sons, Chichester 1982, 1989, 1991.

FR refers to the author's *Free Rings and their Relations*, Second Edition, London Math. Soc. Monographs No. 19, Academic Press, London and New York 1985.

UA refers to the author's *Universal Algebra*, Second Edition, Reidel, Dordrecht 1981.

Albert, A. A.
[40] On ordered algebras, *Bull. Amer. Math. Soc.* **46** (1940), 521.   461, 470
[57] A property of ordered rings, *Proc. Amer. Math. Soc.* **8** (1957), 128–9.   462
Amitsur, S. A.   111, 412, 418
[48] A generalization of a theorem on differential equations, *Bull. Amer. Math. Soc.* **54** (1948), 937–41.   133, 150
[54] Non-commutative cyclic fields, *Duke Math. J.* **21** (1954), 87–105.   133, 137, 139, 150
[55] Finite subgroups of division rings, *Trans. Amer. Math. Soc.* **80** (1955), 361–86.   5, 148, 150
[56] Algebras over infinite fields, *Proc. Amer. Math. Soc.* **7** (1956), 35–48.   118, 419
[65] Generalized polynomial identities and pivotal monomials, *Trans. Amer. Math. Soc.* **114** (1965), 210–26.   332, 365
[66] Rational identities and applications to algebra and geometry, *J. Algebra* **3** (1966), 304–59.   306, 308, 335, 340
Amitsur, S. A. and Small, L. W.
[78] Polynomials over division rings, *Israel J. Math.* **31** (1978), 353–8.
Artin, E.
[28] Über einen Satz von Herrn J. H. Maclagan Wedderburn, *Abh. Math. Sem. Hamb. Univ.* **5** (1928), 245–50.   150
[65] *The collected papers of E. Artin* (Ed. S. Lang and J. Tate), Addison-Wesley, Reading, Mass. 1965   276

478

Artin, E. and Schreier, O.
[26] Algebraische Kennzeichnung reeller Körper, *Abh. Math. Sem. Hamb. Univ.* **5**
(1926), 85–99. 470
Bacsich, P. D. and Hughes, D. Rowlands
[74] Syntactic characterizations of amalgamation, convexity and related properties,
J. Symb. Logic **39** (1974), 433–51. 206
Baer, R.
[27] Über nicht-archimedisch angeordnete Körper, *Sitzungsber. Heidelberg. Akad.*
Wiss. Math. Nat. Kl. **8** (1927), 3–13. 418
Bedoya, H. and Lewin, J.
[77] Ranks of matrices over Ore domains, *Proc. Amer. Math. Soc.* **62** (1977), 233–6.
Benz, W.
[69] Zur Geometrie der Körpererweiterungen, *Canad. J. Math.* **21** (1969), 1097–1122.
Bergman, G. M. 41, 46, 107, 139, 239, 326, 329, 378
[64] A ring primitive on the right but not on the left, *Proc. Amer. Math. Soc.* **15**
(1964), 473–5. 82
[70] Skew fields of noncommutative rational functions, after Amitsur, Sém.
Schützenberger–Lentin–Nivat, Année 1969/70 No. 16 (Paris 1970) 335, 339f., 365
[71] A weak Nullstellensatz for valuations, *Proc. Amer. Math. Soc.* **28** (1971), 32–8.
470
[72] Hereditarily and cohereditarily projective modules, *Proc. Park City Conf. Ring*
Theory (Ed. R. Gordon), Academic Press 1972, 29–62. 191
[74] Modules over coproducts of rings, *Trans. Amer. Math. Soc.* **200** (1974), 1–32.
202, 219, 276
[74'] Coproducts and some universal ring constructions, *Trans. Amer. Math. Soc.* **200**
(1974), 33–88. 23, 46, 249f., 276
[76] Rational relations and rational identities in division rings, I, II, *J. Algebra* **43** (1976),
252–66, 267–97. 345, 347f., 355f., 365
[90] Ordering coproducts of groups and semigroups, *J. Algebra* **133** (1990), 313–39.
76
Bergman, G. M. and Dicks, W.
[75] Universal derivations, *J. Algebra* **36** (1975), 193–211. 276
[78] Universal derivations and universal ring constructions, *Pacif. J. Math.* **79** (1978),
293–337. 276
Bergman, G. M. and Small, L. W.
[75] PI-degrees and prime ideals, *J. Algebra* **33** (1975), 435–62. 344f., 355, 365
Bilo, J.
[80] A geometrical problem depending on the equation $ax - xa = 2$ in skew fields,
Simon Stevin **54** (1980), 27–62. 418
Blanchard, A.
[72] *Les corps non-commutatifs*, Presses Univ. de France, Collection SUP, Paris 1972.
Boffa, M. and Van Praag, P.
[72] Sur les corps génériques, *C. R. Acad. Sci. Paris Ser. A* **274** (1972), 1325–7. 317
[72'] Sur les sous-champs maximaux des corps génériques dénombrables, *C. R. Acad. Sci.*
Paris Ser. A **275** (1972), 945–7.
Bokut', L. A.
[63] On a problem of Kaplansky (R), *Sibirsk. Mat. Zh.* **4** (1963), 1184–5.
[67] The embedding of rings in skew fields (R), *Doklady Akad. Nauk SSSR* **175** (1967),
755–8. 330
[69] On Malcev's problem (R), *Sibirsk. Mat. Zh.* **10** (1969), 965–1005. 330
[81] *Associative Rings* 1, 2 (R), NGU, Novosibirsk 1981. 276, 330
[87] Embedding of rings (R), *Uspekhi Mat. Nauk* **42**:4 (1987), 87–111. (Engl. transl. in
Russian Math. Surveys **42** (1987), 105–38.)
Boone, W. W.
[59] The word problem, *Ann. Math.* **70** (1959), 265–9. 330

Bourbaki, N.
[64] *Algèbre commutative, Ch. 6 Valuations*, Hermann, Paris 1964.   470
Bowtell, A. J.
[67] On a question of Malcev, *J. Algebra* **6** (1967), 126–39.   330
Brauer, R.
[29] Über Systeme hyperkomplexer Zahlen, *Math. Zeits.* **30** (1929), 79–107.
[32] Über die algebraische Struktur von Schiefkörpern, *J. reine u. angew. Math.* **166**
(1932), 241–52.   107
[49] On a theorem of H. Cartan, *Bull. Amer. Math. Soc.* **55** (1949), 619–20.   150
Bray, U. and Whaples, G.
[83] Polynomials with coefficients from a division ring, *Canad. J. Math.* **35** (1983),
509–15.   150
Brungs, H. H.   82
Brungs, H. H. and Törner, G.
[84] Skew polynomial rings and derivations, *J. Algebra* **87** (1984), 368–79.
Burmistrovich, I. E.
[63] On the embedding of rings in skew fields (R), *Sibirsk. Mat. Zh.* **4** (1963), 1235–40.
Cameron, P. J. (verbal communication)   238
Carcanague, J.
[71] Idéaux bilatères d'un anneau de polynômes non commutatifs sur un corps, *J. Algebra*
18 (1971), 1–18.
Cartan, H.
[47] Théorie de Galois pour les corps non commutatifs, *Ann. Ecole Normale Supérieure*
64 (1947), 59–77.   150
Cauchon, G. and Robson, J. C.
[78] Endomorphisms, derivations and polynomial rings, *J. Algebra* **53** (1978), 227–38.
Chang, C. C. and Keisler, H. J.
[73] *Model Theory*, North-Holland, Amsterdam 1973.   46
Chehata, C. G.
[53] On an ordered semigroup, *J. London Math. Soc.* **28** (1953), 353–6.   92
Cherlin, G.
[72] The model companion of a class of structures, *J. Symb. Logic* **37** (1972), 546–56.
[73] Algebraically closed commutative rings, *J. Symb. Logic* **38** (1973), 493–9.
Chevalley, C.
[51] Introduction to the Theory of Algebraic Functions of one Variable, *Amer. Math.
Soc.*, New York 1951.   470
Cohn, P. M.
[54] An invariant characterization of pseudo-valuations, *Proc. Camb. Phil. Soc.*
50 (1954), 159–77.   470
[58] On a class of simple rings, *Mathematika* **5** (1958), 103–17.   276
[59] On the free product of associative rings, *Math. Zeits.* **71** (1959), 380–98. 276
[59'] Simple Lie rings without zero-divisors, and Lie division rings, *Mathematika* **6**
(1959), 14–18.   242
[60] On the free product of associative rings II. The case of (skew) fields, *Math. Zeits.*
73 (1960), 433–56.   202, 276f.
[61] On a generalization of the Euclidean algorithm, *Proc. Camb. Phil. Soc.* **57**
(1961), 18–30.   202
[61'] On the embedding of rings in skew fields, *Proc. London Math. Soc.* (3) **11**
(1961), 511–30.   45, 92, 276
[61''] Quadratic extensions of skew fields, *Proc. London Math. Soc.* (3) **11** (1961),
531–56.   91, 150, 276f.
[62] Eine Bemerkung über die multiplikative Gruppe eines Körpers, *Arch. Math.* **13**
(1962), 344–8.   150
[63] Non-commutative unique factorization domains, *Trans. Amer. Math. Soc.* **109**
(1963), 313–32.   46

[64] Free ideal rings, *J. Algebra* 1 (1964), 47–69. 46

[66] Some remarks on the invariant basis property, *Topology* 5 (1966) 215–28. 20, 46

[66'] On a class of binomial extensions, *Ill. J. Math.* 10 (1966), 418–24. 150, 276

[67] Torsion modules over free ideal rings, *Proc. London Math. Soc.* (3) 17 (1967), 577–99. 79, 92

[69] Dependence in rings II. The dependence number, *Trans. Amer. Math. Soc.* 135 (1969), 267–79. 46, 276

[71] The embedding of firs in skew fields, *Proc. London Math. Soc.* (3) 23 (1971), 193–213. 200

[71'] Rings of fractions, *Amer. Math. Monthly* 78 (1971), 596–615. 46

[71"] Un critère d'immersibilité d'un anneau dans un corps gauche, *C. R. Acad. Sci. Paris Ser. A* 272 (1971), 1442–4. 200

[72] Universal skew fields of fractions, *Symposia Math.* VIII (1972), 135–48. 200

[72'] Skew fields of fractions, and the prime spectrum of a general ring, *Lectures on Rings and Modules*, Lecture Notes in Math. No. 246, Springer Verlag, Berlin 1972, 1–71. 200, 418

[72"] Generalized rational identities, *Proc. Park City Conf., Ring Theory* (Ed. R. Gordon), Academic Press 1972, 107–15. 329

[73] The similarity reduction of matrices over a skew field, *Math. Zeits.* 132 (1973), 151–63. 276, 418

[73'] Free products of skew fields, *J. Austral. Math. Soc.* 16 (1973), 300–8. 276

[73"] The word problem for free fields, *J. Symb. Logic* 38 (1973), 309–14. Correction and addendum, ibid. 40 (1975), 69–74. 330

[73'''] The range of derivations on a skew field and the equation $ax - xb = c$, *J. Indian Math. Soc.* 37 (1973), 1–9. 419

[74] The class of rings embeddable in skew fields, *Bull. London Math. Soc.* 6 (1974), 147–8. 330

[75] Presentations of skew fields I. Existentially closed skew fields and the Nullstellensatz, *Math. Proc. Camb. Phil. Soc.* 77 (1975), 7–19. 329, 419

[75'] Equations dans les corps gauches, *Bull. Soc. Math. Belg.* 27 (1975), 29–39. 419

[76] The Cayley–Hamilton theorem in skew fields, *Houston J. Math.* 2 (1976), 49–55. 418f.

[77] A construction of simple principal right ideal domains, *Proc. Amer. Math. Soc.* 66 (1977), 217–22. Correction, ibid. 77 (1979), 40. 91

[77'] Zum Begriff der Spezialisierung über Schiefkörpern, *Proc. Symp. über geom. Algebra*, Birkhäuser, Basel 1977, 73–82. 419

[79] The affine scheme of a general ring, *Applications of Sheaves, Durham Research Symposium, July 1977*, Lecture Notes in Math. No. 753, Springer Verlag, Berlin 1979, 197–211. 46, 200

[79'] Skew fields with involution having only one unitary element, *Resultate d. Math.* 2 (1979), 119–23. 56

[80] The normal basis theorem for skew field extensions, *Bull. London Math. Soc.* 12 (1980), 1–3. 329

[82] The universal field of fractions of a semifir I. Numerators and denominators, *Proc. London Math. Soc.* (3) 44 (1982), 1–32. 201, 365

[82'] On a theorem from *Skew field constructions, Proc. Amer. Math. Soc.* 84 (1982), 1–7. 329

[83] Determinants on free fields, *Contemp. Math.* 13 (1983), 99–108.

[84] Fractions, Presidential address to the London Mathematical Society, *Bull. London Math. Soc.* 16 (1984), 561–74.

[85] The universal field of fractions of a semifir III. Centralizers, *Proc. London Math. Soc.* (3) 50 (1985), 95–113. 277, 419

[85'] Principles of non-commutative algebraic geometry, *Rings and Geometry* (Ed. R. Kaya), Reidel, Dordrecht 1985, 3–37. 419

[85"] On copowers of an ordered skew field, *Order* 1 (1985), 377–82. 469f

[86] The construction of valuations on skew fields, Lecture Notes, University of Alberta, Edmonton, Alberta 1986.   470

[88] Is there a non-commutative analogue of the resultant? *Bull. Soc. Math. Belg.* **40** (1988), 191–8.   419

[88'] Valuations on free fields, *Algebra, some Current Trends*, Proc. Conf. Varna 1986 (Ed. L. L. Avramov and K. B. Tchakerian), Lect. Notes in Math. No. 1352, Springer Verlag, Berlin 1988, 75–87.   470

[89] Around Sylvester's law of nullity, *The Math. Scientist* **14** (1989), 73–83.   171

[89'] The construction of valuations on skew fields, *J. Indian Math. Soc.* **54** (1989), 1–45.   449, 470

[89"] Generalized polynomial identities and the specialization lemma, *Israel Math. Conf. Proc.* **1** (1989), 242–6.   292

[90] An embedding theorem for free associative algebras, *Math. Pannon.* 1/1 (1990), 49–56.   183

[90'] Rank functions on rings, *J. Algebra* **133** (1990), 373–85.   201

[91] *Algebraic Numbers and Algebraic Functions*, Chapman and Hall, London 1991.
                                                                              66, 420f., 424

[92] A brief history of infinite-dimensional skew fields, *The Math. Scientist* **17** (1992), 1–14.   45

[92'] A remark on power series, *Publicacions Mat.* **36** (1992), 481–4.   41

[93] One-sided localization in rings, *J. Pure Appl. Algebra* **88** (1993), 37–42.   34

Cohn, P. M. and Dicks, W.

[80] On a class of central extensions, *J. Algebra* **63** (1980), 143–51.   125, 329

Cohn, P. M. and Mahdavi-Hezavehi, M.

[80] Extensions of valuations on skew fields, *Proc. Ring Theory Week Antwerp 1980* (Ed. F. van Oystaeyen), Lecture Notes in Math. No. 825, Springer Verlag, Berlin 1980, 28–41.   470

Cohn, P. M. and Reutenauer, C.

[94] A normal form for free fields, *Canad. J. Math.* **46** (1994), 517–31.   329

Conrad, P.

[54] On ordered division rings, *Proc. Amer. Math. Soc.* **5** (1954), 323–8.

Cozzens, J. H.

[72] Simple principal left ideal domains, *J. Algebra* **23** (1972), 66–75.

Cozzens, J. H. and Faith, C. C.

[75] *Simple Noetherian Rings*, Cambridge Tracts in Math. No. 69, Cambridge University Press 1975.   91, 419

Cramer, G.

[1750] *Introduction à l'analyse des lignes courbes*, Cramer et Philbert, Geneva 1750.   160

Dauns, J.

[70] Embeddings in division rings, *Trans. Amer. Math. Soc.* **150** (1970), 287–99.   92

[70'] Integral domains that are not embeddable in division rings, *Pacif. J. Math.* **34** (1970), 27–31.   92

[82] *A Concrete Approach to Division Rings*, Heldermann Verlag, Berlin 1982.

Dehn, M.

[22] Über die Grundlagen der projektiven Geometrie und allgemeine Zahlensysteme, *Math. Ann.* **85** (1922), 184–94.   365

Desmarais, P. C. and Martindale, W. S., III

[80] Generalized rational identities and rings with involution, *Israel J. Math.* **36** (1980), 187–92.

Dicker, R. M.

[68] A set of independent axioms for a field and a condition for a group to be the multiplicative group of a field, *Proc. London Math. Soc. (3)* **18** (1968), 114–24.   45

Dicks, W. *see also* Bergman, Cohn   259

[83] The HNN construction for rings, *J. Algebra* **81** (1983), 434–87.   91

Dicks, W. and Sontag, E. D.   91
   [78] Sylvester domains, *J. Pure Appl. Algebra* **13** (1978), 143–75.   201, 250
Dickson, L. E.
   [05] Definitions of a group and a field by independent postulates, *Trans. Amer. Math.
       Soc.* **6** (1905), 198–204.   7, 45
   [23] *Algebras and their Arithmetics*, Univ. of Chicago Press, Chicago 1923. (Repr
       G. E. Stechert, New York 1938.)
Dieudonné, J.
   [43] Les déterminants sur un corps non-commutatif, *Bull. Soc. Math. France* **71** (1943),
       27–45.   367
   [52] Les extensions quadratiques des corps non-commutatifs et leurs applications, *Acta
       Math.* **87** (1952), 175–242.   150
Djokovic, D. Ž.
   [85] Inner derivations of division rings and canonical Jordan forms of triangular
       operators, *Proc. Amer. Math. Soc.* **94** (1985), 383–6.
Doneddu, A.
   [71] Sur les extensions quadratiques des corps non-commutatifs, *J. Algebra* **18** (1971),
       529–40.
   [72] Structures géometriques d'extensions finies des corps non-commutatifs, *J. Algebra* **23**
       (1972), 18–34.
Dumas, F.
   [91] Sous-corps de fractions rationnelles des corps gauches de séries de Laurent, *Topics in
       Invariant Theory, (Paris 1989/90)*, Lecture Notes in Math. No. 1478, Springer Verlag,
       Berlin 1991, 192–214.
Eilenberg, S.   276
Eilenberg, S. and Niven, I.
   [44] The 'Fundamental theorem of algebra' for quaternions, *Bull. Amer. Math. Soc.* **50**
       (1944), 246–8.   418
Elizarov, V. P.
   [60] On the ring of fractions of an associative ring (R), *Izv. Akad. Nauk SSSR, Ser. Mat.*
       24 (1960), 153–70.
Elliger, S.
   [66] Über galoissche Körpererweiterungen von unendlichem Rang, *Math. Ann.* **163**
       (1966), 359–61.
   [67] Potenzbasiserweiterungen, *J. Algebra* **7** (1967), 254–62.
van den Essen, A.
   [86] Algebraic micro-localization, *Comm. in Algebra* **14**(6) (1986), 971–1000.   92
Everett, C. J.
   [42] Vector spaces over rings, *Bull. Amer. Math. Soc.* **48** (1942), 312–16.   46
Faith, C. C. *see also* Cozzens
   [58] Conjugates in division rings, *Canad. J. Math.* **10** (1958), 374–80.
   [81] *Algebra I, Rings, Modules and Categories*, Second edition, Grundlehren d. math.
       Wiss. No. 190, Springer Verlag, Berlin 1981. (First Edition 1973.)   419
Faith, C. C. and Menal, P.
   [92] A counter-example to a conjecture of Johns, *Proc. Amer. Math. Soc.* **116** (1992),
       21–6.   393, 419
Farber, M. and Vogel, P.
   [92] The Cohn localization of the free group ring, *Math. Proc. Camb. Phil. Soc.* **111**
       (1992), 433–43.
Farkas, D. R. and Snider, R. L.
   [76] K_0 and Noetherian group rings, *J. Algebra* **42** (1976), 192–8.   91
Faudree, J. R.
   [66] Subgroups of the multiplicative group of a division ring, *Trans. Amer. Math. Soc.* **124**
       (1966), 41–8.

[69] Locally finite and solvable subgroups of sfields, *Proc. Amer. Math. Soc.* **22** (1969), 407–13.

Felgner, U.
[74] *Einführung in die Theorie der Schiefkörper*, Inst. f. angew. Math., Univ. Heidelberg 1974.

Ferrero, G.
[68] Classificazione e costruzione degli stems *p*-singolari, *Ist. Lombardo Accad. Sci. Lett. Rend. A* **102** (1968), 597–613.   7

Fisher, J. L.
[71] Embedding free algebras in skew fields, *Proc. Amer. Math. Soc.* **30** (1971), 453–8.
                                                                                        54
[74] The poset of skew fields generated by a free algebra, *Proc. Amer. Math. Soc.* **42** (1974), 33–5.

Fitting, H.
[36] Über den Zusammenhang zwischen dem Begriff der Gleichartigkeit zweier Ideale und dem Äquivalenzbegriff der Elementarteilertheorie, *Math. Ann.* **112** (1936), 572–82.   46

Fuchs, L.
[58] Note on ordered groups and rings, *Fund. Math.* **46** (1958), 167–74.   470

Gelfand, I. M. and Retakh, V. S.
[92] A theory of non-commutative determinants and characteristic functions of graphs, *Funct. Anal. Appl.* **26** (1992), 1–20.

Gerasimov, V. N.
[79] Inverting homomorphisms of rings (R), *Algebra i Logika* **18**:6 (1979), 648–63.   200
[82] Localizations in associative rings (R), *Sibirsk. Mat. Zh.* **23** (1982), 36–54.

Goldie, A. W.
[58] The structure of prime rings under ascending chain conditions, *Proc. London Math. Soc. (3)* **8** (1958), 589–608.   46

Golod, E. S.
[64] On nilalgebras and finitely approximable *p*-groups (R), *Izv. Akad. Nauk SSSR Ser. Mat.* **28** (1964), 273–6.   387

Golod, E. S. and Shafarevich, I. R.
[64] On class field towers (R), *Izv. Akad. Nauk SSSR, Ser. Mat.* **28** (1964), 261–72.
                                                                                       387

Goodearl, K. R.
[79] *von Neumann regular Rings*, Pitman, London 1979. (Repr. Krieger, Malabar, Fla 1991.)   190f., 201
[83] Centralizers in differential, pseudo-differential and fractional differential operator rings, *Rocky Mtn. J. Math.* **13** (1983), 573–618.

Gordon, B. and Motzkin, T. S.
[65] On the zeros of polynomials over division rings, *Trans. Amer. Math. Soc.* **116** (1965), 218–26. Correction, ibid. **122** (1966), 547.   150, 365

Gräter, J.
[93] Central extensions of ordered skew fields, *Math. Zeits.* **213** (1993), 531–55.

Green, J. A. (unpublished)   150

Greenfield, G. R.
[81] Subnormal subgroups of *p*-adic division algebras, *J. Algebra* **73** (1981), 65–9.   151

Greenwood, N. G. (verbal communication)   45, 161

Gruenberg, K. W.
[67] Profinite groups, *Algebraic Number Theory* (Ed. J. W. S. Cassels and A. Fröhlich), Academic Press, London 1967, 116–28.   140

Hahn, H.
[07] Über die nichtarchimedischen Größensysteme, *Sitzungsber. Akad. Wiss. Wien IIa* **116** (1907), 601–55.   91

Hall, M., Jr
   [59] *Theory of Groups*, Macmillan, New York 1959.   76
Hamilton, Sir W. R.
   [1853] *Lectures on Quaternions*, Dublin 1853.   45
Harris, B.
   [58] Commutators in division rings, *Proc. Amer. Math. Soc.* **9** (1958), 628–30.
Hartley, B. and Shahabi Shojaei, M. A.
   [82] Finite groups of matrices over division rings, *Math. Proc. Camb. Phil. Soc.* **92**
      (1982), 55–64.   151
Hensel, K.
   [08] *Theorie der algebraischen Zahlen*, B. G. Teubner, Leipzig 1908.   469
Herstein, I. N.
   [53] Finite subgroups of division rings, *Pacif. J. Math.* **3** (1953), 121–6.   117, 150
   [54] An elementary proof of a theorem of Jacobson, *Duke Math. J.* **21** (1954), 45–8.
   [56] Conjugates in division rings, *Proc. Amer. Math. Soc.* **7** (1956), 1021–2.   150
   [61] Wedderburn's theorem and a theorem of Jacobson, *Amer. Math. Monthly* **68** (1961),
      249–51.
   [68] *Noncommutative Rings*, The Carus Math. Monographs No. 15, Math. Assn. of
      America, J. Wiley and Sons 1968.   332
   [76] *Rings with Involution*, Univ. of Chicago Press, Chicago 1976.   332
Herstein, I. N. and Scott, W. R.
   [63] Subnormal subgroups of division rings, *Canad. J. Math.* **15** (1963), 80–3.
Higman, G.
   [40] The units of group rings, *Proc. London Math. Soc. (2)* **46** (1940), 231–48.   285, 418
   [52] Ordering by divisibility in abstract algebras, *Proc. London Math. Soc. (3)* **2**
      (1952), 326–36.   91
   [61] Subgroups of finitely presented groups, *Proc. Roy. Soc. Ser. A* **262** (1961), 455–75.
                                                                                   316
Higman, G., Neumann, B. H. and Neumann, H.
   [49] Embedding theorems for groups, *J. London Math. Soc.* **24** (1949), 247–54.   202,
                                                                                   231
Hilbert, D.
   [1899] *Grundlagen der Geometrie. Festschrift zur Feier der Enthüllung des Gauss–Weber
      Denkmals in Göttingen*, B. G. Teubner, Leipzig 1899.   45
   [03] *Grundlagen der Geometrie*, 2. Aufl., B. G. Teubner, Leipzig 1903.   45
Hill, D. A.
   [92] Left serial rings and their factor skew fields, *J. Algebra* **146** (1992), 30–48.
Hirschfeld, J. and Wheeler, W. H.
   [75] *Forcing, Arithmetic, Division Rings*, Lecture Notes in Math. No. 454, Springer
      Verlag, Berlin 1975.   311, 419
Hochster, M.
   [69] Prime ideal structure in commutative rings, *Trans. Amer. Math. Soc.* **142** (1969),
      43–60.   200
Hodges, W. A.
   [93] *Model Theory*, Encyc. of Math. and its Applns. No. 42, Cambridge University
      Press 1993.   206
Hua, L. K.
   [49] Some properties of a sfield, *Proc. Nat. Acad. Sci. USA* **35** (1949), 533–7.   150
   [50] On the multiplicative group of a sfield, *Science Record Acad. Sinica* **3** (1950), 1–6.
                                                                                   144, 150, 308
Hughes, D. R. *see* Bacsich
Huzurbazar, M. S.
   [60] The multiplicative group of a division ring (R), *Doklady Akad. Nauk SSSR* **131**
      (1960), 1268–71.   151

Ikeda, M.
 [62] Schiefkörper unendlichen Ranges über dem Zentrum, *Osaka Math. J.* **14** (1962), 135–44.
 [63] On crossed products of a sfield, *Nagoya Math. J.* **22** (1963), 65–71.   125
Jacobson, N.
 [37] Pseudo-linear transformations, *Ann. Math.* **38** (1937), 484–507.   46
 [40] The fundamental theorem of Galois theory for quasi-fields, *Ann. Math.* **41** (1940), 1–7.   150
 [43] *Theory of Rings*, Amer. Math. Soc., Providence, R. I. 1943.   46, 379f., 382
 [44] An extension of Galois theory to non-normal and non-separable fields, *Amer. J. Math.* **66** (1944), 1–29.
 [45] Structure theory for algebraic algebras of bounded degree, *Ann. Math.* **46** (1945), 695–707.   145, 150
 [47] A note on division rings, *Amer. J. Math.* **69** (1947), 27–36.   150
 [50] Some remarks on one-sided inverses, *Proc. Amer. Math. Soc.* **1** (1950), 352–5.   24
 [55] A note on two-dimensional division ring extensions, *Amer. J. Math.* **77** (1955), 593–9.   132, 150
 [56] *Structure of Rings*, Amer. Math. Soc., Providence, R. I. 1956. (Repr. 1964.)   82, 101, 140, 150, 332, 419
 [62] *Lie Algebras*, Interscience, New York 1962.   89, 307
 [75] *PI-algebras, an Introduction*, Lecture Notes in Math. No. 441, Springer Verlag, Berlin 1975.   343
Jacobson, N. and Saltman, D. J.
 [a] Finite-dimensional division algebras, to appear.   45
Jategaonkar, A. V.
 [69] A counter-example in homological algebra and ring theory, *J. Algebra* **12** (1969), 418–40.   78, 92
 [69'] Ore domains and free algebras, *Bull. London Math. Soc.* **1** (1969), 45–6.   54
Johns, B.
 [77] Annihilator conditions in Noetherian rings, *J. Algebra* **49** (1977), 222–4.   393, 419
Johnson, R. E.
 [44] On the equation $x\alpha = \gamma x + \beta$ over an algebraic division ring, *Bull. Amer. Math. Soc.* **50** (1944), 202–7.   419
 [52] On ordered domains of integrity, *Proc. Amer. Math. Soc.* **3** (1952), 414–16.   470
Kamen, E. W.
 [75] On an algebraic theory of systems defined by convolution operators, *Math. Systems Theory* **9** (1975), 57–74.
 [78] An operator theory of linear functional differential equations, *J. Diff. Equations* **27** (1978), 274–97.
Kaplansky, I.
 [48] Rings with a polynomial identity, *Bull. Amer. Math. Soc.* **54** (1948), 575–80.   332, 365
 [51] A theorem on division rings, *Canad. J. Math.* **3** (1951), 290–2.   150
 [58] On the dimension of modules and algebras X. A right hereditary ring which is not left hereditary, *Nagoya Math. J.* **13** (1958), 85–8.   92
Keisler, H. J. *see* Chang
Kerr, J. W.
 [82] The power series ring over an Ore domain need not be Ore, *J. Algebra* **75** (1982), 175–7.   224
Kharchenko, V. K.
 [77] Galois theory of semiprime rings (R), *Alg.i Logika* **16** (1977), 313–63. (Engl. transl. *Algebra and Logic* **16** (1978), 208–58)
 [91] *Automorphisms and Derivations of Associative Rings*, Kluwer Acad. Publs., Dordrecht 1991.   150

Kirezci, M.
[82] On the structure of universal non-IBN rings $V_{n,m}$, *Karadeniz Univ. Math. J.* **5** (1982), 178–98. 24
Klein, A. A.
[67] Rings nonembeddable in fields with multiplicative semigroups embeddable in groups, *J. Algebra* **7** (1967), 100–25. 330
[69] Necessary conditions for embedding rings into fields, *Trans. Amer. Math. Soc.* **137** (1969), 141–51. 23, 40
[70] Three sets of conditions on rings, *Proc. Amer. Math. Soc.* **25** (1970), 393–8.
[70'] A note about two properties of matrix rings, *Israel J. Math.* **8** (1970), 90–2. 23
[71] Matrix rings of finite degree of nilpotency, *Pacif. J. Math.* **36** (1971), 387–91.
[72] A remark concerning embeddability of rings into fields, *J. Algebra* **21** (1972), 271–4.
[72'] Involutorial division rings with arbitrary centers, *Proc. Amer. Math. Soc.* **34** (1972), 38–42.
[80] Some ring-theoretic properties implied by embeddability in fields, *J. Algebra* **66** (1980), 147–55.
Klein, A. A. and Makar-Limanov, L. G.
[90] Skew fields of differential operators, *Israel J. Math.* **72** (1980), 281–7.
Knight, J. T.
[70] On epimorphisms of non-commutative rings, *Proc. Camb. Phil. Soc.* **68** (1970), 589–600. 153
Kohn, S. and Newman, D. J.
[71] Multiplication from the other operations, *Proc. Amer. Math. Soc.* **27** (1971), 244–6. 8
Koo, H. K.
[90] The injectivity of a division ring D as a left $D \otimes_k D^{op}$-module, *Comm. in. Algebra* **18** (1990), 453–61.
Koshevoi, E. G.
[70] On certain associative algebras with transcendental relations (R), *Alg. i Logika* **9**: 5 (1970), 520–9. 54
Köthe, G.
[31] Schiefkörper unendlichen Ranges über dem Zentrum, *Math. Ann.* **105** (1931), 15–39. 69
Krasner, M. unpublished 470
Krob, D. and Leclerc, B.
[a] Minor identities for quasideterminants and quantum determinants, to appear
Kropholler, P. H., Linnell, P. A. and Moody, J. A.
[88] Applications of a new K-theoretic theorem to soluble group rings, *Proc. Amer. Math. Soc.* **104** (1988), 675–84. 91
Krull, W.
[28] Galoissche Theorie der unendlichen algebraischen Erweiterungen, *Math. Ann.* **100** (1928), 687–98. 140, 150
[31] Allgemeine Bewertungstheorie, *J. reine u. angew. Math.* **167** (1931), 160–96. 469
Kurosh, A. G.
[34] Die Untergruppen der freien Produkte von beliebigen Gruppen, *Math. Ann.* **109** (1934), 647–60. 214
Kurshan, R. P.
[70] Rings whose cyclic modules have finitely generated socle, *J. Algebra* **15** (1970), 376–86. 419
Lam, T. Y. and Leroy, A.
[88] Algebraic conjugacy classes and skew polynomial rings, *Perspectives in Ring Theory* (Ed. F. van Oystaeyen and L. Le Bruyn), Proc. Antwerp Conf. Ring Theory, Kluwer, Dordrecht 1988, 153–203. 91

[88'] Vandermonde and Wronskian matrices over division rings, *J. Algebra* **119** (1988), 308–36.

Lam, T. Y., Leung, K. H., Leroy, A. and Matczuk, J.
[89] Invariant and semi-invariant polynomials in skew polynomials rings, *Ring Theory Conf. 1989 in Honor of S. A. Amitsur, Israel Math. Conf. Proc.* **1** (1989), 247–61.

Lambek, J.
[73] Noncommutative localization, *Bull. Amer. Math. Soc.* **79** (1973), 857–72.

Lawrence, J.
[87] Strongly regular rings and rational identities of division rings, *Periodica Math. Hung.* **18** (3) (1987), 189–92.

Lawrence, J. and Simons, G. E.
[89] Equations in division rings– a survey, *Amer. Math. Monthly* **96** (1989), 220–32.
                                                                                            150

Lazerson, E. E.
[61] Onto inner derivations in division rings, *Bull. Amer. Math. Soc.* **67** (1961), 356–8.
                                                                                            71

Leavitt, W. G.
[57] Modules without invariant basis number, *Proc. Amer. Math. Soc.* **8** (1957), 322–8.
                                                                                            46

[60] The module type of a ring, *Trans. Amer. Math. Soc.* **103** (1960), 113–20.

Le Bruyn, L.
[93] Rational identities and a theorem of G. M. Bergman, *Comm. in Algebra* **21** (7) (1993), 2577–81.   365

Leclerc, B. *see* Krob

Leissner, W.
[71] Eine Charakterisierung der multiplikativen Gruppe eines Körpers, *Jahresber. D. M. V.* **73** (1971), 92–100.   45

Lenstra, H. W., Jr
[74] Lectures on Euclidean rings, University of Bielefeld 1974.   79

Leroy, A. *see* Lam

Leung, K. H. *see also* Lam
[89] Positive-semidefinite forms on ordered skew fields, *Proc. Amer. Math. Soc.* **106** (1989), 933–42.

[91] A construction of an ordered division ring with a rank 1 valuation, *Pacif. J. Math.* **147** (1991), 139–51.

Lewin, J. *see also* Bedoya
[72] A note on zero divisors in group rings, *Proc. Amer. Math. Soc.* **31** (1972), 357–9.
[74] Fields of fractions for group algebras of free groups, *Trans. Amer. Math. Soc.* **192** (1974), 339–46.

Lewin, J. and Lewin, T.
[78] An embedding of the group algebra of a torsion-free one-relator group in a field, *J. Algebra* **52** (1978), 39–74.   91

Lewin, T. *see* Lewin, J.

Lichtman, A. I.
[63] On the normal subgroups of the multiplicative group of a division ring (R), *Doklady Akad. Nauk SSSR* **152** (1963), 812–15.
[76] On embedding of group rings in division rings, *Israel J. Math.* **23** (1976), 288–97.
[78] Free subgroups of normal subgroups of the multiplicative group of skew fields, *Proc. Amer. Math. Soc.* **71** (1978), 174–8.   151
[a] Valuation methods in division rings   87, 92

Linnell, P. A. *see* Kropholler

Łos, J.
[55] Quelques remarques, théorèmes et problèmes sur les classes définissables d'algèbres, *Math. Interpretation of formal systems*, North-Holland, Amsterdam 1955, 98–113.   46

Lyndon, R. C. and Schupp. P. E.
[77] *Combinatorial Group Theory*, Ergebn. d. Math. No. 89, Springer Verlag, Berlin 1977. 325
Macintyre, A. J. 237, 323, 330
[72] On algebraically closed groups, *Ann. Math.* **96** (1972), 53–97.
[73] The word problem for division rings, *J. Symb. Logic* **38** (1973), 428–36. 330
[77] Model completeness, *Handbook of Math. Logic* (Ed. J. Barwise), Studies in Logic Vol. 90, North-Holland, Amsterdam 1977. 329
[79] Combinatorial problems for skew fields: I. Analogue of Britton's lemma and results of Adjan–Rabin type, *Proc. London Math. Soc.* (3) **39** (1979), 211–36.
Mahdavi-Hezavehi, M. *see also* Cohn
[79] Matrix valuations on rings, *Antwerp Conf. Ring Theory*, Dekker, New York 1979, 691–703. 470
[80] Totally ordered rings and fields, *Bull. Iran. Math. Soc.* **7** (1980), 41–54.
[82] Matrix valuations and their associated skew fields, *Resultate d. Math.* **5** (1982), 149–156. 470
Mahler, K. 442, 470
Makar-Limanov, L. G. *see also* Klein
[75] On algebras with one relation (R), *Uspekhi Mat. Nauk* **30**: 2 (1975), 217. 418
[83] The skew field of fractions of the Weyl algebra contains a free non-commutative subalgebra, *Comm. in Algebra* **11**(7) (1983), 2003–6.
[85] On algebraically closed skew fields, *J. Algebra* **93** (1985), 117–35. 418
[89] An example of a skew field without a trace, *Comm. in Algebra* **17**(9) (1989), 2303–7. 259
[91] On subalgebras of the first Weyl skew field, *Comm. in Algebra* **19** (7) (1991), 1971–82.
Malcev, A. I.
[37] On the immersion of an algebraic ring in a skew field, *Math. Ann.* **113** (1937), 686–91. 9, 330
[39] Über die Einbettung von assoziativen Systemen in Gruppen, I, II (R, German summary), *Mat. Sbornik NS* **6** (48) (1939), 331–6, **8**(50) (1940), 251–64. 9
[48] On the embedding of group algebras in division algebras (R), *Doklady Akad. Nauk SSSR* **60** (1948), 1499–1501. 91
[73] *Algebraic Systems*, Grundl. math. Wiss. No. 192, Springer Verlag, Berlin 1973. 46, 327
Malcolmson, P.
[78] A prime matrix ideal yields a skew field, *J. London Math. Soc.* (2) **18** (1978), 221–33. 200
[80] On making rings weakly finite, *Proc. Amer. Math. Soc.* **80** (1980), 215–18. 201
[80'] Determining homomorphisms into a skew field, *J. Algebra* **64** (1980), 399–413.
[84] Matrix localizations of firs I, II. *Trans. Amer. Math. Soc.* **282** (1984), 503–18, 519–27.
[93] Weakly finite matrix localizations, *J. London Math. Soc.* (2) **48** (1993), 31–8. 171
Martindale, W. S., III, *see* Desmarais
Matczuk, J. *see* Lam
Mathiak, K.
[77] Bewertungen nichtkommutativer Körper, *J. Algebra* **48** (1977), 217–35. 470
[81] Zur Bewertungstheorie nichtkommutativer Körper, *J. Algebra* **73** (1981), 586–600. 470
[86] *Valuations of Skew Fields and Projective Hjelmslev Spaces*, Lecture Notes in Math. No. 1175, Springer Verlag, Berlin 1986.
May, M.
[92] Universal localization and triangular rings, *Comm. in Algebra* **20**(5) (1992), 1243–57.
May, W.
[72] Multiplicative groups of fields, *Proc. London Math. Soc.* (3) **24** (1972), 295–306. 150

Menal, P. *see* Faith

Michler, G. O. and Villamayor, O. E.

[73] On rings whose simple modules are injective, *J. Algebra* **25** (1973), 185–201.

Mikhailova, K. A.

[58] The occurrence problem for direct products of groups (R), *Doklady Akad. Nauk SSSR* **119** (1958), 1103–5.   330

Montgomery, M. S. and Passman, D. S.

[84] Outer Galois theory of prime rings, *Rocky Mtn. J. Math.* **14** (1984), 305–18.

Moody, J. A. *see* Kropholler

Motzkin, T. S. *see* Gordon

Moufang, R.

[37] Einige Untersuchungen über geordnete Schiefkörper, *J. reine u. angew. Math.* **176** (1937), 203–23.   77, 92

Mumford, D.

[76] *Algebraic geometry I. Complex projective varieties*, Grundl. Math. Wiss. No. 221, Springer Verlag, Berlin 1976.   416

Nagahara, T. and Tominaga, H.

[55] A note on Galois theory of division rings of infinite degree, *Proc. Japan Acad.* **31** (1955), 655–8.

[56] On Galois theory of division rings, *Proc. Japan Acad.* **32** (1956), 153–6.

Nakayama, T.

[53] On the commutativity of certain division rings, *Canad. J. Math.* **5** (1953), 242–4.

Neumann, B. H., *see also* Higman

[49] On ordered division rings, *Trans. Amer. Math. Soc.* **66** (1949), 202–52.   91

[49'] On ordered groups, *Amer. J. Math.* **71** (1949), 1–18.   91f.

[51] Embedding non-associative rings in division rings, *Proc. London Math. Soc. (3)* **1** (1951), 241–56.   276

[54] An essay on free products of groups with amalgamations, *Phil. Trans. Roy. Soc. Ser. A* **246** (1954), 503–54.   210, 235, 275

[73] The isomorphism problem for algebraically closed groups, *Word Problems* (Ed. W. W. Boone *et al*.), North-Holland, Amsterdam, 1973, 553–62.   313

Neumann, H. *see* Higman

Newman, D. J. *see* Kohn

Niven, I. *see also* Eilenberg

[41] Equations in quaternions, *Amer. Math. Monthly* **48** (1941), 654–61.   401, 418

Nobusawa, N.

[55] An extension of Krull's Galois theory to division rings, *Osaka Math. J.* **7** (1955), 1–6.

[56] On compact Galois groups of division rings, *Osaka Math. J.* **8** (1956), 43–50.

Noether, E.

[33] Nichtkommutative Algebra, *Math. Zeits.* **37** (1933), 514–41.   106, 150

Novikov, P. S.

[55] On the algorithmic unsolvability of the word problem in group theory (R), *Trudy Mat. Inst. im. Steklov* **44** (1955), 143 pp. (Engl. transl. *AMS Transl. Ser. 2* **9** (1958), 1–122.)   330

Ore, O.

[31] Linear equations in noncommutative fields, *Ann. Math.* **32** (1931), 463–77.   14, 46

[32] Formale Theorie der Differentialgleichungen, *J. reine u. angew. Math.* **167** (1932), 221–34, II, ibid. **168** (1932), 233–52.   91

[33] Theory of non-commutative polynomials, *Ann. Math.* **34** (1933), 480–508.   46f.

[33'] On a special class of polynomials, *Trans. Amer. Math. Soc.* **35** (1933), 559–84.   56

Passman, D. S. *see* Montgomery

Pfister, A.

[65] Darstellung von −1 als Summe von Quadraten in einem Körper, *J. London Math. Soc.* **40** (1965), 159–65.   470

Pickert, G.
   [51] *Einführung in die höhere Algebra*, Vandenhoeck & Ruprecht, Göttingen 1951.
                                                                                    470
   [59] Eine Kennzeichnung desarguesscher Ebenen, *Math. Zeits.* **71** (1959), 99–108.   45
Platonov, V. P. and Yanchevskii, V. I.
   [92] *Finite-dimensional Division Algebras* (R), Algebra 9 Itogi Nauki i Tekhniki VINITI,
        Moscow 1992. (English transl., Springer Verlag 1995)
Podderyugin, V. D.
   [54] Conditions for the orderability of an arbitrarily ring (R), *Uspekhi Mat. Nauk* **9**: 4
        (1954), 211–16.   470
Van Praag, P. *see also* Boffa
   [71] Groupes multiplicatifs des corps, *Bull. Soc. Math. Belg.* **23** (1971), 506–12.
   [72] Sur les groupes multiplicatifs des corps, *C. R. Acad. Sci. Paris A* **271** (1972), 243–4.
   [75] Sur les centralisateurs de type fini dans les corps existentiellement clos, *C. R. Acad.
        Sci. Paris A* **281** (1975), 891–3.
Procesi, C.
   [68] Sulle identità delle algebre semplici, *Rend. Circ. Mat. Palermo, Ser. 2* **XVII** (1968),
        13–18.   365
   [73] *Rings with polynomial identities*, Dekker, New York 1973.   46, 365
Quebbemann, H.-G.
   [79] Schiefkörper als Endomorphismenringe einfacher Moduln über einer Weyl-Algebra,
        J. Algebra **59** (1979), 311–12.   395
Resco, R. D.
   [87] Division rings and V-domains, *Proc. Amer. Math. Soc.* **99** (1987), 427–31.   391, 419
Retakh, V. S. *see* Gelfand
Reutenauer, C. *see also* Cohn
   [92] Applications of a noncommutative Jacobian matrix, *J. Pure Appl. Algebra* **77** (1992),
        169–81.
Révész, G.
   [81] Universal fields of fractions, their orderings and determinants, Ph.D. Thesis, London
        University 1981.
   [83] Ordering epic *R*-fields, *Manuscr. Math.* **44** (1983), 109–30.   469f.
   [83′] On the abelianized multiplicative group of a universal field of fractions, *J. Pure
        Appl. Algebra* **27** (1981), 277–97.
Richardson, A. R.
   [27] Equations over a division algebra, *Messenger of Math.* **57** (1927), 1–6.   150
Riesinger, R.
   [82] Geometrische Überlegungen zum rechten Eigenwertproblem für Matrizen über
        Schiefkörpern, *Geom. Dedicata* **12** (1982), 401–5.
Roberts, M. L.   201
   [82] Normal forms, factorizations and eigenrings in free algebras, Ph.D. Thesis, London
        University 1982.   419
   [84] Endomorphism rings of finitely presented modules over free algebras, *J. London
        Math. Soc. (2)* **30** (1984), 197–209.   329
Robinson, A.
   [63] *Introduction to Model Theory and the Metamathematics of Algebra*, Studies in Logic,
        North-Holland, Amsterdam 1963.   329f.
   [71] On the notion of algebraic closedness for non-commutative groups and fields, *J.
        Symb. Logic* **36** (1971), 441–4.
   [71′] Infinite forcing in model theory, *Proc. 2nd Scandinav. Logic Symp.* (Ed.
        J. E. Fenstad), North-Holland, Amsterdam 1971, 317–40.
Robson, J. C. *see* Cauchon
Rogers, H., Jr
   [67] *Theory of Recursive Functions and Effective Computability*, McGraw-Hill, New York
        1967.   319

Rosenberg, A.
 [56] The Cartan–Brauer–Hua theorem for matrix and local matrix rings, *Proc. Amer. Math. Soc.* **7** (1956), 891–8.
Rowen, L. H.
 [88] *Ring Theory* I, II, Academic Press, New York 1988.   92
Sabbagh, G.
 [71] Embedding problems for modules and rings with application to model companions, *J. Algebra* **18** (1971), 390–403.   329
Saltman, D. J. *see* Jacobson
Samuel, P. *see* Zariski
Scharlau, W. and Tschimmel, A.
 [83] On the level of skew fields, *Arch. Math.* **40** (1983), 314–15.   470
Schenkman, E. V.
 [58] Some remarks on the multiplicative group of a field, *Proc. Amer. Math. Soc.* **9** (1958), 231–5.
 [59] On a theorem of Herstein, *Proc. Amer. Math. Soc.* **10** (1959), 236–8.
 [61] Roots of centre elements of division rings, *J. London Math. Soc.* **36** (1961), 393–8.
 [64] On the multiplicative group of a field, *Arch. Math.* **15** (1964), 282–5.   150
Schilling, O. F. G.
 [45] Noncommutative valuations, *Bull. Amer. Math. Soc.* **51** (1945), 297–304.   470
 [50] *The Theory of Valuations*, Math. Surveys No. IV, Amer. Math. Soc., New York 1950.   470
Schlesinger, L.
 [1897] *Handbuch der Theorie der Differentialgleichungen*, B. G. Teubner, Leipzig 1897.
                                       91
Schofield, A. H.   223, 250, 259, 268, 329
 [85] *Representations of Rings in Skew Fields*, LMS Lecture Notes No. 92, Cambridge University Press 1985.   96, 201, 275ff.
 [85'] Artin's problem for skew fields, *Math. Proc. Camb. Phil. Soc.* (97) (1985), 1–6.
                                       262, 276
Schreier, O. *see also* Artin
 [27] Über die Untergruppen der freien Gruppen, *Abh. Math. Sem. Hamb. Univ.* **5** (1927), 161–83.   204, 231, 275
Schröder, M.
 [87] Bewertungen von Schiefkörpern, Resultate und offene Probleme, *Results in Math.* **12** (1987), 191–206.
Schupp, P. E. *see* Lyndon
Schur, I.
 [04] Über vertauschbare lineare Differentialausdrücke, *Berl. Math. Ges. Sitzungsber.* **3** (1904), 2–8.   91
Schwarz, L.
 [49] Zur Theorie des nichtkommutativen Polynombereichs und Quotienten-ringes, *Math. Ann.* **120** (1947/49), 275–96.
Scott, W. R. *see also* Herstein
 [57] On the multiplicative group of a division ring, *Proc. Amer. Math. Soc.* **8** (1957), 303–5.   150
Serre, J.-P.
 [49] Extensions des corps ordonnés, *C. R. Acad. Sci. Paris* **229** (1949), 576–7.   470
Shafarevich, I. R. *see* Golod
Shahabi Shojaei, M. A. *see* Hartley
Shepherdson, J. C.
 [51] Inverses and zero-divisors in matrix rings, *Proc. London Math. Soc.* (3) **1** (1951), 71–85.
Shimbireva, E. P.
 [47] On the theory of partially ordered groups (R), *Mat. Sbornik* **20** (1947), 145–78.   91

Shirvani, M. and Wehrfritz, B. A. F.
[86] *Skew Linear Groups*, LMS Lecture Notes No. 188, Cambridge University Press
1986.   148
Simons, G. E. *see* Lawrence
Sizer, W. S.
[75] Similarity of sets of matrices over a skew field, Ph.D. Thesis, London University
1975.   418
[77] Triangularizing semigroups of matrices over a skew field, *Lin. Algebra and its Appls.*
16 (1977), 177–87.   386
Small, L. W. *see also* Amitsur, Bergman
[66] Hereditary rings, *Proc. Nat. Acad. Sci. USA* **55** (1966), 25–7.   92
Smith, D. B.
[70] On the number of finitely generated O-groups, *Pacif. J. Math.* **35** (1970), 499–502.
                                                                                237
Smith, K. C.
[74] Theory and application of noncommutative polynomials, Lecture Notes, University
of Oklahoma 1974.
Smits, T. H. M.
[69] The free product of a quadratic number field and a semi-field, *Indag. Math.* **31**
(1969), 145–69.
[82] On the coproduct of quadratic algebras, *Indag. Math.* **44** (1982), 461–73.
Snider, R. L. *see* Farkas
Sontag, E. D. *see* Dicks
Stafford, J. T.
[83] Dimensions of division rings, *Israel J. Math.* **45** (1983), 33–40.
Steinitz, E.
[10] Algebraische Theorie der Körper, *J. reine u. angew. Math.* **137** (1910), 167–308.
(Repr. W. de Gruyter, Berlin 1930, Chelsea Publ. Co., New York 1950.)   5
Sushkevich, A. K.
[36] On the embedding of semigroups in groups of fractions (R), *Zap. Khark. Mat. O-va*
4: 12 (1936), 81–8.
Sylvester, J. J.
[1884] On involutants and other allied species of invariants to matrix systems, *Johns
Hopkins Univ. Circulars* **III** (1884), 9–12, 34–5.   201
Szele, T.
[52] On ordered skew fields, *Proc. Amer. Math. Soc.* **3** (1952), 410–13.   470
Tits, J.
[72] Free subgroups in linear groups, *J. Algebra* **20** (1972), 250–70.   151
Tominaga, H. *see* Nagahara
Törner, G. *see* Brungs
Treur, J.
[76] A duality for skew field extensions, Ph.D. Thesis, University of Utrecht 1976.
[88] On duality for skew field extensions, *J. Algebra* **119** (1988), 1–22.
[89] Separate zeros and Galois extensions of skew fields, *J. Algebra* **120** (1989), 392–405.
                                                                                407
Tschimmel, A. *see* Scharlau
Valitskas, A. I.
[87] An example of a non-invertible ring embeddable in a group (R), *Sibirsk. Mat.
Zh.* **28** (1987), 35–49.   330
[a] P. Cohn's theorem on embedding certain filtered rings in skew fields (R),   92
Van Deuren, J.-P.   66
Vidal, R.
[81] Anneaux de valuation discrète complets non-commutatifs, *Trans. Amer. Math. Soc.*
267 (1981), 65–81.
Villamayor, O. E. *see also* Michler   419

Vogel, P. *see* Farber
van der Waerden, B. L.
    [30] *Moderne Algebra* I, Springer Verlag, Berlin 1930. 330
    [48] Free products of groups, *Amer. J. Math.* **70** (1948), 527–8. 205, 275
Wagner, W.
    [37] Über die Grundlagen der projektiven Geometrie und allgemeine Zahlensysteme,
    Math. Ann. **113** (1937), 528–67. 365
Wähling, H.
    [87] *Theorie der Fastkörper*, Thales Verlag, Essen 1987. 5, 45
Warfield, R. B., Jr
    [78] Stable equivalence of matrices and resolutions, *Comm. in Algebra* **6** (17) (1978),
    1811–28.
Wedderburn, J. H. M.
    [05] A theorem on finite algebras, *Trans. Amer. Math. Soc.* **6** (1905), 349–52. 150
    [21] On division algebras, *Trans. Amer. Math. Soc.* **22** (1921), 129–35. 150
Wehrfritz, B. A. F. *see also* Shirvani
    [73] *Infinite Linear Groups*, Springer Verlag, Berlin 1973. 151
    [92] On Cohn's embedding of an enveloping algebra into a division ring, *Ukrain. Mat.*
    Zh. **44** (1992), 729–35. 92
Westreich, S.
    [88] Matrix localization and embedding, *Comm. in Algebra* **16** (1) (1988), 75–102.
    [91] On Sylvester rank functions, *J. London Math. Soc.* (2) **43** (1991), 199–214.
Whaples, G. *see* Bray
Wheeler, W. H. *see also* Hirschfeld
    [72] Algebraically closed division rings, forcing and the analytical Hierarchy, Ph.D.
    Thesis, Yale University 1972. 312, 316f., 329, 419
Witt, E.
    [31] Über die Kommutativität endlicher Schiefkörper, *Abh. Math. Sem. Hamb. Univ.* **8**
    (1931), 413. 118
Wood, R. M. W.
    [85] Quaternionic eigenvalues, *Bull. London Math. Soc.* **17** (1985), 137–8. 418
Yanchevskii, V. I. *see* Platonov
Yen, C. T.
    [93] A condition for simple ring implying field, *Acta Math. Hungar.* **61** (1993), 51–2.
Zalesskii, A. E.
    [67] The structure of some classes of matrix groups over a skew field (R), *Sibirsk. Mat.*
    Zh. **8** (1967), 1284–98. 151
Zariski, O. and Samuel, P.
    [60] *Commutative Algebra* II, van Nostrand, Princeton, N. J. 1960. 411
Zassenhaus, H.
    [36] Über endliche Fastkörper, *Abh. Math. Sem. Hamb. Univ.* **11** (1936), 187–220.
    150

SUBJECT INDEX

For any property P, left P, right P or non-P are usually listed under P. An S-object is often listed as -object, e.g. t-adic is listed as -adic, but p-adic comes under p.

abelian valuation, 427
abelianization, 437
absolute GRI, 340
absolute presentation, 317
absolute value, 422
abstract support system, 362
AC = algebraically closed, 308
addition, 3
additive group of a field, 4
-adic topology, 66
-adic valuation, 85, 425
admissible matrix block, system, 159
-algebra, 41
algebraic closure, 308, 329
algebraic dependence, 406
algebraic element, 111
algebraic extension, 140
algebraic matrix, 379
algebraically closed, 308, 367
amalgamation property, 310
Amitsur–Levitzki theorem, 334, 343
Amitsur's GPI-theorem, 332, 365
Amitsur's theorem on rational identities, 339, 344
annihilator ring, 393
antichain, 72
Artin–Schreier condition, 461
Artin's theorem, 101, 103f.
associated matrices, 24
atom, 27
augmentation ideal, 44
augmentation mapping, 55
augmented ring, 44

base ring in a coproduct, 211
basic formula, 309

basic module, 211
Bezout domain, 13
bicentralizer, 110
binomial field extension, 121
Birkhoff–Witt theorem, 89
block, 162
Bokut' classification, 330
bound, of an element, 28
bound module, 34
bounded element, 28
BW-algebra, 90

CAC = characteristically algebraically closed, 370, 418
Cartan–Brauer–Hua theorem, 144
Cayley–Hamilton theorem, 385
central, 62
central field extension, 121
central localization, 387
centralizer, 4
centralizing extension, 339
centre, 4
characteristic of a field, 4
Chevalley's extension lemma, 426, 470
cofinite subset, 11
cogenerator, 389
comaximal, 25
comma category, 203
commutator, 143, 238, 259, 441
compactness theorem of logic, 327, 329
companion matrix, 370
complete variety, 416
concatenation of valuations, 433
cone, 458
conical monoid, 73, 249
conjugate, 112, 233

consistent system of equations, 309
convex subgroup, 432
coprime, 26, 29
coproduct
    in a category, 203f.
    of fields, 223, 268, 449
    of rings, 42, 222, 276
core, 159
Cramer's rule, 160
critically skew field, 149
crossed product, 123ff.
cyclic extension, 133
cyclic matrix, 381

Dedekind's lemma, 101f.
defect of a matrix over $D_k \langle x \rangle$, 409
degenerate, 336, 344
degree
    of a field extension, 94
    of a monomial, 215
    of a polynomial, 48, 84
denominator
    of an admissible matrix block, 162
    of an admissible system, 159, 201
dependable field extension, 318
dependence relation, 406f.
depth, 196
derivation, 49
derived group, 144, 427
Desarguesian plane, 1, 365
determinantal sum of matrices, 163
diagonally closed matrix set, 445
Dieudonné determinant, 437, 466
differential equation, 133
differential polynomial, 46, 91
distributive laws, 4
divergence of a skew polynomial, 59
divisibility ordering, 73
division algebra, 5, 45, 150
division ring, 4
domain
    of a function, 336, 344
    of a subvaluation, 444
dominate, 428

E-associated, 436
EC-field, 309ff.
E-matrix, 437
effective construction, 318
eigenring, 28
eigenspace, 378
eigenvalue, 370, 375, 378, 403
eigenvector, 375
elementary divisor, 382
elementary mapping, 311
elementary sentence, 308
elimination theory, 416f.
embedding condition, 326
epic -field, 154, 336
epi-final, -initial, 177

epimorphism, 153
equivalence
    of homomorphisms, 101
    of local homomorphisms, 154
    of matrix blocks, 163
    of places, valuations, 424
essential extension, 387
essential index (term), 349
essential set, 354
Euclidean algorithm, 79
evaluation map, 333
existential sentence, 308
existentially closed, 309ff.

FAC = fully algebraically closed, 371
factor ring in a coproduct, 211
factorial duality, 27
faithful A-ring, 41, 205
faithful coproduct, 42, 204
field, 3, 154, 336
    coproduct, 223
    of fractions, 8
    spectrum, 173, 200
filter, 11
filtered ring, 83, 228
filtration, 83, 206f.
finite extension, 94
finite field element, 404
finite topology, 98
finitely generated, 280
finitely inert, 284
finitely presented, 280
    point, 417
finitely related, 280
fir, 36f., 40, 46, 222, 248
flat module, 228
flat subset, 337
forcing companion, 311
formally real field, ring, 459
fraction, 15
free k-algebra, 224
free field, 224, 235, 301, 344
free ideal ring, 36f.
    see also fir
free module type, 21
free point, 402
free product
    with amalgamation, 275
    of groups, rings, 42, 204f.
free set in a field, 235, 279f.
free transfer isomorphism, 220
full map, 193
full matrix, 22, 168, 179
fully algebraically closed, 200, 371
fully inverting map, 177
function field, 338
fundamental theorem of Galois theory for
    skew fields, 109

Galois connexion, 99, 109, 142

Galois extension, group, 100, 129ff.
generic division algebra, 344
generic field, 329
generic matrix ring, 46, 343
generic point, 414
Gerasimov–Malcolmson localization
    theorem, 40
global dimension, 39, 218, 244, 247
GPI = generalized polynomial identity, 283,
    332, 339
graded ring, 83
GRI = generalized rational identity, 337,
    340f.
Grothendieck group of projectives, 186

HCLF = highest common left factor, 32
Hahn–Banach theorem, 188, 420, 442
height, 207
    theorem for integral domains, 208
hereditary ring, 39, 246
Hermite ring, 457
Higman's theorem, 316, 325
Higman's trick, 284
Hilbert basis theorem, 90
Hilbert field, 281, 339, 469
Hilbert Nullstellensatz, 411ff., 415, 419
Hilbert 'Theorem 90', 135f.
HNN-extension
    of a field, 231f., 276
    of a ring, 239ff.
Hochster's axioms, 200
hollow matrix, 179, 299
homogeneous element, 84
homogeneous field, 233
homological dimension, 39, 214
honest homomorphism, 177
Horn sentence, 9
Hua's identity, 335
Hua's theorem, 144, 150

I-atom, 29
IBN = invariant basis number, 19, 46, 249,
    328
idealizer, 28
idempotent, 184f., 243
identity, 9, 331f., 335
indecomposable, 31
index of a matrix, 24, 296
induced homomorphism, 219
induced module, 186, 211
inductive class, 311
inertia lemma, 284
inner derivation, 50
inner eigenvalue, 378
inner Galois group, 109
inner order of an automorphism, 61
inner rank, 179, 192
integral domain, 8
internal modification, 35
invariant basis number, 19

invariant element, 28, 57f.
invariant factor, 380, 384
invariant subring, 423
inverse eigenvalue, 403
inversive ring, 33
inverting homomorphism, 14, 156
involution, 56
irreducible algebraic set, 337
irreducible element, 27
isomorphic idempotents, 185
iterated skew polynomial ring, 78ff.

J-ring, J-skew polynomial ring, 79
Jacobson–Bourbaki correspondence, 99,
    150
Jacobson radical, 10, 82, 92
Jacobson–Zassenhaus formula, 137
Jategaonkar's condition, 79
Jordan matrix, normal form, 384

Kaplansky's PI-theorem, 332, 365
Kaplansky's theorem on projective modules,
    191
key term, 216
Klein's nilpotence condition, 23, 40, 250
Klein's theorem, 23, 40

LCLM, LCRM, least common left, right,
    multiple, 33
large submodule, 387
Laurent polynomial, 44, 55
Laurent series, 45, 66
leading term, 84, 215
length of an element in a UFD, 28
level
    of an affine space, 403
    of a field, 470
Lie algebra, 88
linear companion, 294, 370
linear matrix, 293
linearization by enlargement, 284
linearly disjoint, 96, 302
local homomorphism, 154
local ring, 154, 169, 344
localization, 15, 156
locally cyclic group, 146
locally finite algebra, 387
locus of a point, 403
lower central series, 76

magic lemma, 438
Malcev conditions, 9, 23
Malcev–Neumann construction, 76, 79, 91
Malcolmson's criterion, 167, 183
matrix-algebraic algebra, 387
matrix block, 159, 162
matrix cone, 463
matrix functor, 42
matrix-homogeneous, 234, 391
matrix ideal, 172

matrix local ring, 344
matrix preideal, 173
matrix reduction functor, 43, 46, 247
matrix subvaluation, 443
matrix units, 42
matrix valuation, 435
metacyclic group, 148
metro-equation, 369, 418
minimal invariant element, 59
minimal polynomial, 113
monic matrix, 293
monic polynomial, 48
monoid of projectives, 186, 221, 244, 249
monoid ring, 55
monomial matrix, 291
monomial unit, 208
multiplication, 3
multiplication algebra, 258
multiplicative commutator, 143, 238
multiplicative function, 444
multiplicative group of a field, 4, 143ff.,
    150f.
multiplicative set, 15
    of matrices, 157

N-group, 106, 150
N-invariant subgroup, 109
near field, ring, 5, 7, 45
negative element, 457
von Neumann regular ring, 10, 190
nilpotent group, 144
non-singular at infinity, 370
norm, 422
normal basis theorem, 137, 290
normal field extension, 306
normal form in a free ring, 294
normalizer, 263
nullity condition, 180
Nullstellensatz, 411ff., 415, 419
numerator, 159, 201

one, 3
one-unit, 424
opposite ring, 97
order of a block, 196
order-unit, 188
ordered field, 457
ordered group, 75, 421
ordered ring, 457
Ore condition, 15
Ore domain, 16
Ore set, 16
outer cyclic extension, 133
outer derivation, 50
outer Galois group, 110

PAC = polynomially algebraically closed,
    371, 418
p-adic valuation, 469
PI-algebra, 332

PI-degree, 343
P(R)ID = principal (right) ideal domain, 49
PWO = partly well-ordered, 73
partition lemma, 36
partly well-ordered, 73
perfect closure of a commutative field, 316
place, 423
point singularity, 408
pointed bimodule, 227
polynomial, 48
polynomially algebraically closed, 371
positive cone, 458
positive element, 457
Posner's theorem, 343
power series, 38, 66ff.
presentation of a field, 279
prime avoidance lemma, 351, 358
prime (left, right) matrix, 197
prime matrix ideal, 172
prime ring, 343
prime subfield, 4
primitive element (theorem), 110, 137
primitive ring, 10, 82, 92, 332
principal valuation, 424
profinite group, 140
projective-free ring, 39, 185
proper factorization, 298
proper matrix, 368
proper (matrix) cone, 458, 463
proper valuation on a ring, 421
pseudo-linear field extension, 119
pseudo-valuation, 83, 442, 470
pure element, 207, 215
pure field extension, 121
pure matrix block, 162

quadratic field extension, 118, 126, 268ff.
quasi-commutative valuation, 430
quasi-free point, 403
quasi-identity, 9, 46
quasi-variety of algebras, 9
quaternions, 45, 56
quaternion algebra, 118, 272f., 373

Rabinowitsch trick, 412
radical matrix ideal, 174
radical matrix subvaluation, 443f.
rank, of a free module, 19
rank factorization, 179
rank function on projectives, 187
rational closure, 157, 348
rational expression, function, 335ff.
rational identity, 335ff., 345f.
rational meet, 348
rational relation, 344
rational topology, 337, 405
rationality criterion, 69
ray singularity, 408
recursive, recursively enumerable, 318f.
reduced admissible system, 200

reduced automorphism set, 289
reduced centre, 68
reduced element, 208
reduced matrix block, 197
reduced order, 106
reduced product, 12
reduced ring, 10
regular field extension, 96
regular matrix subvaluation, 444
regular ring, 10, 190
regular subset, 16
regularization, 444
representation of a Lie algebra, 89
residue-class field, 154
resultant, 416
retract, 37
reversible Ore set, 16
Rg = category of rings, 15, 41
rigid domain, 33
-ring, 41, 335
ring epimorphism, 153
root of a (matrix) subvaluation, 443f.

Schreier's theorem, 204, 275
Seifert–van Kampen theorem, 275
semifir, 34, 40, 46
semihereditary ring, 39
semilocal ring, 350
semiprime ring, 13
semiprimitive ring, 10
semiuniversal EC-field, 314, 414
separable matrix, 386
separating coproduct, 204
similar elements, matrices, 27, 46
simple ring, 61, 241f., 391
singular eigenvalue, 370
singular ideal, 393
singular kernel, 156, 439, 452
singular matrix, 22, 309
singularity support, 301, 405
skew cyclic matrix, 384
skew field, 4
skew polynomial ring, 49
Skolem–Noether theorem, 52, 105
small cancellation theory, 276
small matrix ideal, 454
socle of a ring, 393
specialization, 154, 344f., 401
specialization lemma, 287, 306, 320, 329
spectrum, 375
split null extension, 153, 394
stably associated matrices, 24
stably isomorphic modules, 186
staircase lemma, 334
standard identity, 343
state, 189
Steinitz number, 146
strict X-ring, 348
strictly cyclic module, 27
strongly regular ring, 10

submultiplicative function, 442f.
subordinate projective module, 186
subordinate valuation, 432
subvaluation, 83, 442, 470
superficial matrix ideal, 457
superfluous block, 163
supernatural number, 146
support, 74, 215
support relation, 351ff., 362
supporting a family, 352, 355
Sylvester domain, 180, 185, 201
Sylvester rank function, 193
Sylvester's law of nullity, 180

tensor ring, 38, 48, 226
topology of simple convergence, 98
torsion-free, 30, 229
torsion group, 145
torsion module, 30
total (matrix) cone, 458, 465
total divisor, 380
total subring, 423
totally algebraically closed, 125
totally coprime, 29
totally transcendental, 234, 379
totally unbounded, 29
trace in an outer cyclic extension, 137
trace ideal, 187
transvection, 220
triangle inequality, 422
triangulable, 371
trivial relation, 34
trivial ring = zero ring, 20
trivial support relation, 354
trivial valuation, 422
trivializable, 34

UFD = unique factorization domain, 28
UGN = unbounded generating number, 19,
    46, 190, 249
ultrafilter, ultraproduct, 11f., 46
unit, 14
universal class, 9
universal denominator, 199, 299
universal derivation bimodule, 251
universal EC-field, 314, 415
universal field of fractions, 176f.
universal localization, 15, 177
universal sentence, 9
universal S-, Σ-inverting ring, 15, 156
unramified extension, 449

valuation, 83, 421, 469
valuation ring, 423
value group, 421
variety
    of algebras, 8
    over a skew field, 408
V-ring, 389, 419

WF = weakly finite, 20, 190, 249
weak algorithm, 38, 232, 276, 293, 297
weakly finite, 20, 46, 190, 249
weakly semihereditary ring, 191f.
Wedderburn's theorem on finite fields, 115, 145, 150
well-positioned family, 216

Weyl algebra, field, 282
word problem, 317, 330

Zariski topology, 337, 405
zero, 3
zig-zag lemma, 312